Nuclear Structure
and Function

Nuclear Structure and Function

Edited by

J. R. Harris

North East Thames Regional Transfusion Centre
Brentwood, Essex, England

and

I. B. Zbarsky

N. K. Koltzov Institute of Developmental Biology
Academy of Sciences of the USSR
Moscow, USSR

Plenum Press • New York and London

Library of Congress Cataloging-in-Publication Data

Nuclear Workshop (11th : 1989 : Suzdal, R.S.F.S.R.)
    Nuclear structure and function / edited by J.R. Harris and I.B.
Zbarsky.
        p.   cm.
    "Proceeding of the Eleventh Nuclear Workshop, organized by the
N.K. Koltzov Institute of Developmental Biology, Academy of Sciences
of the USSR, under the auspices of the European Cell Biology
Organization, held September 18-23, 1989, in Suzdal, USSR"--T.p.
verso.
    Includes bibliographical references.
    Includes index.
    ISBN-13:978-1-4612-7918-1   e-ISBN-13:978-1-4613-0667-2
    DOI:10.1007/978-1-4613-0667-2

    1. Cell nuclei--Congresses.   2. Nucleoproteins--Congresses.
3. Biochemical genetics--Congresses.   I. Harris, James R.
II. Zbarskiĭ, I. B.   III. N.K. Koltzov Institute of Developmental
Biology.   IV. European Cell Biology Organization.   V. Title.
    [DNLM: 1. Cell Nucleus--congresses.   2. Genetics, Biochemical-
-congresses.   3. Nuclear Proteins--congresses.   QH 595 B9643n]
QH595.N837 1989
574.87'32--dc20
DNLM/DLC
for Library of Congress   90-14252
                                                               CIP

Proceedings of the Eleventh Nuclear Workshop, organized by the N. K. Koltzov Institute of
Developmental Biology, Academy of Sciences of the USSR, under the auspices of the European
Cell Biology Organization, held September 18–23, 1989, in Suzdal, USSR

ISBN-13:978-1-4612-7918-1

PREFACE

    This collection of 101 short communications, submitted by some of the
participants at the 11th Nuclear Workshop held in Suzdal, USSR, 18-23
September, 1989, provides a representative survey of the material presented
at the Workshop. Articles have been submitted by both those who delivered
lectures and those who had poster presentations. The order of presentation
at the Nuclear Workshop is roughly maintained within this proceedings book,
but the session titles within the scientific program have not been utilized
as discrete subdivisions within the book, because of the considerable
overlap of subject matter. The overall sequence is as follows: Genome
structure, Gene Structure and Expression, Nucleolar Genes, Structure and
Proteins, Chromatin and Nuclear Granules, Nuclear Matrix and Nuclear
Proteins, Replication and Transcription and finally Nuclear Envelope and
Nuclear Cytoplasmic Transport. Several articles on Nuclear Lipids are also
included, stemming from an evening round-table discussion on lipids.

    The third Wilhelm Bernhard Lecture was delivered in Suzdal by
Professor Harris Busch, who can be seen in the photograph above (on the
left) in the presence of Professor Ilya B. Zbarsky, President of the
organizing committee for the 11th Nuclear Workshop. (Previous Wilhelm
Bernhard lecturers have been Ronald H. Reeder, in Krakow, Poland, 1985 and
Oscar L. Miller, Jr., in Stevensbeek, The Netherlands, in 1987).

    Despite some unforseen delay in the production of this proceedings
book, it is hoped that it will be of widespread interest to all those
concerned with the 'Cell Nucleus' investigated by the numerous technical
approaches available to present-day biological and biomedical science.

                                                        Robin Harris
                                                        Brentwood, UK
                                                        May, 1990

CONTENTS

THE PUTATIVE TRANSPOSABLE ELEMENT T14 IS LOCATED ON
A Y CHROMOSOMAL LAMPBRUSH LOOP AND OTHER GENOMIC SITES
AND IS TRANSCRIBED IN THE GERM LINE OF *DROSOPHILA HYDEI*

R. C. Brand and W. Hennig

Catholic University of Nijmegen, Faculty of Sciences
Department of Molecular and Developmental Genetics
Toernooiveld, 6525 ED Nijmegen
The Netherlands

INTRODUCTION

In *Drosophila* the final differentiation of male germ cells into mature
spermatozoa is dependent on the presence of the Y chromosome. Although the
fertility genes on the Y chromosome are required for the correct completion
of spermatogenesis, these genes are transcribed premeiotically. In
D. hydei, five of the Y chromosomal fertility genes are visible as giant
lampbrush loops in the nucleus of primary spermatocytes (reviews: Hennig,
1987; Hackstein, 1987).

The molecular analysis of the structure of Y chromosomal lampbrush loop
DNA may provide insight into the function of the genes forming these loops.
Genomic DNA clones were isolated and found to represent moderately repetitive
DNA (Vogt et al. 1982; Lifschytz et al., 1983; Hennig et al. 1983; Vogt and
Hennig, 1983; Huijser and Hennig, 1987). Y chromosomal DNA sequences are
constructed from two types of repeated DNA sequences (Vogt and Hennig, 1983).
The first type represents tandemly repeated DNA sequences with a complexity
of several hundred base pairs, which are only found on the Y chromosome and,
therefore, called Y-specific. This DNA constitutes most of the loop DNA.
The second type contains repeated DNA sequences several kb in length, which
are found on the Y chromosome as well as in other genomic sites and, therefore, called Y-associated. Y-associated sequences are more heterogeneous
than Y-specific DNA. Both types of DNA sequences occur interspersed within
the fertility gene forming the lampbrush loop nooses on the short arm of the
Y chromosome (Vogt and Hennig, 1986a,b) and also within the DNA from other
loops (Huijser et al., 1988; Trapitz et al., 1988; Wlaschek et al., 1988).

Y chromosomal lampbrush loop DNA is transcribed into large transcripts
(Meyer and Hennig, 1974; Hennig et al., 1974; Glaetzer and Meyer, 1981).
Miller spreading of the lampbrush loop nooses revealed a single transcription
unit of at least 260 kb in length (Grond et al., 1983). Thus the Y-specific
and interspersed Y-associated DNA sequences within this loop are transcribed
together into a single large primary transcript. Other Y chromosomal
lampbrush loop forming genes code for primary transcripts of a similar size
as found for the nooses or even longer sizes (de Loos et al., 1984). The
cellular function and fate of these giant primary transcripts slowly emerged
from different experimental approaches (Hennig, 1987; Hennig et al., 1989).
The distinction between two types of Y chromosomal DNA sequences may also be

important when the function of lampbrush loop RNA is envisaged. With respect to the Y-specific type of repeated DNA sequences, a model has been proposed in which the transcripts of these sequences have a function in binding chromosomal proteins (Hennig, 1987). These proteins might be involved in the reorganization and replacement of chromosomal proteins during or after meiosis. In this paper we focus on the function of the Y-associated DNA found in lampbrush loops. It is possible that the transcripts from Y-associated DNA sequences (1) have no specific function in spermatogenesis but represent transcripts from transposable elements, (2) encode structural or regulatory RNA, or (3) serve as regulatory RNA. The analysis of a cDNA clone representing an abundant testis RNA species transcribed from Y-associated DNA sequences suggests that a family of transposable elements is involved.

MATERIALS AND METHODS

The origin of *D. hydei* strains, collection of embryos, isolation of nucleic acids, construction of a cDNA library from poly(A)$^+$ testis RNA and screening for sequence similarity to Y chromosomal DNA, cloning of genomic DNA, blot and *in situ* hybridizations are described by Brand and Hennig (1989).

RESULTS AND DISCUSSION

The T14 Family of Y-Associated Sequences

One way to experimentally study the function of Y chromosomal tran-scripts is their analysis using cDNA clones. The partial cDNA clone cDhT14 was selected from a cDNA library constructed from poly(A)$^+$ testis RNA isolated from wild type males of *D. hydei* (Brand and Hennig, 1989). It represents a testis RNA with sequence similarity to Y chromosomal sequences. The insert of clone cDhT14 hybridized to a 2.3 kb PstI fragment which is length specific for DNA from males of *D. hydei* strain Tübingen (Figure 1a). Therefore, these fragments of 2.3 kb must be derived from the Y chromosome. In addition, this cDNA clone hybridized to a number of PstI fragments present in DNA from both males and females. These DNA fragments may originate from either autosomes and/or from both the X and Y chromosome. This complete series of repeated DNA sequences – the T14 family – belongs to the type of Y-associated sequences (Vogt and Hennig, 1983; *cf*. Hennig et al., 1987), because some of these sequences are located on the Y chromosome while others are located on other genomic sites (see also below).

The 8.3 kb PstI DNA Fragments; Characterization of a Genomic Clone

Out of the T14 family of repeated sequences represented by the PstI fragments shown in Figure 1a, the fragments of 8.3 kb are particularly interesting because only these fragments form stable hybrids with the insert of cDht14 after a wash under high stringency conditions (Figure 1b). This indicates a high degree of sequence similarity. Thus the genomic origin of the RNA represented by clone cDhT14 is most likely contained within the PstI fragments of 8.3 kb.

The copy number of these fragments was analyzed by titration of genomic DNA digested with PstI relative to the insert of cDhT14 and with the same insert DNA used as probe (Figure 2). Approximately 10 - 15 copies of the cDhT14 sequence are represented by the PstI fragments of 8.3 kb. The remainder of PstI fragments represents 35 - 40 copies of the cDhT14 sequence. Thus the 8.3 kb PstI DNA fragments constitute 20 - 30% of the total T14 family.

2

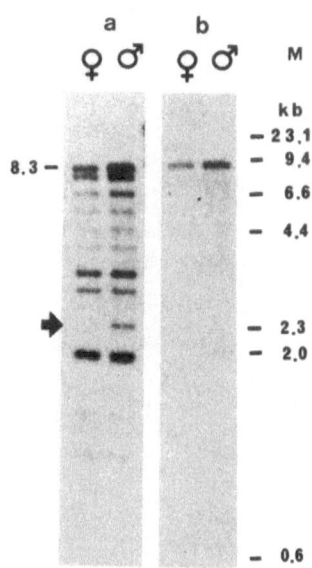

Figure 1a,b. A cDNA clone of a testis poly(A)$^+$RNA shows similarity to Y chromosomal DNA sequences. A Southern blot of PstI digested DNA from *Drosophila hydei* females versus males was hybridized with cDhT14 insert DNA (a, b). Each lane contained approximately 3µg genomic DNA. The nitrocellulose blot was washed at 60°C in 2 x SSC (a) and, after autoradiography, in 0.1 x SSC (b). SSC is 0.15 M NaCl, 0.015 M sodium citrate. Positions of marker DNA fragments (HindIII digested lambda DNA) are indicated on the right. The 2.3kb PstI fragment, which is length specific for DNA of male flies, is marked by an *arrow*. (From Brand and Hennig, 1989).

A genomic clone for one of the 8.3 kb PstI DNA fragments was isolated (Brand and Hennig, 1989). This clone, designated DhT14-8.3 was used as a probe for *in situ* hybridization (Figure 3). Mitotic metaphase chromosomes are strongly labelled at the distal end of the long arm of the Y chromosome (Figure 3a). Therefore most of the 10 - 15 members of the T14 family represented by the DNA fragments of 8.3 kb are located on the Y chromosome. Figure 3b, c shows the localization of sites with a low copy number using polytene chromosomes. Two positions are labelled; region 12D/3A of the X chromosome (Figure 3b) and region 112 of the fifth chromosome (Figure 3c). Transcript in situ hybridization to nuclei of primary spermatocytes strongly labels the lampbrush loop region 'cones' (Figure 3d and schematically shown in Figure 3e). These results show that the majority of the 8.3 kb DNA stretches are located on the Y chromosome while the other copies are on the X and fifth chromosome. In addition the hybridization to transcripts in primary spermatocyte nuclei shows that the Y chromosomal location containing repetitive DNA sequences of the T14 family is transcribed.

Transcription of cDhT14-Related Sequences

Hybridization of the insert of cDhT14 to total RNA from testes of wild type *D. hydei* males reveals an abundant RNA transcript of 5.0 kb (Figure 4: t^{xy}). This RNA is also expressed in testis of XO flies (Figure 4: t^{xo}). Therefore at least one of the copies of the 8.3 kb PstI DNA fragment located on the X and fifth chromosome is actively transcribed in testis. In

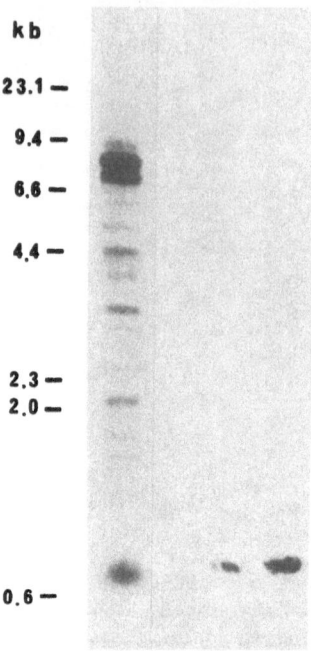

Figure 2.   Titration of sequences with sequence similarity to cDhT14 in
             DNA from *D. hydei* males. Nick translated insert DNA of cDhT14
             (in the presence of an excess of unlabelled vector DNA) was
             hybridized to 5µg genomic DNA digested with PstI and a series
             of cDhT14 DNA digested with PstI. The amount of this control
             DNA corresponds to 1, 5 or 10 copies of the T14 sequence in
             genomic DNA. The genomic size of *D. hydei* males is 4.57 x
             10^8 bp (Zacharias et al., 1982). The DNA blot on a Gene
             Screen Plus membrane was washed at 60°C in 0.3 M Na_2HPO_4
             (pH 7.2) and, after autoradiography, in 0.02 M Na_2HPO_4. The
             latter condition is shown. The extreme reduction in number of
             PstI fragments hybridizing at high stringency as seen on a
             nitrocellulose membrane (see Figure 1b) is less prominent on
             Gene Screen Plus membranes. Ps, PstI.

addition, the labelling of the lampbrush loop region cones by transcript *in
situ* hybridization (Figure 3d) shows that at least some Y chromosomal
sequences with similarity to DhT14-8.3 are transcribed. Whether or not these
Y chromosomal transcripts are processed into the 5.0 kb RNA is hard to prove.
The absence of long transcripts other than the RNA species of 5.0 kb in total
RNA from XY testes (Figure 4: t^{xy}), even after a long exposure (Figure 6c in
Brand and Hennig, 1989), suggests that processing does occur. In male
somatic tissue (the fly carcass after testis dissection: c^{xy} and c^{xo} in
Figure 4) the 5.0 kb RNA was not detected. However, this RNA is also
expressed in ovaries (Figure 4: o^{xx}) and contributed to early and mid
embryogenic stages as a maternal transcript (Figure 4). The transcription of
the 5.0 kb RNA might, therefore, be restricted to the male and female germ
line. *In situ* hybridization to transcripts in tissue sections of testis with
the insert of cDht14 as probe shows that the 5.0 kb RNA is present in the
cytoplasm of primary spermatocytes (Figure 5). As expected, the primary
spermatocyte nucleus is labelled at the same lampbrush loop as in Figure 3d
(the cones are considered to be part of the lampbrush loop pseudonucleolus,
see Figure 3e).

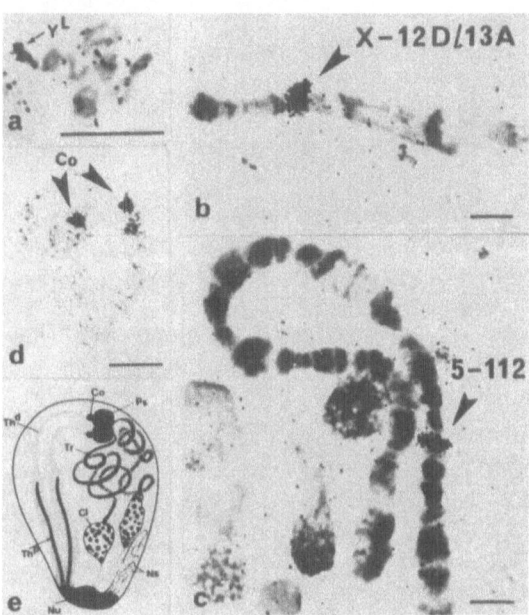

Figure 3 a-e. In situ hybridization of ^{3}H-labelled DhT14-8.3 to neuroblast
metaphase chromosomes (a), polytene chromosomes from salivary
glands (b, c) and transcripts in the nucleus of primary
spermatocytes (d). In all cases wild-type *D. hydei* was used
for the slides. A schematic representation of the Y
chromosomal lampbrush loops (Th, threads; Ps, pseudonucleolus;
Tr, Tubular ribbons; Cl, clubs; Ns, nooses; Co, cones; Nu,
nucleolus) in a primary spermatocyte nucleus is shown (e).
Panel (d) is a bright field picture. Bar represents 10μm.
(From Brand and Hennig, 1989).

The Function of Transcripts From Y-Associated DNA
_____

    The data obtained so far allow us to make a first distinction between
the possibilities for the function of transcripts from Y-associated DNA (see
Introduction). The possibility that the Y chromosomal DNA sequences detected
with clone DhT14-8.3 encode a structural protein involved in spermatogenesis
is unlikely. Indirect evidence does suggest that the primary transcripts of
the lampbrush loop region cones are processed into the 5.0 kb RNA species
(see above). However, a major Y chromosomal contribution to the total amount
of 5.0 kb RNA in testis is excluded, because of the almost identical
concentration of this transcript in total RNA from XO and XY testes (Figure
4). Thus in testis most of the 5.0 kb RNA transcripts originates from the
loci on the X and/or fifth chromosome. In ovaries the latter loci are also
transcribed into the 5.0 kb RNA (Figure 4). An exclusive role for this
mature RNA species or a presumed translation product during spermatogenesis
is therefore unlikely.

    The alternative possibility that the Y chromosomal DNA sequences with
similarity to DhT14-8.3 encode a regulatory RNA or protein present in
catalytic amounts cannot be excluded. As discussed above, the Y chromosomal
contribution to the total amount of 5.0 kb RNA in testis can only be minor
compared to the contribution of loci on the X and/or fifth chromosome. When
this possibility holds true, the latter loci may encode a structural protein,

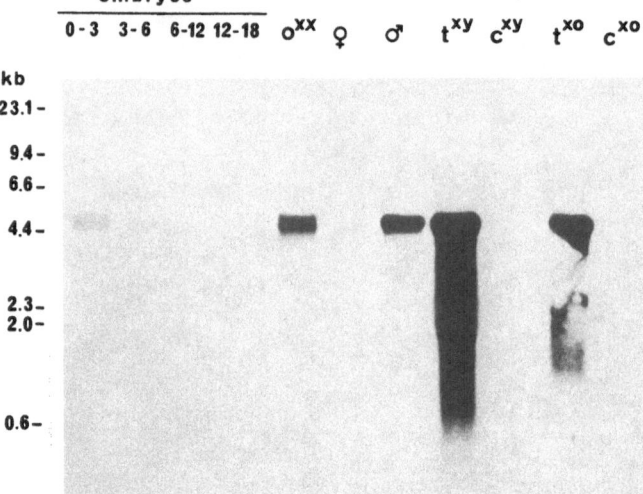

Figure 4.   Northern blot hybridization of 20μg total RNA from *D. hydei*
embryos of 0-3h, 3-6h, 6-12h and 12-18h, ovaries from wild-
type females (o^{xx}), wild-type female flies (♀), male flies (♂),
somatic tissue of wild-type males (c^{xy}), wild-type testes
(t^{xy}), XO male somatic tissue (c^{xo}), and XO testes (t^{xo}) with
the labelled insert of clone cDhT14. Autoradiography with two
intensifying screens was for 16h. In control experiments we
tested whether identical amounts of total RNA were loaded in
different lanes. For this purpose we compared the hybridization with a labelled rDNA repeat probe (clone dhPR21, see
Figure 2 in Huijser and Hennig, 1987). The hybridization
signals were similar in all lanes (not shown). Molecular
weight markers were HindIII digested lambda DNA (on the left).

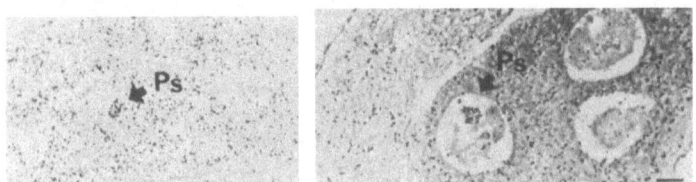

Figure 5.   In situ hybridization of ^{3}H-cDhT14 on transcripts in tissue
sections of testes. Note the absence of label in cells of the
testis wall. Ps, pseudonucleolus; bar represents 5μm.

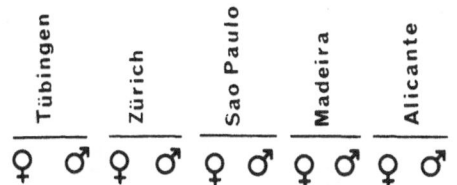

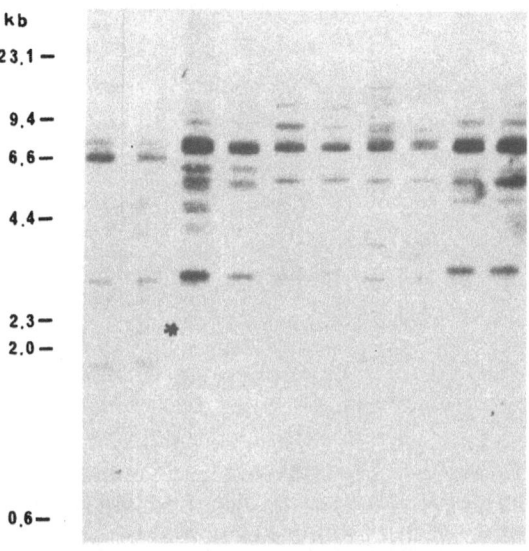

Figure 6.   Comparison of hybridization profiles with cDhT14 insert DNA as
a nick-translated probe on a Southern blot of PstI digested
DNA from different *D. hydei* wild-type strains: Tübingen,
Zürich, Sao Paulo, Madeira and Alicante. For each strain the
DNA from females and males was analyzed separately. Each lane
contained approximately 3µg genomic DNA. The 0.8% agarose gel
was pretreated with 0.25 M HCl for 15 minutes to facilitate
the transfer of high molecular weight DNA fragments. After
hybridization, the DNA blot (on Hybond-N membrane was washed
at 60°C in 0.3 M Na_2HPO_4 (pH 7.2). Molecular weight markers
were HindIII digested lambda DNA (on the left). PstI fragments, which are length specific for DNA of male flies, are
marked by an *asterisk*.
(From Brand and Hennig, 1989).

while a Y chromosomal control region may encode a regulatory RNA or protein.
This distinction between the function for transcripts form DNA sequences on
the Y chromosome and similar sequences located elsewhere in the genome has
been proposed for the Stellate locus in *D. melanogaster* (Hardy et al., 1984;
Livak et al., 1984).

The possibility that the T14 family of repeated DNA sequences represents
a family of transposable elements is supported by several data. Earlier
observations (see Hennig et al., 1987) suggested that Y-associated DNA
sequences belong to the transposable element class. The recently described
micropia family (Huijser et al., 1988; Lankenau et al., 1988), a new family
of retrotransposons, confirms this hypothesis. To test this possibility for
the T14 family of repeated DNA sequences we carried out Southern blot
ybridizations with DNAs from different wild type isolates of *D. hydei*

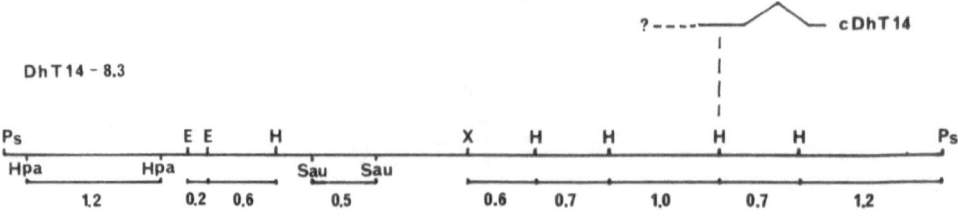

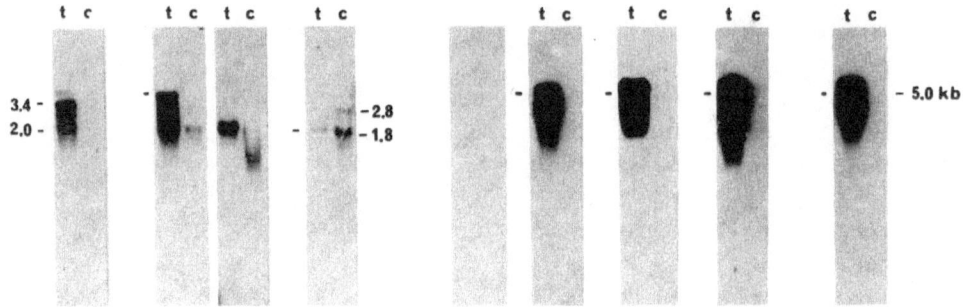

Figure 7.   Restriction map of DhT14-8.3 and Northern blot hybridization
of 20μg total RNA from D. hydei wild type testes (t^{xy}) and
somatic tissue of wild type males (c^{xy}) with labelled subfragments of DhT14-8.3 as probe. Autoradiography with two
intensifying screens was for 4 days. The position of the
partial cDNA clone cDhT14 is aligned with the corresponding
regions in the genomic clone. Ps, PstI; E, ECoRI; H, HindIII;
Hpa, HpaII; Sau, Sau 3AI.

(Zürich, Sao Paulo, Madeira, Alicante and our laboratory strain from
Tübingen). The different PstI fragments hybridizing using cDhT14 as probe
support the possibility that the T14 family of repeated DNA represents
another family of transposable elements, although restriction length
polymorphism cannot be ruled out as an explanation. The germ line specific
expression (Figure 4) and the presence of several short open reading frames
in DhT14-8.3 (data not shown) suggest similarity between this putative
transposable element T14 in D. hydei and the P-element in D. melanogaster
(O'Hare and Rubin, 1983). This cannot be confirmed, however, by the
preliminary DNA sequence analysis of clone DhT14-8.3 obtained so far. It is
clear that the coding region for the 5.0 kb RNA is not fully contained in
this genomic clone. Some subfragments of DhT14-8.3 hybridize to this 5.0 kb
testis RNA while others do not or detect testis and even carcass RNA species
of another length (Figure 7 shows an overexposure of the Northern blots).
Therefore the flanking regions of the genomic DNA sequence and the full
length cDNA clone need to be analyzed to clarify this point. It also remains
to be established whether Y chromosomal copies of this putative transposable
element T14 are merely tolerated within the lampbrush region cones, because
their co-transcription with Y-specific DNA sequences does not interfere with
the function of this particular fertility gene during spermatogenesis, or
play a functional role.

REFERENCES

Brand R. C., Hennig W. (1989) An abundant testis RNA species shows sequence similarity to Y chromosomal and other genomic sites in *Drosophila hydei*. Mol Gen Genet 215:469.

deLoos F., Dijkhof R., Grond C. J., Hennig W. (1984) Lampbrush loop-specificity of transcript morphology in spermatocyte nuclei of *Drosophila hydei*. EMBO J 3:2845.

Glätzer K. H., Meyer G. F. (1981) Morphological aspects of the genetic activity in primary spermatocyte nuclei of *Drosophila hydei*. Biol Cell 41:165.

Grond C. J., Siegmund I., Hennig W. (1983) Visualization of a lampbrush loop-forming fertility gene in *Drosophila hydei*. Chromosoma 88:50.

Hackstein J. H. P. (1987) Spermatogenesis in *Drosophila*, in: Results and Problems in Cell Differentiation, volume 15, Spermatogenesis: Genetic aspects. Hennig W. (ed) Springer-Verlag, Berlin, Heidelberg.

Hardy R. M., Lindsley D. L., Livak K. J., Lewis B., Silversten A. L., Joslyn G. L., Edwards J., Bonaccorsi S. (1984) Cytogenetic analysis of a segment of the Y chromosome of *Drosophila melanogaster*. Genetics 107:591.

Hennig W. (1987) The Y chromosomal lampbrush loops of *Drosophila*, in: Results and Problems in Cell Differentiation, Volume 14, Structure and function of eukaryotic chromosomes. Hennig W. (ed) Springer-Verlag, Berlin, Heidelberg, New York.

Hennig W., Brand R. C., Hackstein J., Huijser P., Kirchhoff C., Kremer H., Lankenau D-H, Vogt P. (1987) Structure and function of Y chromosomal genes in *Drosophila*. Chromosomes Today 9:48.

Hennig W., Brand R. C., Hackstein J., Hochstenbach R., Kremer H., Lankenau D-H, Meyer S., Miedema K., Pötgens A. (1989) Y chromosomal fertility genes of *Drosophila*: a new type of eukaryotic genes. Genome 31:

Hennig W., Huijser P., Vogt P., Jäckle H., Edström J-E (1983) Molecular cloning of microdissected lampbrush loop DNA sequences of *Drosophila hydei* EMBO J 2:1741.

Hennig W., Meyer G. F., Hennig I., Leoncini O. (1974) Structure and function of the Y chromosome of *Drosophila hydei*. Cold Spring Harbor Harb Symp Quant Biol 38:673.

Huijser P., Hennig W. (1987) Ribosomal DNA-related sequences in a Y chromosomal lampbrush loop of *Drosophila hydei*. Mol Gen Genet 206:441.

Huijser P., Kirchoff C., Lankenau D-H, Hennig W. (1988) Retrotransposonlike sequences are expressed in Y chromosomal lampbrush loops of *Drosophila hydei*. J Mol Biol 203:689.

Lankenau D-H, Huijser P., Jansen E., Miedema K., Hennig W. (1988) Micropia: a retrotransposon of *Drosphila* combining structural features of DNA viruses, retroviruses and non-viral transposable elements. J Mol Biol 204:233.

Lifschytz E., Hareven D., Azriel A., Brodsly H. (1983) DNA clones and RNA transcripts of four lampbrush loops from the Y chromosome of *Drosophila hydei*. Cell 32:191.

Livak K. J. (1984) Organization and mapping of a sequence on the *Drosophila melanogaster* X and Y chromosomes that is transcribed during spermatogenesis. Genetics 107:611.

Meyer G. F., Hennig W. (1974) Molecular aspects of the fertility factors in *Drosophila* in: The functional anatomy of the spermatozoon, Afzelius, B. A., ed., Pergamon Press, Oxford, New York.

O'Hare K., Rubin G. M. (1983) Structures of P transposable elements and their sites of ensertion and excision in the *Drosophila melanogaster* genome. Cell 34:25.

Trapitz P., Wlaschek M., Bünemann H. (1988) Structure and function of Y chromosomal DNA. II. Analysis of lampbrush loop associated transcripts in nuclei of primary spermatocytes of *Drosophila hydei* by in situ hybridization. Chromosoma 96:159.

Vogt P., Hennig W. (1983) Y chromosomal DNA of *Drosophila hydei*. J Mol Biol 167:37.

Vogt P., Hennig W. (1986a) Molecular structure of the lampbrush loops nooses of the Y chromosome of *Drosophila hydei*. I. The Y chromosome-specific repetitive DNA sequence family ay1 is dispersed in the loop DNA. Chromosoma-94:449.

Vogt P., Hennig W. (1986b) Molecular structure of the lampbrush loops nooses of the Y chromosome of *Drosophila hydei*. II. DNA sequences with homologies to multiple genomic locations are major constituents of the loop. Chromosoma 94:459.

Vogt P., Hennig W., Siegmund I. (1982) Identification of cloned Y chromosomal DNA sequences from a lampbrush loop of *Drosophila hydei*. Proc Natl Acad Sci USA 79:5132.

Wlaschek M., Awgulewitsch A., Bünemann H. (1988) Structure and function of Y chromosomal DNA. I. Sequence organization and localization of four families of repetitive DNA on the Y chromosome of *Drosophila hydei*. Chromosoma 96:145.

Zacharias H., Hennig W., Leoncini O. (1982) Microspectrophotometric comparison of genome sizes of *Drosophila hydei* and some related species. Genetica 58:153.

DNA SEQUENCE-SPECIFIC LOCATION OF COVALENT DNA-POLYPEPTIDE COMPLEXES IN

EUKARYOTIC GENOMES

Dieter Werner and Marita Pfütz

Institute of Cell and Tumor Biology
German Cancer Research Center
D-6900 Heidelberg, F.R.G.

Work on specific gene expression has not yet identified the factors determining the portions of genomes that are selectively expressed in different cell types. A number of genes are known that are regulated by *transient* interactions of cell-type-specific polypeptides with regulatory DNA sequences. However, this cell-specific regulation cannot be regarded as the basic mechanism determining the phenotype-specific patterns of gene expression because the regulatory polypeptides involved are, on their own, products of cell-type-specifically regulated genes. Alternatively, it has been suspected that the expression of specific sets of genes in various cell types could be induced by *fixed* interactions between nonhistone nuclear olypeptides and regulatory DNA sequences. Theoretically, a highly ordered pattern of *permanent* and site-specific DNA-polypeptide complexes with regulatory functions could encode the portion of the genome that is selectively expressed in each cell type. Slight variations in the locations of such complexes in variant cell types could induce the expression of different sets of genes. This hypothesis is far from being experimentally proved, however, covalent DNA-polypeptide complexes found in all eukaryotic genomes investigated show characteristics that have to be expected from such hypothetical regulatory elements: eg, they are site-specifically located in eukaryotic genomes (Neuer-Nitsche and Werner, 1987; Neuer-Nitsche et al., 1988; Werner and Neuer-Nitsche, 1989), they are associated with the nuclear matrix (Werner and Rest, 1987) and they are metabolically stable (Tsanev et al., 1989).

Nuclear polypeptides that are not released from DNA by alkali, SDS, moderate concentrations of proteases, and phenol have been found in DNA of different species, including mammals (Krauth and Werner, 1979; Neuer et al., 1983; Bodnar et al., 1983; Avramova and Tsanev, 1987; Razin et al., 1988), insects (Plagens, 1978) and plants (Capesius et al., 1980; Avramova et al. 1988). Covalent bonds between these resistant nuclear polypeptides and DNA were first shown by rigorous analysis of DNA from Ehrlich ascites cells. A combined and prolonged nuclease/protease treatment released the linking groups consisting of residual peptides bound to 3' and 5' sites of nucleo-tides (Neuer et al., 1983). The alkali-stability and the phosphodiesterase-sensitivity of the isolated linking groups are in agreement with phosphodiester bonds between O^4-hydroxyl group of tyrosine and DNA. More recently, this experimental design has been also applied by other authors to demonstrate covalent bonds between the resistant nuclear polypeptides and DNA (Avramova and Tsanev, 1987; 1988). Although this procedure indicates covalent DNA-peptide bonds impressively, it gives no information about the size and the chemical structure of the native complexes, because the macromolecular components become denatured and degraded on purpose during the treatments. The size of apparently undegraded covalent DNA-polypeptide complexes could be revealed by ^{32}P-radiolabelling during nik-translation-arrest and SDS-polyacrylamide gel analysis (Werner and Rest, 1987).

The residual DNA-peptide-phosphodiester compounds released by the prolonged treatment of denatured DNA with nucleases/proteases could reflect remnants of phosphodiester linkages between nuclear polypeptides and internal ends of native DNA. However, the frequency of the covalent complexes in genomes is much higher than that of internal DNA ends detectable either by alkali or by S1 nuclease (Werner and Neuer-Nitsche,1989a). Consequently, it was suggested that the complexes could function as polypeptide-linkers joining adjacent ssDNA strands (linker-model). Alternatively, the residual DNA-peptide-phosphodiesters could reflect remants of phosphotriester bonds between polypeptides and the uninterrupted phosphodiester backbone of native DNA (triester-modes). Both, the linker-structure and the phosphotriester complexes could by hydrolyzed during the treatments resulting in residual DNA-peptide-phosphodiesters. Consequently, none of these models could be proved or rejected experimentally until now (Werner and Neuer-Nitsche, 989a). However, computer graphics indicate that the triester-model is stereochemically more likely than the linker-model (Figure 1).

The fraction of DNA fragments associated with the resistant poly-peptides can be isolated by a nitrocellulose filter binding assay (Neuer-Nitsche and Werner, 1985). This opened the possibility to prepare radiolabelled probes for the DNA fractions associated with the resistant polypeptides. Southern hybridisations of unfractioned DNA, DNA passing nitrocellulose filters and retained DNA with such probes, revealed that DNA-polypeptide complexes reside on a subset of DNA sequences which is widely identical in the genomes of different cell types of one species (Neuer-Nitsche and Werner, 1987; Neuer-Nitsche et al., 1988; Werner and Neuer-Nitsche, 1989b). However, minor, cell-type-related variations in the locations of the covalent DNA-polypeptide complexes are not ruled out.

Since the location of covalent DNA-polypeptide complexes could reflect remnants of transcription complexes we investigated the transcriptionally active and inactive ovalbumin gene for its potential association with the residual polypeptides. DNA isolated from hen oviduct cells and from chicken erythrocytes was digested with various restriction enzymes followed by filtration through nitrocellulose filters. The partitioning of DNA fragments of the ovalbumin gene region was investigated by mapping ovalbumin gene-specific DNA probes in the DNA fractions resolved by nitrocellulose filtration. We found that all fragments of the transcriptionally active and inactive ovalbumin gene are not associated with covalent DNA-polypeptide complexes because all fragments could pass the nitrocellulose filters indicating that the covalent DNA-peptide complexes are not directly involved in transcription complexes (Werner and Neuer-Nitsche, 1989b).

By differential screening of recombinant DNA libraries with radio-labelled probes of the DNA fractions retained on nitrocellulose filters and those passing the filters, we detected and cloned DNA sequences from mouse (Neuer-Nitsche et al., 1988), chicken (Werner and Neuer-Nitsche, 1989b) and man (Pfütz and Werner, unpublished) mapping exclusively in the DNA fraction associated with the covalent DNA-polypeptide complexes. Consequently, these sequences reflect or comprise signals for the site-specific locations of the complexes in eukaryotic genomes.

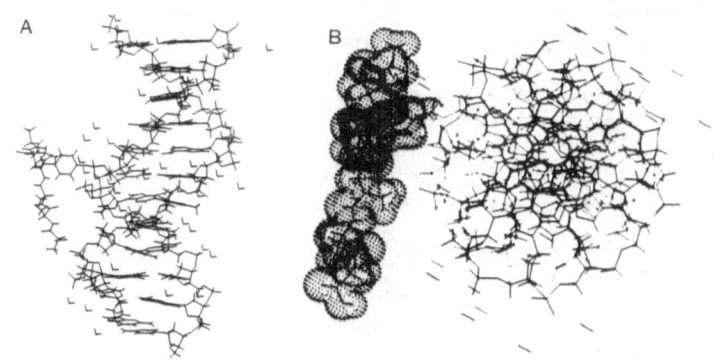

Figure 1.   Computer graphics of the triester-model for covalent DNA-
peptide complexes. A model tripeptide comprising a tyrosine
residue was linked to DNA by formation of a triester bond
between the O^4-hydroxyl group of the tyrosine residue in the
model peptide and the phosphodiester backbone of DNA, and the
resulting complex was energy optimized. Both, the side view
(A), and the view along the DNA axis (B) indicate that the
triester-complex would not distort the DNA structure. In (B)
van der Waals radia are shown for the model peptide. (With
kind permission from S. Suhai, Institute of Epidemiology and
Biometry, German Cancer Research Center, Heidelberg).

REFERENCES

Avramova Z., and Tsanev R. 1987, Stable DNA-protein complexes in eukaryotic
chromatin, J. Mol. Biol., 196:437.
Avramova Z., Ivanchenko M., and Tsanev R. 1988, A protein fraction stably
linked to DNA in plant chromatin, Plant Mol. Biol., 11:401.
Bodnar J. W., Jones C. J., Coombs D. H., Pearson G., and Ward D. C. 1983,
Proteins tightly bound to HeLa cell DNA at nuclear matrix attachment
sites. Mol. Cell. Biol., 3:1567.
Capesius I., Wrauth W., and Werner D., 1980, Proteinase K-resistant and
alkalistably bound proteins in higher plant DNA, FEBS Lett., 110:184.
Krauth W., and Werner D., 1979, Analysis of the most tightly bound proteins
in eukaryotic DNA, Biochim. Biophys. Acta, 564:390.
Neuer B., Plagens U., and Werner D., 1983, Phosphodiester bonds between
polypeptides and chromosomal DNA, J. Mol. Biol., 164:213.
Neuer-Nitsche B., and Werner D., 1985, Screening of isolated DNA for
sequences released from anchorage sites in nuclear matrix, J. Mol.
Biol., 181:15.
Neuer-Nitsche B., and Werner D., 1987, Sub-set characteristics of DNA
sequences involved in tight DNA-polypeptide complexes and their
homology to nuclear matrix DNA, Biochem. Biophys. Res. Commun.
147,335-339.
Neuer-Nitsche B. Lu, X., and Werner D., 1988, Functional role of a highly
repetitive DNA sequence in anchorage of the mouse genome, Nucleic
Acids Res., 16:8361.
Plagens U., 1978, Effect of salt-treatment on manually isolated polytene
chromosomes, Chromosoma (Berl.), 68:1.
Razin S. V., Chernokhvostov V. V., Vassetzky Jr, E. S., Razina M. V., and
Georgiev G. P., 1988, The distribution of tightly bound proteins along
the DNA chain reflects the type of cell differentiation, Nucleic Acids
Res., 16:3617.

Tsanev R., Avramova Z., and Tasheva B., 1989, DNA-bound nonhistone chromosomal proteins and loop organization of chromatin, in: 'Nonhistone Protein Research', I. Bekhor, ed., CRC Press, Boca Raton.

Werner D., and Neuer-Nitsche B., 1989a, Discontinuities of peptide nature in DNA, in: 'Nonhistone Protein Research', I. Bekhor, ed., CRC Press, Boca Raton.

Werner D., and Neuer-Nitsche B., 1989b, Site-specific location of covalent DNA-polypeptide complexes in the chicken genome, Nucleic Acids Res., 17:6005.

Werner D., and Rest R., 1987, Radiolabelling of DNA-polypeptide complexes in isolated bulk DNA and in residual nuclear matrix DNA by nick-translation, Biochem. Biophys. Res. Commun., 147:340.

MINK Sau REPEAT: STRUCTURAL ANALYSIS AND GENOME DISTRIBUTION

M. V. Lavrentieva, M. I. Rivkin, A. G. Shilov, M. L. Kobetz,
I. B. Rogozin and O. L. Serov

Institute of Cytology and Genetics
Academy of Sciences of the USSR
Siberian Department
630090
Novosibirsk-90
USSR

Alu-like interspersed repetitive sequences are represented widely in
primates and rodents (Jelinek and Schmid, 1982). We have identified
similar repeats in the mink genome (*Mustela vison*). The sequence,
designated as mink Sau3A repeat, was found in one of the phage clones
picked out from the American mink X-chromosome library (Lavrentieva et al.,
1986). The mink Sau3A repeat has a close structural resemblance to the
known mouse B2. This 195bp sequence is flanked by short direct repeats of
14bp, contains two split promoter regions for RNA-polymerase III (Krayev et
al., 1982) and 3'A-rich region with overlapping polyadenylation signals
(AATAAA), the sequence that may serve as a polymerase III termination signal
(TCTTT) (Haynes and Jelinek, 1981) and oligo(dA)sequence (A_8) (Figure 1).
An interesting feature is the presence of a perfect polypyrimidine tract
22bp long absent from the known *Alu*- and *Alu*-like sequences. Taking into
account 9 sub-situations, the polypyrimidine tract is 62bp long (positions
136-197, Figure 1). Such tracts were found to be located precisely in
nuclease hypersensitive chromatin regions (Nickol and Felsenfeld, 1983).
Furthermore, *in vitro* experimental data indicate, that the structure of
this kind prevents nucleosome formation (Simpson and Kunzler, 1979).
Noteworthy are also the pyrimidine rich 5'-flank region and the obvious
homology between the segments within the repeat and the 3'-flanking region
(see Figure 1). It cannot be ruled out that this structural resemblance was
a reason why the repeat inserted into this genomic region. Alignment of
mink Sau3A and mouse B2 sequences allowed us to estimate their homology
degree as 55% (Figure 2).

Dot-hybridization analysis of the Sau3A repeat demonstrated that it is
present in $1-2 \cdot 10^5$ copies per mink genome and about 10^4 copies per mink X
chromosome. According to the cytological observations relying on
chromosome length measurements, the X chromosome constitutes 5% of the mink
genome. The good agreement between cytological and dot hybridization
analysis shows that there should be no appreciable difference between the X
and autosomes in copy number per unit DNA.

Analysis of *in situ* hybridization of mink metaphase chromosomes with
[3]H-labelled DNA of the Sau3A repeat has demonstrated that the repeats are

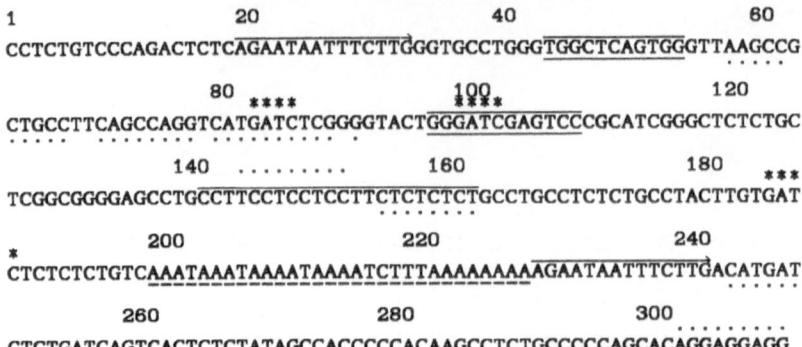

```
1              20              40              60
CCTCTGTCCCAGACTCTCAGAATAATTTCTTGGGTGCCTGGGTGGCTCAGTGGGTTAAGCCG
                                                           . . . . .

    80  ****        100 ****           120
CTGCCTTCAGCCAGGTCATGATCTCGGGGTACTGGGATCGAGTCCCGCATCGGGCTCTCTGC
. . . . .  . . . . . . . . .  .

    140  . . . . . . . . .  160            180  ***
TCGGCGGGGAGCCTGCCTTCCTCCTCCTTCTCTCTCTGCCTGCCTCTCTGCCTACTTGTGAT
                       . . . . . . . .

*       200            220            240
CTCTCTCTGTCAAATAAATAAAATAAAATCTTTAAAAAAAAAGAATAATTTCTTGACATGAT
                                                            . . . . . .

    260            280            300
CTCTGATCAGTCACTCTCTATAGCCACCCCCACAAGCCTCTGCCCCCAGCACAGGAGGAGG
. . . .      . . . . . . . . .       . . . . .  . . . . . . . . .
```

Figure 1.   Nucleotide sequence of mink Sau3A repeat with flanking regions
             cloned in pBR327. Arrows show positions of direct repeats;
             regions homologous to the RNA polymerase III promoter elements
             are boxed; the perfect polypyrimidine tract is indicated by
             the line above; the broken line represents 3'A-rich region;
             dots below indicate the regions of homology between the repeat
             and 3'-flanking region (at positions 57-77 and 282-303, 79-89
             and 243-253, 153-161 and 259-267, respectively); dots above
             indicate repeat (TCC)$_3$ and (AGG)$_3$ complementary to it;
             asterisks denote Sau3A sites.

Sau3A   GGTGCCTGGG----TGGCTCAGTGGGTTAAGCCGCTGCCTTCAGCCAGGTCATGATCT

        ** * ****   *********** ***** ** * * **   ** *   *

B2   GGGG-CTGGAGAGATGGCTCAGTGG-TTAAGA-GCACCTGACTGC----TCTT---CC

Sau3A   CGGGGTACTGGGATCGAGTCCC-GCATCG-------GGCTC-------TCTGCTC-GG

        *** * * * ** * **** *** *   *****   ** * *

B2   GAAGGTCC-GAGTTCAATTCCCAGCAACCACATGGTGGCTCACAACCATCCG-TAAT-

Sau3A   CGGG----GA-G-CCTGCCTTCCTCCTCCTTCTCTCTCTGCCTGCCT-CTCTGCCTAC

        * *   ** * *** **** *   * ***   * * ** * ** ***

B2   -GAGATCTGATGCCCT--CTTC-TGGAG--TGTCTGAA-GACAG-CTACAGTG--TAC

Sau3A   TTGTGATCTCTCTCTGTCAAATAAATAAAATAAA A   TCTTT   (A)$_8$

        **   ** *   ********** **** *   *****   *

B2   TTAC-ATAT----------AATAAATAAA-TAAA(A)$_{0-8}$ TC(T)$_{2-5}$ (A)$_{7-14}$
```

Figure 2.    A comparison of the mink Sau3A sequence and mouse B2 consensus
             sequence (Krayev et al., 1982). The sequences are aligned in
             accordance with the best matching. Asterisks denote nucleo-
             tide matches.


distributed uniformly along the length of all chromosomes including the X
(Figure 3). The telomeric regions are heavily and the centrometric weakly
labelled. The metaphase spreads were examined more thoroughly to determine
how the grains are distributed along the length of the X; 26 of 48 silver
grains (54%) are located on the end of the long arm in the region Xq14-qter
comprising 17% of the total chromosome length. It should be noted that the
silver grain number on the short arm of the mink X is less than expected
```

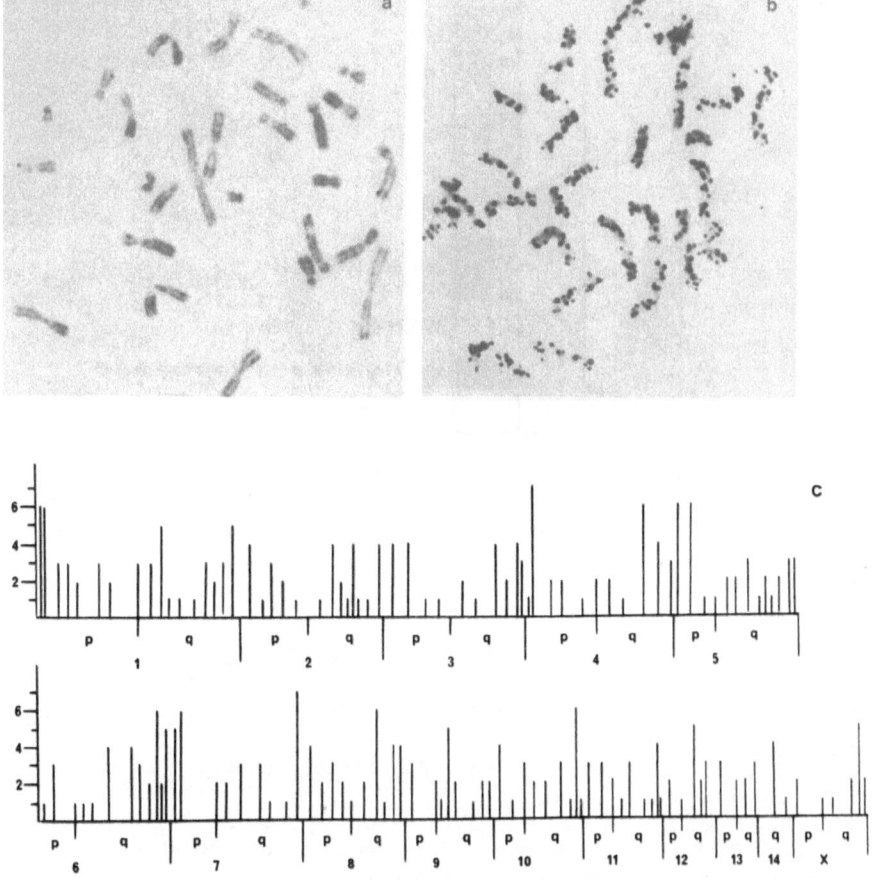

Figure 3.   *In situ* hybridization of a mink metaphase spread with [3]H-
             labelled Sau3A repeat performed under conditions of high (a)
             and low (b) stringency. (c) histogram showing the distri-
             bution of 373 silver grains among mink chromosomes from 25
             metaphase spreads. The abscissa represents the mink
             chromosomes in their relative size proportions, and the
             ordinate number of silver grains.

from measurements of chromosome length. However, this does not seem to be
specific to this region because short arms of other chromosomes also
labelled weakly, for example, those of chromosome 6 or 9 (Figure 3c).

     Closely resembled repeated sequences have been revealed in the genomes
of other *Carnivorae* species *(Mustela lutreola, Mustela erminea, Vulpes
fulvus, Alopex lagopus)* by means of dot-hybridization analysis of their
DNAs under stringent conditions.

     To our knowledge, this is the first evidence for the presence of *Alu*-
like repeats in the carnivore genome.

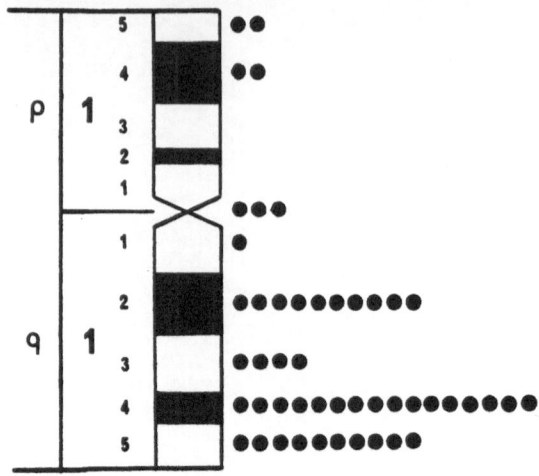

Figure 4.   Regional distribution of silver grains on 35 labeled mink X
chromosomes after *in situ* hybridization with Sau3A repeat.
26 of 48 grains (54%) were located in the region Xq14-qter.

REFERENCES

Haynes S. R., and Jelinek W. R. 1981, Low molecular weight RNAs transcribed
in vitro by RNA polymerase III from *Alu*-type dispersed repeats in
Chinese hamster DNA are also found *in vivo*, Proc. Natl. Acad. Sci.,
USA, 78:6130.
Jelinek W. R., and Schmid C. W. 1982, Repetitive sequences in eukaryotic
DNA and their expression, Ann. Rev. Biochem., 51:813.
Krayev A. S., Kramerov D. A., Skryabin K. G., Ryskov A. P., Bayev A. A., and
Georgiev G. P. 1980, The nucleotide sequence of the ubiquitous
repetitive DNA sequence B1 complementary to the most abundant class of
mouse fold-back RNA, Nucleic Acids Res., 8:1201.
Lavrentieva M. V., Rivkin M. I., Shilov A. G., Karasik G. I., Gradov A. A.,
Kumarev V. P., and Serov O. L. 1986, New strategy for obtaining
chromosome libraries based on the X-chromosome, Dokl. Akad. Nauk SSSR
(Russian), 290:982.
Nickol J. M., and Felsenfeld G. 1983, DNA conformation at the 5' end of the
chicken adult β-globin gene, Cell, 35:467.

STRUCTURE OF THE PLANT MITOCHONDRIAL GENOME AND LIGHT-REGULATED

TRANSCRIPTION OF THE MITOCHONDRIAL GENES

R. I. Salganik, N. A. Dudareva, A. V. Popovsky,
E. V. Kiseleva, and S. M. Rozov

Institute of Cytology and Genetics
Siberian Branch of the USSR Academy of Sciences
630090 Novosibirsk, USSR

INTRODUCTION

The higher-plant mitochondrial (mt) genome differs from the animal genome in many respects. The mt genomes of plants attain 2500 kb, but those of animals do not exceed 18 kb. The plant mt genomes are organised as multiple circular molecules and they contain, in addition to high molecular weight (HMW) DNA, low molecular weight (LMW) DNA, including plasmid-like molecules (Newton, 1988).

However, the structural organization of the plant mt genome and, particularly, the regulation of its transcription remain unclear. In this paper, we present the results of biochemical and electron microscopy studies of the organization of the sugarbeet (*Beta vulgaris L.*) mt genome and data pertaining to the regulation of its transcription.

MATERIALS AND METHODS

Preparation of sugarbeet mitochondria and mtDNA, as well as methods of their electron microscopic analysis, were previously described (Dudareva et al., 1988; Kiseleva et al., 1989). Clones pBN6601 (cytochrome c oxidase subunit II), pZMEH 680 (apocytochrome b), and copy V (the alpha subunit of F_1 ATPase) were kindly provided by Dr C. J. Leaver, the clones were used as probes in the hybridization experiments. The antibodies against the - subunit of *E. coli* RNA polymerase were kindly provided by Dr V. V. Roshke and Dr P. P. Laktionov.

RESULTS AND DISCUSSION

During 1.5% agarose gel electrophoresis the mtDNA separates into a broad band which contains more than 90% of the mt DNA representing the HMW fraction, and three discrete bands representing the LMW fraction. Judging from electrophoretic behaviour after S_1-nuclease treatment, the LMW mtDNA are supercoiled circular molecules of 1.3, 1.4, and 1.6 kb. Electron microscopy analysis reveals that the deproteinized multipartite sugarbeet mtDNA consists of large circular molecules with contour lengths corresponding to 62-496 kb, supercoiled subgenomic DNA (3.1-31 kb), and plasmid-like minicircular

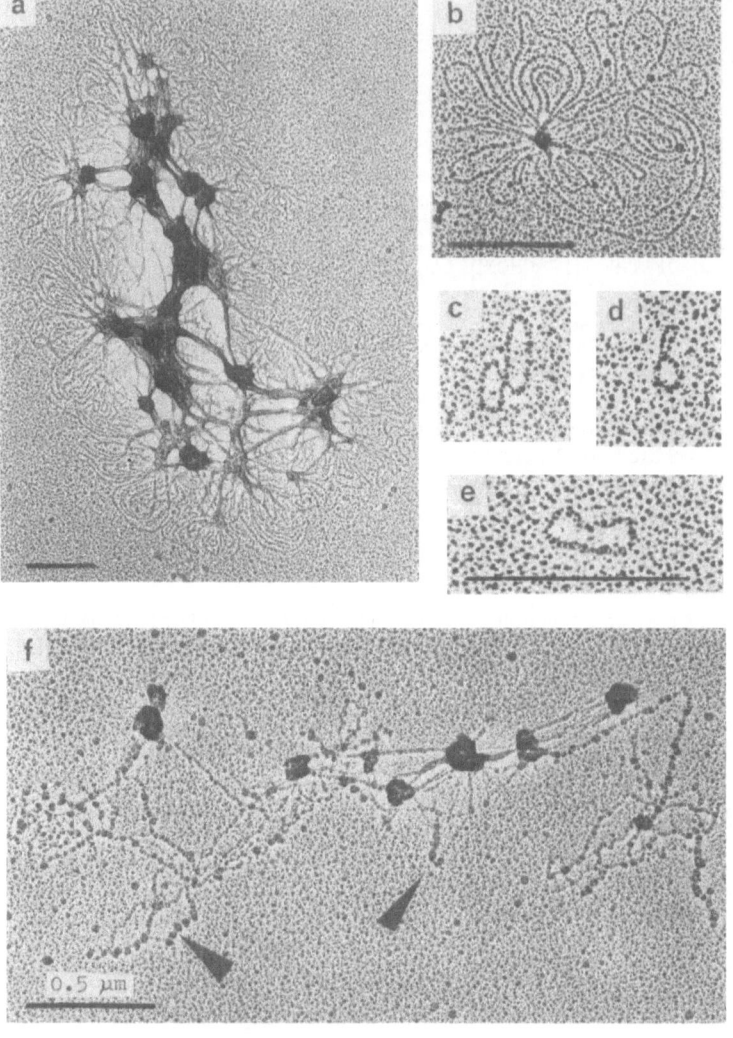

Figure 1.   Electron micrographs of sugarbeet mtDNA spreads: (a) a high
molecular weight mtDNA consisting of interconnected rosettes;
(b) a circular molecule of mtDNA forming a rosette; (c – e)
relaxed forms of minicircular mtDNA; (f) a high molecular
weight mtDNA containing fibrils with regularly distributed
granules of different sizes (arrows).

molecules with contour lengths corresponding to 0.6-2.4 kb (Figure 1a – e).
However, after a single phenol deproteinization without proteinase K treatment, the proteins remain attached to the mtDNA molecules. Electron
microscopy identified deoxynucleoprotein fibrils with regularly distributed
nucleosome-like globules (d=10-12 nm), larger globules (d=25-30 nm)
resembling nucleomeres, and rarely very large chromomere-like globules
(d=150-200 nm) (Figure 1f). These data for the first time demonstrate a
similarity between the molecular organization of the plant mitochondrial and
nuclear genomes. Recently, a very similar chromosome organization was
revealed in the chloroplast (Kiseleva et al., 1989).

Histones and histone-like basic proteins were extracted from the sugar-beet nuclei, mitochondria and chloroplasts by perchloric and sulphuric acids. Their electrophoretic patterns were compared. As Figure 2 shows, the nuclei contain the usual histone set, H1, H2a, H2b, H3, and H4. The histone-like proteins of sugarbeet mitochondria differ from the chloroplast histone-like proteins, and both differ from the nuclear histones in molecular weight and electrophoretic pattern. The data support the idea of the different evolutionary origin of the three cellular compartments.

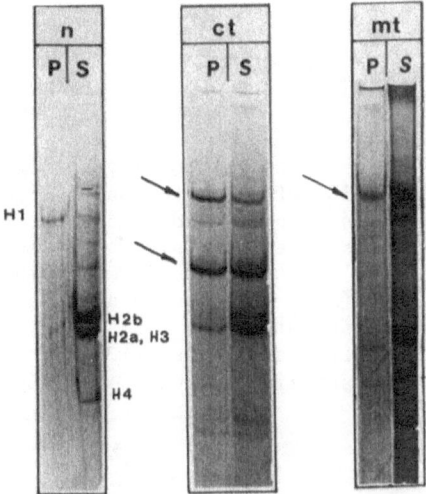

Figure 2.   Basic proteins of Beta vulgaris nuclei (n), chloroplasts (ct), and mitochondria (mt) extracted with perchloric (P) and sulphuric (S) acids.

Little is known about the regulation of mt gene transcription. Mito-chondria function predominantly in the dark, when photosynthesis cannot meet the energy demands of the cell. It was reasonable, hence, to suggest that the mt genes are expressed mainly in the dark. To test this suggestion, we compared the transcription level in the mitochondria from 5-day-old sugarbeet seedlings grown in the dark and additionally exposed to 10 h daylight. MtRNA from these seedlings were dot-blot-hybridized using cloned mt genes as probes. Figure 3a shows that all these genes are well expressed in the dark; in contrast, their transcription is suppressed in the case of daylight exposure. These data suggest that the expression of the plant mt genes may be light-regulated. It seems hardly conceivable that mitochondria possess their own photoreceptors; the chloroplasts receive photons and may then transmit the light signal through an unknown mediator to the mito-chondria.

In the studies of the possible mechanisms of light-regulated trans-cription, we found that RNA polymerase of plant mitochondria has antigens in common with the β-subunit of E. coli RNA polymerase. Electron microscopy with the use of monoclonal antibodies against the β-subunit of E. coli RNA polymerase and the protein A-colloidal gold complex as a marker, made it possible to visualize the location of RNA polymerase and, accordingly, the transcriptionally active sites of the mitochondrial chromosomes. In the

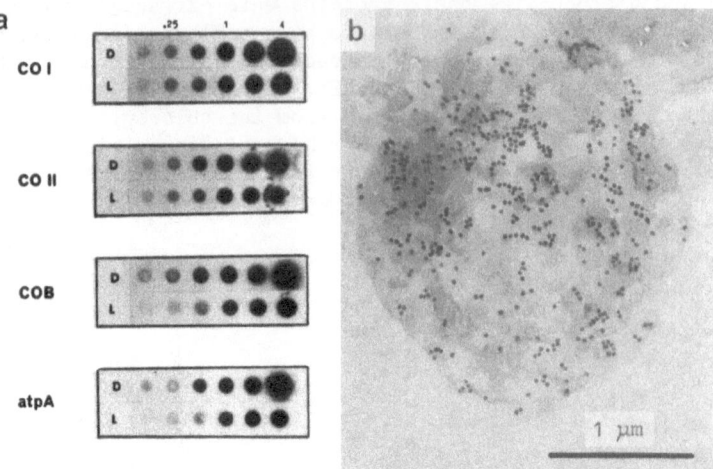

Figure 3.   Analysis of the transcriptional activity of sugarbeet mtDNA.
(a) transcription of the mitochondrial genes: D – mitochondria
isolated from sugarbeet seedlings grown in the dark, L –
mitochondria isolated from sugarbeet seedlings exposed to
daylight after growth in the dark. Numbers are the amounts of
mtRNA in dots (mg); (b) binding of monoclonal antibodies
against the β-subunit of *E. coli* RNA polymerase to
mitochondria after a mild osmotic shock. The mitochondria were
isolated from seedlings grown in the dark.

dark, about 70% of all the mitochondria have a large number of RNA polymerase molecules (Figure 3b); in the light, the number of mitochondria with
RNA polymerase molecules is as low as 10%. It seems reasonable to assume
that the synthesis of mtRNA polymerase may also be induced in the dark,
whereas the enzyme molecules are abolished by hydrolysis or in some other way
in the light. This may be one of the important mechanisms of transcription
regulation.

REFERENCES

Dudareva, N. A., Kiseleva E. V., Boyarintseva A. E., Maystrenko A. G.,
    Khristolyubova N. B., and Salganik R. I. 1988, Structure of mito-
    chondrial genome of *Beta vulgaris L.,* Theor. Appl. Genet., 76:753.
Kiseleva E. V., Dudareva N. A., Dikalova A. E., Khristolyubova N. B.,
    Salganik R. I., Laktionov, P. P., Roshke V. V., and Zaichikov E. F.
    1989, The chloroplast genome of *Beta vulgaris L.*: structural
    organization and transcriptional activity, Plant Sci., 62:93.
Newton K. J. 1988, Plant mitochondrial genomes: organization, expression and
    variation, Ann. Rev. Plant Physiol., 39:503.

ROLE OF DAMAGE OF SKELETAL STRUCTURES IN THE INDUCTION OF CHROMOSOME

ABERRATIONS

S. I. Zaichkina, O. M. Rosanova, G. F. Aptikaeva, and
E. E. Ganassi

Institute of Biological Physics
Acad. Sci. USSR Pushchino
Moscow Region, 142292, USSR

INTRODUCTION

Studies of the mechanism of induction of chromosome aberrations by ionizing radiation and endonucleases treatment are means of determining the native organization of nuclear structure. The modern radial loop model for chromosome organization received supporting evidence from these investigations. In the present work we compared the effects of radiation and nucleases at both molecular and cytogenetic levels. It was found that the damage of skeletal structures of chromosome play a decisive role in production of chromosome aberrations.

MATERIALS AND METHODS

An asynchronous culture of Chinese hamster fibroblasts (clone 431) were cultured by a standard technique. 24 h after seeding, exponentially growing cells were permeabilized by the technique of Obe et al. (1985), and cultured on glass slides under standard growth conditions. Cells were fixed after 7.5 and 18 h to analyze cells in metaphase and after 24 h to analyze cells with micronuclei (Zaichkina, and Ganassi, 1984). Restriction enzymes Sal I and Hind III (Amersham), Eco RV and Hae III (USSR) and DNA-ase I (Boehringer, Mahnheim, GFR) were used.

Cells were exposed to γ-rays ^{60}Co at a dose rate of 2.7 Gy/min. The yield of DNA damage was determined spectrophotometrically by a specially developed cytochemical method *in situ* (Zaichkina et al., 1988).

RESULTS AND DISCUSSION

Analysis of the repair kinetics in the post-irradiation period indicated that nearly 1% of the DNA lesions produced remained non-repaired (Sakai and Okada, 1984: Zaichkina et al., 1988). Comparison of the distribution of non-reparable lesions among cells with cytogenetic damage indicated that the yield of DNA damage per aberrant cell is similar for different doses, and approximately 1 - 2% of these lesions initiated chromosome aberrations (Figure 1). It is seen that size of the DNA radiation target of chromosome mutagenesis and hence the cell death, is

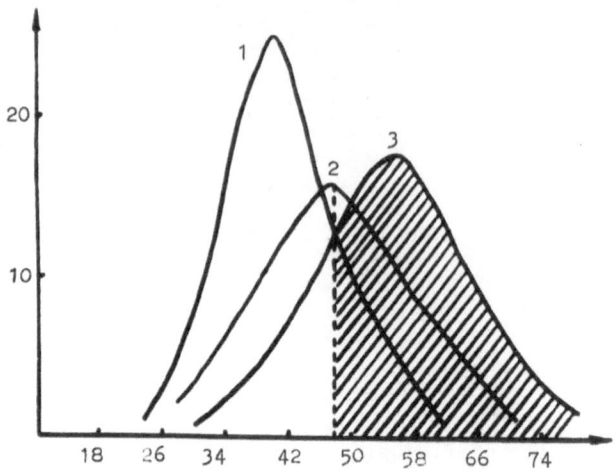

Figure 1.   Distribution of non-reparied DNA lesions among individual
             cells irradiated with a dose of 5 Gy (1), 10 Gy (2), 15 Gy
             (3). On the ordinate gives the number of cells (%); on the
             abscissa is the amount of single-strand breaks (%), (the
             dashed area represents cells with micronuclei).

Table I. Molecular and cytogenetic damages induced
          by 100 unit of various nucleases.

Enzyme	Number of single-strand breaks	Cells with micronuclei, %	Yield of aberrations per cell in S stage
Eco RV	1700±200	16,5±2.0	0.58±0.06
Hae III	1700±200	7.5±1.0	0.33±0.04
Hind III	1850±200	18.0±2.0	0.50±0.05
Sal I	1250±150	14.7±1.5	0.53±0.06
DNA-asa I	1460±150	11.5±1.5	0.49±0.05

small and is compatible with the amount of DNA conected with the protein
scaffold (Agutter and Richardson, 1980).

     Study of DNA lesions induced by enzyme treatment *in vito* (Table I)
indicates that the yield of cytogenetic damage is different for various
nucleases, while the numbers of the induced DNA breaks is approximately the
same. In contrast to the available literature (Natarajan and Obe, 1984;
Bryant et al., 1987) the effeciency of aberration induction was found to

Table II. Relative efficiency of the cytogenetic damage induced by various restriction enzymes.

Enzyme	Object	Times after treatment, h	Relative efficiency	
Sau 3A ↓GATC	CHO	22	1	Winegar and Preston, 1988
Eco RV GAT↓ATC	Chinese hams.fibr.	18	0.76	Own data
Pvu II CAG↓CTC	V79	18	0.76	Bryant, 1984
Alu I AG↓CT	CHO	18	0.79	Winegar and Preston 1988
Sal I G↓TCCAC	Chinese hams.fibr.	18	0.68	Own data
Hind III A↓AGCTT	—"—	18	0.61	—"—
Msp I G↓CCC	CHO	18	0.47	Winegar and Preston, 1988
Hae III GG↓CC	Chinese hams.fibr.	18	0.34	Own data
Bam HI G↓GATCC	CHO KI	16	0.19	Bryant et al. 1987
Eco RI G↓AATTC	CHO KI	16	0.08	—"—

depend not on the type of the cut ends but rather on the presence of a specific sequence in the DNA target. Table II presents our own and literature data on the cytogenetic effect of various restriction enzymes. The maximal yield of cytogenetic damage induced by Sau 2A, which produces cohesive-end double-strand breaks is assumed to be 1. The restriction enzyme Hae III producing blunt-end breaks is the least efficient. Inefficiency in producing chromosomal aberrations of Bam H_1, which exerts a similar effect at molecular level as Eko RV (Bryant, 1984) is consistent with the data on the absence of the sequence recognized by Bam H_1 in the anchor DNA (Mirkovitch et al., 1984). Comparison of the distribution of nuclease-induced DNA lesions among cells with cytogenetic damage indicated that the average yield of DNA damage was similar, the yield of breaks per aberrant cell being different for different nucleases. As in γ-irridiated cells, a very small number of DNA lesions initiate chromosomal aberrations. This suggests that the probability of damage by nucleases depends on the presence of specific DNA sequences in the DNA target.

Non-reparable damage in the anchor DNA may produce lesions in the axial structure of the chromosome. According to the current ideas, nonhistone acidic proteins are the essential component of this structure (Gasser et al.,

1986). We showed the disturbance of this axial structure on cytologic preparations irradiated with 8 Gy in the G_1 stage. Metaphase preparations were treated with DNAase 1 for 2 h at $37^{\circ}C$ to remove DNA and stained with fast acid green, which is specific at pH 5 for acidic proteins. It was found that the yield of chromosome aberrations per cell in the presence of DNA is similar to that in cells treated with DNAase I. These data indicate that the damage of the protein component may lead to a disturbance of chromosome structure.

Thus, the results presented demonstrate that metaphase chromosomes do have an axial structure, and the lesions of specific DNA sequences anchored in the scaffold play a key role in the induction of chromosome aberrations.

REFERENCES

Aggutter P. C., and Richardson J. C. W. 1980, Nuclear non-chromatin proteinaceous structure - their role in the organization and function of the interphase nucleus, J. Cell Sci., 44:395.

Bryant P. E. 1984, Enzymatic restriction of mammalian cell DNA using Pvu II and Bam H$_1$: evidence for the doublestrand break origin of chromosomal aberration, Internat. J. Radiat. Biol., 46:57.

Bryant P. E., Birch D. A., and Yeggo P. A. 1987, High chromosomal sensitivity of Chinese hamster xrs 5 cells to restrection endonuclease induced DNA doublestrand breaks, Internat. J Radiat. Biol., 52:537.

Gasser S. M., Laroche T., Falquet J., Boy E., de la Tour, and Laemmli U. K. 1986, Metaphase chromosome structure Involvement of Topoisomerasa II, J. Mol. Biol., 188:613.

Mirkovitch J., Mirault M. E., and Laemmli U. K. 1984, Organization of the higher-order chromatin loop: specific DNA attachment sites on nuclear scaffold, Cell, 39:223.

Nafarajan A. T., and Obe G. 1980, Molecular mechanism involved in production of chromosomal aberrations. III. Restriction endonucleases, Chromosoma, 90:120.

Obe G., Palitti F., Tanzarella C., Degrassi F., and De Silva R. 1985, Chromosomal aberrations induced by restriction endonucleases, Mut. Res., 150:359.

Sakai K., and Okada S. 1984, Radiation-induced DNA damage and cellular lethality in cultured mammalian cells, Radiat. Res., 98:479.

Winegar R. A., and Preston R. I. 1988, The induction of chromosome aberrations by restriction endonucleases that produce blunt-end or cohesive-end double-strand breaks, Mutat. Res., 197:141.

Zaichkina S. I., and Ganassi E. E. 1984, Micronuclear test to estimate chromosomal aberrations induced by different agents, Studia Biophysica, 99:203.

Zaichkina S. I., Livanova I. A. Akhmadieva A. Kh., Antipov A. V., and Ganassi E. E. 1988, On the contribution of irreparable injury to DNA to initiation of chromosome mutagenesis in chinese hamster cells exposed to γ - and 70 GeV proton-radiation, Radiobiologia, 28:438.

POSSIBLE FUNCTIONAL STRUCTURES IN THE CHROMOMERE

N. A. Reznik, G. P. Yampol, E. V. Kiseleva,
N. B. Khristolyubova, and A. D. Gruzdev

Institute of Cytology and Genetics
Siberian Branch of the USSR Academy of Sciences
630090 Novosibirsk, USSR

INTRODUCTION

The basis of each chromosome is a DNA molecule lying between its
telomeres (Gruzdev and Reznik, 1981). The chromosome after partial
deproteinization exhibits a set of long (60 - 90 kbp) DNA loops (Paulson and
Laemmli, 1977) which in milder conditions are organized into rosette
structures composed of short (about 5 kbp) loops (Okada and Comings, 1979).
As the long loops are considered to represent the DNA molecules of the
chromomeres (Okada and Comings, 1979) the short ones may correspond to the
nucleomeres. Noted variations in the length of the loops and in the number
of the rosette loops (Okada and Comings, 1979) may reflect variations in the
genetical content of the chromomere. This means that there should exist a
correspondence between the genetical units (genes, exons, introns) and the
structural units of the chromomere.

To establish the correspondence we compared the lengths of structural
versus genetical units. Average values for the loops lengths were taken
from Glazkov (1988). The lengths of the exons and introns for 315 genes
with the well defined exon-intron boundaries were obtained from the data
bank of gene sequences (Nucleotide Sequences, 1985). Because of the data
bank constitution the sample mostly contains relatively small genes without
their regulatory sequences. These drawbacks of the sample were kept in mind
during the following discussion.

RESULTS AND DISCUSSION

Comparison of the gene average size 1.5 - 2 kbp (Table I) to the long
loop size clearly demonstrates that the average chromomere may contain more
than one average-sized gene. Comparison of the gene average size to the
short loop size (about 5 kpb) indicates a possible correspondence between
them. An additional evidence in favour of the correspondence may be derived
from the rosette structure. Figure 1 demonstrates that the rosette core
consists of the separate granulae bound to the base of short loops. Moreover, the rosette core is clearly different from the fastener of the loop
base to the nuclear matrix, the fastener is more stable to deproteinization
than the core.

Table I. Lengths of Eukaryotic Genes, Exons, and Introns
(Base Pairs)

Taxon[*]	Gene length		Exon length		Intron length	
	$\bar{x} \pm \sigma$	N	$\bar{x} \pm \sigma$	N	$\bar{x} \pm \sigma$	N
Primates	1968 ± 1778	58	216 ± 242	302	436 ± 679	159
Rodents	2202 ± 1405	27	166 ± 139	341	403 ± 518	164
Mammals	1445 ± 921	21	137 ± 83	84	415 ± 502	41
Vertebrates	1538 ± 1504	15	152 ± 146	108	390 ± 447	67
Invertebrates	1983 ± 1729	23	347 ± 393	86	398 ± 441	63
Plants	1580 ± 1196	16	214 ± 349	86	243 ± 297	59
Total:	genes -	160	exons -	1007	introns - 553	

Table I contains data extracted from the published sequencies
of 160 complete genes and 155 gene fragments (5). * Taxons
are classified as in the original data bank (5).

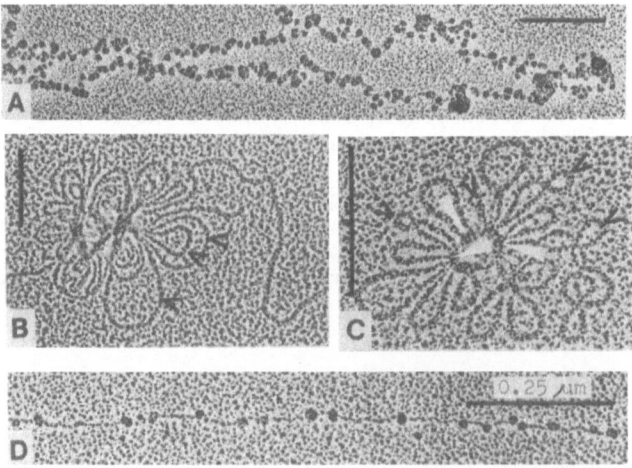

Figure 1.   Structural levels of DNA packaging in chromatin of eukaryotes.
(A) Unravelled chromomere loop with the preserved nucleosomal
structure (about 400 nucleosomes on the 20 μm long loop).
(B,C) The loops of variable size (black arrows) compose the
rosette-like structure through interacting granulae at their
bases (white arrows). (D) Nucleosomes are not evenly distributed along the DNA molecule. (A,D) polytene chromosomes of
Chironomus thummi; (B,C) rat liver chromatin.

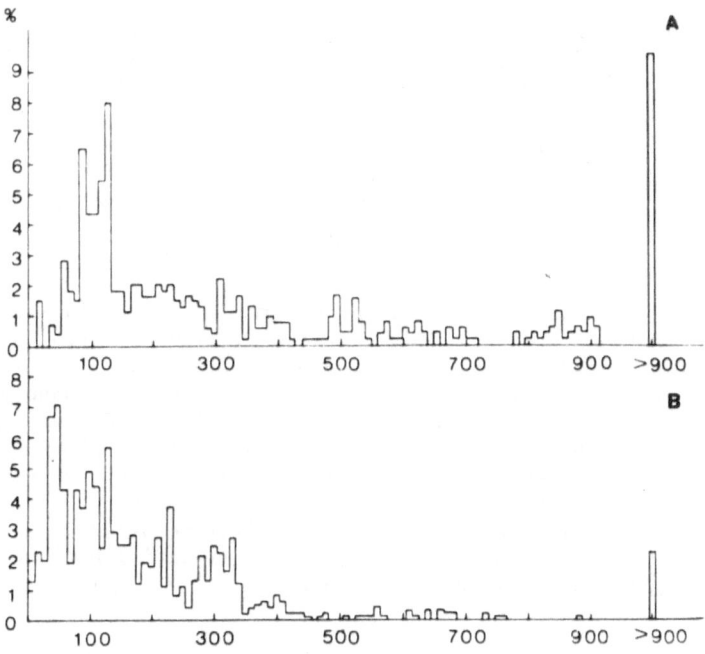

Figure 2.   Distribution frequencies (in percent) of introns (A) and
             exons (B) length (in base pairs).

        In search of the correspondence between small structural and functional
units one may easily note that both the average (Table I) and modal (Figure
2) length values of the introns and exons are close to the nucleosomal repeat
length 150 - 200 bp. However, nucleosomes are not identified, neither with
introns nor with exons. An argument against the identity may be that the
variation in the nucleosomal repeat lengths is 100 - 1,000 times smaller than
the variation in the exon and intron lengths (Figure 2). To clarify the
situation we took advantage of the fact that specific small nuclear RNP
(snRNP) particles are present in the chromatin and possess high affinity to
the exon-intron boundaries. We supposed that these particles, when bound to
the chromosomal DNA, hinder the formation of nucleosomes. It means that they
contribute to the variability of the nucleosomal repeat lengths. Particularly, small introns and exons (145 bp length) can not form nucleosomes. On
the other hand, the very long exons and introns can easily form nucleosomes
and nucleomeres. The suggestion that fasteners of the DNA microloops should
be close to the location of the snRNP particles gives us a possibility to
make a plausible model of any gene packaging. 88 gene models were built,
some of them are presented in Figure 3. Statistical analysis of the models
reveals that the average gene is composed of two nucleomeres. This value is
close but not equal to the conventional value: one nucleomere per one rosette
loop. The last value of one nucleomere per gene was found in 57 percent of
genes analysed (2 in 15%; 3 in 16%; 4 in 6%; 5 in 4%, and 6 in 2% of genes).
The majority of the exons adjacent to the rosette loop bases (78% of the
first and 53% of the last ones) are too short to be packaged into nucleosomes.
Another interesting feature of the models is that many genes contain long
(nucleosome binding) pieces in series with short (nucleosome free) pieces of
DNA. Long packaged in nucleosomes pieces are predominantely introns in
vertebrates- they are exons in invertebrates and plants. As a summary to the
above reasoning, the chromomere structure with indication of the gene-exon-
intron subdivisions is presented schematically in Figure 3.

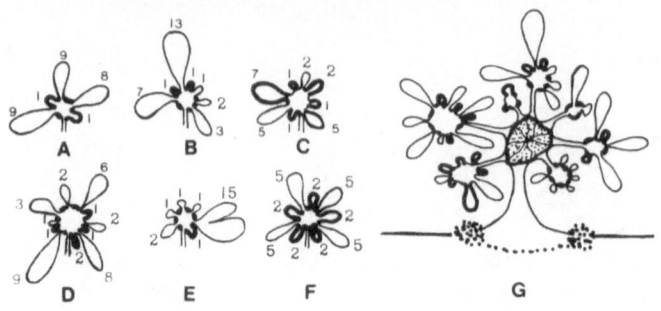

Figure 3.   Models of DNA folding for several genes. Thin lines – for
             introns, thick lines – for exons. The expected number of the
             nucleomeres in a given gene is indicated. The values in
             brackets are, respectively, the gene size in base pairs and
             the number of nucleomeres formed. (A) Human preproglucagon
             gene (6,455 bp; 3); (B) Mouse MHC class II H2-IA-beta (haplo-
             type b) gene (5,801 bp; 3); (C) Rat beta-actin gene (2,758
             bp; 4); (D) Chicken ovalbumin gene (5,280 bp; 6); (E) Rabbit
             Ig germline kappa isotype K1 (allotype b4) gene (4,545 bp;
             3); (F) Sea urchin (*Psammechinus miliaris*) histone complex:
             genes h1, h4, h2b, h3, and h2a.

REFERENCES

Glazkov, M. V. 1988, Structural and functional organization of DNA in the
     interphase nucleus, Mol. Biol. (USSR), 22:16.
Gruzdev, A. D., and Reznik, N. A. 1981, Evidence for the uninemy of eukary-
     otic chromatids, Chromosoma, 82:1.
Okada, T. A., and Comings, D. E. 1979, Higher order structure of chromosomes,
     Chromosoma, 1:1.
Paulson, J. R., and Laemmli, U. K. 1977, The structure of histone-depleted
     metaphase chromosomes, Cell, 12:817.
Nucleotide Sequences 1985, IRL Press, Oxford Washington DC.

RAT DNA FINGERPRINTING FOR *RATTUS NORVEGICUS*: A NEW APPROACH IN GENETIC

ANALYSIS

P. L. Ivanov and A. P. Ryskov

Institute of Molecular Biology
USSR Acad. of Sciences
117984 Moscow B-334, USSR

Recent findings with highly effective DNA probes for RFLP testing (of hypervariable minisatellite DNA type) has led to the invention of DNA fingerprinting - the new technique of great value for individual identification, establishing biological kinship and studies in population genetics (Jeffreys et at., 1985a; Jeffreys et al., 1985b: Vassart et al., 1987). More recently, we have demonstrated a potential range of M13 phage DNA applications as a probe for DNA fingerprinting of organisms of different taxonomic groups including animals, plants and microorganisms, specifically for the determination of species and inbred lines, stock, variety and strain distinction (Ryskov et al., 1988). We anticipate now that the DNA fingerprinting procedure with M13 phage DNA as a probe will make it possible to realize some new approach in genetic analysis, by establishing whether or not a particular locus is associated with the inheritance of genetic disease (Lewin, 1986).

A good way to realize this approach is by using appropriate animal models, which widens the possibilities of genetic analysis.

In this study we characterized by DNA fingerprinting the strain of Krushinsky-Molodkina rats with the hereditary predisposition for epileptic attacks and performed the comparative fingerprint analysis of these defective and normal rat genomes by comparing the whole restriction fragment data from affected and unaffected animals. To determine the feasibility of using DNA fingerprints for linkage analysis in rats we have investigated the DNA fingerprints of the rat strain with audiogenic epilepsia. These inbred rats were obtained in Moscow State University in the early 50s by Krushinsky and Molodkina as a result of the selection of white laboratory rats *R. norvegicus* with hereditary persistence of extraordinarily high sensitivity to the sound of the electric bell (Krushinsky, 1960). If the intensity of the bell ring is higher than 100 db, these rats (so called KM rats) have severe cramps and finally develop a convulsive shock. This pathology is steadily inherited and the segregation picture for the trait in rat pedigrees makes it possible to suggest that the condition is due to several defective alleles combined in one genome to produce defective phenotype. No genetic markers have yet been linked to this disease.

So, one may try to trace these defective and probably structurally changed loci in the genome by means of multiple markers provided by a hypervariable probe. In our work we used M13 phage DNA as a probe.

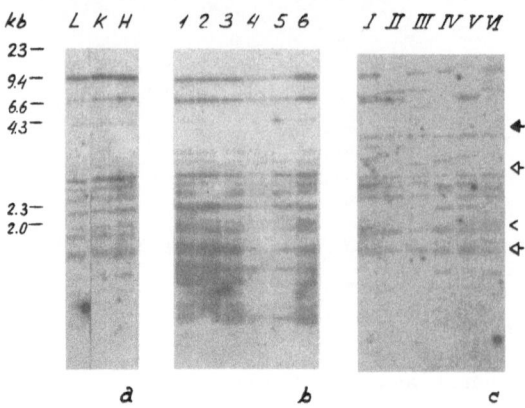

Figure 1.   DNA fingerprints of individual albino rats: somatic stability;
for liver, kidney and heart (a); inbred strain (b); outbred
laboratory stock (c).

Resolution of DNA fragments was maximized by electrophoresis in 20-cm-long
agarose gels.

First of all we have estimated the genetic homogeneity of KM rat strain.
In these experiments DNA isolated from rat liver or kidney was fragmented by
BspRI restriction endonuclease which yields a wide spectrum of hybridizable
bands, though as we have shown, HinfI and MvaI were good choices as well;
note that the DNA fingerprint is largely dependent on the restriction enzymes
involved.

Figure 1b shows fingerprints of six KM rats: the hybridization patterns
are identical, so the genetic constitution of this inbred strain is highly
homogeneous. For comparison, on the right panel (Figure 1c) there are six
randomly taken white rats from laboratory stock: each individual exhibits its
own distinctive pattern of hybridization. On the left panel (Figure 1a)
there is an illustration of comparisons within an individual. The autoradiograph shows an analysis of DNA extracted from samples of liver, kidney
and heart taken from the same KM rat. There are, as expected, no instances
of difference between these fingerprints. The data obtained one can regard
as the basis for the determination and registration of this particular
inbred stock. Figure 2 shows the computer aided analysis of hybridization
data. Computerised scanning densitomery does provide fast track scanning and
image processing for accurate band recognition and vacilities for correlation
analysis.

Our subsequent work was comparative fingerprint analysis of these
defective and normal rat genomes, since comparing the whole restriction
fragment data from affected and unaffected animals is the first step in
establishing whether or not a particular hypervariable locus is associated
with the inheritance of genetic disease.

Figure 3 demonstrates segregation of hypervariable fragments detected by
M13 phage DNA in rat genomes. In these experiments DNA's were digested with
MvaI, HinfI and BspRI which proved to be most informative restrictases for
this sort of analysis. The patterns of outbred albino rats reveal a marked
individual polymorphism. The frequency of band sharing between individual
animals in outbred laboratory population was approximately 0.35 for MvaI and
0.61 for BspRI. Meanwhile, some bands were shared by all the rats studied.

32

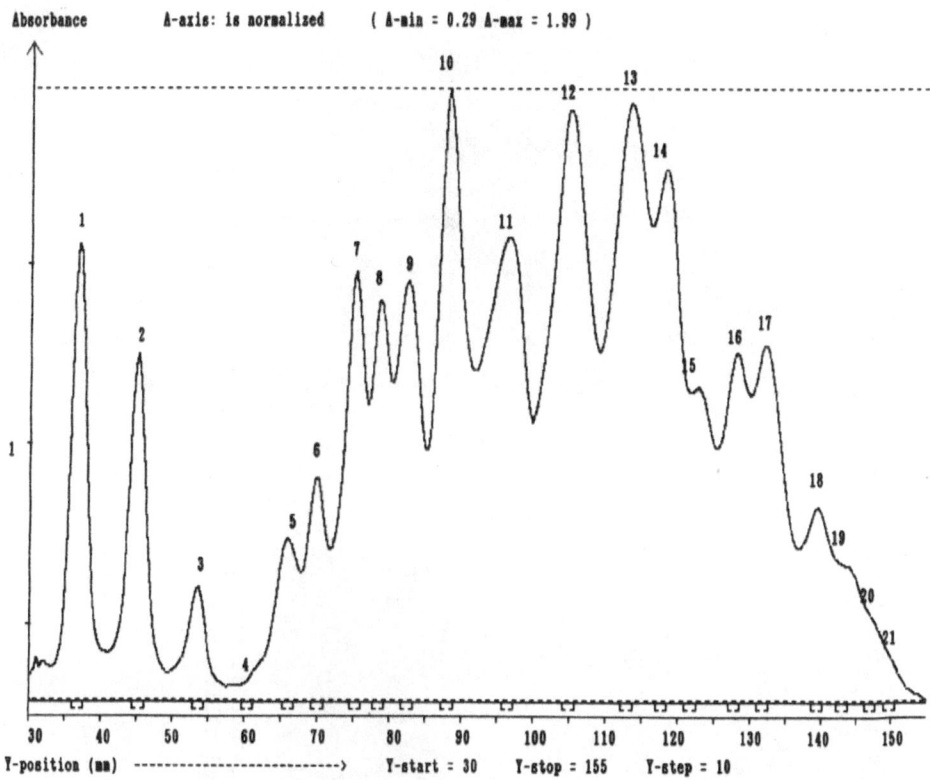

Figure 2.   Densitogram of KM individual fingerprinting (track 'l', Figure
            1b) and computer aided quantitative interpretation.

PEAK #	POSITION mm	HEIGHT AU	AREA AU*mm	REL. AREA %	PEAK #	POSITION mm	HEIGHT AU	AREA AU*mm	REL. AREA %
1	37.0	1.27	2.21	6.1	12	105.0	1.64	3.20	8.8
2	45.4	0.96	1.70	4.7	13	113.0	1.66	3.25	9.0
3	53.8	0.31	0.56	1.5	14	117.8	1.47	2.88	7.9
4	60.6	0.05	0.11	0.3	15	121.8	0.85	1.72	4.8
5	66.2	0.45	0.86	2.4	16	127.8	0.96	1.87	5.2
6	70.2	0.62	1.15	3.2	17	131.8	0.99	1.92	5.3
7	75.4	1.19	2.21	6.1	18	139.4	0.54	1.04	2.9
8	78.6	1.11	2.09	5.8	19	143.0	0.38	0.76	2.1
9	82.6	1.16	2.25	6.2	20	147.0	0.24	0.47	1.3
10	88.2	1.69	3.20	8.8	21	149.8	0.12	0.25	0.7
11	96.6	1.29	2.53	7.0					

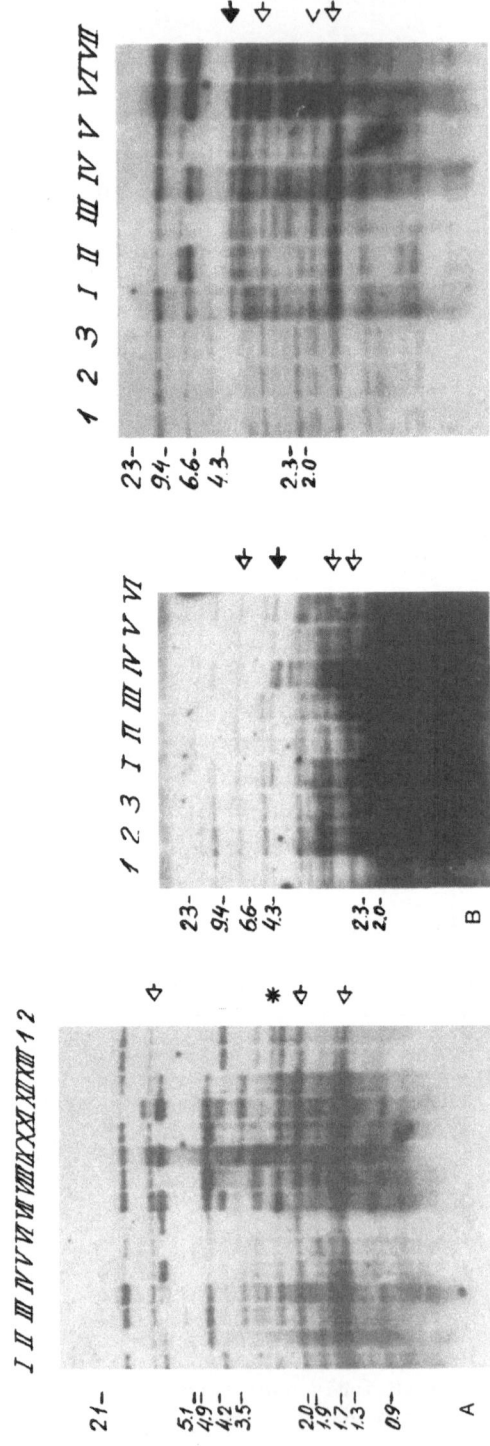

Figure 3. Segregation of hypervariable fragments detected by M13 phage DNA in rat genomes: 1-3 KM inbred individuals; I-Xiii outbred albino rats. Rat DNAs were digested with MvaI (A), Hinf I (B), BspRI (C).

34

We refer to these alleles as population markers characterizing definite groups of organisms within species (such as lines, stocks, varieties, strains etc.) and probably species themselves.

In this context, it is significant that in KM rats DNA fingerprint distinctions were found, comprising loss of some marker alleles present in all normal albino rats studied. At the same time there is an example of the appearance of a novel band which we could not find in normal rats. These findings seem promising, when considering the possibilities of pathogenetically relevant alterations in DNA fingerprints. However, we realize the necessity of further segregation analysis of multiple markers in rat pedigrees to obtain evidence of linkage in coupling with audiogenic epilepsia. The question is now being investigated further.

REFERENCES

Jeffreys A. J., Wilson V., Thein S. L. 1985a, Hypervariable 'minisatellite' regions in human DNA, Nature, 314:67.
Jeffreys A. J., Wilson V., Thein S. L. 1985b, Individual-specific 'fingerprints' of human DNA, Nature, 316:76.
Krushinsky L. V., 1960, 'Animal behavioural reactions in normal state and pathology', Moscow State University, Moscow (in Russian).
Lewin R., 1986, DNA fingerprints in health and disease, Science, 233:521.
Ryskov A. P., Jincharadze A. G., Prosnyak M. I., Ivanov P. L., Limborskaya S. A. 1988, M13 phage DNA as a universal marker for DNA fingerprinting of animals, plants and microorganisms, FEBS Lett., 233:388.
Vassart G., Georges M., Monsier R., Brocas H., Lequarre A. S., Christophe D. 1987, A sequence in M13 phage detects hypervariable minisatellites in human and animal DNA, Science, 235:683.

THE EXTRACELLULAR LYMPHOCYTE AND BLOOD PLASMA DNAs CONTAIN THE DISCRETE

SIZE MOLECULES HOMOLOGOUS TO THE C_k FRAGMENT OF THE Ig GENE

V. I. Vasyukhin, L. A. Lipskaya, A. G. Tsvetkov, and
O. I. Podgornaya

Institute of Cytology of AS USSR
Leningrad, 194064, USSR

A spontaneous release of a DNA-containing complex by human lymphocytes has been reported (Anker et al., 1975). The release process was unrelated to cell death and the released DNA consists of newly synthesized DNA. Possibly, part of the human blood plasma DNA represents the DNA released by lymphocytes. It was shown in the work by Anker et al., (1984), that the purified DNA released by antigen stimulated human lymphocytes, when injected into the nude mice is capable of inducing the formation of specific antibodies expressing human characteristics. In the present paper we attempt to show by the technique of the Southern blot analysis that extracellular lymphocyte and human blood plasma DNA contain molecules homologous to the human C_k fragment of the Ig gene.

Human peripheral blood lymphocytes were isolated under sterile conditions by the Ficoll-Isopaque gradient technique with the use of heparinized whole blood samples from healthy donors (Böyum, 1968). The lymphocytes were cultured at a concentration of 5×10^6 cells/ml in sterile Hanks solution for 30 min at $37^\circ C$. After the incubation, the lymphocytes were removed by the centrifugation for 10 min at 400xg and then the supernatant was centrifuged for 10 min at 1,500 g and finally for 20 min at 12,000 g. This supernatant (usually about 50 ml) was concentrated by ultrafiltration (The Immersible CX-30 Ultrafilter, Millipore) to 1 ml. The extracellular DNA was obtained by a standard technique with two cycles of RNAse A and proteinase K treatment and phenol deproteinization (Maniatis et al., 1982). Total cellular DNA was obtained from the lymphocytes by the same technique. The extrachromosomal lymphocyte DNA was obtained according to Hirt (1967). Human blood plasma DNA was obtained as for the extracellular DNA, except that there was no ultrafiltration. Conditions of agarose gel electrophoresis, nick translation and Southern blotting were described by Maniatis et al., (1982). We used nylon membranes (Zeta-probe, Bio-Rad) and performed hybridization under the conditions specified by the supplier. The plasmid that contains the constant part of human k chain Ig gene (Hieter et al., 1982) was a gift of Dr V. A. Pospelov.

Cell death counts by trypan blue did not show any differences at the beginning or after thirty minutes of incubation. The quantity of the extracellular DNA was usually about 0.02% and never more than 0.05% of the cellular DNA. The quantity of the human blood plasma DNA was about 0.1 µg DNA per ml plasma. As shown in Figure 1, the size of the cellular DNA was more than 48 kb; the size of the extrachromosomal, extracellular lymphocyte

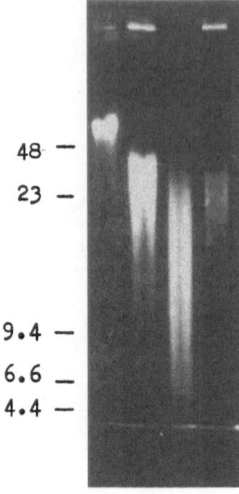

Figure 1.   Electrophoresis of 0.3 μg total cellular DNA (1), 0.5 μg
extrachromosomal DNA (2), 0.4 μg extracellular lymphocyte DNA
(3) and 0.15 μg blood plasma DNA (4), in 0.3% agarose gel.
Numbers on the left represent the positions of the λ DNA in kb.

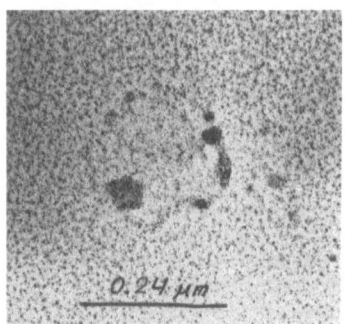

Figure 2.   Covalently closed circular DNA from extracellular lymphocyte
DNA. The bar represents 0.24μm.

and blood plasma DNAs was between 40 to 2 kb (Figure 1). These DNAs were
sensitive to the DNase I and resistent to the RNase and pronase treatment.

The samples of the extracellular lymphocyte DNA were examined by
electron microscopy according to Davis et al., (1971). They contained
covalently closed circular (ccc) DNA molecules. Contour lengths ranged
widely, from 0.1 to 2.8 μm, with mean length of 0.6 μm. A representative
electron micrograph of ccc DNA is shown in Figure 2.

We used the C_k human Ig region probe to study by the Southern blot
hybridization whether extracellular lymphocyte and plasma DNA contained some
molecules homologous to the Ig gene. Extracellular lymphocyte and plasma
blood DNA were applied to electrophoresis without restriction, but after

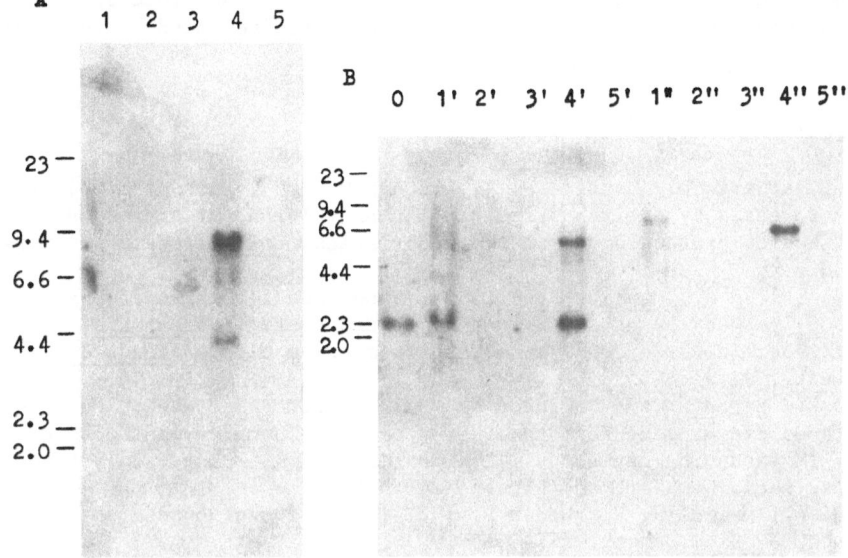

Figure 3.   Southern blot hybridizations. The samples of the DNA were
electrophoresed in 0.4% (A) or 0.8% (B) agarose gels, Southern
transferred onto nylon and hybridized with a nick translated
C_k Ig gene probe. Numbers on the left represent the positions
of restricted fragments of the λ DNA.
0. C_k fragment Ig gene DNA (20 pg).
1. Total lymphocyte DNA (5 µg); 1', after the EcoR I digestion
    (10 µg); 1'', after the Bgl II digestion (10 µg).
2. Extrachromosomal lymphocyte DNA (5 µg); 2', after the EcoR
    I digestion (5 µg); 2'', after the Bgl II digestion (5 µg).
3. Extracellular lymphocyte DNA (0.3 µg); 3', after the EcoR
    I digestion (0.3 µg); 3'', after the Bgl II digestion
    (0.3 µg).
4. Blood plasma DNA (5 µg); 4', after the EcoR I digestion
    (5 µg); 4'', after the Bgl II digestion (5 µg).
5. Lymphocyte DNA broken by cell endonuclease (5 µg); 5',
    after the EcoR I digestion (5 µg); 5'', after the Bgl II
    digestion (5 µg).

hybridization discrete bands of DNA, homologous to the probe, were visible
(Figure 3A), and the differences between the total lymphocyte and blood
plasma DNA were revealed, when the samples of the DNA were digested with
EcoR I and Bgl II before electrophoresis (Figure 3B). Unfortunately, the
quantity of the extracellular lymphocyte DNA in this experiment was not
enough to produce an answer after the hybridization. The result obtained
after the Bgl II digestion apparently suggested that the Ig gene in the
sample of blood plasma DNA is rearranged. The new band after the EcoR I
digestion of the blood plasma DNA that is absent in the sample of total
lymphocyte DNA was unexpected, as we do not know some recombination signal
sequences in 2.5 kb EcoR I-EcoR I region that we used as a probe. It may be
that in cell damage the molecules of genome DNA can be broken by endonucleases at certain definite points. This supposition can explain the
presence of discretely sized fragments of the Ig gene in extracellular DNA.
To study this possibility we homogenized the lymphocytes, incubated for 30
min at 37°C and obtained the DNA from this solution. After the Southern
blotting no discrete bands of DNA, homologous to the probe, could be

discerned. The data obtained apparently suggest that discrete size molecules homologous to the C_k fragment Ig gene are released by the living cells.

REFERENCES

Anker P., Jachertz D., Maurice P. A., Stroun M. 1984, Nude mice injected with DNA released by antigen stimulated human T lymphocytes produce specific antibodies expressing human characteristics, Cell Biochem. and Funkt., 2:33.
Anker P., Stroun M., Maurice P. A. 1975, Spontaneous release of DNA by human blood lymphocytes as shown in an in vitro system, Cancer Res., 35:2375.
Böyum A. 1968, Isolation of lymphocytes from human blood, Scand. J. Clin. Invest., 21:31.
Davis R. W., Simon M. and Davidson N. 1971, Electron microscope heteroduplex methods fro mapping regions of base sequence homology in nucleic acids, in: 'Methods in Enzymology', L. Grossman and K. Moldav, ed., Academic Press Inc., New York and London, 21D:413.
Hieter P. A., Maizel J. V., Leder P. 1982, Evolution of human Ig_k region genes, J. Biol. Chem., 257:1516.
Hirt, 1967, Selective extraction of polyoma DNA from infected mouse cell cultures, J. Mol. Biol., 26:365.
Maniatis T., Fritch E. E., Sambrook J. 1982, 'Molecular Cloning. A Labaratory Manual', Cold Spring Harbor Laboratory Press.

NON-RANDOM DISTRIBUTION OF ALU-FAMILY DNA REPEATS IN HUMAN CHROMOSOMES

L. V. Filatov, S. E. Mamayeva, and N. V. Tomilin

Institute of Cytology
The Academy of Sciences of the USSR
Leningrad 194064
Tikchoretskii Av. 4, USSR

Human retrotransposons, the Alu-family DNA repeats (AFRs), are members of an abundant heterogeneous family of short intersperced repeats containing promoter for RNA polymerase III, which is active in vitro (Perez-Stable et al., 1984) but is silent in HeLa cells in vivo (Poulson and Schmid, 1986). We have studied long-range distribution of AFRs in human chromosomes using tritium-labeled (Filatov et al., 1984; Filatov et al., 1987; Filatov et al., 1988) and biotin-labeled Alu-probes and have found that the distribution is non-random: some chromosome bands reproducibly show a more intensive hybridization signal compared to other bands.

In experiments with tritium-labeled Alu-probes (DNA of genomic lambda clone containing multiple Alu-inserts) two types of cells were studied: phytohaemagglutinin-stimulated lymphocytes from peripheral blood of four normal donors (SPBL), and non-stimulated bone marrow cells of five independent acute leukaemia patients (NSBMC). For each individual 20 metaphases were analysed, and Figure 1 shows a typical cumulative plot for chromosome 6 of one normal donor (SPBL). The data obtained with 9 independent individuals are summarized in Table I. Two types of Alu-clusters were found:- 1) present in all individuals studied either in SPBL or in NSBMC: these clusters were named as conservative (C-clusters); 2) present only in one of the two groups: these clusters were named as variable (V-clusters). Some clusters were unambiguously assigned to G-negative (Reverse) bands: 6p21, 10o13-p14, 3q26.2, 14q24, 16p13, but some C-clusters were over centromeres of chromosomes 14, 16 and 21. However, the resolution of the tritium method is not high enough to conclude definitely that the clusters are actually over centromeres and not over nearest R-bands. Two V-clusters detected over G-bands (6q22 and 15q21) may in fact belong to R-positive subbands (6q22.2 and 15q21.2) seen by a high-resolution G-banding.

Non-random character of the distribution of AFRs in human chromosomes was confirmed recently by workers in the USA using a biotin-labeled Alu-probe and a fluorescent method for detection of biotin (Korenberg and Rykowski, 1988), which suggested that Alu maps to Reverse bands. With this fluorescent method we have obtained similar results, suggesting that many R-bands actually contain an increased concentration of AFRs. Figure 2 shows the distribution of Alu repeats in chromosomes 6 and 7 of acute myelomonocytic leukaemia cells (line U-937): a strong fluorescence signal (streptavidin-phycoerythrin conjugate) is seen over bands 6P21, 6q25-q27,

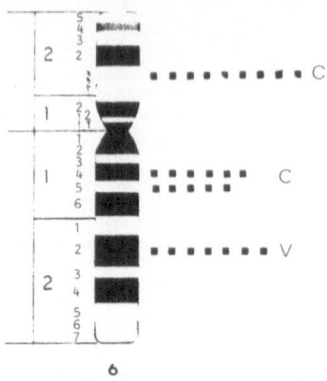

Figure 1.   Clusters of Alu-family DNA repeats in chromosome 6 of PHA-
stimulated lymphocytes of one normal donor. Tritium-
labeled DNA of genomic lambda clone CAR42, containing
multiple Alu repeats is used as a probe. (For other details
see Filatov et al., 1987).

7q11.2, 7q22 and 7q32-q34, all of which are R-positive. With the tritium
method we have also detected Alu clusters in some leukaemia patients over
7q22 (Table I) and 7q32-q34. However, in U937 cells we never detected an
increased fluorescence over 7pter and 7qter, in agreement with the data
(Table I) obtained with a tritiated Alu-DNA. It should be noted that many
distal telomeric R-bands (5pter, 10qter, 11qter, 13qter) were found to be
free of Au1 by both the fluorescent (Korenberg and Rykowski, 1988) and
tritium method (Table I). The data are consistent with the view that the
major fraction of Alu-clusters might be associated with R-bands, but many
do not contain Alu-clusters. Some R-bands positive for Alu in the fluorescent method were found to be Alu-negative in the tritium method (3p21,
2qter, 16q22, etc), which might be explained by relatively low abundance of
AFRs in these bands.

    We have studied also the distribution of AFRs in the human genome by
biochemical method. Figure 3 shows a Southern blot of human placental DNA
cleaved with different restriction nucleases and probed with Alu-DNA labeled
with ^{32}P. Samples were electrophoretically separated in a 1.8% agarose gel
and transferred to nitrocellulose membrane. It is seen that most Alu
sequences in human DNA are associated with HpaII tiny fragments of 1kb or
less. Islands of non-methylated HpaII sites (HTF-islands) in vertebrate DNA
are known to be associated with regulatory regions of housekeeping genes
available for interactions with nuclear regulatory proteins (Bird, 1986).
Alu-repeats are HFT-like sequences (Korenberg and Rykowski, 1988) which seem
to be non-methylated (Figure 3) and concentrated in some euchromatic R-bands,
containing mostly expressed housekeeping geens. Interestingly, Alu-repeats
were suggested to have a mobile enhancer of transcription by RNA polymerase
II (Bak and Jorgensen, 1984).

    The existence of V-clusters of Alu-repeats (Table I) suggest site-
specific amplification or directed transposition of Alu-repeats in some
omatic cells (both SPBL and NSBMC represent different lineages of blood stem
cells). These data as well as silence of the RNA polymerase III promoter in
HeLa cells (Poulson and Schmid, 1986) suggest the existence of an efficient
mechanism of modulation of Alu transposition/amplification. RNA polymerase
II-dependent transcription is modulated by proteins which binds to specific
sequences conserved in promoters. In HeLa cells we have detected nuclear

Table I. Clusters of Alu-family DNA repeats in human chromosomes.

Chromosome band	Type of band	Presence (+) or absence (−) of Alu-cluster over indicated band as detected with tritiated Alu-probes										Presence of Alu-cluster over indicated band as detected with biotinylated Alu-probe (Korenberg and Rykovski, 1988)	Type of cluster
		PHA-stimulated lymphocytes of normal donors				Bone marrow cells of acute leukemia patients							
		1	2	3	4	1	2	3	4	5		
1p32-p35	R	+	+	+	+	+	+	+	+	+	+	C
1q21-q25	R/G	+	+	+	+	+	+	+	+	+	+/−	C
2q22-q23	G/R	−	−	−	+	+	+	+	+	+	+/−	C
3q26.2	R	−	−	−	−	+	+	+	+	+	−	V
6p21	R	+	+	+	+	+	+	+	+	+	+	C
6q14-q15	G/R	+	+	+	+	+	+	+	+	+	+/−	C
6q22	G	−	−	−	−	−	−	−	−	−	−	V
7q21-q22	G/R	−	−	−	−	+	+	+	+	+	+	V
8p11-p12	R/G	+	+	+	+	+	+	+	+	+	+	V
10p13-p14	R	+	+	+	+	+	+	+	+	+	+	C
10q21-q22	G/R	+	+	+	+	+	+	+	+	+	+/−	C
11p14-p15	G/R	−	−	−	−	+	+	+	+	+	+/−	C
11q13-q14	R/G	−	−	−	−	−	+	+	+	+	+/−	V
14q24	R	+	+	+	+	+	+	+	+	+	+/−	V
15q21	G	+	+	+	+	+	+	+	+	+	+	C
16p13	R	+	+	+	+	+	+	+	+	+		
14cen	cen	+	+	+	+	+	+	+	+	+	−	C
16cen	cen	+	+	+	+	+	+	+	+	+	−	C
21cen	cen	+	+	+	+	+	+	+	+	+	−	C

R − reverse band, G − Giemza-positive band; C − conservative, V − variable cluster.

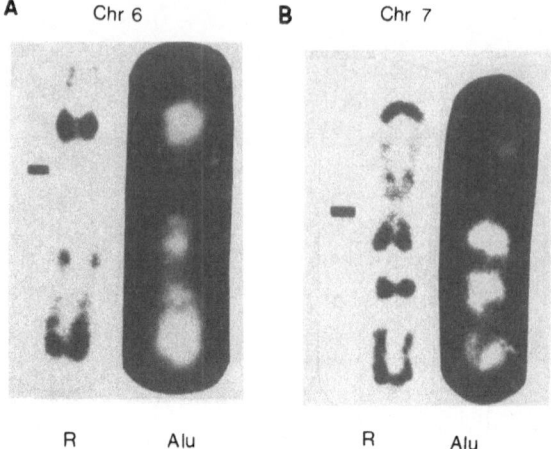

Figure 2.   Distribution of Alu repeats in chromosomes 6 (A) and 7 (B) of the permanent human cell line U-937. Biotin-labelled DNA of plasmid BLUR8 was used as a probe, biotin was detected using streptavidin-phycoerythrin conjugate kindly gifted to us by Dr Y. Lasebnik. Photograph was made on Opton inverted microscope with epifluorescence. R-bands are taken from Atlas.

Figure 3.   Southern blot of human placental DNA cleaved with indicated restriction endonucleases, electrophoresed in a 1.8% agarose gel and probed with ^{32}P-labelled DNA of lambda CAR42, containing multiple Alu inserts. The major band in the HpaII lane is less than 0.5kb in size.

protein which binds to the conservative sequence motif (GGAGGC hexamer) of Alu-repeats (Tomilin et al., 1989; Tomilin and Bozhkov, 1989). One of the GGAGGC hexamers is situated in a spacer between enhancing and directing elements of bipartite internal RNA polymerase III promoter of Alu-repeats, and we suggested that the protein is a repressor of Alu transcription and transposition.

Since many hot spots of chromosome rearrangements coincide with clusters of Alu-repeats, our data may have important implications in understanding genome stability and genome mobilization during genetic stress. Alu mapping in the interphase nuclei may help to map spatial distribution of HTF-island and regulatory sequences.

REFERENCES

Bak A. L., and Jorgensen A. L. 1984, RNA polymerase III control regions in retrovirus LTR, Alu-type repetitive DNA, and papovavirus, J. Theor. Biol., 108:339.
Bird A. 1986, CpG-rich islands and the function of DNA methylation, Nature, 321:209.
Filatov L. V., Mamayeva S. E., and Tomilin N. V., 1987, Non-random distribution of Alu-family repeats in human chromosomes, Mol. Biol. Rep., 12:117.
Filatov L. V., Mamayeve S. E., and Tomilin N. V., 1988, 'Conservative' and 'variable' clusters of Alu-family DNA repeats in human chromosomes, Mol. Biol. Rep., 13:79.
Korenberg J. R., and Rykowski M. C. 1988, Human genome organization: Alu, LINES, and the molecular structure of metaphase chromosome bands, Cell, 53:391.
Perez-Stable C., Ayres T. M., Shen C. K. J. 1984, Distinctive sequence organization and functional programming of an Alu repeat promoter, Proc. Natl. Acad. Sci. USA, 81:5291.
Poulson K. E., and Schmid C. W. 1986, Transcriptional inactivity of Alu repeats in HeLa cells, Nucleic Acids Res., 14:6145.
Tomilin N. V., and Bozhkov V. M. 1989, Human nuclear protein interacting with a conservative sequence motif of Aly-family DNA repeats, FEBS Letters, in press.
Tomilin N. V., Filatov L. V., Mamayeva S. E., and Svetlova M. P. 1984, Cloning and localyzation of some repeated sequences of the human genome, Molekuljarnaya genetika (Russian), 3:16-20.
Tomilin N. V., Perelygina L. M., and Podgornaya O. I. 1989, Sequence-specific binding of a human nuclear protein to the Alu-family repeated DNA, in: 'Macromolecules in the functioning cell', Proceedings of the 6th Soviet-Italian Symposium, Moscow, Nauka, Ed. by A. Bayev, 1988, in press.

ISOLATION AND PROPERTIES OF A NOVEL SPECIFIC FRACTION OF CHROMOSOMAL DNA

FROM HUMAN, *DROSOPHILA* AND PLANT CELLS

Nickolai A. Tchurikov and Natalia A. Ponomarenko

Engelhardt Institute of Molecular Biology
The Academy of Sciences of the USSR
Vavilov str., 32, Moscow B-334, 117984, USSR

Pulsed field gel (PFG) electrophoresis allows the separation of large DNA molecules and permits resolution of intact chromosomal DNAs from unicellular eukaryotic genomes (Schwartz and Cantor, 1984). PFG electrophoresis is currently unable to separate much longer chromosomes. We have used PFG electrophoresis of intact undigested DNA from human, *Drosophila* and plant cells and found a resolvable DNA fraction. The procedure used allows one to isolate DNA *in situ* by cell lysis in solid agarose with proteinase K-detergent treatment with a minimum of shearing, as described by Schwartz and Cantor (1984). DNA-agarose plugs containing total uncleaved DNA were separated in LKB Pulsaphor system. Figure 1 (A and B) shows the ethidium bromide (Et Br) staining pattern of gels. We tested DNA samples from human sperm, *D. melanogaster* cells (from Oregon RC flies or Schneider tissue culture cells) and from *Arabidopsis thaliana* protoplasts. Pulse times of 100 seconds suffice to separate yeast chromosomal DNA and concatamers of bacteriophage lambda. In human DNA lane (Figure 1A), two discrete bands migrated to a position of about 2 and 1 Mb in the region of little or no resolution. The lane also contains fairly abundant DNA of a heterogeneous size around 100-200 kb where concatamers of λ-DNA are separated. Changing the PFG pulse length from 100 to 4500 seconds (Smith et al., 1987) considerably affected the pattern of the human DNA sample (Figure 1B). There is only one DNA band which should correspond to the 100-200 kb region, whereas *S. pombe* chromosomes and *S. cerevisiae* chromosome XII are resolved. Thus, the two regions of lower resolution at a switching interval of 100 seconds appear as 2-Mb and 1-Mb bands which originate from smeared DNA. The results presented here indicate that only one heterogeneous fraction of DNA can be separated in PFG, mainly in the region of 100-200 kb. This DNA which makes up 5% or rather more of the total cellular DNA was denoted as '<u>forum</u>' DNA (fDNA). Similar results were obtained with *Drosophila* and *Arabidopsis,* DNA samples.

The above observation has prompted us to use either conventional agarose mini-gel electrophoresis or even direct electroelution of a 100-200 kb DNA migrating fraction into dialysis bags in order to isolate fDNA from DNA-agarose plugs. Figure 1C demonstrates the Et Br staining pattern in a mini-gel. DNA bands from several species are migrating in the region above 20 kb and correspond to the 100-200 kb DNA fraction DNA remaining in agarose plugs was isolated by homogenization in a Dounce homogenizer (f⁻DNA). For further experiments, we cloned EcoRI-BamHI fragments of human fDNA in the plasmid pUC12. Four randomly cloned DNA fragments without Alu or Kpn repeats (Houck

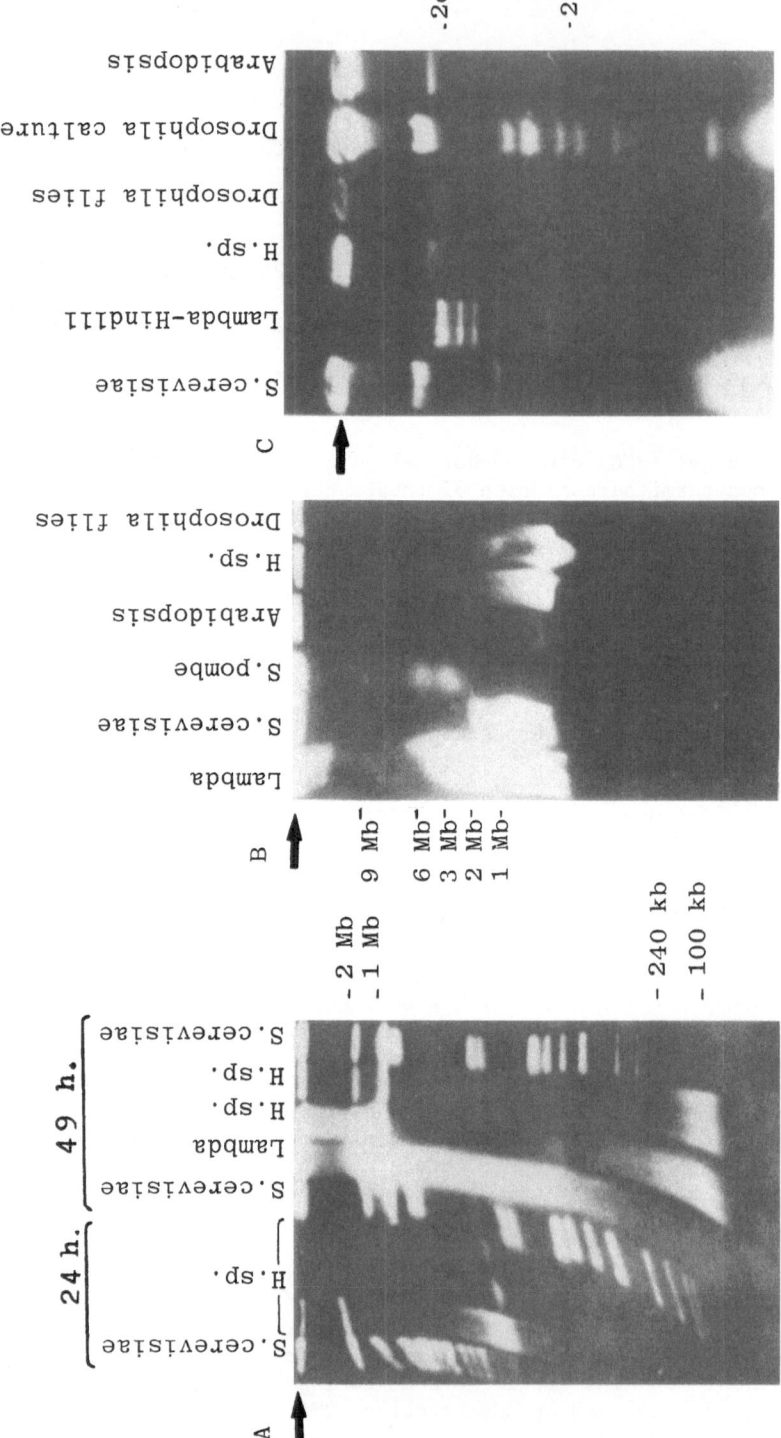

Figure 1. Electrophoretic separation of migrating DNA from a DNA-agarose plug containing total uncleaved DNA. Arrows indicate the slots. PFG was run with 100 s pulse time (A) or 75 min pulse time (B). (C) separation of a mini gel.

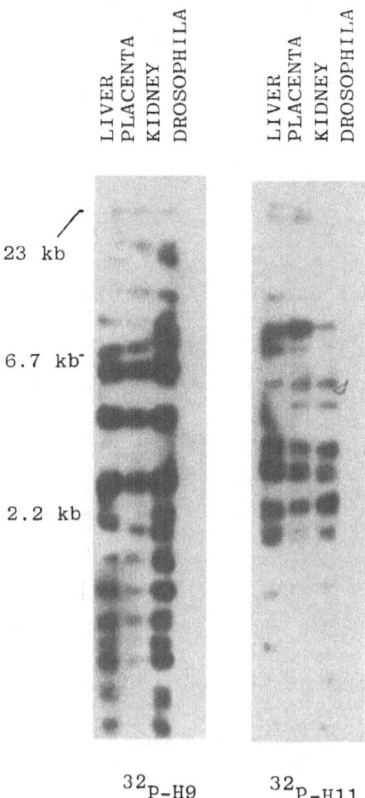

Figure 2.   Southern blot analysis of cloned DNA. Human genomic DNA
                samples from different sources of *D. melanogaster* D DNA were
                digested by EcoRI and BamHI endonucleases. After separation,
                DNA blots were prepared and probed with H9 or H11 cloned
                human sequences containing 2.8 kb EcoRI and 1.4 kb EcoRI-
                BamHI DNA fragments, respectively.

et al., 1979; Shafit-Zagardo et al., 1982) were used to probe human genomic
DNA blots (H1, H9, H10 and H11). The results suggest that repetitive DNA
sequences are present in the cloned DNA fragment (Figure 2A). All the clones
analysed revealed different patterns of genomic hybridization and minor
differences in the hybridization pattern given by DNA from different tissues
of different individuals. This may be a result of both restriction fragment
length polymorphism and/or evidence of DNA sequence mobility in the genome.
Thus it seems that fDNA is enriched with repetitive sequences.

    Samples of fDNA and f⁻DNA were used in a straight forward Southern
transfer experiment to study the distribution of sequences between two DNA
fractions (Figure 3). The same blot was probed successively with human
cloned fDNA fragment H10 and cDNA synthesized from human poly(A)$^+$RNA with an
oligo(dT)-primer. H10 gives a much stronger signal with fDNA. On the other
hand, nucleotide sequences corresponding to the 3'-ends of mRNA are
preferentially located in the non-migrating DNA (f⁻DNA). Therefore, the
above observations strongly argue in favour of fDNA specificity. For an
independent confirmation of the distribution of sequences between fDNA and
f⁻DNA, we used DNA fingerprinting analysis (Vassart et al., 1987). As shown
in Figure 4, almost all hypervariable regions of the human genome are
present in fDNA, although some differences are detected. We conclude,

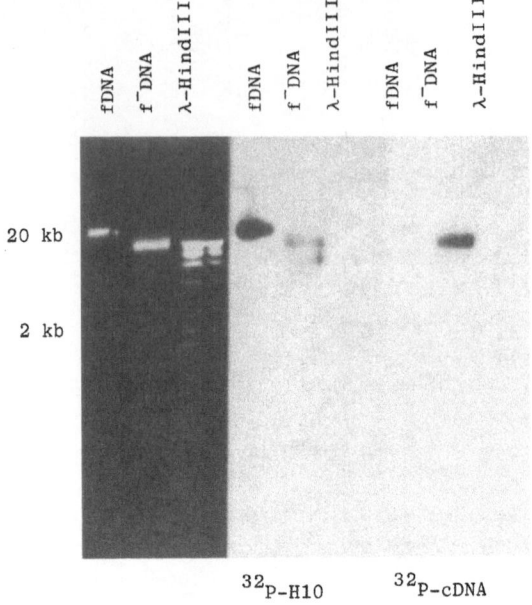

Figure 3.   Southern-blot analysis of two human DNA fractions. The
             autoradiograms show hybridization with the (^{32}P)-labelled H10
             clone containing 0.8 kb EcoRI-BamHI DNA fragment cloned from
             cDNA preparation.

therefore, that fDNA is heterogeneous, ie it contains DNA fragments from
many different regions of the genome.

The results suggest that non-random DNA degradation occurs during
preparation of the sample and that breaks in DNA are mostly separated from
one another by distances of about 100-200 kb. Both the length of fDNA and
the partition of sequences between fDNA and f⁻DNA strongly suggest that fDNA
is excised in a non-random way. These data may indicate the existence in
eukaryotic chromosomes of some higher-order structures posessing 100-200 kb
DNA stretches. The novel fraction of eukaryotic chromosomal DNA reported
here may assist the study of higher structures in chromosomes and,
apparently, to identify a number of new repetitive families in mammalian and
plant genomes. We expect that cloning of 100-200 kb fDNA stretches will
allow us to address some of the questions raised by this report.

ACKNOWLEDGEMENTS

We thank G. P. Georgiev, C. L. Smith and Ch. R. Cantor for their
encouragement; N. I. Barbakar, A. J. Leigh Brown, L. G. Airich and D. R.
Beritashvily for their help.

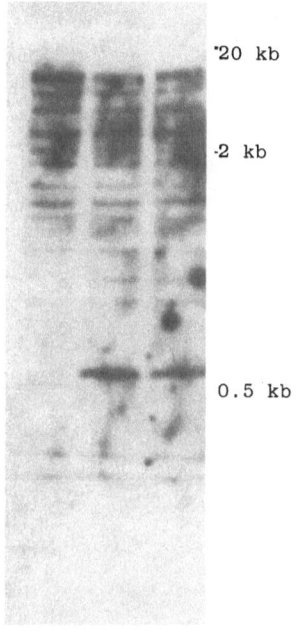

^{32}P-M13

Figure 4.   DNA fingerprinting of human DNA fractions. DNA preparations
             were treated with HaeII endonuclease. The blot was probed
             with (^{32}P)-labelled DNA of M13 phage.

REFERENCES

Houck C. M., Rinehard F. P., and Schmid C. W. 1979, A ubiquitous family of
    repeated DNA sequences in the human genome, J. Mol. Biol., 132:289.
Schwartz D. C., and Cantor Ch. R. 1984, Separation of yeast chromosome-
    sized DNA by pulsed field gradient gel electrophoresis, Cell, 37:67.
Shafit-Zagardo B., Maio J. I., and Brown F. L. 1982, KpnI families of long,
    interspersed repetitive DNAs in human and other primate genomes,
    Nucleic Acids Res., 10:3175.
Smith C. L., Matsumoto T., Niwa O., Kleo S., Fan J.-B., Yanagida M., and
    Cantor Ch. R. 1987, An electrophoretic karyotype for Schizoccharomyces
    pombe by pulsed field electrophoresis, Nucleic Acids Res., 15:4481.
Vassart G., Georges M., Monsieur R., Brocas H., Lequarre A. S., and
    Chistophe D. 1987, A sequence in M13 phage detects hypervariable
    minisatellites in human and animal DNA, Science, 235:683.

STRUCTURAL ELEMENTS OF BALBIANI RING BRa OF *CHIRONOMUS THUMMI*

N. N. Kolesnikov, S. S. Bogachev, S. V. Scherbic,
A. V. Taranin, S. I. Baiborodin, A. P. Donchenko,
T. E. Sebeleva, and I. I. Kiknadze

Institute of Cytology and Genetics
Siberian Department of the USSR Academy of Sciences
Novosibirsk, USSR

To study the molecular-cytological organisation of the tissue specific puff Balbiani ring BRa of *C. thummi* a number of clones from a microlibrary of a part of the fourth chromosome were hybridized *in situ* to polytene chromosomes. Clones, containing insertions homologous to the DNA of BRa, were identified. The nucleotide sequence of some of them (F6.2; C1.2; C6-10) were determined. The length of the F6.2 is 2333 bp, the content of AT pairs is 63%; the fragment C1.2 is 2040 bp long, with 67% AT pairs. Comparative computer analysis revealed some similarity between these fragments. They include the next structural elements: a) coding regions (ORF) with promoter zone, b) tandem direct imperfect repeats of two types, c) pseudogene, d) mobile element (MEC). Schematically these elements are represented in Figure 1. The length of the coding region α is 723 bp in the F6.2 and 738 bp in the C1.2. Both regions are preceded by zones containing signal sequences for RNA-polymerase II and occupying the same topographic positions. At the nucleotide level these zones are to some degree homologous to the promoter of the globin gene of *C. thummi* (Trewitt et al., 1988). The α regions are flanked by direct repeats R1, 21 bp each. Units are repeated 8 times in the F6.2 and 13 times in the C1.2 fragments. In the case of the F6.2 the open reading frame (ORF) is terminated into the first of these R1 repeats, but it can be extended from the β region. With the assumption of the splicing coding regions α and β may be regarded as an exons, tandem repeats R1 as an intron. A noteworthy feature of the consensus of these repeats (AGGTTATGTTGCCAAATTTG) is the conservative AG$^{\downarrow}$GT sequence, which may serve as donor and acceptor sites in splicing (Keller and Noon, 1984). The C1.2 has an analogous sequence arrangement but interpretations here are made difficult by insertion of a mobile element into the β-region (Figure 1A).

In the first third of the F6.2 three overlapping ORF were established (Figure 1A). However insertions of only one nucleotide at two positions restores ORF 654 bp long. The polyadenylation signal is at a distance 38 bp away from the last codon. The amino acid sequence is to a certain extent homologous to the α-part and more than 50% to the β. Such homology suggests that the pseudogene Ψβ is located in the beginning of the F6.2 fragment.

Between the pseudogene and the promoter another type of tandem repeat - R2 is situated (Figure 1A); the consensus sequence is TCCCCCTTCCCA, which is repeated 13 times. Such repeats, under certain conditions, may change the

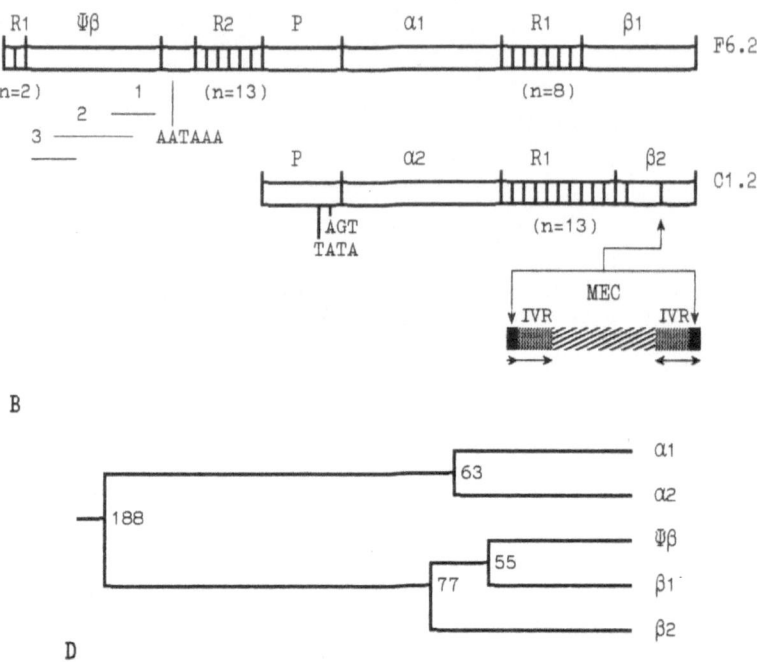

Figure 1.   A scheme showing the organization of cloned fragments F6.2,
C1.2 and C6-10 from Balbiani ring BRa DNA of *Chironomus thummi*.
(A) Structural elements of the F6.2 and C1.2. R1, R2 –
tandem direct imperfect repeats, n – the number of repeated
units. Ψβ – pseudogene, lines with numbers 1, 2, 3 – open
reading frames (ORF) found in this part, AATAAA – signal of
polyadenilation. P – promoter zone with signal sites for RNA-
polymerase II. α and β-coding parts (ORFs). MEC – mobile
element inserted into the β-part. IVR – inverted repeats.
(B) The phylogenetic tree constructed on the basis of comparison and alignment of amino acid sequences deduced by
corresponding nucleotide sequences – α, β and Ψβ.
(C) The C6-10 fragment, a, b, c – one long ORF, b – nucleotide
sequence, which corresponds to the deduced amino acids
containing the Zn^{++}-finger motif, d – AT-rich region.
(D) A value of relative evolutionary distance (Feng et al.,
1985).

conformational state of DNA and may become involved in the regulation of gene
activity.

The results of the analysis of the derived amino acid sequences of
α- and β-parts of both fragments demonstrated that at the N-ends there were
16 amino acids with the properties of a signal peptide (Heijne, 1984).
Potential glycosylation and phosphorylation sites were identified. Post-
translation modifications of this kind are the feature of the secretion
glycoproteins encoded by the BRs (Hamodrakas, Kaftos, 1984). Comparative
analysis demonstrated more than 45% homology for the amino acid sequences
deduced from the α- regions of the F6.2 and C1.2 and homology of about 40%
for the β- regions (Figure 1B). These similarities between the coding parts
and the flanking R1 repeats suggest their derivation from a common ancestral
sequence (Figure 1B). Thus sequence analysis showed that the F6.2 contains
coding elements with signal sites of RNA-polymerase II. In addition, the
F6.2 clone strongly hybridized to the DNA strands of the transcriptionally
active puff - BRa, and also with poly A$^+$ m-RNA isolated from the salivary
glands.

To show whether these coding sequences were actually expressed in cells
the hybrid gene β-galactosidase-F6.2α was constructed. It was expressed in
a bacterial system and fusion protein appeared. This fusion protein was
extracted and injected into rabbits to raise antibody. The specificity of
the anti-(β-gal/F6.2α) antibody was tested in control experiments. When
preabsorbed with β-gal, antibody was incubated with a Western blot
containing salivary gland proteins, and a single band with an apparent
molecular weight of 67 kDa, reacted. We concluded that the α element in the
F6.2 encoded a part of a 67 kDa protein (p-67).

By Western blot analysis p-67 was detected in the protein spectrum of
isolated secretion of salivary glands and unexpectedly in other tissues with
secretory activity, such as body fat, malpigian tubules, midgut, neural
ganglia and hemolymph. Then samples of all tissues were prepared for
electron microscopy and p-67 was localized using preabsorbed anti-(β-gal/
F6.2α) antibody coupled to particles of colloidal gold. Gold coupled antibody was observed in all tissues examined. While the particles appeared
slightly concentrated in the peritrophic membranes, they were also observed
in the nucleus and cytoplasm of all cells and the lumen of salivary glands.
Gold particles were generally not observed in mitochondria, Golgi apparatus,
secretory granules and vacuoles. It thus appears that p-67 is a ubiquitous
component of the cell and may be equivalent to a house-keeping protein.

These results were surprising because the F6.2 clone hybridized in situ
with BRa, whose activity appeared only in four special lobe cells of
salivary glands and correlated with the synthesis of the additional
secretory polypeptide - ssp-160 (Kolesnikov et al., 1981). But, it is
necessary to note that the secretory proteins of the salivary gland (sp-180,
-150,-35) give a weak positive result on Western blot, and ssp-160 also
cross-reacts with antibody against fusion protein. There is no reaction
with the secretory proteins of the sp-1 group. Whether such immunological
similarity exists in the structure of some secretion protein genes and the
one of p-67 (or not) needs further experimental support. It is now clear
that BRa contains not only the structural gene coding for the tissue-
specific protein ssp-160, but also the house-keeping gene encoding p-67,
which may be involved in the secretion process.

A mobile element of Chironomus genome was detected in the C1.2 sequence
(Figure 1A). The whole length of MEC is 596 bp, with AT content - 71%.
There are two inverted repeats by 107 bp in the ends of MEC. The inverted
sequences are in their turn flanked by direct repeats of 5 bp, which are a
target-duplication of the host DNA. The inverted repeats contain CAAT and

TATA — boxes. Within the body of MEC there are three overlapping ORFs. Thus, molecular properties of MEC are similar to those known for the transposable elements of the eucariotic genome, however MEC has some features of its own. MEC was hybridized *in situ* to the polytene chromosomes from natural and laboratory populations of *C. thummi*. The total number of hybridization sites was 90; MEC was localized in various chromosome regions, the centromeric and telomeric, and the puffs. In *C. th. pieger*, a species closely related to *C. th. thummi*, the number of hybridization sites was found to be twice smaller. In hybrid larvae of these species the localization of MEC was altered, compared to the parental. A new puff, normally not observed, develops from a band in one of the homologous chromosomes, which contains MEC, while the other band, without MEC, on another homoloque remains condensed. In separate cases, insertion of MEC may lead to the activation of genetic material, however the functional meaning of the MEC presence in BRa is still unknown.

Apart from MEC we found another clone, C6-10, which also hybridized to BRa and exhibited a dispersed localization. It has more than 90 localization sites in the genome. The C6-10 clone hybridized intensivly with poly A$^+$ mRNA from larvae and indeed we found long ORF — 2043 bp in this fragment. The whole length of the C6-10 is 2384 bp. When we made the same experiments as in the case of the F6.2, Western blot analysis showed that C6-10 encodes a protein of molecular weight ca 75kDa, detected in different tissues. Immunocytochemical analysis demonstrated that p-75 localizes in the nucleolus and in a few separate bands of polytene chromosomes. Searching for homology by protein data base showed the inferred amino acid sequence of the C6-10 has a partial homology to DNA-binding proteins (Figure 1C).

Thus, BRa is a complicated locus, which contains at least three genes: one is tissue-specific, the other two probably belong to house-keeping genes.

REFERENCES

Feng D. F., Johnson M. S., and Doolittle R. F. 1985. Aligning amino acid sequences: Comparison of commonly used methods, J. Mol. Evol., 21:112.
Homodrakas S., and Kafatos F. C. 1984. Structural implications of primary sequences from a family of Balbiani ring-encoded proteins in Chironomous, J. Mol. Evol., 20:296.
Heijne G. 1985. Signal sequences. The limit of variation, J. Mol. Biol., 184:99
Keller E. B., and Noon W. A. 1985. Intron splicing: a concerved internal signal in introns of *Drosophilia* pre-m-RNA, Nucl. Acids Res., 13:4971.
Kolesnikov N. N., Karakin E. I., Sebeleva T. E., Meyer L., and Serfling E. 1981. Cell-specific synthesis and glycosylation of secretory proteins in larval salivary glands of *Chironomus thummi*, Chromosoma, 83:661.
Trewitt P. M., Saffarini D. A., and Bergtrom G. 1988. Multiple clustered genes of the haemoglobin VIIB subfamily of *Chironomus thummi thummi* (Diptera), Gene, 69:91.

ROLE OF SPECIFIC PROTEIN S-S BONDS IN THE QUASISUBUNIT STRUCTURE OF

CHROMOSOMAL DNA

V. A. Struchkov, N. B. Strazhevskaya, and Yu. D. Blokchin

All-Union Cancer Research Center
Academy of Medical Sciences of the USSR
Kashirskoye shosse 24
115478 Moscow, USSR

The study of DNA-bound residual protein (RP) is essential for understanding the structural and functional arrangement of eukaryotic chromosomal
DNA. Of particular interest is the research (Andoch and Ide, 1972; Dounce
et al., 1973; Lange, 1974) demonstrating the role of S-S bonds of DNA-
covalently linked RP in a tandem arrangement of chromosomal DNA subunits.
Recently, Struchkov and Strazhevskaya (1989) have found the existence in DNA
of several types of RP specific S-S bonds which probably determine different
levels of eukaryotic as well as prokaryotic chromosomal DNA. This work was
undertaken to study the mechanism of thiol-induced DNA degradation and the
role of S-S bonds of DNA-linked RP in the quasisubunit and loop structure of
DNA. The experiments were performed with a DNA supramolecular complex (DNA
SMC) obtained from loach sperm and hen erythrocytes by the mild phenol-
extraction method (Georgiev and Struchkov, 1961). DNA SMC were found to
contain minor amounts of DNA-bound RP, RNA and lipids (Strazhevskaya et al.,
1979; Struchkov and Strazhevskaya, 1988). These DNA structures are highly
intact, since they retain the elements of looped organization, possess
biological properties such as immuno- and tissue specificity, hormone-
dependence and transforming activity, and are the target of ionizing radiation damage *in vivo* (Strazhevskaya and Struchkov, 1977). DNA SMC possesses
1% RP which contains cysteine, and is 40% acidic aminoacid; it is composed
of 4-5 peptides (12-70 kD). We used 2-mercaptoethanol (ME), dithiothreitol
(DTT) and $NaBH_4$ as S-S cleaving agents.

Figure 1 shows sedimentation data on the DNA SMC of loach sperm after
incubation with ME and DTT as a function of pH, time of treatment and thiol
nature (the experimental conditions excluded the effect of nucleases). It
appeared that during long-term incubation (5 and 10 days) with thiols, DNA
SMC was cleaved into double-strand subunits of different size. For example,
ME treatment at pH 4.4 cleaved DNA SMC into subunits of size $4-6 \times 10^5 D$; ME
treatment at pH 8 in the presence of EDTA cleaved DNA complex into subunits
$10^8 D$ in size, while incubation with DTT in the same condition resulted in
subunits of size $1-2 \times 10^7 D$. However, $NaBH_4$ at pH 8 failed to induce
degradation of DNA SMC. We believe that thiol-induced DNA SMC degradation
is due to the cleavage of S-S bonds of RP. To understand the mechanism of
thiol-induced DNA SMC degradation we studied the heat denaturation of ME-
induced subunits at acid pH values.

It is seen (Figure 2) that ME-induced degradation of hen erythrocyte

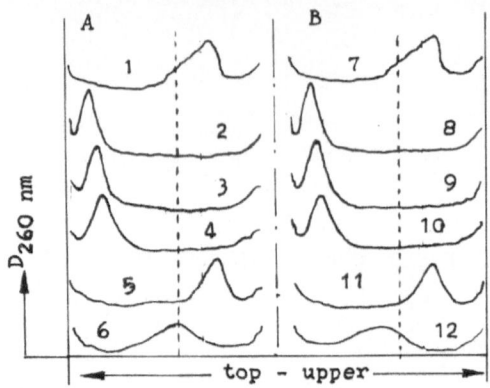

Figure 1.   UV-sedimentation of thiol-induced subunits of loach sperm DNA
             SMC on neutral 5-20% sucrose gradient (2mkg DNA, SW-41 rotor,
             20°C, 39,000 rpm, 2.5h). Incubation with 0.1% thiols was
             performed at 37°C for 5 days (A) and 10 days (B). Prior to
             sedimentation DNA was dialysed against 0.14M NaCl-0.01M SSC at
             pH 7. (1,7) Control, incubation without thiol; (2,8) ME at pH
             4.4; (3,9) ME at pH 5.3; (4,10) ME at pH 5.6; (5,11) ME at pH
             8 with 0.025M EDTA; (6,12) DTT at pH with 0.025M EDTA. Dotted
             line indicates lambda phage DNA, 31×10^6D.

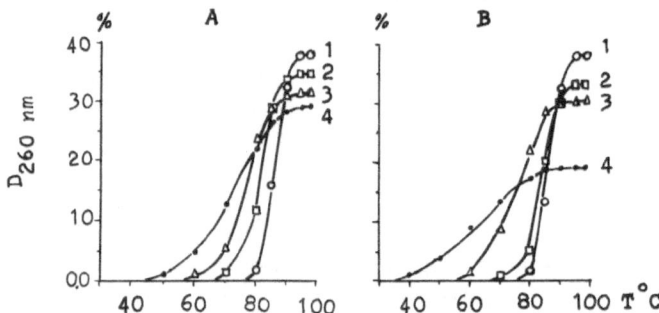

Figure 2.   Melting of hen erythrocyte DNA SMC in norm and after incubation
             with 0.1% ME for 5 days (A) and 10 days (B). (1) Control,
             incubation without thiol; (2-4) incubation with ME at pH 5.9,
             5.3 and 4.4, respectively. The ordinate represents hyper-
             chromatic effect in %. Prior to melting DNA was dialysed
             against 0.14M NaCl-0.01M SSC.

DNA SMC is accompanied by a decrease of hyperchromatic effect of subunits,
indicating the presence of 'sticky' ends. These facts allow us to conclude
that ME-induced degradation of DNA SMC results from a 'slanting' double-
strand break. We have calculated that this break involves, on average,
180bp. This value is characteristic for all DNA SMC preparations and is
reliably tested for subunits of size $0.5-2 \times 10^6$D. Thus, not only S-S bonds
of RP, but also complementary 'sticky' ends of subunits take part in tandem
linking of eukaryotic chromosomal DNA subunits.

      It is suggested (Figure 3) that the tandem linkage of two subunits is
formed by 4 peptides arranged in part through S-S bonds and these pairs of

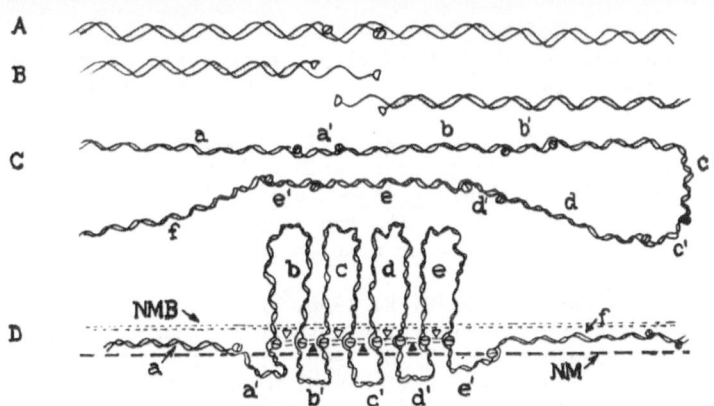

Figure 3.   Possible quasisubunit structure of eukaryotic chromosomal DNA.
(A) A tandem linking of subunits by S-S bonds of peptides
occuring in antiparallel DNA strands at a distance of 180bp.
(B) Thiol-induced 'slanting' double strand break resulting in
the formation of 'sticky' ends. (C) A tandem quasisubunit
organization of DNA by peptide S-S bonds and complementary
'sticky' ends. (D) Looped organization of DNA by clustered
peptides. (b-e) Big loops; (a'-e') small loops; (NM) nuclear
matrix; (NMB) nuclear membrane; (▲) topoisomerases; (Δ) lipid.

peptides occur in antiparallel DNA strands at a distance of 180bp. If we
assume that DNA SMC has, on average, 1% of RP, then the molecular weight of
the peptides which tandemly bond DNA subunits of size 5×10^5D must be
0.15kD. Such peptides were detected (Neuer et al., 1983; Welsh and Vyska,
1981). However, according to our hypotheses these low-molecular weight
sulphur-containing peptides must be linked by weak peptide bonds (Figure 3)
to form larger aggregates of size 12-70kD, which are usually detected by the
electrophoresis of DNA-bound RP (Neuer et al., 1983; Struchkov and
Strazhevskaya, 1989). Such clustered peptides can serve as a base for the
formation of small and big loops. Small loops, ie the parts of 'sticky'
ends of quasisubunits may be the sites of interaction with topoisomerases 1
and 11, and the lipids localized in metabolically active parts of chromatin:
topoisomerases - at the initiation points of transcription and replication
(Wang, 1985); chromatin cardiolipin - wholly in DNA SMC (Struchkov and
Strazhevskaya, 1988), and activates the dnA protein initiating replication
(Sekimizu and Kornberg, 1988). Big loops carry genetic information. S-S
bridges may possibly play a flanking role between genes, through which an
exchange of DNA fragments can occur. After histonization, these loops may be
the basis for nucleosome→nucleomere→ chromomere supercoiling. In this case,
clustered and nonclustered DNA-bound peptides will represent chromomeric and
nonchromomeric sites of the chromatid, respectively.

REFERENCES

Andoch T., and Ide T. 1972, Disulfide bridges in protein linking DNA in
    cultured mouse fibroblasts, strain L-P3, Exp. Cell Res., 74:525.
Georgiev G. P., and Struchkov V. A. 1961, On the polymeric deoxyribonucleic
    acid of animal origin (In Russian), Biofisika, 6:745.
Dounce A. L., Chandra S. K., and Townes P. L. 1973, The structure of higher
    eukaryotic chromosomes, J. Theor. Biol., 42:275.
Lange C. S. 1974, The organization and repari of mammalian DNA, FEBS Letters,
    44:153.
Neuer B., Plagens U., and Werner D. 1983, Phosphodiester bonds between poly-
    peptides and chromosomal DNA, J. Mol. Biol.,164:213.

Sekimizu K., and Kornberg A. 1988, Cardiolipin activation of dnA protein, The initiation protein of replication in Escherichia coli, J. Biol. Chem., 263:7131.

Strazhevskaya N. B., and Struchkov V. A. 1977, Organization of supramolecular DNA complexes of the eukaryotic chromatin and their role in the radiation effect (In Russian), Radiobiologiya, 17:163.

Strazheveskaya N. B., Krasichkova Z. I., and Kruglova N. L. 1979, Radiation damage and repair of DNA-membrane complex in mammalian cells, Studia Biophys., Berlin, 3:205.

Struchkov V. A., and Strazhevskaya N. B. 1988, The composition of DNA-bound lipids in regenerating rat liver (In Russian), Biochimiya, 53:1449.

Struchkov V. A., and Strazhevskaya N. B. 1989, Specific disulfide bonds of DNA-residual protein complex, Dokl. Acad. Nauk SSSR, 307:755.

Wang J. C. 1985, DNA-topoisomerases, Annu. Rev. Biochem., 54:665.

Welsh R. S., and Vyska K. 1981, Organization of highly purified calf thymus DNA. 1. Cleavage into subunits and release of phosphopeptides, Biochem. Biophys. Acta, 655:291.

AN ASPECT OF GENOME VARIABILITY: RETROTRANSPOSONS (MDG1 COPIES) IN THE

EUCHROMATIN AND HETEROCHROMATIN OF *DROSOPHILA MELANOGASTER*

V. A. Gvozdev, Yu. A Shevelyov, and M. D. Balakireva

Institute of Molecular Genetics
USSR Academy of Sciences
Moscow, USSR

INTRODUCTION

Regions of pericentromeric heterochromatin enriched with satellite DNA
and retrotransposons scattered over the eukaryotic genome are usually
regarded as components of selfish DNA whose expansion in the genome is
determined by their own internal properties (Orgel et al., 1980). It is
believed that selfish DNA has no effect on the individual's phenotype or
biological properties so long as no vital genes are altered. But even if
these components of the eukaryotic genome have no functional role, it is
still interesting to examine their variability (amplification, transpositions
and divergence) in the course of the genome's evolution.

We shall look into the variability of those sites in the genome that
carry inserted copies of MDG1 (a variety of copia-like elements - retrotransposons) in pericentromeric heterochromatin and along polytene
chromosomes. The interest in the genomic pattern of retrotransposons, and
specifically MDG1, has been aroused by indications of non-random distribution
of MDG1 along chromosomes both in the laboratory stocks and in a natural
population (Gvozdev, 1986). Furthermore, an inbred stock of *D. melanogaster*
characterized by low fitness and poor competitive ability exhibited
spontaneous transpositions of several retrotransposons, including MDG1, into
preferred 'hot spots' along chromosomes (Figure 1). These transposons were
accompanied by a remarkable biological effect: an increase of the stock's
fitness (Pasyukova et al., 1986).

The hot spots along polytene chromosomes, in the euchromatin, often
prove to belong to the so-called regions of intercalary heterochromatin (IH),
which show a preference for conjugation with each other (Figure 1) and, like
pericentromeric heterochromatin (PH), are distinguished by late replication
(Zhimulev et al., 1982). IH regions often associate with PH saturated with
repeats, including satellite DNA and mobile elements. In order to
characterize the sites of MDG1 insertions along polytene chromosomes (ie in
the euchromatin and in IH), as in PH, it was necessary to clone extended
genomic regions carrying inserted copies of MDG1.

RESULTS AND DISCUSSION

From genome libraries obtained in cosmid vectors pHC79 and pJB8,

Nuclear Structure and Function, Edited by J. R. Harris and
I. B. Zbarsky, Plenum Press, New York, 1990

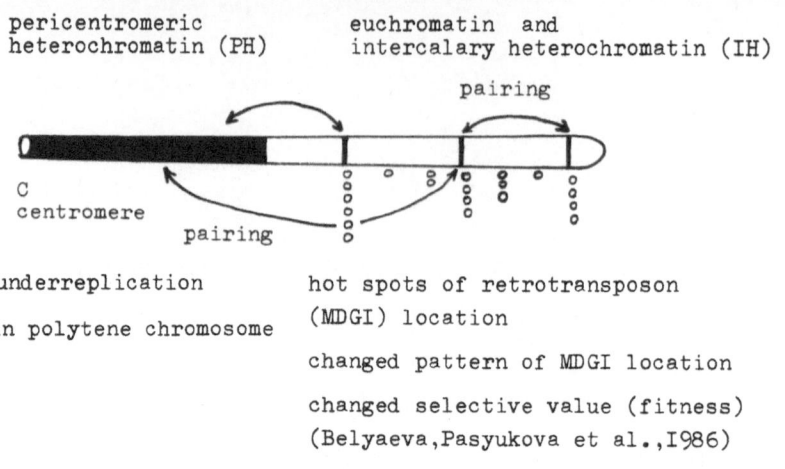

pericentromeric   euchromatin and
heterochromatin (PH)   intercalary heterochromatin (IH)

pairing

C
centromere
        pairing

underreplication   hot spots of retrotransposon
                              (MDGI) location
in polytene chromosome
                              changed pattern of MDGI location

                              changed selective value (fitness)
                              (Belyaeva,Pasyukova et al.,I986)

Figure 1.   Heterochromatin and enchromatin in *D. melanogaster* genome
              (Scheme).

RECOMBINANT COSMIDS ∼ 40 kb
CARRYING MDGI COPIES

cosmid I   ▨   } no cross homology (except MDGI)
cosmid 2   ▢   between cosmids

chromosome

                    IH   IH

C
centromere   MDGI   MDGI

              additional sites of hybridization
              intercalary heterochromatin (IH),
              hot spots of MDGI insertions

REGIONS OF INTERCALARY HETEROCHROMATIN CONTAIN
VARIABLE SETS OF REPEATS (MOBILE ELEMENTS)

Figure 2.   Recombinant cosmids (40kb) carrying MDGI copies.

enabling fragments of 40-50kb to be cloned, we selected recombinant cosmids
that carried MDGI copies. Upon *in situ* hybridization with polytene
chromosomes, a considerable number of the recombinant cosmids exhibited
'additional' hybridization sites, other than the MDGI location sites (Figure
2). These recombinant cosmids also frequently hybridized with the chromo-
centre and the β-heterochromatin adjacent to it. As a rule, the additional
hybridization sites varied from stock to stock. Consequently, many MDGI
insertion sites consist of repeat clusters (including other mobile elements)
both scattered over the genome and grouped in regions of pericentromeric

heterochromatin. It has also been established that, for instance, two
cosmids that carry no homologous sequences (except MDG1) hybridize with one
and the same set of additional sites in polytene chromosomes (Figure 2). A
number of such sites coincide with potential MDG1 hot spots identified in
other stocks. Hence, these cosmids carry different repeats that can be
located in the same sites, including hot spots (IH regions). Since the
additional hybridization sites in cosmids vary from stock to stock, we can
assume that a substantial part of IH sites is represented by structurally
variable genome regions, largely made up of mobile elements. The functional
role of these regions is totally unclear, and their molecular characterization is so far rather hazy, It has recently been demonstrated that IH
regions, at least in the polytene nuclei, are predominantly associated with
the nuclear membrane (Hochstrasser et al., 1986). It is still not clear
how these regions relate to the loop-like domain organization of the
chromosomal structure.

The results of *in situ* hybridization suggest that IH and PH regions may
contain homologous repeats, including identically or similarly structured
MDG1 copies. It was interesting therefore to try to characterize the MDG1
copies and their surroundings in the PH regions, which are known to be
largely under-replicated in polytene chromosomes.

Of the cosmids containing fragments under-replicated in polytene
chromosomes, ie genomic PH fragments, we chose two that carried homology
regions other than MDG1 for a more detailed characterization (Figure 3).
The sequencing of the DNA fragments adjacent to MDG1 demonstrated that the
MDG1 insertion sites are represented by a new family of imperfect tandem
heterochromatic repeats with a repeat unit of 1.0 or 1.5kb in two different
regions of heterochromatin. Repeats from different PH regions show zones of
homology that cannot be accounted for by the insertion of mutually homologous
mobile elements into different tandems. These repeats seem to be subject to
an untrivial mechanism of variability, which is responsible for the tandem
structure described above. *In situ* hybridization with appropriate radioactive probes failed to reveal such repeats along polytene chromosomes in
the vicinity of MDG1 copies, with the exception of one IH site 12E, where
MDG1 copies occur but rarely.

The bulk of under-replicated MDG1 copies in heterochromatin were found
to present a distinctive structural variant characterized by the absence of
the EcoRI restriction site in LTR. It was shown that the targets of insertions of many heterochromatic MDG1 copies are the same in different stocks.
Thus, while MDG1 copies are mobile in euchromatin and IH, they stably remain
in their insertion sites in PH in the course of evolution.

The stability of MDG1 in heterochromatin may seem to contrast with the
variability of heterochromatic regions due to amplification and diminutions,
revealed by cytogeneticists. However, we observed these properties of PH as
well. As it turned out, the regions of heterochromatic tandem repeats are
amplified along with the inserted MDG1 copies in cell cultures (the data of
A. Kalmykova). Interestingly enough, two established independent *Drosophila*
cell cultures showed a similar (ten-fold) degree of amplification of these
repeats.

Analysis of the sequences surrounding two inserted MDG1 copies revealed
the nucleotide motifs that, according to the data found in literature, ensure
the local amplification of *Drosophila* genes. Moreover, these regions were
found to contain adenine nucleotide blocks spaced by multiples of the
nucleotide number in a turn of the DNA helix and are probably responsible
for DNA bending. The current view is that 'bent' DNA regions in eukaryotes
are located in the regions of replication origins (Williamson et al., 1988).

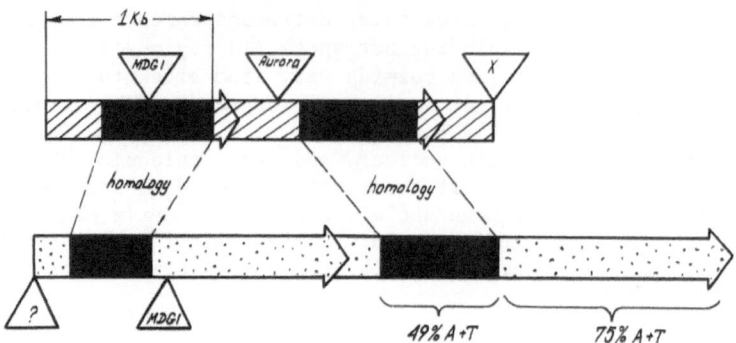

Figure 3. Tandem divergent repeats in two different heterochromatic
regions (50 clusters in the genome). The regions of homology
in two different tandem (black boxes) contain deletions
(insertions). The origin of the homology region is obscure.
No duplications of flanking sequences are detected. AT-rich
regions (75% A+T) are adjacent to homology regions (49% A+T).

The study of IH variability is of general interest because it correlates, according to our data for an inbred *Drosophila* stock, with changed
biological properties of individuals. Of course, one can hardly claim that
the spontaneous sudden changes in selective value are a direct consequence
of the insertion of retrotransposon copies. The other type of variability,
local amplification of a certain type of tandem repeat in PH can underly the
meiotic drive phenomenon - the unequal transfer to the offspring of two types
of gametes produced by a heterozygous individual (Wu et al., 1988). Thus,
the local replication of tandem repeats in PH may determine an important
biological property that plays an essential role in population processes. To
elucidate the biological role and the molecular mechanisms underlying these
types of variability, more profound studies of adequate genetic models are
required.

REFERENCES

Belyaeva E. Sp., Ananiev E. V., and Gvozdev V. A. 1984, Distribution of
    mobile dispersed genes (mdgl and mdg3) in the chromosomes of Drosophila
    melanogaster, Chromosoma, 90:16.
Gvozdev V. A. 1986, Mobile genetic elements in Drosophila melanogaster: a
    study of distribution and saltatory transpositions coupled with fitness
    changes, Soviet. Scient. Review (Section D) Physicochemical Biology
    Reviews, vol. 6:107, Harwood Press, London.
Hochstrasser M., Mathog D., Gruenbaum Y., Saumweber H., and Sedat J. W. 1986,
    Spatial organization of chromosomes in the salivary gland nuclei of
    D. Melanogaster, J. Cell. Biol., 102:112.
Orgel L. E., Crick F. H. C. and Sapienza C. 1980. Selfish DNA, Nature,
    288:645.
Pasykova E. G., Belyaeva E. Sp., Kogan G. L., Kaidanov L. Z. and Gvozdev
    V. A. 1986, Concerted transpositions of mobile genetic elements coupled
    with fitness changes in Drosophila melanogaster, Mol. Biol. Evol.,
    3:299.

Williams J. S., Eckdahl T. T. and Anderson J. N. 1988, Bent DNA functions
    as a replication enhancer in Saccharomyces cerevisiae, <u>Molec. Cell.
    Biol.</u>, 8:2763.

Wu C.-I., Lyttle T. W., Wu M.-L., and Lin G.-F. 1988, Association between a
    satellite DNA sequence and the responder of Segregation Distorter in
    D. melanogaster, <u>Cell.</u>, 54:179.

Zhimulev I. F., Semeshin V. F., Kulichkov V. A. and Belyaeva E. Sp. 1982,
    Intercalary heterochromatin in Drosophila. I. Localization and general
    characteristics, <u>Chromosoma</u>, 87:197.

IS ELEMENTS IN *AGROBACTERIUM TEMEFACIENS* STRAINS AND THEIR PUTATIVE

IMPLICATIONS IN HORIZONTAL GENE TRANSFER TO PLANTS

C. De Meirsman, J. Vanderleyden and A. Van Gool

F. A. Janssens Memorial Laboratory of Genetics
University of Leuven
Willem de Croylaan 42
Leuven, Belgium

IS ELEMENTS IN PLANT-INTERACTIVE BACTERIA

Recent elucidation of molecular mechanisms in plant-bacteria inter-actions gave rise to the discovery of transposable elements. Predominantly, they were identified in bacterial insertion mutants impaired in a successful interaction with plants.

Among these elements several insertion sequences (IS) have now been thoroughly characterized in particular in *Agrobacterium tumefaciens*, *Pseudomomas savastanoi* and several *Rhizobium* species (Galas and Chandler, 1989).

OCCURRENCE OF IS ELEMENTS IN *AGROBACTERIUM TUMEFACIENS*

In Table I, IS and isoIS elements are listed that occur in various *A. tumefaciens* strains. Only the best characterized sequences are mentioned and their overall genomic location indicated.

DISTRIBUTION OF THESE IS ELEMENTS AMONG *A. TUMEFACIENS* STRAINS

Using nick translated internal DNA fragments of IS elements as probes, interaspecies distribution of IS426, IS427, IS66 and IS866 has been analyzed by Southern hybridization with restricted total DNA or pTi DNA, derived from various *A. tumefaciens* strains.

In the case of IS427 (De Meirsman et al., 1989) the internal *AccI* restriction fragment only gave hybrids on restricted total DNA of *A. tumefaciens* T37 and no signals were observed in Southern hybridization with restricted total DNA of the biotype I C58 nopaline strains, the 15955 octopine strain, or C58 derived A208 and A114 strains. The distribution of IS866 has been examined by Bonnard et al. (1989). Out of the nine *A. tumefaciens* strains examined, IS866 only occurred in biotype I grapevine isolates CG401, 2654 and 2655. In biotype II strains it only occurred in the AB2/73 strain, while in the biotype III group it could only be detected among strains of the vitopine and octopine/cucumopine subgroup. Table II shows the distribution of IS426 among the chromosomes and Ti plasmids of

Table I. Occurrence of IS and isoIS Elements in *A. Tumefaciens* and Their Genomic Location.

Insertion Sequence	Strain	Biotype	IS Genomic Location chromosome	pTi Plasmid	Reference
IS66	A66, spontaneous A6 variant	octopine biotype I	2isoIS66 identified	T region *iaaH* locus 3isoIS66 identified	Machida et al. (1984)
IS426	A208 (C58 background with pTiT37 plasmid)	nopaline biotype I	present	T region in between 6a and 6b loci	Vanderleyden et al. (1986)
IS427	T37 obtained from M.D. Chilton	nopaline biotype I	n.d.	besides IS427 2isoIs427 were detected	De Meirsman et al. (1989)
IS866	Tm4	octopine/ cucumopine biotype III	n.d.	T_A region *iaaH* locus 5isoIS866 detected	Bonnard et al. (1989)

n.d. : IS element was not detected by Southern hybridization

various *A. tumefaciens* and *A. rhizogenes* strains. IS66 was shown to occur only in octopine and not in nopaline strains (Machida et al., 1987). In the octopine strain A66, IS66 was detected both in the chromosome and in pTi A66, including its occurrence within the T region.

SEQUENCE HOMOLOGY AMONG *A. TUMEFACIENS* IS ELEMENTS

Intraspecies IS homology was examined both by cross hybridization on Southern blots and through homology search in case IS nucleotide sequence data were available. Table III summarizes this intraspecies homology search. Since very little intraspecies IS sequence homology could be detected, an interspecies comparison was carried out taking in account IS elements that had been previously characterized in other plant interactive bacteria. As shown in Figure 1 a striking sequence homology could be observed between the inverted repeat sequences of IS426 and those of IS51. The latter insertion sequence has been shown to impair virulence in phytopathogenic *Pseudomonas savastanoi* strains through its insertion in the pIAA iaaM gene that has a strikingly homologous counterpart in gene 1 carried by the *A. tumefaciens* pTi T-DNA (Yamada et al., 1986).

IS426 AND IS51 SEQUENCE HOMOLOGY WITH T_CDNA IN pTi PLASMIDS WITH SPLIT T REGIONS

Yamada et al. (1986) observed sequence homology between IS51 and a 527bp stretch in the T_C region of the octopine plasmid pTi15955. As shown in Figure 2, the examination of sequence homology between IS426 and T_C region of pTi15955 also led to the identification of a highly homologous 105bp nucleotide sequence. When the T_C-DNA homologous regions of IS51 and of IS426 were located on the T_C-DNA nucleotide sequence (Figure 3) it became apparent that they were flanked by a four bp direct repeat. T_C-DNA homologous sequences were also found in IS866 and in IS66 (Bonnard et al., 1989).

Table II. Distribution of IS426 among Various *A. Tumefaciens* and *A. Rhizogenes* Strains[1].

Strain	IS426 occurrence	
	Plasmid	Chromosome
Agrobacterium tumefaciens		
nopaline types		
A136	0	+
A208	-	+
C58	-	+
T37$_{BRAUN}$	-	+
T37$_{LIPP.}$	-	+
AT181	±	-
octopine types		
A6NC	-	.
ACH5	-	.
B6-806	-	±
15955	-	.
leucinopine types		
542$_{TEMPE}$	-	+
Agrobacterium rhizogenes		
A4PC	-	.
HR1	-	.
TR101	-	.
TR105	-	.
8196	-	.
15834	-	.

[1] Data provided by G. Jen, CIBA-GEIGY, Research Triangle Park, Chapel Hill, N.C., U.S.A.

0, pTi cured strain ; -, no hybridization; ±, weak hybridization; +, hybridization

DISCUSSION

The above homologies, together with the flanking 4bp direct repeat, may point towards a composite transposon structure that gave rise to at least part of the T_C region through deficiencies in its translocation performance. This is in line with the observation that within the T_C DNA no characteristic plant-like gene loci could be mapped in the T region. However, the composite transposon structure could have introduced border or pseudoborder sequences in the T region and consequently might affect the transfer and integration of T-DNA in the plant through the creation of an alternative pattern of T strand generation (Zambryski, 1988). Different IS elements or iso-IS sequences could become subjected to a concerted translocation under appropriate selection pressures and result in the formation of composite transposons as illustrated above for the T_C DNA in split (pTi) T regions.

Table III.   Intraspecies IS Sequence Homology as Detected by
Cross Hybridization and Nucleotide Sequence
Analysis.

	IS426	IS427	IS66	IS866
IS426	+	-	-	n.d.
IS427	-	+	-	n.d.
IS66	-	-	+	±
IS866	n.d.	n.d.	±	+

+, strong overall homology; ±, weak and partial homology; -, no homology

observed; n.d., hybridization was not carried out.

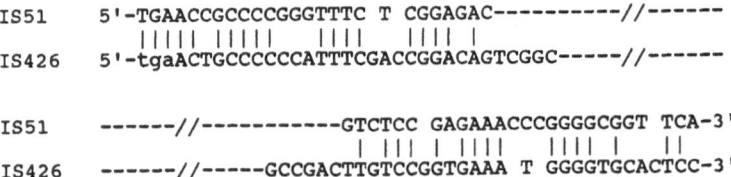

```
IS51    5'-TGAACCGCCCCGGGTTTC T CGGAGAC----------//------
           ||||| |||||    ||||   ||||| |
IS426   5'-tgaACTGCCCCCCATTTCGACCGGACAGTCGGC-----//------


IS51    ------//-----------GTCTCC GAGAAACCCGGGGCGGT TCA-3'
                          | ||| | ||||   |||| |    ||
IS426   ------//-----GCCGACTTGTCCGGTGAAA T GGGGTGCACTCC-3'
```

Figure 1.   DNA homology between the inverted repeats of IS51 (Yamada et
al., 1986) and IS426 (Vanderleyden et al., 1986). For
comparison of the left inverted repeats (top) 3 bases
(indicated by small letters) of the duplicated target sequence
of IS426 have been included. Base matches are indicated with
dashes.

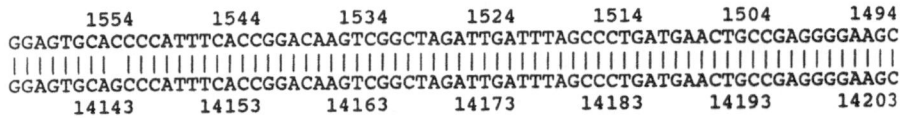

```
        1554      1544      1534      1524      1514      1504      1494
GGAGTGCACCCCATTTCACCGGACAAGTCGGCTAGATTGATTTAGCCCTGATGAACTGCCGAGGGGAAGC
||||||||| |||||||||||||||||||||||||||||||||||||||||||||||||||||||||||||
GGAGTGCAGCCCCATTTCACCGGACAAGTCGGCTAGATTGATTTAGCCCTGATGAACTGCCGAGGGGAAGC
        14143     14153     14163     14173     14183     14193     14203
```

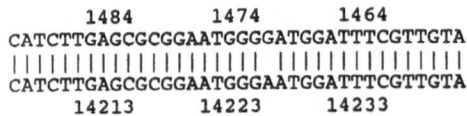

```
        1484      1474      1464
CATCTTGAGCGCGGAATGGGGATGGATTTCGTTGTA
|||||||||||||||||||||| |||||||||||||||
CATCTTGAGCGCGGAATGGGAATGGATTTCGTTGTA
        14213     14223     14233
```

Figure 2.   Homology between the published nucleotide sequences of IS426
(Vanderleyden et al., 1986) and the T_C region of T-DNA of
octopine plasmid of A. tumefaciens 15955 (Barker et al., 1983).
Numbers on the first line refer to the nucleotide sequence of
the lower strand of IS426, beginning with the right inverted
repeat, over a distance of 105 bases. Numbers on the bottom
line refer to the T_C region of pTi15955. Numbers have been
taken from the original published sequences. Base matches are
indicated with dashes.

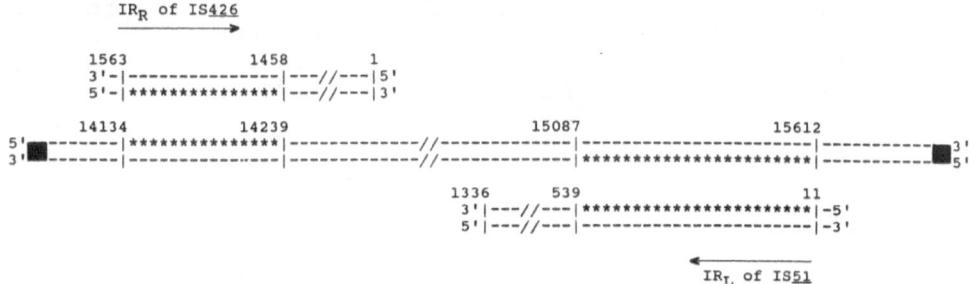

Figure 3.   Localization of the matching sequences of IS51 (Yamada et al., 1986) and IS426 (Vanderleyden et al., 1986) with the T_C region of pTi15955 (Barker et al., 1983). The compared DNA regions are represented by double interrupted lines (top and bottom strand). Matching sequences are represented by asterisks. The border sequences which define the edges of T_C-DNA are represented by black boxes. Numbers have been taken from the original published sequences. The orientations of the inverted repeats of IS426 (IR_R) and IS51 (IR_L) and for which DNA homology is found within T_C-DNA, are indicated by arrows. The figure is not drawn to scale.

In addition our intra- and interspecies homology data show that *A. tumefaciens* IS elements are useful markers to trace genetic diversification among strains and species, and that in the case of IS elements which are only located on a particular plasmid, routes of genetic exchange among strains and species can be reconstructed. The occurrence of these IS elements in plant interactive bacteria qualifies them as even more important markers suitable for ecologic and evolutionary genetic analysis. In *A. tumefaciens* vitopine and octopine strains carrying a pTi plasmid with split T region there seems to be a prevalence of IS insertions at or near to the right borders of T_L, T_A or T_B regions (Bonnard et al., 1989). Consequently, these insertions would not only affect the mechanism but also the outcome of T-DNA transfer to the plant. Since the relaxation of right border sequences by the plant induced *virD* products is of crucial importance to trigger a differential replication preceding the polarized ss DNA transfer to the plant.

REFERENCES

Bonnard G., Vincent F. and Otten L. 1989. Sequence and distribution of IS866, a novel T-region-associated insertion sequence from *Agrobacterium tumefaciens*. Plasmid 22, 70.
De Meirsman C., Croes C., Desair J., Verreth C., Van Gool A. and Vanderleyden J. 1989. Identification of insertion sequence element IS427 in pTiT37 plasmid DNA of an *Agrobacterium tumefaciens* T37 isolate. Plasmid 21, 129.
De Meirsman C., Desair J., Vanderleyden J., Van Gool A. P. and Jen G. C., 1987. Similarities between nucleotide sequences of insertion elements of *Agrobacterium tumefaciens* and *Pseudomonas savastanoi* in relation to *Agrobacterium tumefaciens* T_C-DNA. Nucleic Acids Res. 15, 10591.
Galas D. J. and Chandler M., 1989. Ch. 4 Bacterial Insertion Sequences. In: Berg D. E. and Howe M. M., Mobile DNA, pp 109-162, American Society for Microbiology, Washington D. C.
Vanderleyden J., Desair J., De Meirsman C., Michiels K., Van Gool A. P., Chilton M.-D. and Jen G. C., 1986. Nucleotide sequence of an insertion

sequence (IS) element identified in the T-DNA region of a spontaneous varient of the Ti-plasmid pTiT37. Nucleic Acids Res. 14, 6699.

Zambryski P., 1988. Basic processes underlying *Agrobacterium* - mediated DNA transfer to plant cells. Annu. Rev. Genet. 22, 1.

FUNCTIONAL ANALYSIS OF THE NUCLEOLIN GENE PROMOTER FROM THE MOUSE

Barbara J. Stevens, Valerie Housset, Patrick Calvas,
Francois Amalric and Henri-Marc Bourbon

Centre de Recherches de Biochimie et de Genetique du C.N.R.S.
118 route de Narbonne
31062 Toulouse, France

INTRODUCTION

The gene coding for a major, nonribosomal protein of the nucleolus, nucleolin, has been isolated and characterized in three rodent species (Bourbon et al., 1988a and b). A sequence comparison of the 5' terminal regions indicated a remarkable conservation within around 800bp upstream from the start sites and extending about 825bp downstream into the first intron. It was determined that this highly conserved region constituted an extended CpG island (Bird, 1987), whose boundaries and G+C content are likewise conserved in the three species (Bourbon et al., 1988b). In an attempt to define DNA elements of the 5' terminal regions which are potentially active in the transcriptional regulation of the gene, we made a compilation of the conserved sequences and identified stretches, termed homology blocks, in which 7 out of 8 nt had been conserved among the species. This evolution-oriented approach to define motifs important for promoter function provided us with a series of putative cis-acting regulatory elements. In this work, we have tested the effectiveness of the 5' upstream region which contains these conserved motifs to act as a promoter using a functional assay and we have determined the relative promotion strength of the region.

RESULTS AND DISCUSSION

Nucleolin is encoded by a single gene in the three rodents examined. The coding sequences of these genes extend from about 9kb for rat and mouse to over 11kb for hamster and are split into 14 exons that encode between 706-713 amino acid residues (Figure 1). The overall sequence and structure of the gene has been well conserved among these species and the exon-intron junctions have also been strictly maintained during evolution (Bourbon, 1988; Bourbon et al., 1988a and b).

The nucleotide sequence of the mouse promoter region is shown in Figure 2. It contains a number of features typical of genes fulfilling housekeeping or growth control functions in the cell (Dynan, 1986). In the proximal promoter, consensus TATA and CCAAT sequences are lacking. However, a conserved, AT-rich sequence flanked by GC-rich segments (GATTACTG) which may serve as a TATA-box related element (Breathnach and Chambon, 1981) is

Nuclear Structure and Function, Edited by J. R. Harris and
I. B. Zbarsky, Plenum Press, New York, 1990

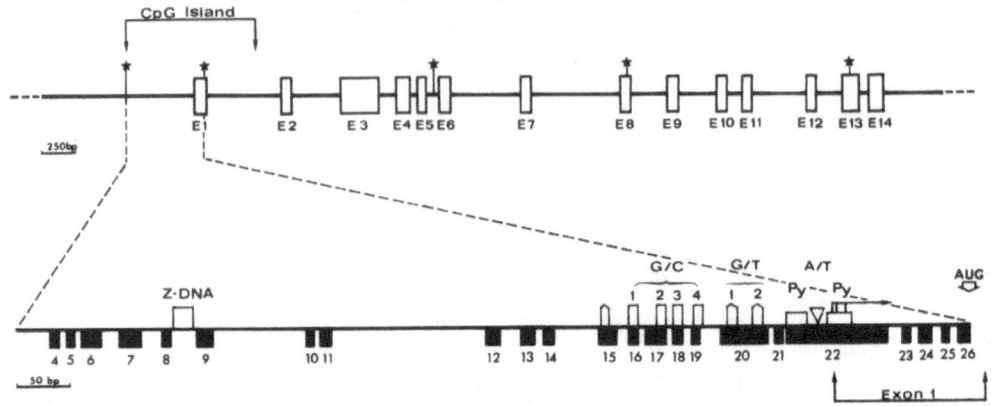

Figure 1.   Physical map of the mouse nucleolin gene (above) and the NcoI
             fragment from the 5' terminal region (below), shown in 5' to
             3' orientation. Exons are denoted by numbered clear boxes.
             * Indicates NcoI restriction sites. Homology boxes (see text)
             are denoted by numbered filled-in boxes. G/C:G/C boxes; G/T:
             inverted CCAAT-like boxes; Py: pyrimidine-rich stretches;
             A/T: TATA-related sequence. Start sites are indicated →.

located at −23 to −16 upstream from the first of three mRNA start sites.
Likewise, two inverted CCAAT-related elements (Benoist et al., 1980) with
strong homologies between them are located at positions between −74 and −96.
In the start site region, a palindromic sequence motif presents homology to
the consensus glucocorticoid-response element (GRE; Ringold, 1985). Three
conserved pyrimidine-rich stretches also appear in this region: the Py box
motif (−30 to −50), the cap site motif, and nearly the whole leader
sequence. These stretches show clear homologies to corresponding regions in
vertebrate ribosomal protein genes (Dudov and Perry, 1984; Mager, 1988).
The other notable feature is the presence of four CCCGCC hexanucleotide
sequences (GC boxes) which are potential binding sites for the transcription
factor Sp1 (Dynan and Tijan, 1985; Kadonaga et al., 1986), between −130 and
−240; one of these boxes is perfectly consensus (GCCCCGCCCC) in each species.
Finally, in the distal region, a stretch of alternating purine and pyrimidine
residues which has the potential of adopting a Z-DNA conformation has been
strongly conserved.

    We tested the ability of the 5' flanking region from the mouse nucleolin
gene to initiate transcription correctly in transient expression assays using
the chloramphenicol acetyltransferase (CAT) reporter gene in the plasmid
pSV2CAT (Gorman et al., 1982). The SV40 early promoter region was removed
from the plasmid by digestion with HindIII and AccI and the ends were filled
in with Klenow polymerase. An 881bp fragment (NcoI/NcoI fragment, see
Figure 1) was purified from a gel, treated with Klenow and blunt-end ligated
immediately upstream from the CAT coding sequence. The CAT gene was linked
downstream to the SV40 splice and polyadenylation sequences. The resultant
plasmid, pNuCATDir, contained the initiator codon ATG, the mRNA leader and a
730bp upstream region which included all the putative control elements of the
nucleolin gene described above. To maximize translational efficiency, the
nucleolin initiation codon was maintained in frame with that of the CAT gene.
Two other plasmids, pSVO'CAT, produced by re-ligation of AccI/HindIII
digested pSV2CAT without an insert, and pNuCATInv, in which the nucleolin
fragment was orientated in the 3' to 5' direction, were constructed for
comparative purposes.

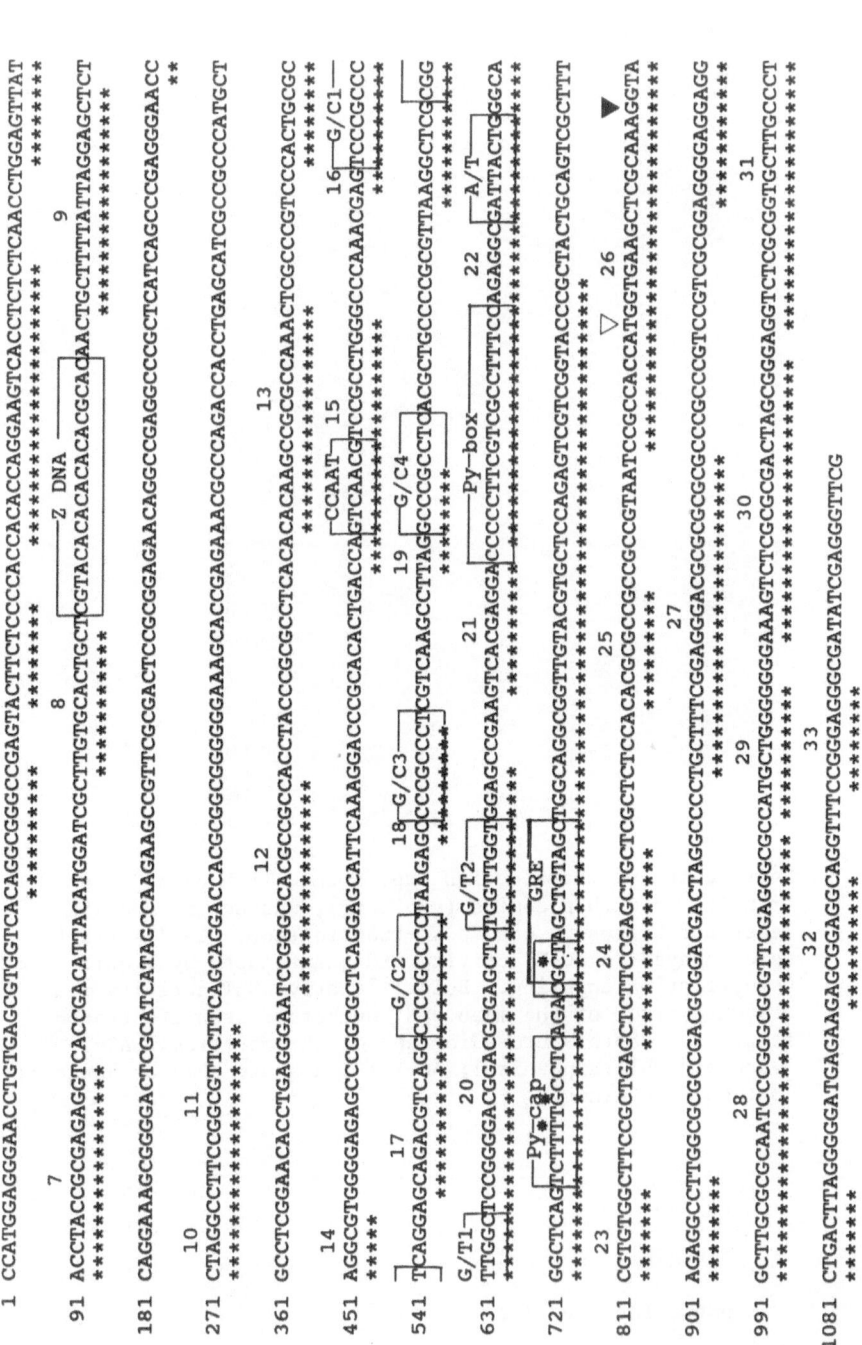

Figure 2. Nucleotide sequence of the mouse 5' terminal region extending from the NcoI site at −730 to +410 in the first intron. Homology boxes (see text) are underlined by * and numbered. Promoter elements are boxed. * transcriptional start sites; ▽ ATG codon; ▼ end of exon 1.

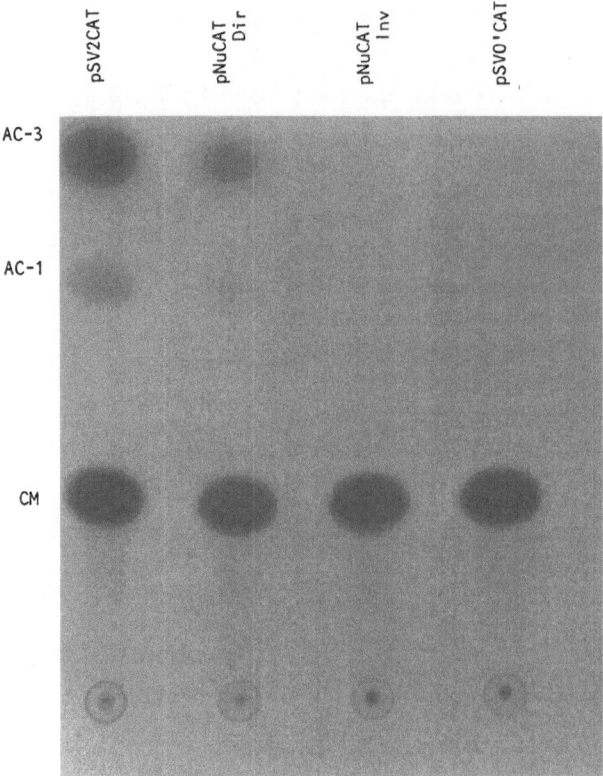

Figure 3.   Representative transfection experiment. BHK cells were
transfected with plasmids (see text), washed and fed 15h.
later and harvested 48h. post-transfection. Cell extracts
were assayed for CAT activity and quantitated by liquid
scintillation counting. Below: relative CAT activity ±
standard error of the mean. N: number of separate trans-
fections. Photo: Autoradiograph of chromatogram. AC-3 :
3-acetate chloramphenicol; AC-1 : 1-acetate chloramphenicol;
CM : chloramphenicol.

	%	SEM	N
pSV2CAT	100	–	5
pNuCATDir	77.2	± 20	5
pNuCATInv	3.2	± 1.6	2
pSVO'CAT	0.15	± 0.06	3

Plasmids were transfected into hamster BHK cells using the calcium phosphate precipitation technique. Following transient expression of the plasmid, cell extracts were prepared, incubated with ^{14}C-chloramphenical and acetyl coenzyme A and assayed for the presence of acetylated ^{14}C-chloramphenicol by thin-layer chromatography. A representative transfection experiment is presented in Figure 3. Under our conditions, the efficiency of the nucleolin fragment to direct transcription and translation of CAT compares favorably with that of the SV40 early promoter (77% relative to 100% for pSV2CAT). We can affirm, therefore, that the nucleolin insert in pNuCATDir functions as an effective and strong promoter. The inverse orientation of the insert in pNuCATInv promoted a weak CAT activity (3.2%) which is nonetheless significant with regard to the negligible value (0.15%) found by transfection with the promoter-less plasmid PSVO'CAT. In the pNuCATInv construction, an ATG codon is present in both the nucleolin and CAT sequences, and a TATA box as well as a pyrimidine-rich stretch are located within 90bp upstream (see Figure 2). These motifs could account for the minimal, correct initiation of transcription in this plasmid. Furthermore, this suggests that the region above the A/T-rich sequence and the pyrimidine-rich stretches in the correctly oriented NcoI fragment in pNuCATDir confers the main promotion strength of the nucleolin promoter. As in the case of other housekeeping genes, eg, CHO *aprt* (Park and Taylor, 1988), mouse *dhfr* and *hprt* (see Dynan, 1986), it is likely that the four potential Sp1 binding sites located in this upstream region are functionally important for the nucleolin promoter. Work is in progress to elucidate this question.

CONCLUSIONS

Our results show that the 5' flanking region of the mouse nucleolin gene contains a functional promoter which confers correct and efficient expression of the CAT reporter gene. We further conclude that the nucleolin genes cloned from mouse, rat and hamster (Bourbon, 1988; Bourbon et al., 1988a and b) represent functional genes. In addition to features found in housekeeping gene promoters, other elements of the nucleolin promoter show homology rather to sequences found in rRNA genes (GRE) and in ribosomal protein genes (CpG island; Py-rich stretches). Indeed, the major elements of the promoter and first intron of r protein genes required for transcription (Dudov and Perry, 1984; Chung and Perry, 1989) are also present in the nucleolin promoter and first intron. Efforts to examine mechanisms for a possible co-regulation of expression for genes involved in ribosome biogenesis should take into account these homologies.

ACKNOWLEDGEMENT

This work was supported by grants from the Association pour la Recherche sur le Cancer (ARC).

REFERENCES

Benoist C., O'Hare K., Breathnach R., and Chambon P. 1980, The ovalbumin gene-sequence of putative control regions, Nucl. Acids Res., 8:127.
Bird A., 1987, CpG islands as gene markers in the vertebrate nucleus, Trends Gen. 3:343.
Bourbon H-M., 1988, Characterisation du gene codant pour lat nucleoline; Evolution structurale chez les rongeurs, Thesis, Doctorat d'Etat-Sciences, Université Paul Sabatier, Toulouse.
Bourbon H-M., Lapeyre B., and Amalric F. 1988a, Structure of the mouse nucleolin gene. The complete sequence reveals that each RNA binding

domain is encoded by two independent exons, J. Mol. Biol., 200:627.

Bourbon H-M., Prudhomme M., and Amalric F. 1988b, Sequence and structure of the nucleolin promoter in rodents: characterization of a strikingly conserved CpG island, Gene, 68:73.

Breathnach R. and Chambon P. 1981, Organization and expression of eucaryotic split genes coding for proteins, Ann. Rev. Biochem., 50:349.

Chung S. and Perry R. P. 1989. Importance of introns for expression of mouse ribosomal protein gene rpL32, Mol. Cell. Biol. 9:2075.

Dudov K. P. and Perry R. P. 1984, The gene family encoding the mouse ribosomal protein L32 contains a uniquely expressed intron-containing gene and an unmutated processed gene, Cell, 37:457.

Dynan W. S. 1986, Promoters for housekeeping genes, Trends Gen. 2:196.

Dynan W. S. and Tijan R. 1985, Control of eukaryotic messenger RNA synthesis by sequence-specific DNA-binding proteins, Nature, 316:774.

Gorman C., 1985, High efficiency gene transfer into mammalian cells, p. 143 - 190 in 'DNA Cloning' Vol. 2, D. M. Glover, ed., IRL Press, Oxford.

Gorman C. M., Moffat L. F., and Howard B. H. 1982, Recombinant genomes which express chloramphenicol acetyltransferase in mammalian cells, Mol. Cell. Biol., 2:1044.

Kadonaga J. T., Jones K. A., and Tijan R. 1986, Promoter-specific activation of RNA polymerase II transcription by Sp1, Trends Biochem., 11:20.

Mager W. H. 1988, Control of ribosomal protein gene expression, Biochim. Biophys. Acta, 949:1.

Park J-H. and Taylor M. W. 1988, Analysis of signals controlling expression of the Chinese hamster ovary aprt gene, Mol. Cell. Biol., 8:536.

Ringold G. M. 1985, Steroid hormone regulation of gene expression, Ann. Rev. Pharmacol. Toxicol., 25:529.

EXPRESSION OF RIBOSOMAL PROTEIN GENES IN *XENOPUS* DEVELOPMENT

P. Pierandrei-Amaldi and B. Cardinali

Istituto di Biologia Cellulare C. N. R.,
Via Marx 43 - Rome 00137 - Italy.

INTRODUCTION

Ribosome production is one of the major projects of developmental systems, such as *Xenopus* oocytes and embryos, which thus are particularly suitable for the investigation of the regulation of ribosome biogenesis. A typical feature of the embryo is its utilization of maternal stored material accumulated during the oogenesis. In particular the *Xenopus* embryo is completely dependent on maternal gene products until the 'midblastula transition' (4000 - 8000 cells); at this stage transcription is activated and the embryo begins to use the products of its own genes (Newport and Kirschner, 1982). The stored material includes, beside proteins and mRNAs, a huge amount of ribosomes (10^{12} per oocyte) accumulated as 80S particles, which are sufficient to support protein synthesis for a considerable part of embryogenesis; when they become limited the embryo begins making new ribosomes. We were interested in studying, during this period of development, the regulation of expression of the protein component of the ribosomes, the numerous ribosomal proteins (r-proteins), whose coordinated production implies a fine regulation. A coregulation with ribosomal RNAs, with which they are functionally related, could also be expected.

EXPERIMENTAL SYSTEM AND APPROACHES

The information we have obtained on this problem came at first from the analysis of the expression of r-protein genes in developing oocytes and embryos. These studies have been carried out at various regulatory levels: transcription of r-protein genes, maturation and accumulation of their transcripts, synthesis and stability of the r-proteins (reviewed in Amaldi et al., 1989). As for the influence of rRNA genes on the expression of r-protein genes we had some answers from the anucleolate embryo, one of the few mutants available in Xenopus (Elsdale et al., 1958). This homozygous mutant, which carries a deletion of the rRNA gene cluster, can survive up to the tadpole stage using maternal ribosomes, which allow it to pass through the developmental period concerned with the production of new ribosomes. The third approach we have been using consists of the introduction of molecules (cloned genes, proteins and antibodies) in oocytes and embryos by microinjection, thus interfering with the normal pattern of expression of r-protein genes.

The data obtained from the various experimental approaches indicated that the expression of r-protein genes in *Xenopus* development involves a regulation at the level of stability of the r-protein mRNA, apparently affected by the accumulation of unutilized r-proteins (Pierandrei-Amaldi et al., 1985). A more detailed analysis of the gene for r-protein L1 has shown that the production of L1mRNA is controlled by the L1 protein itself which interferes with the correct processing of the transcripts (Bozzoni et al., 1984; Caffarelli et al., 1987; Pierandrei-Amaldi et al., 1988). Another typical feature of these genes is that they are also controlled at the level of translation. Supporting evidence for this comes from the observation that during development the production and accumulation of r-protein mRNA (rp-mRNA) is uncoupled from its utilization (Pierandrei-Amaldi et al., 1982; Baum and Wormington, 1985). In fact rp-mRNA starts to be synthesized and accumulated at the blastula stage, when many genes become transcriptionally activated. This mRNA remains for several hours mostly in mRNPs, and is mobilized onto polysomes around the tailbud stage. The mobilization, which leads to the synthesis of new r-proteins and is concomitant with a significant increase of rRNA (Brown and Littna, 1964), appears to respond to a shortage of ribosomes when the maternal store has been used up (Pierandrei-Amaldi et al., 1985). A similar delay in the untilization of rp-mRNA was observed also during oogenesis (Cardinali et al., 1987). It could be expected that this translational control could be autogenously regulated as in *E. coli* where r-proteins, if in excess relative to the rRNA, prevent further translation of their mRNA (Nomura et al., 1984). In order to test this hypothesis, we injected an excess of purified *Xenopus* r-proteins in *Xenopus* oocytes to see if they had any effect on the endogenous synthesis of r-proteins. Similar experiments were carried out by adding excess proteins to an *in vitro* system programmed with mRNA coding for *Xenopus* r-proteins. No inhibition of r-protein synthesis was obtained in these experiments suggesting that the translation of rp-mRNA in *Xenopus* is not feed-back regulated (Pierandrei-Amaldi et al., 1985b). This conclusion was even more strongly supported by experiments in anucleolate embryos, where it has been observed that the absence of rRNA genes does not interfere with the synthesis of rp-mRNA nor with the synthesis of r-proteins; however newly synthesized r-proteins, which do not find rRNA to assemble with, are unstable and are degraded (Pierandrei-Amaldi et al., 1985). A similar behavior was reported in this mutant for 5S RNA (Miller, 1974), thus indicating that the absence of rRNA has no effect on the synthesis of the other ribosomal components but is crucial for their stability.

EXPERIMENTAL ANALYSIS OF TRANSLATIONAL CONTROL

As mentioned above, the distribution of rp-mRNA between the translationally active polysomes and inactive mRNPs changes specifically during development: the percentage of rp-mRNA loaded on polysomes increases progressively from stage 26 to stage 30, when stored ribosomes become limited. A similar pattern is observed in the anucleolate mutant in the early period of r-protein synthesis; however when at later stages the maternal store is finished, and no ribosomes can by synthesized, the rp-mRNA available is completely recruited onto polysomes (Pierandrei-Amaldi et al., 1985). The temporal relationship between shortage of ribosomes and rp-mRNA recruitment suggested to us the idea that the amount of ribosomes available in the cell might interfere with the efficiency of translation of rp-mRNA and consequently with the production of new ribosomes. To test this possibility we experimentally modified the amount of available ribosomes in developing embryos: an increase was obtained by microinjection of purified ribosomes into fertilized eggs, a decrease was induced by treatment with a drug which reduces the amount of 80S and ribosomal subunits. The effect of this

manipulation on the partition of rp-mRNA between polysomes and mRNA was
analyzed in developing embryos: an inverse relationship between the amount
of ribosomes and rp-mRNA loading on polysomes was observed (Pierandrei-
Amaldi et al., submitted). Although these results suggest that the amount
of available ribosomes signals the need for new ones, as also indicated by
experiments carried out in *Drosophila* (Schmidt et al., 1985), we do not know
how this is achieved. Some particular structure in the rp-mRNA, perhaps
together with specific factor(s), could be responsible for the observed
control. Structural elements present at the 5' untranslated region of the
mRNA for r-protein S19, also shared by other rp-mRNAs, were identified as
able to confer the characteristic translational pattern during embryo
development to the mRNA carrying them (Mariottini et al., this volume). On
the other hand it was reported that the dissociation from polysomes and
deadenylation of mRNA for r-protein L1 in maturing oocytes in mediated by
the 3' portion of mRNA (Hyman and Wormington, 1988).

An interesting feature of the translational control emerged from
experiments of gene dosage increase obtained in developing embryos by
microinjection of the cloned gene for r-protein L1 in fertilized eggs. This
manipulation induced in the embryo a tenfold increase of the corresponding
mature mRNA which, although present in a higher amount, maintained a
relative distribution between polysomes and mRNPs similar to controls and
typical of the developmental stage. This indicates that the percent, rather
than the absolute amount, of rp-mRNA to be loaded on polysomes is regulated
(Pierandrei-Amaldi et al., 1988). These results suggest the possibility
that some factor(s), which may act in a positive or negative way on the
rp-mRNA translation, could be involved. Considering that the various
rp-mRNAs share common structural features at their 5' end, one can speculate
on the implication of a particular factor(s) able to recognize similar
structures and interact with this class of mRNAs for a coordinated
translational regulation.

ACKNOWLEDGEMENTS

This research was partially supported by a grant from 'Progetto
Finalizzato Biotecnologie e Biostrumentazione CNR'.

REFERENCES

Amaldi F., Bozzoni I., Beccari E. and Pierandrei-Amaldi P. 1989,
    Expression of ribosomal protein genes and regulation of ribosome bio-
    synthesis in Xenopus development, Trends Biochem. Sci., 14:175.
Baum E. Z., and Wormington W. M. 1985, Cordinate expression of r-protein
    genes during Xenopus development, Dev. Biol., 111:488.
Bozzoni I., Fragapane P., Annesi F., Pierandrei-Amaldi P., Amaldi F., and
    Beccari E. 1984, Expression of two Xenopus laevis r-proteins genes in
    injected frog oocytes. A specific block interferes with the L1 RNA
    maturation, J. Mol. Biol., 180:987.
Brown D. D., and Littna E. 1964, RNA synthesis during the development of
    Xenopus laevis, the South African clawed toad, J. Mol. Biol., 8:669.
Caffarelli E., Fragapane P., Gehering C., and Bozzoni I. 1987, The
    accumulation of mature RNA for Xenopus laevis ribosomal protein L1 is
    controlled at the level of splicing and turnover of the precursor RNA,
    Embo. J., 6:3493.
Cardinali B., Campioni N., and Pierandrei-Amaldi P. 1987, Ribosomal protein,
    histone and Calmodulin mRNA are differently regulated at the
    translational level during oogenesis of Xenopus laevis, Exp. Cell Res.,
    169:432.
Elsdale T. R., Fishberg M., and Smith S. 1958, A mutation that reduces

nucleolar number in Xenopus laevis, Exp. Cell. Res., 14:642.

Hyman L. E., and Wormington W. M. 1988, Translational inactivation of ribosomal protein mRNA during Xenopus oocyte maturation, Genes Dev., 2:598.

Miller L. 1974, Metabolism of 5S RNA in the absence of ribosome production, Cell, 3:275.

Newport J., and Kirschner M. 1982, A major developmental transition in the early Xenopus embryo: II - Control of the onset of transcription, Cell, 30:687.

Nomura M., Gourse R., and Baugham G. 1984, Regulation of the synthesis of ribosomes and of ribosomal components, Annu. Rev. Biochem., 53:75.

Pierandrei-Amaldi P., Campioni N., Beccari E., Bozzoni I., and Amaldi F. 1982, Expression of ribosomal-protein genes in Xenopus development, Cell, 30:163.

Pierandrei-Amaldi P., Beccari E., Bozzoni I., and Amaldi F. 1985, Ribosomal protein production in normal and anucleolate Xenopus embryos: regulation at the post-transcriptional and translational levels, Cell, 42:317.

Pierandrei-Amaldi P., Campioni N., Gallinari P., Beccari E., Bozzoni I., and Amaldi F. 1985b, Ribosomal protein synthesis is not autogenously regulated at the translational level in Xenopus laevis, Dev. Biol., 167:281.

Pierandrei-Amaldi P., Bozzoni I., and Cardinali B. 1988, Expression of the gene for ribosomal protein L1 in Xenopus embryo: alteration of gene dosage by microinjection, Genes Dev., 2:23.

Schmidt T., Chen P. S., and Pellegrini M. 1985, The induction of ribosome biosynthesis in a non mitotic secretory tissue, J. Biol. Chem., 260:7645.

RIBOSOMAL PROTEIN GENES IN *XENOPUS LAEVIS*: ORGANIZATION, STRUCTURE AND
IDENTIFICATION OF THE cis-ELEMENT RESPONSIBLE FOR THEIR TRANSLATIONAL
CONTROL

P. Mariottini, F. Annesi, C. Bagni, Q. M. Chen.
A. Francesconi, C. D. Pesce, J. J. Serra and
F. Amaldi.

Dipartimento di Biologia,
II Università di Roma 'Tor Vergata',
via E. Carnevale,
00173 Roma, Italy.

The biosynthesis of ribosomes requires the coordinate expression of the
genes coding for their structural components, a few rRNA molecules and
numerous ribosomal proteins (r-proteins). In particular about 70 different
r-proteins are coordinately synthesized during *Xenopus* oogenesis and embryo
development (Pierandei-Amaldi et al., 1982). A number of different
experimental approaches (reviewed by Amaldi et al., 1989) have shown that
the control of r-protein synthesis in *Xenopus* involves at least two types
of regulation: 1) a post-transcriptional regulation, operated by feedback of
the r-proteins themselves, that controls the processing and stability of
r-protein transcripts and consequently the amount of the corresponding mRNA
present in the cell; and 2) a translational regulation that controls the
efficiency of utilization of r-protein mRNA (rp-mRNA) in response to
cellular need for new ribosomes.

The coregulation of the genes for different r-proteins implies that
they share some structural features responsible for the common regulatory
mechanisms. Thus we consider interesting the analysis and comparison of the
organization and structure of the genes for different r-proteins. These
genes are not reiterated; in general two copies (a and b) per haploid genome
are present in *Xenopus laevis*. Comparison of the sequences of the cDNA
corresponding to the two copies of r-proteins L1 (Loreni et al., 1985), L14
(Beccari et al., 1986) and S8 (Mariottini et al., 1988) showed that these
originated from a duplication of the whole genome which occurred in this
species about 30 million years ago. Accordingly, we have found only one
copy of the gene for r-protein L1 in another frog species, *Xenopus
tropicalis,* which did not undergo the genome duplication. As in all other
eukaryotes the genes for the various r-proteins are not clustered but
rather dispersed in the genome, and contain several introns. At present the
nucleotide sequence is available for the entire genes of r-protein L1a
(Loreni et al., 1985) and of r-protein L14a (Beccari and Mazzetti, 1987),
for most of the gene of r-protein S8a and for the 5' region of the gene of
r-protein S19a.

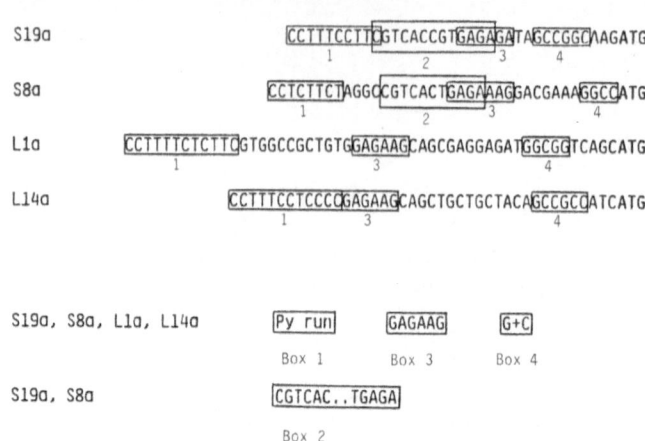

Figure 1.   Sequence comparison of the 5' UTR of different *Xenopus* r-protein genes. Similarities among sequences are evidenced.

STRUCTURAL SIMILARITIES AMONG r-PROTEIN GENES

The comparison of the structure of these genes shows several similarities, which probably reflect the fact that they are members of a class of coregulated genes:

1)   An interesting typical feature is that in all cases transcription start sites are located within a 12-20 pyrimidine stretch preceded by a non-canonical TATA-box. Similar transcription start sites have also been observed in mouse r-protein genes and it has been noted that this type of 5' end is common to several 'housekeeping' genes in vertebrates (Wagner and Perry, 1985).
2)   In all cases the first intron is localized exactly after the initiation ATG codon, as in the genes for L1a, L14a and S19a, or very close to it, as in the gene for r-protein S8a, thus separating the 5'UTR (5' untranslated region) from the coding portion of the gene.
3)   The first exons, which code for the 5'UTR of the mRNAs are always quite short, between 35 and 50bp (Figure 1). Besides starting with a run of 8-12 pyrimidines (due to the above mentioned initiation in the middle of a run of pyrimidine), they display several other sequence similarities (Mariottini et al., 1988).
4)   Evident sequence homology extends also for about 100 nucleotides in the 5' portion of the first introns of the genes for L1a, L14a and S19a. It is possible, although not proved, that this region is involved in the transcription of this class of genes, as it has been shown for two mouse r-protein genes (Moura-Neto et al., 1989).
5)   The 3'UTR of the *Xenopus* r-protein genes analyzed are also typical: they are short, 40-60 nucleotides, and display a number of sequence similarities between each other (Mariottini et al., 1988). It can be supposed that this region is involved in the control of transcript stability, which has been shown to be one of the regulations operating in this class of genes (Pierandrei-Amaldi et al., 1985a; Hyman and Wormington, 1988).

REPEATED SEQUENCES

A computer search has revealed the presence of repeated sequences in the introns and in the 5' flanking regions of these genes (Figure 2).

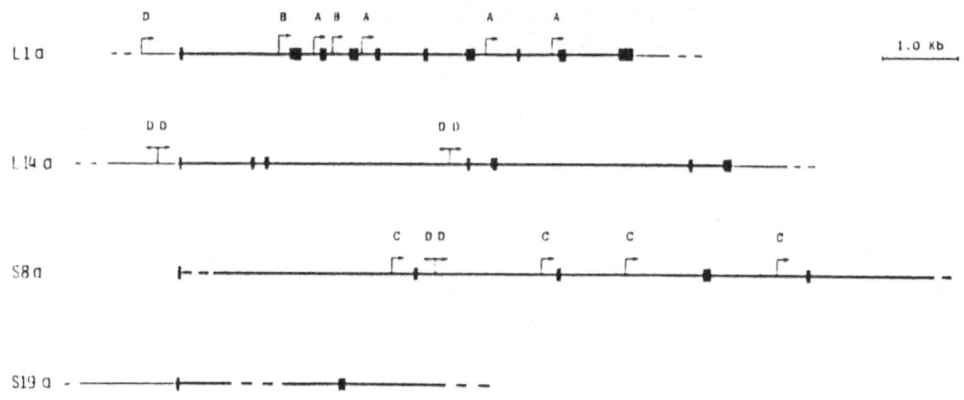

Figure 2.   Comparison of the genes for different *Xenopus* r-proteins.
Black boxes: exons. Thick lines: introns. Thin lines:
flanking regions. Letters and arrows indicate location and
orientation of repeated sequences (see text).

A)   A sixty nucleotide sequence (A in Figure 2) has been found to be
present, with a homology of about 80%, in introns 2, 4, 7 and 8 of the gene
for r-protein L1a (Loreni et al., 1985). Curiously the central part of this
sequence, a 13 nucleotide box, is perfectly complementary to the most
conserved region of the 28S ribosomal RNA (Cutruzzolà et al., 1986). This
repeated sequence has not been found in other r-protein genes, nor anywhere
elso by a computer search of gene data banks.
B)   More recently another sequence, 80 nucleotides long, has been found to
be repeated twice in the same gene for r-protein L1a, in introns 1 and 3 (B
in Figure 2). Also, this has not been found anywhere else.
C)   A different but analogous repeated sequence is present in the gene for
r-protein S8a (C in Figure 2). This sequence, about 90 nucleotides long, is
present with an 85% homology in introns 1, 2, 3, and 4 of this gene, but
could not be found in the other r-protein genes nor in the gene banks.
D)   Somewhat different is the situation with another repeated sequence,
about 150 nucleotides long, which has been found, always arranged in two
inverted copies, in various positions of three r-protein genes (D in Figure
2). It is present: 1) upstream of the gene for r-protein L14a, and in its
3rd intron, 2) in the 2nd intron of the gene for r-protein S8a, and 3)
upstream of the gene for r-protein L1a, although here, due to the limit of
sequence data, only a portion of the sequence has been detected. This
sequence has not been found anywhere else in the gene data banks.

    It is not clear yet if these repeated sequences, some of which appear
to be gene specific and others common in the class of r-protein genes, have
some functional relevance for gene activity or regulation, or not. They
appear to be members of the so called short interspersed repeated sequences.

'CpG ISLANDS'

    A computer analysis has been carried out to determine the CpG
dinucleotide distribution along these genes. It has been observed that in
all cases the frequency of the CpG is strongly repressed with respect to the
statistically expected value, with the exception of the regions, about one
thousand nucleotide long, encompassing the 5' end of the genes. This
observation is in line with the notion that 'CpG islands' on the 5' is a

typical feature of vertebrate housekeeping genes (Bird, 1986). The gene for r-protein S8a has a second 'CpG island' in an internal position of the gene, around the 3rd intron. This might indicate the presence here of the 5' end of an overlapping gene.

ROLE OF THE 5'UTR IN THE TRANSLATIONAL REGULATION OF rp-mRNA

As mentioned above, the genes up to now studies for *Xenopus* r-proteins share a typical first exon (Figure 1). It seems possible that this 5' untranslated portion (5'UTR) of the rp-mRNA (mRNA specific for r-proteins) could be involved in its translational regulation which we had observed during *Xenopus* development (Pierandrei-Amaldi et al., 1982; 1985a), where the production and accumulation of rp-mRNA are uncoupled from its util- ization. In fact mRNA for r-proteins begins to be synthesized and accumulated at blastula stage, when many genes are transcriptionally activated at the end of cleavage, but for over twenty hours it is mostly kept unused as light mRNPs; it is mobilized on to polysomes only at the tail-bud stage, when an active production of new ribosomes is required, the maternal store having been used up. We have also shown that, at variance with prokaryotes, this translational control is not due to an autogenous regulation by r-proteins (Pierandrei-Amaldi et al., 1985b).

To verify the hypothesis that the characteristic 5'UTR of rp-mRNAs is responsible for their translational behavior, we have constructed a precisely designed fused gene. The 5'UTR of the gene for *Xenopus* r-protein S19a has been joined, exactly at the level of the initiation ATG codon, with the coding sequence of CAT gene deprived of its own 5'UTR (Mariottini and Amaldi, submitted for publication). Upon introduction *in vivo* by microinjection in fertilized eggs, the fused gene is transcription- ally activated after the blastula stage but the utilization of S19-CAT mRNA is delayed as it occurs for the endogenous rp-mRNAs. This has been demonstrated by the pattern of appearance of CAT activity during development, and by the distribution of S19-CAT mRNA between polysomes and mRNPs. Thus we could conclude that the 5'UTR of the mRNA for r-protein S19a confers to an unrelated mRNA a translational regulation property, similar to that of rp-mRNA during *Xenopus* development. This 5'UTR can thus be viewed, by analogy with transcriptionally regulated systems, like a 'cis-acting' element for the regulation of mRNA utilization. The structural similarity of this 5'UTR with those of other rp-mRNAs, suggests that our conclusion can be of more general validity for all translationally controlled rp-mRNAs.

ACKNOWLEDGEMENTS

This work has been partially supported by grants from 'Progetto Finalizzato Ingegneria Genetica, C.N.R.', and from 'Ministero Pubblica Istruzione'.

REFERENCES

Amaldi F., Bozzoni I., Beccari E., and Pierandrei-Amaldi P. 1989, Expression of r-protein genes and regulation of ribosome biosynthesis in Xenopus development, Trends Biochen. Sci., 14:175.

Beccari E., and Mazzetti P. 1987, Nucleotide sequence of r-protein L14 gene of Xenopus, Nucl. Acids Res., 15:1870.

Beccari E., Mazzetti P., Mileo A. M., Bozzoni I., Pierandrei-Amaldi P., and Amaldi F. 1986, Sequence coding for r-protein L14 in X. laevis and X. tropicalis; homologies in the 5'UTR are shared with other r-protein mRNAs, Nucl. Acids Res., 14:7633.

Bird A. P. 1986, CpG-rich islands and the function of DNA methylation, Nature, 321:209.

Cutruzzolà F., Loreni F., and Bozzoni I. 1986, Complementarily of sequence elements in 28S rRNA and in r-protein genes of X. laevis and X. tropicalis, Gene, 49:371.

Hyman L. E., and Wormington W. M. 1988, Translational inactivation of ribosomal protein mRNAs during Xenopus oocyte maturation, Genes Dev., 2:598.

Loreni F., Ruberti I., Bozzoni I., Pierandrei-Amaldi P., and Amaldi F. 1985, Nucleotide sequence of the L1 r-protein gene of Xenopus laevis: remarkable sequence homology among introns, EMBO J., 4:3483.

Mariottini P., Bagni C., Annesi F., and Amaldi F. 1988, Isolation and nucleotide sequences of cDNAs for X. Laevis r-protein S8: similarities in the 5' and 3' UTR of mRNAs for various r-proteins, Gene, 67:69.

Moura-Neto R., Dudov K. P., and Perry R. P. 1989, An element downstream of the cap site is required for transcription of the gene encoding mouse r-protein L32, Proc. Nat. Acad. Sci. USA, 86:3997.

Pierandrei-Amaldi P., Campioni N., Beccari E., Bozzoni I., and Amaldi F. 1982, Expression of ribosomal protein genes in Xenopus development, Cell, 30:163.

Pierandrei-Amaldi P., Beccari E., Bozzoni I., and Amaldi F. 1985a, Ribosomal protein production in normal and anucleolate Xenopus embryos: regulation at the post transcriptional and translational levels, Cell, 42:317.

Pierandrei-Amaldi P., Campioni N., Gallinari P., Beccari E., Bozzoni I., and Amaldi F. 1985b, R-protein synthesis is not autogenously regulated at the translational level in Xenopus, Dev. Biol., 107:281.

Wagner M., and Perry R. P. 1985, Characterization of the multigene family encoding the mouse S16 ribosomal protein, Mol. Cell. Biol., 5:3560.

EXPRESSION OF NUCLEOPROTEIN mRNAs DURING RAT SPERMIOGENESIS

Pekka Mali, Antti Kaipia, Marko Kangasniemi, Jorma Toppari,
Minna Sandberg, Eero Vuorio, Pamela C. Yelick, Norman B.
Hecht and Martti Parvinen

Institute of Biomedicine
Departments of Anatomy (PM, AK, MK, JT, MP) and Medical
Biochemistry (MS, EV), University of Turku
SF-20520 Turku, Finland

Department of Biology,
Tufts University, Medford,
Massachusetts 02155, USA (PCY, NBH)

INTRODUCTION

DNA-associated proteins of spermatozoa are different from somatic cell histones. In mammals, arginine-rich protamines replace histones and transition proteins in spermatids during compaction of the chromatin (Poccia, 1986). The mouse protamine mRNAs are expressed in early spermatids by a haploid genome (Kleene et al., 1983). To find the exact stages of differentiation where the expression of spermatidal nucleoprotein mRNAs occur, we used three different cDNA-hybridization techniques to measure protamine 1, protamine 2 and transition protein (TP1) mRNA levels.

MATERIAL AND METHODS

Segments of seminiferous tubules from defined stages of the cycle of the seminiferous epithelium were collected by transillumination-assisted microdissection (Parvinen and Ruokonen, 1982) for Northern blot analysis (Mali et al., 1988), where RNA was isolated and fractionated as described by Chirgwin et al. (1979) and Thomas (1980). For slot blot microanalysis of RNA (Krawczyk et al., 1988), an improved technique was used for separation of 1mm segments of seminiferous tubules at accurately defined stages of the cycle. This technique involves rapid phase contrast microscopic screening of intermediate squash preparations (Toppari et al., 1985) with new criteria for distinction of stages II and VII (Kangasniemi et al., 1989). The *in situ* hybridization analysis was performed as described by Sandberg and Vuorio (1987) using ^{35}S-labelled pMP1, pMP2 and pMTP1 cDNA probes. In some cases, the length of poly(A) chain was analyzed using RNase H digestion of total RNA annealed with oligo (dT) prior to Northern analysis (Mali et al., 1988).

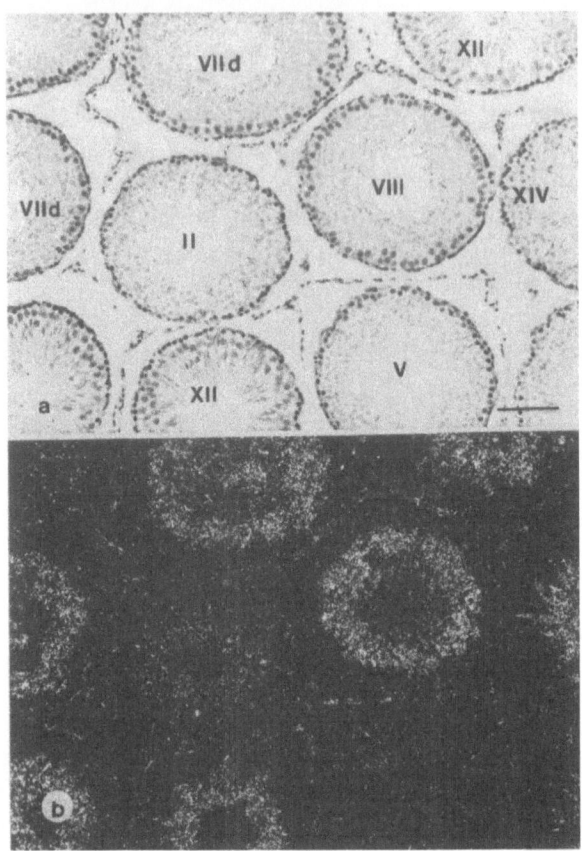

Figure 1.   Normal (a) and dark field (b) photomicrographs of a
            hematoxylin-stained paraffin section of rat testis to show the
            in situ hybridization of the protamine 1 cDNA probe. Stages
            VIId, VIII, XII and XIV have strong hybridization signals over
            the steps 7-14 spermatids, whereas stage V has a background
            level of grain density. An intermediate radioactivity is seen
            above a stage II tuble. Bar: 100μm. Reproduced from the
            Journal of Cell Biology, 1988, vol. 107, p. 408 by copyright
            permission of the Rockefeller University Press.

RESULTS

     The *in situ* hybridizations showed marked differences between individual
seminiferous tubles in the level of mRNA detected by the radioactive
protamine 1 probe. The radioactivity was detected over round, elongating
and elongated spermatids from late stage VII to stage III of the cycle
(Figure 1). Protamine 1 mRNA showed two molecular sizes in Northern blot
analysis. In stages II-III, IV-V and VI there was a low level of the
smaller size protamine 1 mRNA, wheras the level increased during the second
half of stage VII. Only the larger mRNA species was observed from stage VII
to stage XII. The densitometric analysis showed an approximately 10-fold
difference between the minimal (stage VI) and maximal (stages VIII-XIV)
expression of protamine 1 mRNA. The Northern blot analysis after removal of
poly(A) chain revealed that both the large (580 base) and the small (450

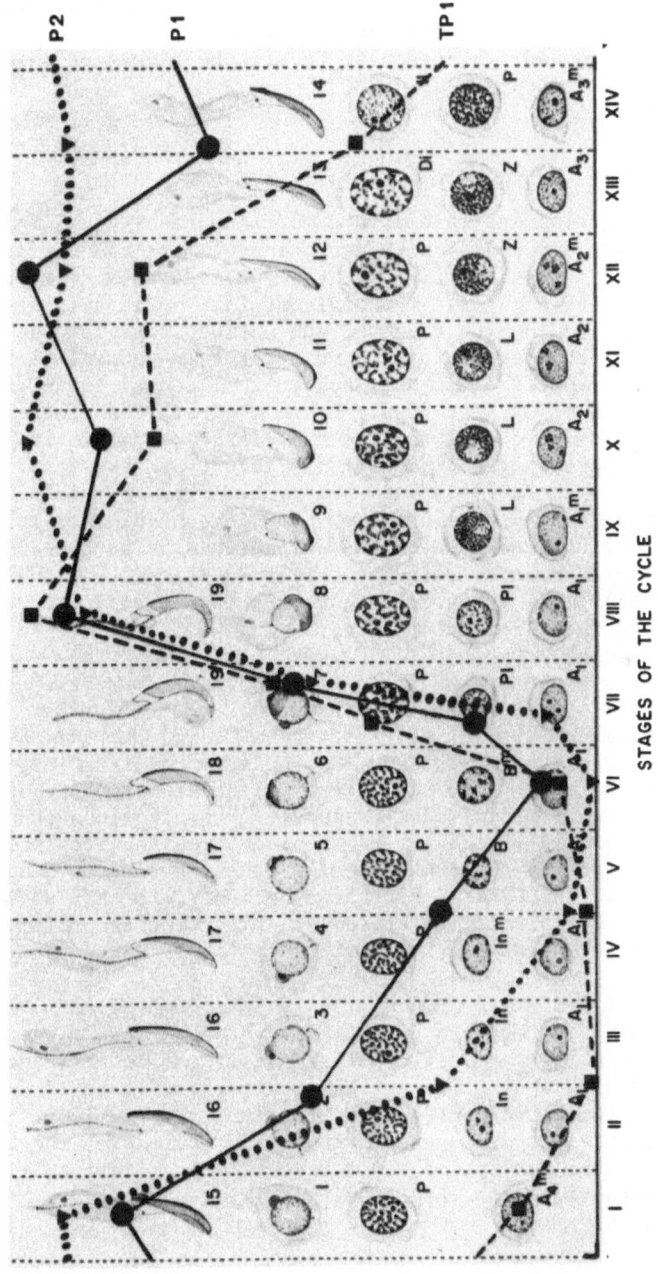

Figure 2.   Densitometric analyses (arbitrary units) of Northern blots of protamine 2 (P2), protamine 1, (P1), and transition protein 1 (TP1) mRNAs, superimposed on the map of spermatogenesis of Dym and Clermont (1970, reproduced by permission of Alan R. Liss, Inc.).

base) forms of protamine 1 mRNA were reduced to an identical length of *ca*. 420 bases (Mali et al., 1988).

The expression of transition protein and protamine 2 genes showed minor temporal differences in expression, although the general picture was similar to that of protamine 1 gene (Figure 2). The slot blot analysis revealed that the level of TP1 mRNA increased at substage VIIb; slightly earlier than mRNAs of the protamines. It also disappeared earlier, during stage XIV of the cycle. All mRNAs showed two molecular forms; the reduction in size was associated with translation.

DISCUSSION

The nuclear elongation and chromatin compaction are typical features of spermiogenesis, and are associated with cessation of most transcription (Monesi, 1964). The RNA must be stored in the spermatids since many proteins are synthesized beyond this point. The nucleoprotein mRNAs are examples of this; they are synthesized during the last 48h period of haploid gene activity in round nucleated spermatids in an accurately sequential fashion. The somatic and testis-specific histones become replaced by transition proteins during steps 9-12 of spermiogenesis in the rat; the time of the nuclear elongation. The beaded ultrastructure of the chromatin in spread preparations is simultaneously replaced by a smooth type that facilitates the packaging of the chromatin (Kierszenbaum and Tres, 1975). Transition proteins become replaced by protamines from step 15 onwards (Meistrich et al., 1978). The final condensation of the chromatin is also reflected by rapidly reducing DNA-fluorochrome binding ability detected by flow cytometry (Toppari et al., 1985). All these events are accurately reflected by levels of specific mRNAs and their molecular sizes.

The time from the beginning of stage VIII to the beginning of stage II when the highest levels of protamine 1 mRNA are present in spermatids is 6.7 days. The lifetime of transition protein 1 mRNA is somewhat shorter, *ca*. 5.5 days. This suggests that these mRNAs undergo post-transcriptional processing before translation that involves a molecular weight reduction. The polysomal form of protamine 1 mRNA is about 130 nucleotides shorter than the stored one (Kleene et al., 1984). We conclude from the stage-specific molecular size analysis that the protamine 1 protein is first synthesized in step 13-14 spermatids. The RNase H digestion demonstrates that a small poly(A) fragment, characteristic for stored mRNA, remains in protamine 1 mRNA during and after translation.

The regulation of protamine mRNA synthesis in spermatids remains to be investigated. Its synthesis occurs during those steps of spermiogenesis that have the poorest ability to differentiate *in vitro* (Toppari and Parvinen, 1985), and also spermatids are the most sensitive cells for the effects of hypophysectomy (Clermont and Morgentaler, 1955). It is therefore possible that the steps of spermiogenesis involving nucleoprotein transitions, are critical for optimal culture conditions, hormonal stimulation and interaction with the Sertoli cells.

ACKNOWLEDGEMENT

This project has been supported by The Academy of Finland (Project no. 200 at the Medical Research Council).

REFERENCES

Chirgwin J. M., Przybyla A. R., MacDonald R. J., and Rutter W. J. 1979.
    Isolation of biologically active ribonucleic acid from sources enriched
    in ribonuclease. Biochemistry, 18:5294.

Clermont Y., and Morgentaler H. 1955. Quantitative study of spermatogenesis
    in the hypophysectomized rat. Endocrinology 57:369.

Dym M., and Clermont Y. 1970. Role of spermatogonia in the repair of the
    seminiferous epithelium following X-irradiation of the rat testis.
    Am. J. Anat. 128:265.

Kangasniemi M., Kaipia A., Mali P., Toppari J., Huhtaniemi I., Parvinen
    I. and M. 1989. Modulation of basal and FSH-stimulated cyclic AMP
    production in rat seminiferous tubules staged by an improved
    transillumination technique. Anat. Rec. (in press).

Kierszenbaum A. L., and Tres L. L. 1985. Structural and transcriptional
    features of the mouse spermatid genome. J. Cell Biol. 65:258.

Kleene K. C., Distel R. J. and Hecht N. B. 1983. cDNA clones encoding
    cytoplasmic poly(A)$^+$ RNAs which first appear at detectable levels in
    haploid phases of spermatogenesis in the mouse. Dev. Biol. 98:455.

Kleene K. C., Distel R. J. and Hecht N. B. 1984. Translational regulation
    and deadenylation of a protamine mRNA during spermiogenesis in the
    mouse. Dev. Biol. 105:71-

Meistrich M. L., Brock W. A., Grimes S. R., Platz R. D., and Hnilica L. S.
    1978. Nuclear protein transitions during spermatogenesis. Federation
    Proc. 37:2522.

Monesi V. 1964. Ribonucleic acid synthesis during mitosis and meiosis in the
    mouse testis. J. Cell Biol. 22:521.

Parvinen M., and Ruokonen A. 1982. Endogenous steroids in rat seminiferous
    tubules. Comparison of the stages of the epithelial cycle isolated
    by transillumination-assisted microdissection. J. Androl. 3:211.

Poccia D. 1986. Remodeling of nucleoproteins during gametogenesis,
    fertilization and early development. Int. Rev. Cytol. 105:1.

Sandberg M., and Vuorio E. 1987. Localization of types I, II and III collagen
    mRNAs in developing human skeletal tissues by in situ hybridization.
    J. Cell. Biol. 104:1077.

Thomas P. S. 1980. Hybridization of denatured RNA and small DNA fragments
    transferred to nitrocellulose. Proc. Natl. Acad. Sci. USA, 77:5201.

Toppari J., Eerola E., and Parvinen M. 1985. Flow cytometric DNA analysis of
    defined stages of rat seminiferous epithelial cycle during in vitro
    differentiation. J. Androl. 6:325.

Toppari J., and Parvinen M. 1985. In vitro differentiation of rat
    seminiferous tubular segments from defined stages of the epithelial
    cycle: morphologic and immunolocalization analysis. J. Androl. 6:334.

SPLICING CONTROL AND NUCLEUS/CYTOPLASM COMPARTMENTALIZATION OF RIBOSOMAL

PROTEIN L1 RNA IN *X. LAEVIS* OOCYTES

Paola Fragapane, Elisa Caffarelli. Paola Mazzetti,
Matteo Lener, Paola Pierandrei-Amaldi and Irene Bozzoni

Centro di Studio per gli Acidi Nucleici del C.N.R. and
Dipartimento di Genetica e Biologia Molecolare
Universita' 'La Sapienza', Roma

INTRODUCTION

In *X. laevis,* the gene encoding for the r-protein L1 is post-transcriptionally regulated. In the presence of excess amounts of free L1 r-protein a specific pre-mRNA which contains the third intron of the gene is accumulated. In the absence of splicing the pre-mRNA is degraded leading to the reduction of the mRNA levels available for translation. A model has been proposed in which the L1 r-protein can feedback regulate its synthesis by controlling the efficiency of splicing of its own precursor RNA (Bozzoni et al., 1984). In order to simplify the analysis of the splicing process we have constructed plasmids in which specific exon-intron-exon portions of the L1 gene have been cloned in front of the CAT gene. When injected in *X. laevis* nuclei they direct the synthesis of the CAT enzyme only if the intron is spliced out. Using this system we have studied the splicing of the third intron and the role of intron removal in the process of mRNA transport to the cytoplasm.

MATERIALS AND METHODS

Plasmid Constructions

A 550bp PstI-BamHI fragment, containing the promoter of the r-protein gene L14 (Beccari et al., 1986), has been cloned in the PstI-BamHI sites of plasmid pSP65CAT to give plasmid L14CAT. 002CAT has been obtained by inserting the BamHI site of L14CAT a 330bp fragment containing 100bp of exon 2, the whole intron 3 and 20bp of exon 4 (Loreni et al., 1985). When the intron is spliced out a leader of 120 bases, with no ATG, is created preceding the ATG of the CAT coding region. 003CAT has been obtained by cloning in the BamHI site of L14CAT a 390bp HinfI fragment covering 90bp of exon 3, the whole intron 3 and 60bp of exon 4. A leader sequence of 150 bases is created before the ATG of the CAT coding region.

CAT Assay and RNA Analysis

2ng of the different plasmid constructs were injected in the nucleus of *X. laevis* oocytes, and after 6 hours incubation at 25°C pools of 10 oocytes

Nuclear Structure and Function, Edited by J. R. Harris and
I. B. Zbarsky, Plenum Press, New York, 1990

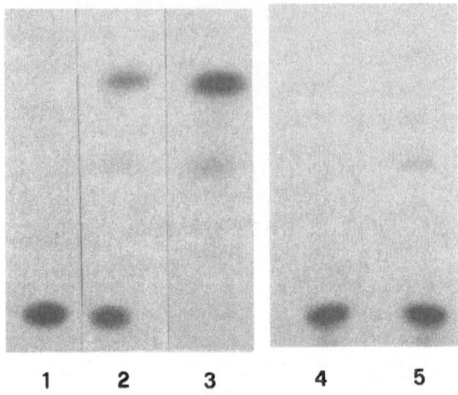

Figure 1.   CAT assay from oocytes injected with: pSP65CAT (1), L14CAT
(2), 002CAT (3), 003CAT (4), antibodies against L1 and 003CAT
(5).

were analysed for CAT activity (Carnevali et al., 1989). ^{32}P-labelled SP6-
transcripts were injected in the nucleus of X. *laevis* oocytes. After 1
hour incubation at 25°C total RNA was extracted from manually dissected
nuclei and cytoplasms. Labelled RNA was run on 8% acrylamide-urea gels and
visualized by autoradiography.

RESULTS AND DISCUSSION

    In order to analyse the process of intron removal from the primary
transcript of the L1 r-protein gene of X. *laevis* we have constructed a
series of plasmids containing in front of the CAT gene different exon-
intron-exon portions of the L1 gene. Transcription is driven from the L14
r-protein gene promoter which is highly active in *Xenopus* oocytes (Carnevali
et al., 1989). The correct removal of the intron allows the formation of an
mRNA coding for the CAT enzyme. Figure 1 shows that while undetectable CAT
activity is obtained with plasmid pSP65CAT (lane 1), high levels of
activity are detected with L14 CAT which has the L14 promoter in front of
the coding sequence (lane 2). Interestingly, plasmid 002CAT, which has the
second intron, produces levels of CAT activity higher than plasmid L14CAT
(lane 3). These results show that the presence of a spliceable intron in a
precursor RNA increases the efficiency of transcript export to the cytoplasm
and its translation. The construct containing the third intron (003CAT),
which is normally not spliced, shows undetectable levels of CAT activity.
Our previous results (Bozzoni et al., 1984; Caffarelli et al., 1987)
indicated that the block of splicing of the third intron could be due to a
direct or indirect effect of the L1 protein. To test this hypothesis we
have coinjected, together with plasmid 003CAT, antibodies against the
r-protein L1. As shown in Figure 1 (lane 5) the presence of antibodies
increases the levels of CAT activity. It can be deduced that partial
depletion of the L1 protein slightly releases the splicing block leading to
the production of increased amounts of mature translatable RNA. These
results support the idea that the L1 protein is indeed involved in the
splicing control of its pre-mRNA and show that the method is suitable for
rapid analysis of intron splicing. The availability of such an easy assay
will make possible the screening of a large number of mutants and the
identification of the intron regions which are responsible for such
regulation.

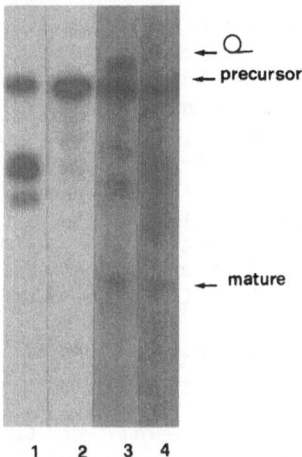

Figure 2.   Gel electrophoresis of RNA extracted from nuclei (N) and
cytoplasms (C) of oocytes injected with SP6-transcripts
containing the third intron (1 and 2) and the fourth intron
(3 and 4) of the L1 gene.

The oocyte system is very suitable for studies concerning the nucleus/
cytoplasm compartmentalization in that the manual dissection of the cell is
very easy. We have taken advantage of this feature and of the RNA microinjection technique to analyse the compartmentalization of different RNA
components. Figure 2 (lanes 1-2) shows that when ^{32}P-labelled SP6-trans-
cripts containing the third intron (003) are injected in the nucleus of X.
laevis oocytes part of the pre-mRNA escapes to the cytoplasm, while all the
products of RNA breakage are retained in the nucleus. The presence of pre-
mRNA in the cytoplasm could be due to the inability to form spliceosome
complexes as recently reported by Legrain and Rosbash (1989). It is also
interesting to observe that the products of RNA turnover are confined to the
nucleus showing that the transport to the cytoplasm is an active process
involving selection of specific molecules. Figure 2 (lanes 3-4) shows the
nucleus/cytoplasm compartmentalization of the RNA components derived from
the injection of a pre-mRNA containing the fourth intron of the L1 gene,
which undergoes normal splicing (004). It appears that the typical product
of RNA splicing, the lariat, is confined to the nucleus whereas precursor RNA
and mature RNA are partitioned in the two compartments. In this case the
presence of pre-mRNA in the cytoplasm could be due to saturation of the
splicing components and therefore once again to the inability of all the
pre-mRNA molecules to form spliceosome complexes. Analysis of nuclear
complexes has shown that while the 003 pre-mRNA does not even form the
prespliceosome, the 004 pre-mRNA is in 30S and 60S complexes (data not
shown).

ACKNOWLEDGEMENTS

This work has been supported by grants form Progetto Finalizzato
'Biotecnologie e Biostrumentazioni' of the C.N.R., by Istituto Pasteur
Fondazione Cenci-Bolognetti and by M.P.I.

REFERENCES

Beccari E., Mazzetti P., Mileo A. M., Bozzoni I., Pierandrei-Amaldi P., and Amaldi F. 1986, Sequences coding for the ribosomal protein L14 in *X. laevis* and *X. tropicalis*; homologies in the 5' untranslated region are shared with other r-protein mRNAs, Nucl. Acids Res., 14:7633.

Bozzoni I., Fragapane P., Annesi F., Pierandrei-Amaldi P., Amaldi F., and Beccari E. 1984, Expression of two *X. laevis* r-protein genes in injected frog oocytes. A specific splicing block interferes with the L1 maturation, J. Mol. Biol., 180:987.

Caffarelli E., Fragapane P., Gehring C., and Bozzoni I. 1987, The accumulation of mature mRNA for the *X. laevis* r-protein L1 is controlled at the level of splicing and turnover of the precursor RNA, EMBO J., 6:3493.

Carnevali F., La Porta C., Ilardi F., and Beccari E. 1989, Nuclear factors specifically bind to upstream sequences of *X. laevis* r-protein gene promoter, Nucl. Acids Res., in press.

Legrain P., and Rosbash M. 1989, Some cis- and transacting mutants for splicing target pre-mRNA to the cytoplasm, Cell, 57:573.

Loreni F., Ruberti I., Bozzoni I., Pierandrei-Amaldi P., and Amaldi F. 1985, Nucleotide sequence of the L1 r-protein gene of *X. laevis*: remarkable sequence homology among introns, EMBO J., 4:3483.

GENETIC ANALYSIS OF SPERMATOGENESIS IN *DROSOPHILA HYDEI:* MALE STERILE
MUTATIONS AFFECTING NUCLEAR DEVELOPMENT DURING THE PRIMARY SPERMATOCYTE
STAGE

Johannes H. P. Hackstein, Heinz Beck[1], Wolfgang Hennig,
Ron Hochestenbach, Hannie Kremer and Helmut Zacharias[2]

Department of Molecular and Developmental Genetics
Catholic University of Nijmegen, Faculty of Science
Toernooiveld, NL-6525 ED Nijmegen, The Netherlands

1) Present address: Cosmital SA, route de Chesalles 21
CH-1723 Marly/Switzerland

2) Institute of Zoology, University of Kiel
D-2300 Kiel/Germany

INTRODUCTION

  Spermatogenesis in *Drosophila* has been of considerable interest since
Bridges demonstrated that the Y chromosome is essential for the formation of
functional spermatozoa, but completely dispensable for somatic development
(Bridges, 1916). It could be shown that only six male fertility genes are
located on the Y chromosome of *D. melanogaster* and less than 16 on the Y
chromosome of *D. hydei*. These genes have sizes up to 1,500 kb and apparently
they do not code for major protein constituents of the spermatozoa (review:
Hackstein, 1987). The deletion or inactivation of a single fertility gene
renders the males carrying such a mutation completely infertile, although
only slight cytological effects on the germ cells can be observed. Even the
deletion of the whole Y chromosome does not prevent a nearly complete
terminal differentiation of the germ cells. In contrast, there exists a
very large number of X chromosomal and autosomal genes which can mutate to
male sterile alleles (see discussion by Hackstein, 1987). Some of these male
sterile mutations cause distinct morphological phenotypes in male germ cells
(Kiefer, 1983). Male sterile mutations on the X chromosome of *D.
melanogaster* have been induced and analyzed cytologically (review: Lifschytz,
1987), but a systematic screen for male sterile mutations on the whole genome
has not been performed. We therefore started a systematic screen for male
sterile mutations on the X chromosome and the autosomes of *D. hydei*. This
species has an especially favourable cytology of the primary spermatocyte
nucleus, which at this stage of development acquires a peculiar shape and
displays five pairs of giant Y chromosomal lampbrush loops which are the
cytological equivalents of male fertility genes (Hackstein et al., 1982;
review: Hennig, 1985).

  In this contribution we will show that the primary spermatocyte nucleus
of *D. hydei* undergoes a characteristic sequence of morphogenetic movements
which can be affected by a number of X chromosomal and autosomal male sterile
mutations. The unfolding of the Y chromosomal lampbrush loops is controlled

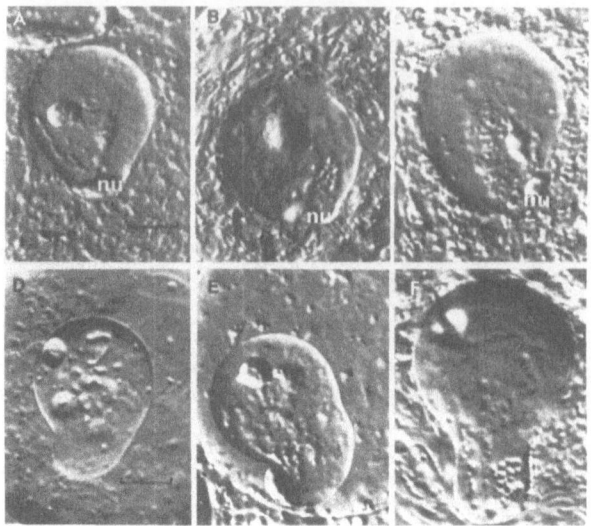

Figure 1.   Mutations affecting the shape of the primary spermatocyte
            nucleus. Nuclei are stage SPC II or III respectively.
            A: *wild-type*, B: *ms(1)808*, C: *ms(3)C16*, D: *ms(2)6*, E: *ms(2)73*,
            F: *ms(3)11*. Differential interference contrast (DIC). Bar
            represents 10µm. nu: nucleous.

by a number of non-allelic male sterile mutations located on chromosome 3.
We will discuss the hypothesis that the Y chromosomal lampbrush loops shape
the primary spermatocyte nucleus (c.f. Lifschytz, 1987) and provide genetical
and cytological evidence for the mechanisms which shape the nucleus and
control the characteristic topology of the Y chromosomal lampbrush loops.

MATERIALS AND METHODS

     Mutagenesis by ethanemethanesulfonate (EMS) and X-rays has been des-
cribed by Hackstein et al. (1982); the screening for male sterile mutations
has been published by Hackstein et al. (1987). Preparation of primary
spermatocytes, cytology, staging and microscopy are described in another
publication (Hackstein et al., 1989).

RESULTS AND DISCUSSION

     We screened more that 16,000 chromosomes of *D. Hydei* for male sterile
mutations and recovered 365 different mutants. Squash preparations of whole
testes were made and the germ cells were inspected with phase-contrast – and
DIC – optics. Seven male sterile mutations (*ms(1)808, ms(2)6, ms(2)73,
ms(3)11, ms(3)C16, ms(3)C20, ms(4)31*, see Figure 1) alter the shape of the
primary spermatocyte nucleus characteristically. Males hemi- or homozygous
for one of the mutations *ms(4)31, ms(1)808* and *ms(3)C20* (arranged in the
order of decreasing strength of phenotype) possess primary spermatocyte
nuclei which remain more or less spherical during the whole primary
spermatocyte stage, whereas *wild-type* nuclei show characteristic changes of
their shape (Figure 2, stages SPC I – IV). Spermatocyte nuclei of *ms(3)11*
males exhibit a normal shape at stage SPC I, show an extremely elongated tip
at stages SPC II and III and become spherical again at stage SPC IV. The tip

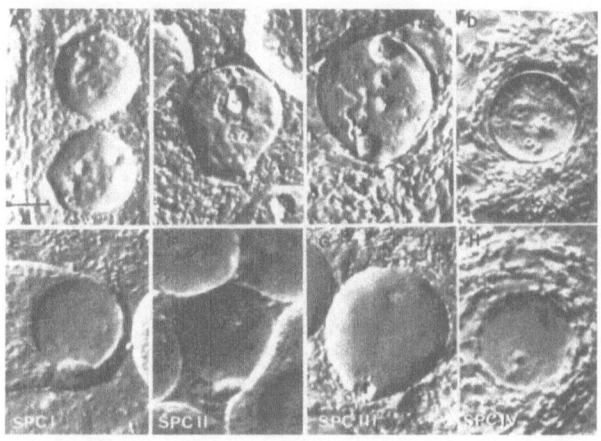

Figure 2.   Development of primary spermatocyte nuclei; upper row: *wild-type X/Y,* lower row: *X/O.* SPC I - IV: Primary spermatocyte stages I to IV. DIC; Bar 10 μm.

of the pearshaped spermatocyte nucleus of *ms(2)6* and *ms(2)73* males is bent characteristically during the stages SPC II and III whereas stages SPC I and IV exhibit a normal shape. The spermatocyte nucleus of *ms(3)C16* males is slightly deformed - with an abnormally positioned nucleolus - during stages SPC II, III, and IV; it only becomes spherical immediately before the onset of the meiotic divisions. Thus, mutations affecting the shape of the primary spermatocyte nucleus only act at distinct periods of the meiotic prophase and they do not block the subsequent meiotic divisions. These mutations neither effect the preceeding gonial stage nor do they interfere with the postmeiotic shaping of the spermatid nucleus. The underlying mechanisms and the function of these morphogenetic movements remain unclear, although the following observations might shed some light on this question. The mutants *ms(2)6* and *ms(2)73* most likely are allelic, and *ms(2)73* appears to be the stronger allele. In both mutants the formation of the nebenkern from the mitochondria fails, but in *ms(2)73* males the tip of the primary spermatocyte nucleus is bent more strongly, and in addition to the above-mentioned phenes, a size-polymorphism of the spermatid nuclei is observed (Hackstein et al., 1989). This is indicative for a distortion of chromosome segregation during the meiotic divisions. In males of the constitution *ms(2)6 / ms(2)73* negative complementation occurs and the germ cells are unable to enter the meiotic divisions. Therefore, it is not unlikely that modifications of the testes-specific tubulin machinery might be the reason for the pleiotropic phenotype of the different combinations. Both, the aberrant shape of the nucleus and the failure of nebenkern formation could be the consequences of structural changes in the distribution or the function of the mucrotubuli. These microtubules might surround the spermatocyte nucleus or, alternatively, structure the nuclear matrix.

The shaping of the primary spermatocyte nucleus does not depend on the presence of the Y chromosome since the nucleus executes its morphogenetic program in the absence of the Y chromosome also (Figure 2). In addition, male sterile mutations which interfere with the unfolding of the Y chromosomal lampbrush loops *(ms(3)5, ms(3)E4, ms(3)E1, ms(3)C16)* do not necessarily affect the shaping of the nucleus, although the shape of *ms(3)C16* nuclei is modified and the nucleolus is abnormally positioned in such primary spermatocyte nuclei (Figure 3). Thus, the shaping of the nucleus and the unfolding of the lampbrush loops are largely independent processes. The

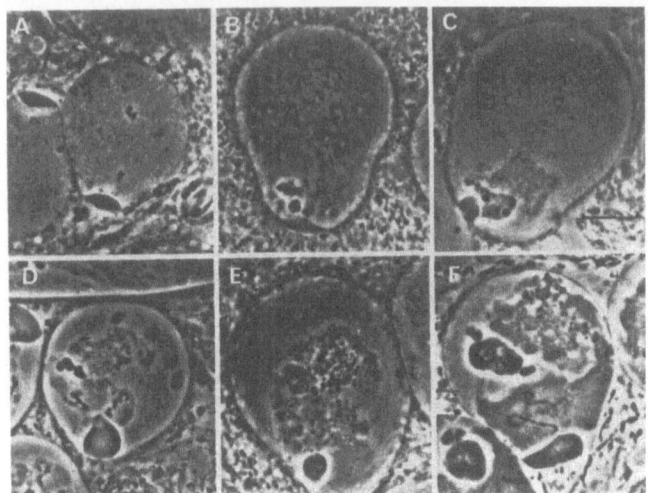

Figure 3.   Unfolding of the Y chromosomal lampbrush loops. A: *X/O, wild-type*; B: *ms(3)5*, C: *ms(3)E4*, D: *ms(3)E1*, E: *ms(3)C16*, F: *X/Y wild-type*. Phase contrast, Bar 10 μm.

temperature-sensitive period for the unfolding of the lampbrush loops in the temperature-sensitive mutant *ms(3)5* coincides with the onset of the gonial development (Hackstein et al., 1987) indicating a regulation of the activity of the Y chromosome at the very beginning of the spermatogenic pathway. The other mutations cause stops in the unfolding of the lampbrush loops at different steps, independent from the developmental stage of the nuclear envelope.

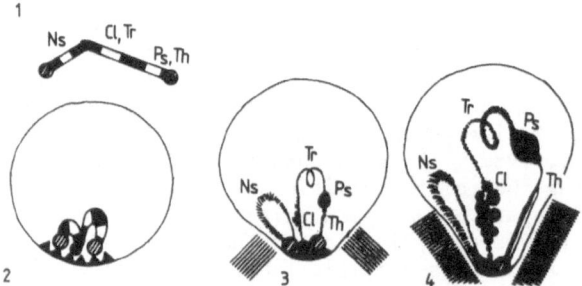

Figure 4.   Schematic representation of the morphogenetic events during the primary spermatocyte stage. 1: Metaphase Y chromosome with terminal nucleolus organizer regions (shaded), centromer (hatched) and locations of the loop-forming fertility genes (open bars). 2: During stage SPC I (or earlier) the Y chromosome attaches with its nucleolus organizer regions and its centromer to a specific site of the nuclear membrane (triangles). 3: During stages SPC I and II the Y chromosome decondenses and the transcription of the lampbrush loops starts. 4: Nearly completed unfolding of Y chromosomal lampbrush loops; Ns: nooses, Cl: clubs, Tr: tubular ribbons, Ps: pseudonucleolus, Th: threads. The shaded bands to the left and the right of the nucleus indicate the (hypothetical) microbubules which shape the nuclear envelope.

The characteristic topological arrangement of the Y chromosomal lampbrush loops is influenced neither by mutations affecting the shaping of the nucleus nor by mutations interfering with the unfolding of the Y chromosomal lampbrush loops. Rather their position in the nucleus is determined by their location on the Y chromosome and the site of attachment of the two terminal nucleolus organizer regions and the centromere of the Y chromosome to the nuclear envelope (Figure 4). The unfolding of the lampbrush loops is the consequence of the gradual decondensation of the Y chromosome and the onset of transcription of the loop-forming fertility genes. Because the Y chromatid becomes completely decondensed, the different lampbrush loops appear as beads on a string, and Y chromosomal deletions alter their topology as expected. The attachment-site between the Y chromosome and the nuclear membrane is abnormal in the mutant *ms(3)C16* and may be the reason for the premature termination of the unfolding of the lampbrush loops. Thus, genetic analysis has proved to be a powerful tool for the analysis of the complex processes which determine the development of the primary spermatocyte nucleus.

REFERENCES

Bridges C. D., 1916, Non-disjunction as proof of the chromosome theory of heredity. Genetics, 1:1 and 107.
Hackstein J. H. P., 1987, Spermatogenesis in *Drosophila*, in: Spermatogenesis, genetic aspects, W. Hennig, ed., Springer Verlag, Berlin p 63.
Hackstein J. H. P., Leoncini O., Beck H., Peelen G., and Hennig W., 1982, Genetic fine structure of the Y chromosome of *Drosophila hydei*. Genetics, 101:257.
Hackstein J. H. P., Hennig W., and Steinmann-Zwicky M., 1987, Autosomal control of lampbrush loop formation during spermatogenesis in *Drosophila hydei* by a gene also affecting somatic sex determination Wilhelm Roux's Arch Dev. Biol., 196:119.
Hackstein J. H. P., Beck H., Hennig W., Hochstenbach R., Kremer H., and Zacharias H., 1989, Spermatogenesis in *Drosophila hydei*: a genetic survey, submitted.
Hennig W., 1985, Y chromosome function and spermatogenesis in *Drosophila hydei*. Adv. Genet. 23:179.
Kiefer B. I., 1973, Genetics of sperm development in *Drosophila* in: Genetic mechanisms in development, E. H. Ruddle, ed., Academic Press, London, p 47.
Lifschytz E., 1987, The developmental program of spermiogenesis in *Drosophila*: A genetic analysis. Int. Rev. Cytol. 109:211.

THE STUDY OF THE INFLUENCE OF cis-REGULATORY SEQUENCES AND trans-REGULATORY

FACTORS OF EUCARYOTES ON CAT GENE EXPRESSION IN *X. LAEVIS* OOCYTES

V. P. Korzh, A. V. Kazansky, I. L. Sleptsova
and S. I. Gorodetsky

Koltsov Institute of Developmental Biology
Vavilov Institute of General Genetics
USSR Academy of Sciences
Cardiological Center of USSR Academy of Medical Sciences
Moscow, USSR

INTRODUCTION

The specificity of expression of eucaryotic genes depends on various
cis- and trans-regulatory sequences and factors. The specificity of
expression of bovine casein genes has long been a question of intensive
interest. A convenient approach to the study of functional sites of a gene
is the use of transformation of animal embryos or cells *in vitro*. But the
question is to move forward more quickly than through transgenic animals.
The study of gene regulation of mammary glands also needs a simple model.
One possible model - *Xenopus laevis* oocytes, has been used recently to study
factors involved in immunoglobulin gene expression (Sweeney and Old, 1988).

The same approach has been used for the study of regulation of bovine
casein genes with special reference to trans- and cis-factor involvement.
The set of bovine Casein genes was isolated earlier by Tkach et al. (1988).
Regulation of casein genes is based on the strict control of the amounts of
specific factors and is time-, tissue- and sex-dependent. This is a question
of great importance for developmental biology and biotechnology.

Xenopus laevis oocytes have been used for evaluating the process of
regulation of casein genes. The effectiveness of a set of regulatory sites
for bovine casein genes has been assayed after microinjection of recombinant
vectors into oocytes. All vectors were constructed using the chloramphenicol
acetyltransferase (CAT) gene; regulatory sites of casein genes and regulatory
sites of some other genes were introduced. The same approach was used for
the evaluation of the effect of regulatory factors of bovine mammary gland
(such as proteins and mRNA) on expression of recombinant vectors.

METHODS

β- and κ-casein genes have been isolated (Tkach et al., 1988).
Flanking elements of some genes were cut from recombinant phages: 5'-flanking
regions of β-casein gene (1.8kb), κ-casein gene (4.5kb), 5'-flanking regions
of TAT gene of rat (1.1kb) (Shinomiya et al., 1984) (gift of Dr G. Schutz)
and TK gene of herpes simplex virus; 3'-flanking regions of β-casein gene

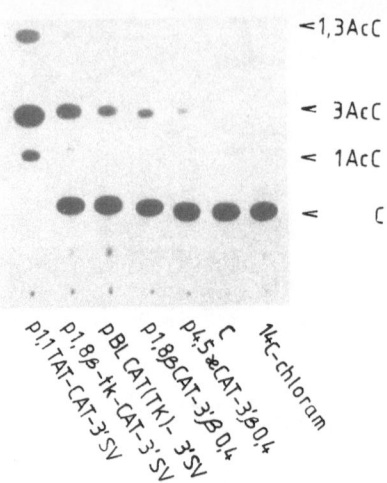

Figure 1.   Comparative analysis of promoter influence on CAT activity in
              extracts of oocytes.

(including non-translatable region and polyadenylation site) and SV-40.   All
the above mentioned fragments were used for construction of a number of
recombinant vectors based on the pBL CAT plasmid.

     Nuclear extracts of bovine mammary gland cells and hepatocytes have
been obtained as previously described (Lobanenkov et al., 1988) on DEAE- and
heparin-sepharose.   The DNA-binding fraction of proteins was identified by
gel retardation and used for microinjection.

     Microinjections have been made as follows.   Oocytes were put into
modified barth solution (pH 7.4) with antibiotics.   20ml of DNA solution
(1mg/ml) with/without nuclear extracts were microinjected into the oocyte
nucleus.   Oocytes were incubated for 40h and fixed at -196°C.   CAT activity
was measured as described by Gorman et al. (1982).

RESULTS AND DISCUSSION

The Influence of cis-Factors on CAT Expression

     The analysis of the influence of 5'-flanking sequence (Figure 1) has
shown that the maximum expression of CAT was obtained with the TAT promoter
(1.1kb).   The decrease of CAT activity was recorded when other promoters
were used in the following order: TAT>TK + β-casein gene promoter (1.8kb)>TK
>β-casein gene promoter (1.8kb)>κ-casein gene promoter (4.5kb).   We found an
additive effect of the two spliced promoters β-casein (1.8kb) + TK on
enhancement of CAT expression.   The analysis of the influence of 3'-non-
translatable regions β-0.4kb>SV40>β-3.5kb (Figure 2) lead us to the
conclusion that the increasing activity of the 3'regulatory site correlates
with the reduction in its length.

The Influence of trans-Factors on CAT Expression

     Preliminary results of analysis of the influence of nuclear extracts on
CAT expression effectivity has shown that the inhibition of CAT activity has
been shown after simultaneous microinjection of DNA and nuclear proteins
into oocyte germinal vesicles (Figure 3).

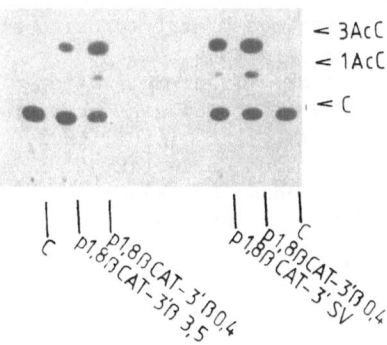

Figure 2.   Comparative analysis of the influence of 3'-non-translatable regions on CAT activity in extracts of oocytes.

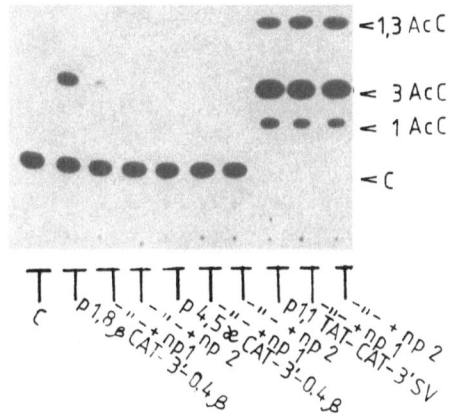

Figure 3.   Comparative analysis of the influence of nuclear DNA-binding proteins on CAT activity in extracts of oocytes.

We can predict from our previous results that the search for possible trans-regulators of specific gene expression in differentiated cells will be more fruitful via the mRNA route. This point is substantiated by the limitations of *X. laevis* oocytes as a test system, due to the cell compartmentalization.

REFERENCES

Gorman C. M., Moffat L. F. and Howard B. H. 1982. Recombinant genomes which express chloramphenicol acetyltransferase in mammalian cells. <u>Mol. Cell. Biol.</u>, 2:1044.

Lobanenkov V. V., Klenova E. M., Gorodetsky S. I. and Adler V. V. 1988. DNA-binding protein factors which interacts in specific way with promoter elements of TAT gene of rat. <u>Dokl. Akad. Nauk. USSR</u>, 298, 3:746.

Shinomiya T., Scherer G., Schmid W., Zentgraf H. and Sclaitz G. 1984.

Isolation and characterization of rat tyrosine aminotransferase gene.
Proc. Nat. Acad. Sci. USA, 81:1344.

Sweeney G. F. and Old R. E. 1988. Trans-activation of transcription, from
promoters containing immunoglobulin gene octamer sequences, by myeloma
cell mRNA in *Xenopus* oocytes, NAR, 16, 11:4903.

Tkach T. M., Kapelinskaya T. V. and Gorodetsky S. I. 1988. Genes of
Caseins of Bos Taurus. Cloning and Restriction Analysis.
Dokl.Akad.Nauk USSR, 229, 6, 1489-1493.

THE DETECTION OF NONCANONICAL CONFORMATION REGIONS IN DNA FROM EUKARYOTIC

CELLS

G. P. Zhizhina, S. A. Navarrete and G. P. Troitskaya

Institute of Chemical Physics
Academy of Sciences
Moscow, USSR

INTRODUCTION

In recent years the DNA double helix has been shown to be polymorphic, its local structure being dependent on nucleotide sequence. The inter-conversion between alternative conformations is probably involved in gene expression and other nuclear processes. The recently discovered left-handed Z-DNA appears to be one of these conformations. Eukaryotic genomes contain $(dC-dG)_n$ and $(dG-dT)_n$ sequences, which are able to transform from B to Z-form. In supercoiled plasmid DNA these regions exist in Z-form under physiological conditions. The methylation of cytosine, the presence of Mg^{+2} or Zn^{+2} and chromatin supercoiling are among the cell factors contributing to Z-DNA stabilization (Rich et al., 1984). Some factors have an increased level in tumour cells (Fyodorov, 1982; Andronikashvili and Mosulishvili, 1980; Luchnik and Glaser, 1981). Therefore, we have assumed that Z-like segments may occur in DNA from tumours under physiological conditions.

MATERIAL AND METHODS

DNA Isolation

DNA from mice liver, Ehrlich ascites carcinoma and calf thymus was isolated after lysis in buffer A (150mM NaCl, 20mM EDTA, pH 8.0) and 1% SDS. Lymphocytes from patients with chronic lymphoid leukaemia were lysed in buffer B containing 100mM NaCl, 10mM Tris-HCl, pH 8.0, 10mM EDTA, 1% SDS and 100µg/ml Proteinase K (Serva). The deproteinization was performed by 1M NaCl and chloroform-isopropanol mixture. After dissolving in 1 SSC DNA was treated with RNAse and Pronase (Serva) and deproteinized. Before experiments DNA was dissolved in buffer C (50mM NaCl, 5mM Tris-HCl, pH 7.8).

DNA Retention by Nitrocellulose Filters

Aliquots of DNA solution (6-8µM nucleotide) and various amount of NaCl (0.5-4.5M) were incubated for 1h at $22^{\circ}C$ and filtered through Chemapol, Synpor filters, 0.23µm pore size, prewashed with the same salt solution. UV-spectra (220-300nm) of each sample were registered before and after filtration on a Beckman DU-50 spectrophotometer. The relative amount of bound DNA was estimated as $\Delta A_{260}/A^{\circ}_{260}$, where A°_{260} is absorbance at 260nm of DNA solution before filtration.

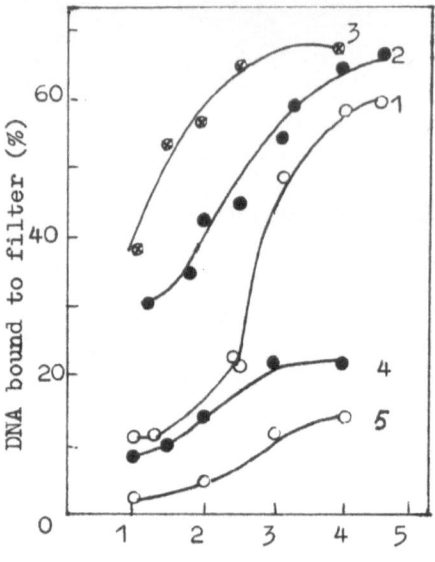

Figure 1.   Retention of DNA by nitrocellulose filters with increasing
NaCl concentration: PM2 DNA (1); phage DNA (2); DNA from
leukaemic lymphocytes in the absence (3) and in presence of
2mM $MgCl_2$ (4) or 5µM EB (5).

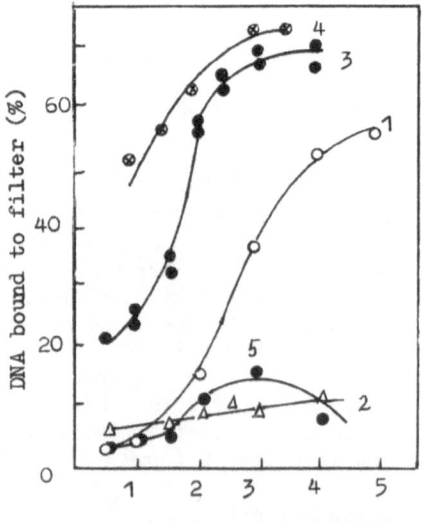

Figure 2.   Retention of DNA by nitrocellulose filters with increasing
NaCl concentration: DNA from mice liver (1); DNA from calf
thymus in the absence (2) and in presence of 2mM $MgCl_2$ (3)
or 5µM EB (4); (dG-dC)$_9$ olygonucleotide (5).

Optical Titration of DNA by Ethidium Bromide

Successive aliquots (3-5μl) of ethidium bromide (EB) solution (50μM) were pipetted at 15 min intervals into 2.5ml of 30-50μM DNA solution and spectra were registered from 360 to 600nm. The concentration of bound EB was determined from the ΔA_{480} value; that of free EB was estimated from A_{460}, since the DNA-EB complex has no absorbance at 460nm, as was shown in our equilibrium dialysis system.

RESULTS AND DISCUSSION

Figures 1 and 2 show the retention of various DNAs by nitrocellulose filters, depending on the NaCl concentration. The binding curves of PM2 DNA, olygonucleotide $(dG-dC)_9$ and DNA from mouse liver, calf thymus and leukaemic lymphocytes are sigmoidal, midpoint varying from 1.9 to 3.0M. It is not the case for phage DNA, single-strand DNA, RNA and serum albumin. As shown by Kuhnlein et al. (1980), the sigmoid binding curves of DNA are due to the salt-induced cooperative transition from B-form passing through the filter to Z-form retained by it. In 1M NaCl the binding of PM2 and mouse liver DNA does not exceed 5-7% and that of DNA from calf thymus and leukaemic cells is equal to 25-30%. The low salt binding level of the latter two can be attributed to the presence of some Z-like segments in the right-handed helix of DNA.

It has been established (Pohl et al., 1972; Rich et al., 1984) that EB induces a Z-B transition and $MgCl_2$ facilitates the reverse. DNA incubation with $MgCl_2$ (2mM for 10 min at $50^{\circ}C$) results in enhance of low salt binding to filter. The incubation with EB (5μM for 1h at $22^{\circ}C$) has an opposite effect and prevents from the salt-induced increase of DNA retention by filter. These results may be also accounted for existance of Z-form regions in natural DNA.

The method of DNA titration by EB was used to verify this conclusion. As Walker et al. (1985) have found, EB binds to the Z-form in a highly cooperative manner inducing Z-B transition of $(dG-dC)_n$ sequences. The Scatchard analysis of drug binding data consists of a plot of r/C_f versus r, where r is the ratio of the bound drug to DNA base concentration and C_f is that of the free drug. A positive slope of the data in a Scatchard plot indicates a cooperative binding. Scatchard analyses of EB binding data to DNAs under investigation are shown in Figure 3. All these EB binding isotherms reveal a positive slope at low r. This is not observed in the case of λphage DNA. These data may be accounted for by a Z-B transition of some Z-like regions in DNA from calf thymus, mouse liver and leukaemic cells.

The circular dichroism (CD) spectra serve as a sensitive indicator of DNA conformation. The CD study established that molar ellipticity at 250nm and 295nm of the poly(dG-dC) B-form remains unchanged after EB binding; in contrast, the ellipticity of the Z-form is a function of the bound drug concentration (Walker et al., 1985). In our difference spectra of mice liver DNA the values of $\Delta(\varepsilon_L - \varepsilon_R)$ at 250 and 290nm vary depending on the bound EB concentration, as seen from Figure 4. The shape of these curves is similar to that reported by Walker et al. (1985) for Z-B transition induced by EB.

The stabilization of Z-like regions may be due to some non-histone tightly bound proteins resistant to protease treatment. DNAs from mouse liver, Ehrlich carcinoma, hepatoma 22A and human leukaemic lymphocytes were digested with DNAse I and exposed to 12.5% polyacrylamide gel electrophoresis. The appearance of three protein bands corresponding to ~44, 50

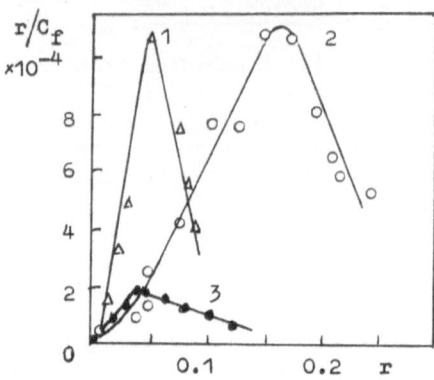

Figure 3.   The optical titration data: Scatchard analyses of EB binding to DNA from calf thymus (1), leukaemic lymphocytes (2) and mice liver (3).

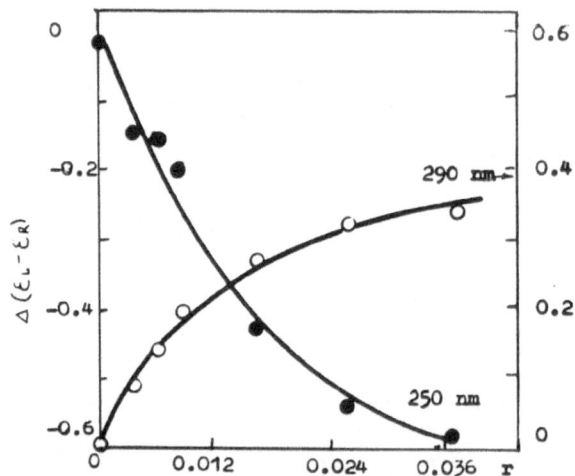

Figure 4.   The CD data: values of $\Delta(\varepsilon_L - \varepsilon_R)$ as function of r for EB binding to mice liver DNA. Spectra were registered on a JASCO-500A spectropolarimeter.

and 70kD was observed in all the samples. These polypeptides are believed to be nuclear matrices proteins.

The analysis of all the data suggests that Z-like segments exist in mammalian and tumour DNA under physiological conditions.

REFERENCES

Andronikashvili E. L., and Mosulishvili L. H. 1980, Human leukaemia and trace elements, in: 'Metal ions in biological systems', H. Sigel, ed., M. Dekker Inc., New York, Basel.

Fyodorov N. A. 1982, A role of DNA enzymatic methylation in cell differentiation and transformation, Exp. oncologiya, 4:3.

Kuhnlein U., Tsang S. S., and Edwards J. E. 1980, Cooperative structural transition of PM2 DNA at high ionic strength and its dependence on DNA damages, Nature (London), 287:363.

Luchnik A. V., and Glaser V. M. 1981, DNA topological linking numbers in malignantly transformed syrian hamster cells, Mol. Gen. Genet., 183:553.

Pohl Z., Jovin T. M., Baehr W., and Holbrook T. 1972, Ethidium bromide as a cooperative effector of DNA structure, Proc. Natl. Acad. Sci. USA, 69:3805.

Rich A., Nordheim A., and Wang A. 1984, Chemistry and biology of left-handed DNA, Ann. Rev. Biochem., 53:791.

Walker G. T., Stone M. P., and Krugh T. R. 1985, Ethidium binding to left-handed (Z) DNAs results in region of right-handed (B) DNA at the intercalation site, Biochemistry, 24:7462.

CHARACTERIZATION OF CYTOPLASMIC LOW-MOLECULAR-MASS RNA COMPLEMENTARY

ASSOCIATED WITH POLY(A)$^+$ RNA FROM RAT LIVER

I. M. Konstantinova, O. A. Petukhova, V. A. Kulitchkova
and V. I. Volkova

Institute of Cytology of the Academy of Sciences of the USSR
Tikhoretsky ave 4, Leningrad
194064, USSR

INTRODUCTION

The fraction of low-molecular-mass RNA (sacc-RNA) associated by complementary interactions with high-molecular-mass poly(A)$^+$ RNA was isolated and studied. The data obtained so far show that small associated cytoplasmic RNA is heterogeneous in size and oligonucleotide composition. The fraction of sacc-RNA is exposed to glucocorticoid control in rat liver. The tight binding of a part of sacc-RNA with peptides has been observed. A close similarity of the oligonucleotide composition of sacc-RNA and RNA-component of specific small nuclear RNP has been found.

METHODS

Poly(A)-containing RNA/poly(A)$^+$ RNA was purified from high-molecular-mass cytoplasmic RNA by oligo(dT)cellulose chromatography (Aviv and Leder, 1972). Electrophoresis of low molecular weight RNA was run in 10% polyacrylamide gel (Loening, 1967).

Low-molecular-mass RNA associated with poly(A)$^+$ RNA was isolated as follows. Purified by two passages through oligo(dT) cellulose, high-molecular-mass poly(A)$^+$ RNA was subjected to denaturation with 90% formamide at 65°C, or in boiling water for 10 min. After denaturation, high-molecular-mass RNA was precipitated with 2.5M NaCl, and the non-precipitated low-molecular-mass RNA was collected by precipitation with ethanol. In some cases, after denaturation, poly(A)$^+$ RNA was applied to an oligo(dT) cellulose column, and the poly(A)$^-$ low-molecular-mass RNA fraction, which did not bind to the column, was eluted in the void volume. The fraction of low molecular weight RNA thus far isolated was termed 'small associated cytoplasmic' RNA (sacc-RNA). T$_1$ RNase fingerprints were prepared according to the method described by Pedersen and Haseltine (1980). Northern blot RNA-RNA hybridization was performed according to Thomas (1980).

RESULTS AND DISCUSSION

The oligonucleotide map of the sacc-RNA separated by denaturation from the total rat liver cytoplasmic poly(A)$^+$ RNA is shown in Figure 1. The

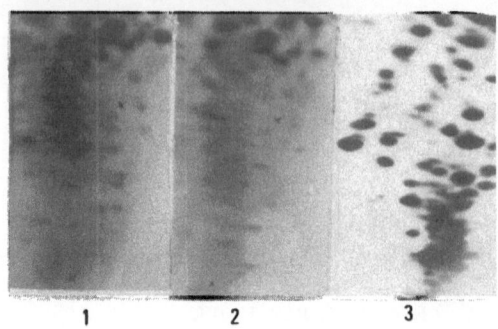

Figure 1.   The T_1-fingerprints of small RNA. 1 - sacc RNA; 2 - α-RNA;
3 - total low molecular weight RNA. x - the position of the
marker dye.

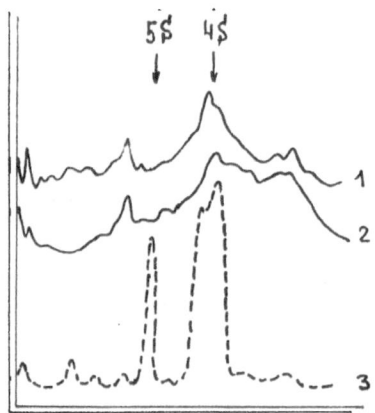

Figure 2.   Electrophoresis in 10% PAGE in the TBE buffer of small
associated RNA (1, 2). 1 - control; 2 - cortisole-induced;
3 - total cytoplasmic low molecular mass RNA.

fingerprint reveals the oligonucleotide heterogeneity of this fraction. The
electrophoretic pattern shows that sacc-RNA contains molecules of different
size classes (Figure 2). The specific character of the fraction of small
associated cytoplasmic RNA is supported by the observation that different
ways of denaturation (heat and formamide) result in the same electrophoretic
and fingerprint pattern (data not shown). Besides, as can be seen in Figure
1, the T_1 fingerprint of the sacc-RNA is totally different from the T_1-
oligonucleotide map of total small cytoplasmic rat liver RNA.

Under the influence of cortisole, a few species of the sacc-RNA are
stimulated (data not shown). This suggests that the few specific kinds of
sacc-RNA are associated with the total diverse population of the
glucocorticoid-stimulated messenger RNA in the liver and allows one to
propose the possibility of co-ordinate hormonal regulation of post-transcriptional events by small associated cytoplasmic RNA.

Of special interest is the close similarity of the oligonucleotide
composition of small associated cytoplasmic RNA and α-RNA - the component of

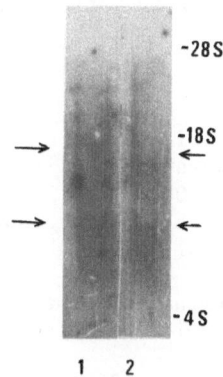

Figure 3.   Northern blot hybridization of poly(A)$^+$ RNA (cortisole-stimulated) with: ^{32}P-sacc RNA (1) and α-RNA (2).

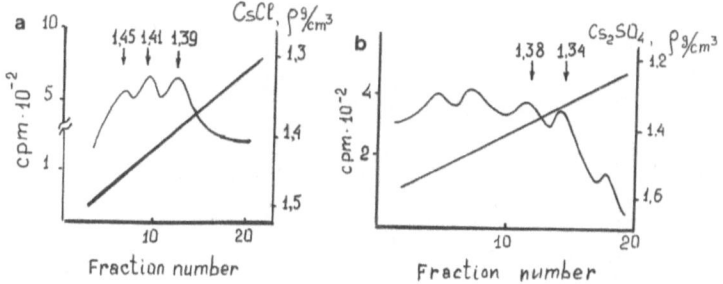

Figure 4.   Separation of ^{32}P-sacc RNA in buoyant density gradient of CsCl (a) and Cs$_2$SO$_4$ (b). The material has not been fixated. RNA was labelled by T$_4$ polynucleotide kinase with ^{32}P-γ-ATP.

small specific nuclear RNP (Figure 1). This RNP is tightly bound to chromatin, as was shown earlier (Konstantinova et al., 1984). The sacc-RNA, as well as α-RNA, is capable of a complementary interaction with poly(A)$^+$ RNA (Figure 3).

Within α-RNP, α-RNA is tightly bound to the protein moiety (Konstantinova et al., 1984). That is why we proposed a similar tight binding of α-RNA to protein for the sacc-RNA preparations. The centrifugation of sacc-RNA-^{32}P in CsCl and Cs$_2$SO$_4$ density gradients shows (Figure 4) that sacc-RNA contains, besides free RNA, some material with buoyant density characteristic for RNP (ρ = 1.39-1.45g/cm^3 in CsCl and ρ = 1.34-1.38g/cm^3 in CsSO$_4$ gradients). The sacc-RNA thus contains some RNA tightly bound to peptides, and this bond withsatnds the phenol-detergent treatments during RNA isolation.

In addition, in the fraction of free cytoplasmic low-molecular-mass RNA, some RNA tightly bound to peptides can be observed. After the ^{14}C-labelled amino-acids were injected *in vitro,* in the purified total low-molecular-mass cytoplasmic RNA, RNA associated with ^{14}C-amino acid labelled material, was observed in SDS-electrophoresis. It had the mobility of 4.5S RNA, and in Cs_2SO_4 gradient displayed the RNP buoyant density (ρ = 1.38g/cm^3).

REFERENCES

Aviv H., and Leder P., 1972, Purification of biologically active globin
    messenger RNA by chromatography on oligothymidylic acid cellulose,
    Proc. Natl. Ac. Sci., 69:1408.
Konstantinova I. M., Turoverova L. V., Petukhova O. A. and Vorob'ev V. I.,
    1984, Cortisole-induced small RNP tightly bound to chromatin, FEBS
    letters, 177:241.
Loening U. E., 1969, The determination of the molecular weight of ribo-
    nucleic acid by polyacrylamide gel electrophoresis, Biochem. J.,
    113:131.
Pedersen F. S., and Haseltine W. A. 1980, A micromethod for detailed
    characterization of high molecular weight RNA, Methods Enzymol.,
    Acad. Press, N.Y., 65:680.
Thomas P. S., 1980, Hybridization of denatured RNA and small DNA fragments
    transferred to nitrocellulose, Proc. Natl. Acad. Sci., 77:5201.

IN VITRO AMPLIFICATION OF THE HUMAN INSULIN GENE

Ryszard Slomski[1], Malgorzata Jungerman[1] and Adam Kraszweski

[1] Institute of Human Genetics
Polish Academy of Sciences
Strzeszyńska 32,
60-479 Poznań, Poland

[2] Institute of Bioorganic Chemistry
Polish Academy of Sciences
Noskowskiego 12/14
61-704 Poznań, Poland

INTRODUCTION

DNA analysis is an increasingly important source of information in modern medicine. Various mutations responsible for genetic disorders have been described by means of direct analysis of human genomic DNA. Studies of a specific DNA sequence have to be preceded by construction and screening of the whole genome library to identify the sequences of interest. The alternative method is the polymerase chain reaction (PCR). PCR is an *in vitro* method for the primer-directed enzymatic amplification of specific DNA sequences using DNA polymerase (Saiki et al., 1985). With repeated cycles of DNA denaturation, annealing of the primers and DNA synthesis, a large number of copies of specific DNA sequences is faithfully synthesized in a few hours. The enzyme used in this procedure is a heat-stable DNA polymerase purified from bacterium *Thermus aquaticus* - Taq DNA polymerase (Chien et al., 1976).

In this paper we describe application of the PCR for amplification of the A and B chains of the human insulin gene. Several point mutations in the insulin gene causing insulin-dependent diabetes (IDD) have been described (Schoelson et al., 1983). We have isolated DNA samples from non-affected individuals and from patients with diabetes mellitus. Patients showed an elevated level of endogenous insulin with probable synthesis of abnormal molecules of proinsulin or insulin. They all required regular administration of exogenous insulin and showed normal response to it. The scope of work was to study the sequence of the insulin gene from patients with IDD based on PCR products synthesized on the template of their genomic DNA.

MATERIALS AND METHODS

Preparation of DNA Samples

Genomic DNA was isolated from 10ml of the whole blood according to the method of Kunkel (1977).

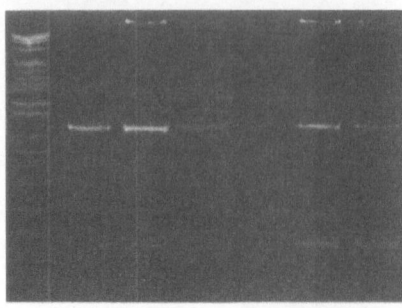

Figure 1.   Electrophoresis in 12% polyacrylamide gel of 132bp PCR product
             containing a sequence coding for the B chain of insulin.

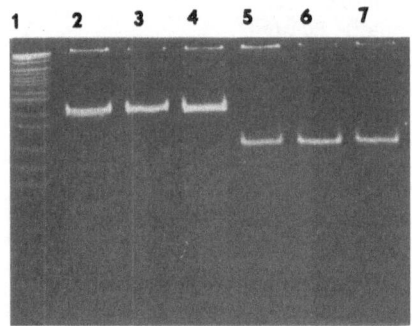

Figure 2.   Electrophoresis in 12% polyacrylamide gel of PCR products
             containing sequence coding for the B chain of insulin
             synthesized after annealing of primers at 40°C. 1, KBL size
             marker; 2,3,4, 132bp PCR product; 5,6,7, 90bp PCR product.

Synthetic Oligonucleotides

     Oligonucleotide primers were designed to direct synthesis of sequences
coding for the A and B chain of human insulin. 5'-ends of primers for
synthesis of the A chain coding sequences have been modified with sequences
enabling direct cloning of the PCR product in vectors after digestion with
restriction endonucleases EcoRI and BamHI.

Sequence Amplification with Taq DNA Polymerase

     Reactions with Taq DNA polymerase (New England Biolabs) were performed
as described by Slomski et al. (1988). Annealing of the primers to the
template was performed at optimal temperatures for each pair of oligonucleotides. Seven pg to 700ng of genomic DNA was used in the reaction to test
the possibility of decreasing the amount of the template. Eight to 48
cycles of PCR were performed for the comparison of the efficiency of DNA
synthesis.

RESULTS

     We have designed three pairs of primers for amplification of unique
sequences in the human genome: one pair for amplification of 100bp fragment

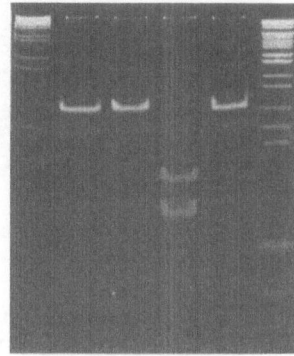

Figure 3.   Electrophoresis in 12% polyacrylamide gel of 90bp PCR product
containing a sequence coding for the B chain of insulin.
1, λDNA digested with HindIII; 2,3,5, 90bp PCR product; 4,
90bp PCR product digested with AluI fragments of 40 and 50bp;
6, KBL size marker.

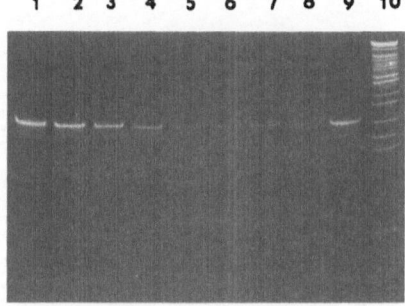

Figure 4.   Electrophoresis in 12% polyacrylamide gel of 90bp PCR product
containing a sequence coding for the B chain of insulin. PCR
was performed using serial dilutions of genomic DNA. The
amount of genomic DNA was ranged from 7pg + 700ng. 1, 700ng;
2, 70ng, 3, 7ng; 4, 700 pg; 5,6,7, 70pg; 8, 7pg; 9, 7ng; 10,
KBL size marker.

containing sequence coding for A chain of insulin and two pairs for
amplification of 90 and 132bp fragments containing sequence coding for B
chain of insulin. We have established experimentally the optimal temperature
of annealing of the primers to the template which is different for each pair
of primers. Unspecific PCR products were synthesized only when annealing
temperature was too low (Figure 1). At a correct temperature of annealing
the length of the PCR product corresponded to the distance framed by primers
(Figures 2 and 3).

    We have studied efficiency of DNA synthesis in PCR with the use of
serial dilutions of genomic DNA. We were able to visualise after
electrophoresis the PCR product obtained with 7pg of genomic DNA
corresponding to the amount of DNA in a single diploid cell (Figure 4). At

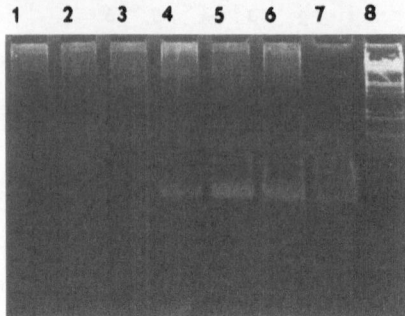

Figure 5.   Electrophoresis in 1.8% agarose gel of 100bp PCR product
              containing sequence coding for the A chain of insulin.
              Various number of cycles were performed: 1, 8 cycles; 2,
              16 cycles; 3, 24 cycles; 4, 32 cycles; 5, 40 cycles; 6,7,
              48 cycles; 8, λDNA digested with HindIII.

least 24 cycles of PCR were needed to visualise the product synthesized with
the use of 0.5µg of genomic DNA (Figure 5).

DISCUSSION

      We have applied the polymerase chain reaction for rapid production of
large number of copies of DNA sequences coding for the A and B chains of
human insulin. The temperature of primer annealing to the template was
established experimentally. At too low a temperature several different
fragments of DNA were synthesized due to unspecific annealing of primers to
partially homologous sequences in genomic DNA. At appropriate temperatures
only one fragment of the length corresponding to the distance between
primers is synthesized. Identification of PCR product was based on its
length revealed after electrophoresis directly after the reaction or after
digestion of the product with an appropriate restriction enzyme. By means
of PCR very small amounts of genomic DNA may be studied. Seven pg of genomic
DNA contained in a single diploid cell is enough to visualise the PCR
product after electrophoresis. However, the standard amount of DNA in the
reaction mixture is 0.5µg, which allows us to visualise the product after
24 cycles of PCR. Amplification of DNA is also possible with primers
bearing additional sequences on their 5' ends, as we have shown by amplifying
a sequence coding for A chain of insulin. These additional sequences enable
cloning of the PCR product in a vector after digestion with BamHI and EcoRI,
and they have no influence on either specificity or efficiency of reaction.
We have established parameters of PCR for several pairs of primers. PCR
enabled us to synthesize with high efficiency specific DNA sequences coding
for insulin in patients with IDD and to study the nucleotide sequence of
their insulin genes without construction of genomic library and screening.

REFERENCES

Chien A., Edgar D. B., Trela J. M. 1976, Deoxiribonucleic acid polymerase
     from the extreme thermophile *Thermus aquaticus*. J. Bacteriol.,
     127:1550.

Kunkel L. M., Smith K. D., Boyer S. H., Borgaonkar D. S., Wachtel S. S., Miller O. J., Breg W. R., Jones H. W., Rary J. M. 1977, Analysis of human Y-chromosome-specific reiterated DNA in chromosome variants. Proc. Natl. Acad. Sci. USA, 74:1245.

Saiki R. K., Scharf S., Faloona F., Mullis K. B., Horn G. T., Erlich H. A., Arnheim N. 1985, Enzymatic amplification of β-globin genomic sequences and restriction site analysis for diagnosis of sickle cell anaemia. Science, 230:1350.

Saiki R. K., Gelfand D., Stoffel S., Scharf S., Higuchi R., Horn G. T., Mullis K. B., Erlich H. 1987, Primer-directed enzymatic amplification of DNA with thermostable DNA polymerase. Science, 239:487.

Slomski R., Reiss J., Jungerman M. 1988, Application of non-radioactive methods of DNA detection in analysis of human genetic disorders. Acta Biochimica Polonica, in press.

CHROMOSOMES E AND S IN THE OOGENESIS OF THE GALL MIDGE

M. N. Gruzova and F. M. Batalova

Institute of Cytology
Academy of Sciences of the USSR
Leningrad, USSR

INTRODUCTION

It has been well documented that the extra chromosomes (L and E) in some insects are characteristic of gamete lines and are eliminated from somatic cells in early embryogenesis (Boswell and Mahowald, 1985). However, the behaviour of these chromosomes in oogenesis, their share in the formation of polar germ plasma which determines the fate of future gametes remains so far obscure. The origin, structural and biochemical organization of these chromosomes are also to be clarified. This paper gives the data on the behaviour of chromosomes S (S-ch) and E (E-ch) during oogenesis, meiotic division and early embryogenesis of the raptorial gall midge *Aphidoletes aphidimyza*. The cytology and cytogenetics of this species, as well as the whole group of zoophagic gall midges, has not been properly studied (Matuszewski, 1982).

RESULTS AND DISCUSSION

In ovarioles of pupae the youngest oocytes are already at diplotene stage: S chromosomes forming the central group of compact bivalents are surrounded by a network of strands (capsule) separating them from E-ch; the latter are diffused and located under the nuclear envelope, in close contact with nuclear pores and perinuclear material (Figures 1, 2). ^{3}H-uridine is incorporated in E-ch during the whole period of oogenesis, however their transcriptional activity is several orders lower than the relative activity of trophocyte nucleoli (Figures 3, 4). S-ch are not active in RNA synthesis before the start of vitellogenesis, when they become despiralized and, incorporate ^{3}H-uridine. The NOR's reaction with $AgNO_3$, first used by us for the analysis of E-ch reveals strong differences between S- and E-ch: at all stages of oogenesis E-ch show intensive 'staining', whereas S-ch are poorly impregnated (Figure 5). The pattern of silver staining of E-ch is comparable with that of nucleoli of follicular and trophic cells (Figure 6). (It should be pointed out that there are no nucleoli in the oocytes of *A. aphidimyza*). These results allow us to believe that E-ch in the oogenesis of *A. aphidimyza* are uncommonly found.

The number of S bivalents in diakinesis and metaphase of meiosis 1 vary from 4 to 5 (Figures 8 - 12) not only in different insects, but even in one and the same female. S-ch are always located inside the spindle, whereas

Nuclear Structure and Function, Edited by J. R. Harris and
I. B. Zbarsky, Plenum Press, New York, 1990

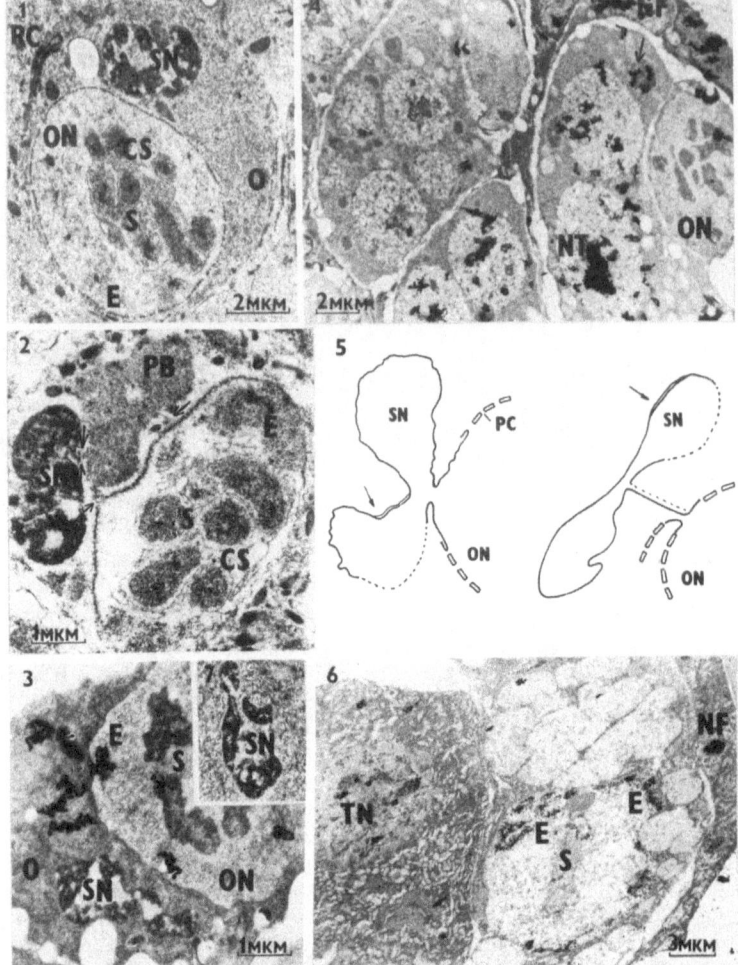

Figure 1.   Germarium oocyte (O); in oocyte nucleus (ON) electron dense
bodies (S chromosomes) are in central position and are
surrounded by net of strands; diffused chromatin masses (E
chromosomes) lie under the nuclear membrane; SN – 'small
nucleus', RC – ring canal, CS – capsule strands.

Figure 2.   Young oocyte with nucleus (ON), 'small nucleus' (SN) and
perinuclear body (PB) are interconnected (arrows); CS –
capsule strands.

Figure 3.   Fragment of oocyte (O) with nucleus (ON) and 'small nucleus'
(SN); E chromosomes (E) incorporate ^{3}H-uridin; S – S chromosomes.

Figure 4. ^{3}H-Uridine incorporation in cells of germarium; nucleoli of
trophocytes (NT) and of follicular cells (NF) incorporate more
actively, tracks in perinuclear bodies are visible (arrows);
ON – oocyte nucleus.

Figure 5.   Scheme showing the 'small nucleus' budding off from the oocyte
nucleus (ON); PC – pore complex.

Figure 6.   AgNO$_3$ Impregnation of egg follicle cells; Ag(+) reaction in
the E chromosomes (E) is similar to that in the nucleoli of
follicular (NF) and trophocyte nuclei (TN); S chromosomes (S)
are weakly stained.

Figure 7.   Chromatin (Ag(+) reaction in 'small nucleus' (SN).

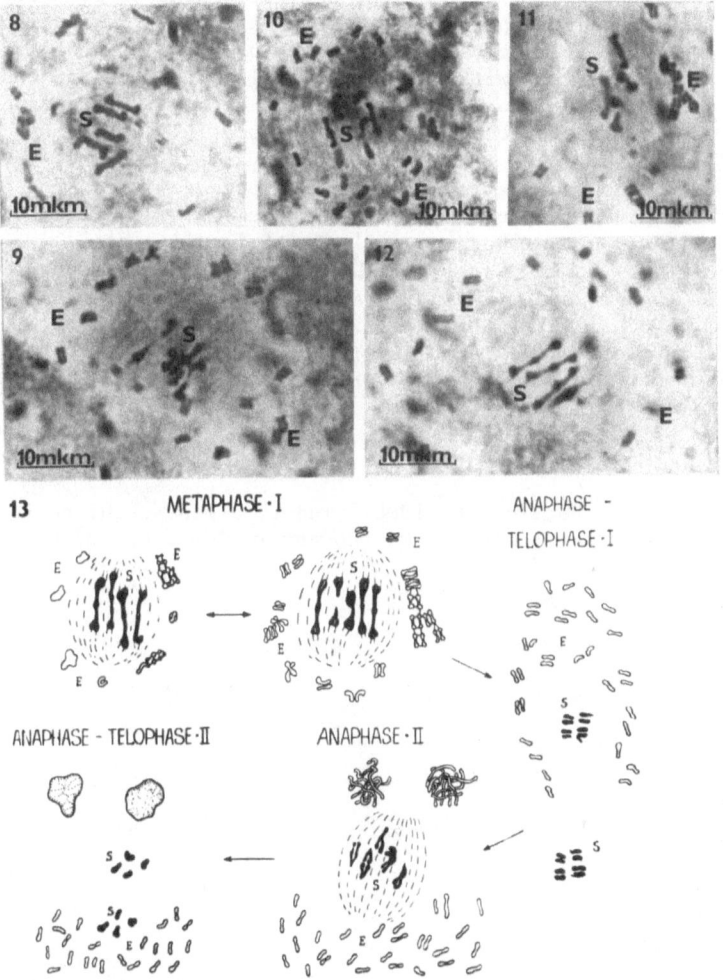

Figures 8–12. Metaphase I of meiosis; S bivalents (S) are situated in
                spindle, E chromosomes lie outside; 8, 9 – 5S bivalents;
                10 – 12 – 4S bivalents.
Figure 13.   Scheme of meiotic division A. *aphidimyza* (for detailed
                explanations see text).

E-ch are on the outside. The number of E-ch is hard to establish, since
they stick together as on aggragates or associate each other (Figure 11).
Where the associations are minimal, it can be seen that E-ch are represented
by both achiasmatic bivalents and univalents and amount to 17 to 26 (Figures
8 - 10 and 12). It can be thought that association, and pasting of E-ch is
due to the high content of rDNA. Anaphase, telophase of meiosis I, takes place
10 to 15 min after oviposition. E-ch are located close to or around one of
the groups of S-ch, whereas the other possibily forms a polar nuclei
(Figure 13). At this stage E-ch are mostly represented by univalents. We
failed to observe a spindle for E-ch, as reported for other species of gall
midges (Mutuszewski, 1982). The metaphase II has been registered 30 min
after oviposition. Sometimes in the zone of division there occur either two
small groups of chromosomes, or two nuclei 5-8mkm in diameter (Figure 13).
In all probability, both are polar nuclei at different stages of formation
rather than 'small nuclei'. The latter have been previously described in

oocytes of some species of gall midges and, according to some authors, are of somatic origin and participate in zygote formation (Matuszewski, 1982). Similar 'small nuclei' have been detected by us in oocytes of *A. aphidimyza*. Our observations disagree with what is generally accepted and come for the following observations. In germarium oocytes of pupar and imagoes, in close proximity ot their nuclei one can see 2-3 chromatin bodies 3-4mkm in diameter ('small nuclei'). They have double nuclear membrane devoid of pores, contain dense chromatin clumps and 'a package' of strand similar to those of the capsule of oocyte nucleus (Figure 1). The pictures observed are interpreted by us as 'budding off' of the 'small nuclei' from the oocyte nucleus (Figure 5). This is also evidenced by a direct association of membranes of these nuclei, and an invariable 'package' of strands. Silver staining of chromatin of 'small nuclei' is similar to that of E-ch (Figure 7). Therefore, we conclude that 'small nuclei' arize due to blebbing of the major nucleus and possibly contain chromatin of E-ch.

REFERENCES

Boswell R.E. and Mahowald A.P. 1985. Tudor, a gene required for assembly of Germ Plasm in *Drosophila Melanogaster, Cell.* 43, 97.

NUCLEAR REPROGRAMMING IN BOVINE EMBRYOS AFTER NUCLEAR TRANSPLANTATION

J. Kanka, J. Jr. Fulka[+] and J. Fulka

The Czechoslovak Academy of Sciences
Institute of Animal Physiology and Genetics
277 21 Libechov, Czechoslovakia

[+] The Institute of Animal Production
Praha - Uhríneves, Czechoslovakia

INTRODUCTION

A fundamental and still unsolved problem concerning the developmental process of eukaryotic organisms is whether differentiated cells ration the genomic totipotency of the zygote nucleus (DiBerardino, 1988). There are probably differences among mammalian species in the extent to which nuclei from early embryos are able to support development after transfer to an enucleated egg. Mouse embryos initiate transcription of heterogenous nuclear RNA (hnRNA) at the early 2 cell stage (Flach et al., 1982) and nuclear totipotency in mouse is lost at the time of embryonic genome activitation (McGrath and Solter, 1984). On the other hand, sheep embryos initiate transcription at the 8 to 16 cell stage (Crosby et al., 1988), and blastomere nuclei of these embryonic stages (Willadsen, 1986) and even of ICM stage (Smith and Wilmut, 1989) are able to support development to term, when fused to enucleated oocytes. Similarly, in cattle embryos transcription starts at the 8 cell stage (Camous et al., 1986), and blastomere nuclei of 8 - 16 cell stage (Prather et al., 1987) and even of 32 cell stage (Marx, 1988) are totipotent. The question arises as to what processes on the ultrastructural and molecular level are responsible for reprogramming the mammalian embryonal nucleus.

MATERIALS AND METHODS

Oocytes and embryos were recovered by flushing the oviducts of superovulated heifers and handled as described by Pavlok et al. (1988). The oocytes containing the first polar body were incubated for 15 min in the medium with cytochalasin B (5μg/ml, Serva) and bisected into two halves (Tarkowski, 1977). Isolated blastomeres from the 8 cell stage were agglutinated to encleated halves in PBS containing phytohaemagglutinin (300μg/ml, Serva). Agglutinated cells were exposed to the fusion conditions of Stice and Robl (1988). Then the products were washed several times in the medium and cultured in 0.1ml droplets of medium under paraffin oil at 37.5°C under 5% CO_2 in air for several periods of time. Developing embryos (2, 4 and 8 cell stages) were incubated for 20 min in medium enriched with 5-^{3}H-urdine (UVVVR Prague, specific activity 765GBq/mol) at a final concentration of 3.7MBq/ml. After incubation, the embryos were

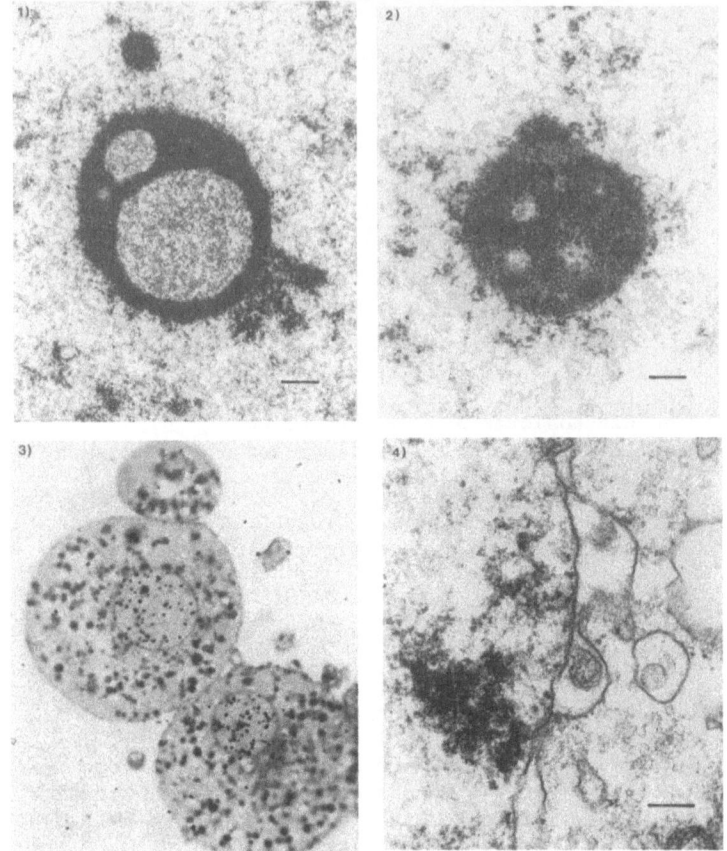

Figure 1.   Nucleolus from the 2nd cell cycle after fusion. Large
             nucleolar vacuole in the center of NPB. Massive penetration
             of chromatin occurred into the dense fibrillar network of NPB.
             Bar represents 200nm.

Figure 2.   Nucleolus at the 4th cell cycle after fusion is formed by
             intermingled filamentous components. Bar represents 200nm.

Figure 3.   Autoradiogram from the 4th cell cycle after fusion. The
             labelling over the nuclear area indicates beginning of hnRNA
             synthesis.

Figure 4.   Blebbing activity of nuclear envelope after fusion. Note
             granular contents which evaginate from the nucleus into the
             perinuclear space. Bar represents 100nm.

washed repeatedly in cold medium, fixed and embedded for electron microscopy
(Crozet et al., 1986). From individual nuclei semi-thin sections (0.5 /um)
were prepared for autoradiography and thin sections for electron microscopy.

RESULTS

     After fusion the nucleolar development is stopped at the stage typical
for 8 cell bovine embryo. The characteristic features of this stage are the
appearance of a large electronlucid area in the center of nucleolar precursor
body (NPB) and association of the NPB with a clump of condensed chromatin

(Figure 1). In reconstituted embryos the ultrastructure was 'frozen' on this stage for the next three cell cycles. Autoradiograms from subsequent semi-thin sections of all these nuclei did not show any signs of hnRNA synthesis.

At the 4th cell cycle of the reconstituted embryo the nucleolus was composed of intermingled filamentous components and secondry small vacuoles (Figure 2). Autoradiograms of these nuclear sections revealed labelling over the nuclear area as proof of hnRNA synthesis (Figure 3).

Blebbing activity of nuclear envelope (Szöllösi and Szöllösi, 1988) appeared in reconstituted embryos at the first and second cell cycle after fusion (Figure 4). During later developmental stages this dynamic activity of nuclear envelope was lacking.

DISCUSSION

Successful development to term of nuclear transplant embryos does not necessarily indicate that the nucleus has been reprogrammed but may simply indicate that the donor and recipient are functionally compatible (Stice and Robl, 1988). However, the information about nuclear changes is lacking in most nuclear transport experiments (Smith and Wilmut, 1989).

In reconstituted bovine embryos nucleolar development and hnRNA transcription are switched on again at the cell cycle which is temporally appropriate for the recipient cytoplasm. A similar situation was found in reconstituted mouse embryos (Howlett et al., 1987). The 8 cell nucleus can be reprogrammed after transfer, with an apparent switch-off of transcription followed by the production of the hsp 68/70kD proteins at the normal developmental stage. However, this limited reprogramming in mouse is not sufficient to allow much more than one cleavage division.

Multiple nuclear envelope blebs were found during the first and second cell cycle of reconstituted bovine embryos, within the perinuclear space. These structures were observed also in somatic (thymocyte) and embryonic (mouse) nuclei following cell hybridization (Szöllösi and Szöllösi, 1988). It is supposed that blebbing has a special role in nucleocytoplasmic transport at the beginning of development and its reappearance after fusion, observed in our experiments, can be considered as a part of reprogramming process of embryonal nuclei.

REFERENCES

Camous S., Kopecný V., and Fléchon J.-E. 1986, Autoradiographic detection of the earliest stage of /^{3}H/-urdine incorporation into the cow embryo, Biol. Cell., 58:95.
Crosby I. M., Gandolfi F., and Moor R. M. 1988, Control of protein synthesis during early cleavage of sheep embryos, J. Reprod. Fertil., 82:769.
Crozet N., Kanka J., Motlik J., and Fulka J. 1986, Nucleolar fine structure and RNA synthesis in bovine oocytes from antral follicles, Gamete Res., 14:65.
DiBerardion M. A. 1988, Genomic multipotentiality of differentiated somatic cells, in: 'Regulatory Mechanisms in Developmental Processes', G. Eguchi, T. S. Okada and L Saxén, eds., Elsevier Sci. Publ. Ireland, Ltd.
Flach G., Johnson M. H., Braude P. R., Taylor R. A. S., and Bolton V. N. 1982, The transition from maternal to embryonic control in the 2-cell mouse embryo, EMBO, 1:681.
Howlett K., Barton Sh. C., and Surani M. A. 1987, Nuclear cytoplasmic

interactions following nuclear transplantation in mouse embryos, Development, 101:915.

Marx J. L. 1988, Cloning in sheep and cattle embryos, Science, 239:463.

McGrath J., and Solter D. 1984, Inability of mouse blastomere nuclei transferred to encleated zygotes to support development in-vitro, Science, 226:1317.

Pavlok A., Torner H., Motlík J., Fulka J., Kauffold P., and Duschinski U. 1988, Fertilization of bovine oocytes in vitro: effect of different sources of gametes in fertilization rate and frequency of fertilization anomalies, Anim. Reprod. Sci., 16:207.

Prather R. S., Barnes F. L., Sims M. M., Robl J. M., Eyestone W. H., and First N. L. 1987, Nuclear transplantation in the bovine embryo: assessment of donor nuclei and recipient oocyte, Biol. Reprod., 37:59.

Smith L. C., and Wilmut L. 1989, Influence of nuclear and cytoplasmic activity on the development in vivo of sheep embryos after nuclear transplantation, Biol. Reprod., 40:1027.

Stice S. L., and Robl J. M. 1988, Nuclear reprogramming in nuclear transplant rabbit embryo, Biol. Reprod., 39:657.

Szöllösi M. S., and Szöllösi D. 1988, 'Blebbing' of the nuclear envelope of mouse zygotes, early embryos and hybrid cells, J. Cell Sci., 91:257.

Tarkowski A. K. 1977, In vitro development of haploid mouse embryos produced by bisection of one-cell fertilized eggs, J. Embryol. exp. Morphol., 38:187.

Willadsen S. M. 1986, Nuclear transplantation in sheep embryos, Nature (Lond), 320:63.

CELL DIFFERENTIATION ACCORDING TO MATURATION OF NUCLEOLAR APPARATUS AND

TOPOGRAPHY OF RIBOSOMAL GENES

T. L. Marshak[1], V. Mares[2], A. A. Karavanov[1],
V. V. Nosikov[3] and V. Ya. Brodsky[1]

[1] N. K. Koltzov Institute of Developmental Biology
USSR Academy of Sciences
Moscow, USSR

[2] Institute of Physiology
Czechoslovak Academy of Sciences
Prague, Czechoslovakia

[3] All-Union Institute for Genetics and Selection of
Industrial Microorganisms
Moscow, USSR

INTRODUCTION

The structure, volume and number of nucleoli reflect the transcription
of rRNA genes, which correlates with the intensity of cell metabolism in
general. Despite the availability of much experimental data concerning the
relation between the organization of the nucleolar apparatus and the rate
of ribosomal gene transcription, this problem still requires further
investigation. In particular, specific features of nucleolar organization
in cells of different transcription rates are practically unknown; there
are no data on the differences in spatial organization of ribosomal genes
between cells of the same type or of cells having different rate of protein
synthesis.

To study these questions we used granule cells (GC) and Purkinje cells
(PC) of rat cerebellum. These cells differ in the rate of protein and RNA
synthesis and in the structure of their nucleoli (Zatsepina and Marshak,
1987).

MATERIALS AND METHODS

We used neurons from the cerebellum of Wistar rats. The number of
nucleoli in PC and GC nuclei was determined in squashes prepared from the
cerebellum of rats of different age (embryonic day 19 (ED 19) – postnatal
day 21 (PD 21) and PD 60) using silver nitrate staining, specific for
nucleolar proteins (Likovsky and Smetana, 1981). For each age group,
nucleoli were counted in 200 – 900 cells of each type. *In situ* hybridization
was performed according to Gall and Pardue (1971). Recombinant DNs (plasmids
pRr 154 and pRr B13) containing genes coding for 18S RNA, transcribed spacer,
28S RNA or 28S RNA only, were used. The specific activity of the probes was

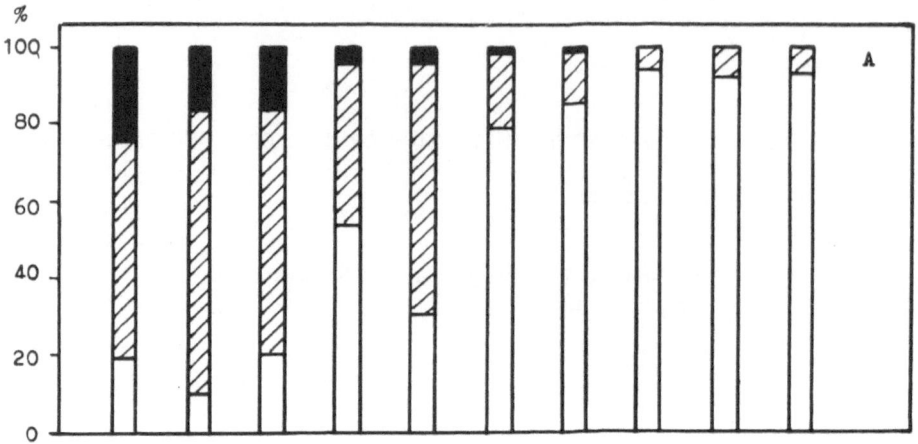

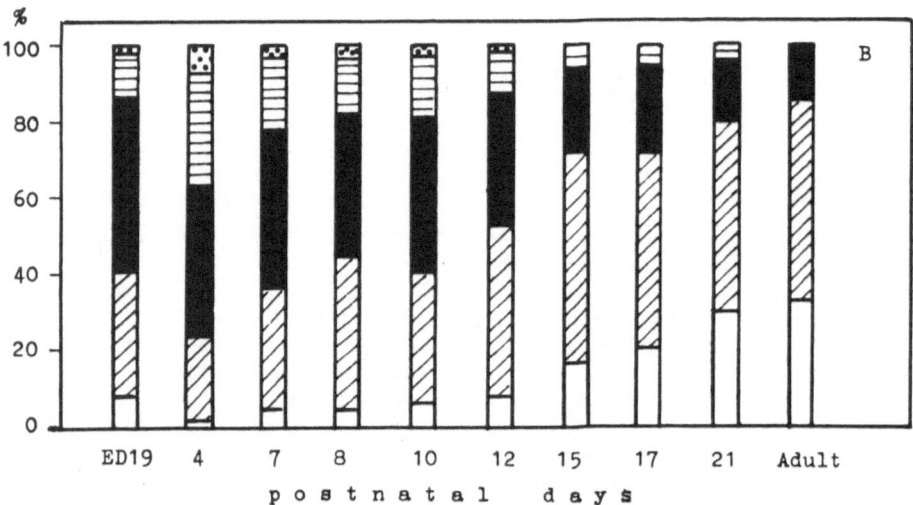

Figure 1.   Ratio of cells with different number of nucleoli in
cerebellum of rats of different age. A - Purkinje cells,
B - granule cells. Abscissa shows the number of cells.
Various cell classes are shown as follows:

☐ - mononucleolar cells   ▨ - binucleolar cells

■ - trinucleolar cells   ▤ - tetranucleolar cells

▦ - pentanucleolar cells

1.5 x 10^6 cpm/µg (pRr 154) and 8 x 10^6 cpm/µg (pRr B13). To reveal unspecific background, a 100 to 200-fold excess of rat rRNA, used as a competitor, was added to the hybridization mixture.

RESULTS AND DISCUSSION

As shown in Figure 1, the number of mononucleolar PC increased during development, and at PD 17 - 21 reached the adult level. Concomitantly, the number of binucleolar cells gradually decreased and approached the adult level also at PD 17 - 21. The kinetics of both processes were non-uniform. On ED 19 a relatively high content of PC nuclei with 3 nucleoli was detected, but already at PD 10 few remained. Single PC nuclei containing 4 nucleoli were present at ED 19, and at later stages they disappeared completely. In animals of all age groups mononucleolar PC nuclei contained a single large nucleolus whose size further increased during development. In some bi- and tri- nucleolar cells the nucleoli were similar and resembled those of mononucleolar cells; other cells, besides containing a large nucleolus, also contained a small one, which was similar to micronucleoli and satellite nucleoli, characteristic of other cell types or to Cajal bodies of nerve cells (Lafarga et al., 1982; Hadjiolov, 1985). By the criterion of nucleolar size, two groups of binucleolar PC nuclei (with 'symmetric' and 'asymmetric' nucleoli) could be distinguished. Similarly, trinucleolar PC nuclei could be classified into three groups - 'symmetric', 'asymmetric type I' (two large and one small nucleoli), and 'asymmetric type II' (one large and two small nucleoli). Changes in the number of multinucleolar PC nuclei are summarized in Table I.

Table I. Changes in ratio between bi- and trinucleolar
Purkinje cells of different types* during
development.

Ade of rats, days	Binucleolar cells		Trinucleolar cells		
	Symmetric	Asymmetric	Symmetric	Asymmetric	
				TypeI	TypeII
ED 19	74,6	25,4	33,3	52,1	14,6
PD 4	4,8	95,2	-	52,0	48,0
PD 7	10,8	89,2	12,1	31,0	56,9
PD 10	3,6	96,4	8,8	23,5	67,7
PD 15	14,1	85,9	-	-	-
PD 21	63,7	36,3	-	-	-
PD 60	13,4	86,6	-	-	-

(*) See text

ED - Embryonic development

PD - Postnatal life

GC nuclei in all age groups contained only small nucleoli, with the number varying from 1 to 5 (Figure 1). In rare cases, up to 6 nucleoli were detected. On ED 19 neurons of this type contained mainly 2 - 3 nucleoli, but on PD 2 neurons with 3 - 4 nucleoli prevailed. During later stages the number of multinucleolar neurons decreased and reached the adult level on PD 21.

Thus, in neurons having different transcription rates, the general organization of nucleolar apparatus was similar, and the number of nucleoli in these cells gradually decreased during development. However, in PC and GC nuclei the rate and the final extent of this decrease are different. Already on ED 19 only single PC nuclei contained 4 nucleoli, and at PD 12 - 15 the cells containing single nucleolus prevailed. On the contrary, more than 50% of GC nuclei contained 3 and more nucleoli on ED 19, and even at PD 21 the number of such cells was still rather high (20%). In adult rats 90% of PC nuclei and only 30 - 40% of GC nuclei contained a single nucleolus. Binucleolar GC nuclei were predominant, and a considerable quantity of trinucleolar GC nuclei were predominant, and considerable quantity of trinucleolar GC nuclei were also detected. Such a diversity may be due to time differences in the neuronal differentiation, since exit from the proliferation cycle has proved to be one of the factors producing a diminishing of the number of nucleoli in the cell (Hadjiolov, 1985).

The presence of PC nuclei with two 'asymmetric' nucleoli, a transient increased number of cells with two 'symmetric' nucleoli at PD 17 - 21 and characteristic changes in number of various types of neurons containing three nucleoli led us to the conclusion that satellite nucleoli appear by dissociation of large nucleoli. It is intriguing as to what happens to the ribosomal genes during this process? An increased volume of PC nucleoli during development is indirect evidence for consolidation of the nucleolar organizers. However, *in situ* hybridization showed that some nucleolar organizers remain dispersed in the karyoplasm. Results of ^{3}H-rDNA-nucleolar DNA hybridization demonstrate that nuclei of adult rat PC and GC show a diffuse label distribution all over the karyoplasms, along with large labelled clusters, whose number corresponds to that of the cell nucleoli. Distribution of silver grains was similar, when either pRr 154 or pRr B13 probes were employed. The excess of unlabelled competitor rRNA added to the hybridization mixture decreased both nucleolar and karyoplasmic labelling. The diffuse labelling indicating the presence of dispersed (inactive) uncleolar organizers was already detected both in PC and GC nuclei on PD 6. In adult rats, the ratio between diffuse label and labelled clusters (corresponding to the nucleoli) varied from cell to cell. Dispersed rRNA genes may be regarded as a 'depot' of temporarely untranscribed genes. Alternatively, these genes may represent the surplus genes not involved in rRNA transcription at all. The variation in number of diffuse silver grains is indirect evidence in favour of the first assumption.

Hybridization *in situ* with ^{3}H-rDNA revealed some features characteristic of cells having an actively transcribing ribosomal genome. Namely, in PC nuclei the distribution of silver grains after *in situ* nuclear DNA-rDNA hybridization is asymmetric, and the number of silver grains in some of the nuclei was 1.5 times higher than in other diploid cells, thus implying rDNA amplification.

REFERENCES

Gall J. G. and Pardue M. L. 1971, Nucleic acid hybridization in cytological preparation, in: 'Methods in Enzymology', L. Groessman, K. Moldave eds., Academic Press, New York, 21:470.

Hadjiolov A. A. 1985, 'The Nucleolus and Ribosome Biogenesis', Springer
    Verlag, New York.
Lafarga M., Crespo D., Villegar J., Berciano M. and Gonzales C. 1982, The
    accessory body of Cajal, Trab. Inst. Cajal., 73:5.
Likovský F. and Smetana K. 1981, Further studies on the cytochemistry of the
    standardized silver staining of interphase nucleoli in smear
    preparation of Yoshida ascitic sarcoma cells in rats, Histochemistry,
    1981, 72:301.
Miale L. and Sidman R. L. 1961, An autoradiographic analysis of histogenesis
    in the mouse cerebellum, Exp Neurol., 4:277.
Zatsepina O. V. and Marshak T. L. 1987, The ultrastructure and number of
    nucleoli in cerebellar nervous cells, Tsitologia, 29:1115.

GENE FAMILIES CODING FOR THE EYE LENS PROTEINS OF CEPHALOPODS

S. I. Tomarev, and R. D. Zinovieva

N. K. Koltzov Institute of Developmental Biology
USSR Academy of Sciences
Moscow, USSR

INTRODUCTION

The lens as neomorph has been formed in at least nine systematic groups of metazoa (De Jong, 1981). It provides a convenient experimental model for the study of mechanisms of organ and tissue formation during evolution. The structural proteins of the vertebrate lens, or crystallins, and genes coding for them have been intensively studied (Wistow and Piatigorsky, 1988). Similar data from other systematic groups are scarce (Siezen and Shaw, 1982; Piatigorsky et al., 1989). This paper deals with characteristics of the major polypeptides of the squid and octopus lenses, obtained by cDNA cloning and sequencing.

METHODS

cDNA libraries form poly(A)$^+$ RNA of the squid *(Ommastrephes sloanei pacificus)* and octopus *(Paractopus defleini)* lenses were obtained (Tomarev and Zinovieva, 1988). From these libraries we isolated recombinant clones showing the strongest hybridization signals with (^{32}P) probes synthesized from squid and octopus poly(A)$^+$ RNAs, respectively.

RESULTS AND DISCUSSION

Eighteen squid clones were selected and at least eight of them coded for homologous polypeptides with calculated M_r values of 23.7kDa (1 clone), 26.2kDa (4 clones), 36.4kDa (2 clones) and 46kDa (1 clone). These polypeptides are probably either glutathione S-transferase (GST) subunits themselves, or share an evolutionary ancestor. Such conclusions follow from the similarity of the identified squid lens polypeptides (SLP) and GST subunits of different species (20 - 30% excluding the central variable regions of SLP, see Figure 1). Moreover, 12 invariant and 24 conservative residues characteristic of GST subunits sequences (Rhoads et al., 1987) were usually present in the SLP sequences; the content of α-helical and β-sheet structures in SLP and GST were also similar (Mannervik and Danielson, 1988). Three known N-terminal sequences of the SLP (25 amino acid residues) are 85 - 96% homologous with the sequence of the major lens protein of the squid *Nototodarus gouldi* (Siezen and Shaw, 1982). According to our results and the published data the apparent M_r's of the major SLPs

Nuclear Structure and Function, Edited by J. R. Harris and
I. B. Zbarsky, Plenum Press, New York, 1990

```
                    *                                                                    *  *
1  MPKYTLYFNSRGRAEICRMLFAAANIPYNDVRIDY----------SEW--DIYRSKMPG--SC--LPVLEINDSIQIPQTMAIARYLARQF
2                        N  GVQ T K FEF----------N --  K ND  S--M---V  D DGQNKM E        EN
3                        S  G       EL----------A --TQFKT     C--HM-- I    DTET V     S     E
6  MSG  PV H   A   M CI W L  GVFFEEKL QSP------EDL--EKLKKDGNL-MFDQW MV    DGM-KLA  R  LN I TKY
7   P  IV   PV   C AT L DQGQSWKEEVTI-------DV --LQGSL STCLYGQ-- KF --DGDITLY SN     L H G SL

                                                                                          *
1  GFYGKHHLDMARVDFICDSFYDIFNDYMRMYHDQKGRVMFELMSQMREWYAARNENSG YEECYMQPSMA PSAQMSQEVDNSDTLADCSEMR
2  Y  NNM  FI Y  C E LH    YF TKN  F-------
3     NNM  FK  CL   LFEL   -----
4    +C E VD  L LF  KE  I YDQ  D  NR  IDMDGRMTFGTVGGYAVQSRGDGGYYVKSRG GGYPVQGRG
6  DL  DMKER LI MYSEGIL LTEMII------
7  L  DQKEA L  MVN GVE LRCK G------

1  SQDSMVEPPSQKLSPELESQSSLCSERPQCGPPDP-----
3  ----------
4  DTGYSSQTRSDDACLGQGRGEVDTGMSYDASTGVCTDINRGDMSSDINSGLYSGGRMDDSCHTSESRRMDDPCGTDESRRLDVPCHSDDHY
6  ----------
7  ----------

                                                                                          *
1                   MMGSDFERLSFNE-GR--MLEM-RRRYDETCRRVLPFLEGTLKQRYGGDRYFMGEYMTM
2                     Q  GTDMSPDMDPTQ-- TSYIQN LD LIS  R EM N  KEF  DQ ML
3                       --AVYNEKDAAK--KT L-QK FQN  L   YM K  EANK  AGW I DQILL
4  RSDNPCTDDSCQAEDRRGHGHSDSHRIDISSEESASRRSRNHA A -  S -- L  M  K   MR  EMQH  HI C DE C
5                          + GSFSK  -  V -- Y R--     M  R   MR EMHHN  NQ I DQ
6                             QLVICPPDQ EAKTALA-K-- R KN Y   AF KV  SH- Q - LV NRL R
7                           TLIYT YENG--KDDY-VKALPGH-LK--     -  TL S NQ  KNFIV NQISF

                   *
1  CDIMCYCALENPLLDNAYLLHPYPKLRGLRDRVSRNQRINSYFTLRNYTDF
2  M  C   M EDQTTFNNF  MS WK  ASHPK TP LKK  N NW
3  T      IQE N KE   AA  T  AAHPK AA EKK  N A
4  V   A  S MQE PS  SN    E  ASQIN SQ IKR YPS
5  V IHLLEL LYVEEFD S   M S +
6  V IHLLEL LYVEEFD S  TSF L KAFKS  I SLPNVKKPLQPGSQRK PAMDAKQIEEARKVFKF
7  A YNLLDL LVHQVLAPGS DNF L SAYVA L ARPK KAFLSSPDHLNRPINGNGKQ
```

Figure 1. Comparison of amino-acid sequences of the SLP (1 - 36.4kDa; 2 - 26.2kDa; 3 - 23.7kDa; 4 - 46kDa) and OLP (5) with the sequences of rat Y_a (6) and Y_b (7) GST subunits. Sequence (1) is written in full; for other sequences only differing amino-acid residues are shown. An asterisk * indicates invariant residues (Rhoads et al., 1987); + shows beginning and end corresponding partial sequences.

are in the range of 27 - 30kDa. We believe that the 26.2kDa polypeptide is the most abundant while the 23.7kDa, 36.4kDa and 46kDa polypeptides are present in smaller amounts (Tomarev and Zinovieva, 1988). The two latter SLP are probably responsible for the formation of high molecular weight aggregates of the squid lens proteins, since their central portions are acidic, while the major SLP contain a basic polypeptide. The coding potential of other clones that we have obtained is still unknown, but available data suggest that they do not code for the major SLP.

At least nine of the 42 selected octopus clones coded for the polypeptide homologous with intermediate filament proteins and lamins. Our present data on the polypeptide structure are not sufficiently conclusive to determine precisely which type of intermediate filament protein the octopus lens polypeptide (OLP) belongs. Clones coding for GST-related polypeptides are not abundant in our library. Only one clone homologous with the major SLP and GST subunits has been identified (Figure 1). We also isolated clones coding for α- and β-tubulins of the octopus. The known part of the octopus α-tubulin amino acid sequence (positions 212-415) is highly homologous with corresponding parts of mammalian and *Drosophila* α-tubulins (90 - 92%). Using immuno-dot titration we have shown that the tubulin content of the octopus lens is by one order of magnitude higher than that of the squid lens and at least two order of magnitude higher than that of the frog lens. Therefore, tubulins and intermediate filament proteins are much more abundant in the octopus lens than in the squid or frog lenses. We have also characterized 3 octopus clones coding for a polypeptide homologous with ferritin.

The SLP and OLP are not homologous with vertebrate crystallins. However, we believe that the strategy of using pre-existing molecules as structural lens proteins in the course of lens formation during evolution is characteristic of cephalopods as well as vertebrates. Protein molecules might be selected as structural lens proteins by criteria such as high thermodynamic stability and ability to accumulate to high concentrations without causing opacification. The formation of the squid lens seems to occur mainly be recruitment of one class of protein molecules (GST subunits), but in the octopus lens other proteins (tubulins, intermediate filament proteins, ferritin) might have been recruited as well. We believe that the presence of various classes of major proteins is not crucial for the formation of a transparent lens.

REFERENCES

De Jong W. W., 1981, Evolution of lens and crystallins. In: 'Molecular and Cellular Biology of the Eye Lens', H. Bloemendal, ed., Wiley, New York.
Mannervik B. and Danielson U. H. 1988, Glutathione transferases - structure and catalytic activity, CRC Critical Rev. Biochem., 23:283.
Piatigorsky J., Horwitz J., Kuwabara T. and Cutress C. E. 1989, The cellular eye lens and crystallins of cubomedusan jellyfish, J. Comp. Physiol. A, 164:577.
Rhoads D. M., Zarlengo R. P. and Tu C.-P. 1987, The basic glutathione S-transferases from human livers are products of separate genes, Biochem. Biophys. Res. Commun., 145:474.
Siezen R. J. and Shaw D. C. 1982, Physicochemical characterization of lens proteins of the squid *Nototodarus gouldi* and comparison with vertebrate crystallins, Biochim. Biophys. Acta, 704:304.
Tomarev S. I. and Zinovieva R. D. 1988, Squid major lens polypeptides are homologous to glutathione S-transferases subunits, Nature, 336:86.
Wistow G. J. and Piatigorsky J. 1988, Lens crystallins: The evolution and expression of proteins for a highly specialized tissue, Ann. Rev. Biochem., 57:479.

LOACH *(MISGURNUS FOSSILIS)* OOCYTE 5S rRNA GENES: HETEROGENEITY OF PRIMARY
STRUCTURE AND LOCATION OF A TRANSCRIPTION STIMULATORY SIGNAL IN THEIR
UPSTREAM SPACER

M. Timofeeva, P. Felgenhauer*, J. Sedman[+], N. Shostak,
N. Kupriyanova, G. Posmogova, A. Lind[+] and A. Bayev

V. Engelhardt Institute of Molecular Biology
*N. Koltsov Institute of Developmental Biology USSR
Academy of Sciences, Moscow, USSR
[+]Biocentre, Estonian Academy of Sciences, Tartu, USSR

INTRODUCTION

The genes coding for ribosomal 5S rRNAs are a suitable object to study
the following aspects of gene expression and regulation:
1)   the regulation and function of the RNA polymerase III-system,
2)   the mechanism of the different ribosomal genes which coordinate
transcription,
3)   the maintanance of a distinct level of rRNA synthesis in different
tissues, in different physiological states, during hormonal induction and in
different stages of development. We used the oocyte 5S rRNA genes of
Misgurnus fossilis in the analysis mechanism of developmental control of
transcriptional activity. The 5S rRNA genes in *Misgurnus fossilis* genome,
as in *Xenopus,* are represented by thousands of copies (Shostak et al., 1984).
Most of them are expressed only during oogenesis and code for oocyte 5S
rRNA which differ from somatic 5S rRNA by primary (Mashkova et al., 1981) and
secondary structures (Serenkova et al., 1984). The oocyte 5S rRNA genes
were cloned in pBR322. In this work we have shown the heterogeneity of
genome copies of oocyte 5S rRNA genes and the existence in upstream spacer
of the specific cis-signal that influences their transcriptional activity.
The role of this signal in developmental regulation of gene expression is
discussed.

RESULTS AND DISCUSSION

The nucleotide sequence of *Misgurnus* cloned 5S rRNA gene copies is
presented in Figure 1. One can see that there are a certain number of
substitutions, not only in spacer but also in the structural region of the
gene. Some of these substitutions are localized inside the internal control
region of gene (ICR), which interacts with the transcriptional factors and
RNA polymerase III. The analysis of transcriptional activity of these
clones by microinjection of DNA into *Misgurnus* oocyte nuclei is shown in
Figure 2. The results indicate that the synthesis of 5S rRNA is active with
the injection of clones 202, 206, 208, 209, 309, 314, lower with the clones
207 and 210 and substantially lower with the clone 317. The lowest activity

a

```
        -60        -50        -40        -30        -20        -10
AAGCTTTCGTTGATGTTTACGCCGTTCTCGCATTAGT-CTTGAAGTGGCTCTCGAAACAAGCCAACACTCgcttacgg...............
             GA  C AT                              TG T GT       ...............
             G   C A                                T T  A            t      ...............
        -C       T    C A                         A T   T T G        ...............
        C        T    C A                           T    T G          ...............
      A C      GT   C AA                             T G TT GT A T    ...............
             GA C A                                 TG T G  A         ...............
             G    TC A                               T T  AT          ...............
             GA  C A                                TG T G A   c      ...............
        T    GA  C A                                TG T                ...............
        T    GA  C AT                               TG T GT A   c     ...............
        T       A G    C A       T                CT TG T            ...............
```

b

```
gcuuacggccacaccaaccugagcaagcccgaucucgucugaucucggaagccaagcagguuugggccugguuaguacuuggaugggagacugccugggaauaccaggugguguuaagcuu oocyte 5SRNA
         u   c    c          c      u      c                                                                                  somatic
```

```
       10        20        30        40        50        60        70        80        90       100       110
GCTTACGGCCACACCACCCTGAGCACGCCCGCTCTCGTCTGATCTCGGAAGCCAAGCAGGGTCGGGCCTGGTTAGTACTTGGATGGGAGACCGCCTGGGAATACCAGGTGCTGTAAGC--pHf5S 209
           C    C    C                       C A            A C CT         C    -- 202
      T    C A  C    C           T           T              C              C    -- 206
           A    C    C             T         C          A   T  A           C    -- 207
           C    C    C  A                    C          A   T  A           C    -- 208
           A    C    C                      T C             T              C    -- 210
           C    C    C                       C C                           C    -- 309
           C    C    C                       C    A A      TT  C           A    -- 317
   C       C    C    C                       C A    T C    C   A           C       314.1
           C    C    C   C T                 C    A A      TT  C           A    -- 314.2
   C       C    C    C                       C A    T C    C   A           C       205.1
       A C      C    C   C T     G           C    A A      TT  C           A    -- 205.2
```

c

```
                                                         '         <-----primer------------->
ATGCTTABGGCCACACCACCCTGAGCACGCCCGCTCTCGTCTGATCTCGGAAGCCAAGCAGGGTTGGGCCTGGTTAGTACTTGGATGGGAGACCGCCTAAGAATATCAGGTGTTGTAAGCA  p5SC1
ATGCTTABGGCCACACCACCCTGAGCACGCCCTCTCTCGTCTGATCTCGGAAGCCAAGCAGGGTCGGGCCTGGTTAGTACTTGGATGGGAGACCGC......................   2
GCTTABGGCCACACCBCCCGCTCTCGTCTGATTTCGGAAGCCAAGCAGGGTCGGGCCTGGTTAGTACTTGAATGGGAGACCGC......................              3
                    CTGGTTAGTACTTGGATGGGAGACTGC......................                                              5
TGCTT------ACACCACCATGAGCATGCCCGCTCTCGTCTGATCTCGGAAGCCAAGCAGGGTTGGGCCTGGTTAGTACTTGGATGGGAGACCTC......................   6
TGCTT------ACACCACCATGAGCATGCCCGCTCTCGTCTGATCTCGGAAGCCAAGCAGGGTTGGGCCTGGTTAGTACTTGGATGGGAGACCTC......................   7
                    CCGCTCTCGTCTGATCTCGGAAGCCAAGCAGGGTCGGGCCTGGTTAGTACTTGGATGGGAGACCGC......................          8
CCACACCACCCTGAGCACGCCCGCTCTCATCTGCTCTCGGAAGCCAAGCAGGGTCGGGCCTGGTTAGTACTTGGATGGGAGACCTC......................             10
TGCTT------ACAAACA-CCTGAGCACGCCCGCTCTCGTCTGATCTCGGAAGCCAAGCAGGGTCGGGCCTGGTTAGTACTTGGATGGGAGACCGC......................  11
                    CCCGCTCTCGTCTGATCTCGGAAGCCAAGCAGGGTCGGGCCTGGTTAGTACTTGGATGGGAGACTGC......................          13
TGCTT------ACACCACCCTGAGCACGCCCGCTCTCGTCTGATCTCGGAAGCCAAGCAGGGTCGGGCCTGGTTAGTACTTGAATGGGAGACCGC......................  14
                    CAGGGTCGGGCCTGGTTAGTACTTGGATGGGAGACCGC......................                                  15
CACCACCCTGAGCACGCCCGCTCTCGTCTGATCTCGGAAGCCAAGCAGGGTCGGGCCTGGTTAGTACTTGGATGGGAGACCTC......................               16
CAAACCCTGAGCACGCCCGCTCTCATCTGATCTTGGAAGCCAAGCAGGGTCGGGCCTGGTTAGTACTTGGATGGGAGACCGC......................               17
CACACCACCCTGAGCACGCCCGCCCTCGTCTGATCTCGGAAGCCAAGCAGGGTCGGGCCTGGTTAGTACTTGGATGGGAGACCGC......................           18
                    TGCCCGCTCTCGTCTGATCTCGGAAGCCAAGCAGGGTCGGGCCTGGTTAGTACTTGGATGGGAGACCGC......................      19
                    AGCAGGGTCGGGCCTGGTTAGTACTTGGATGGGAGACTGC......................                              20
ATGCTTABGGCCACACCACCCTGAGCACGCCCGCTCTCGTCTGATCTCGGAAGCCAAGCAGGGTTGGGCCTGGTTAGTACTTGGATGGGAGACCGC......................  21
           ACCCTGAGCACGCCCGCTCTCGTCTGATCTCGGAAGCCAAGCGGGTCGGGCCTGGTTAGTACTTGGATGGGAGACTGC......................       22
TGCTT------ACACCACCCTGAGCACGCCCGCTTTCGACTGATCTCGGAAGCCAAGCAGGGTCGGGCCTGGTTAGTACTTGGATGGGAGACCGC......................  23
TGCTT------ACACCACCCTGAGCACGCCCGCTCTCGTCTGATCTCGGAAGCCAAGCAGGGTCGGGCCTGGTTAGTACTTGGATGGGAGACCGC......................  24
         ACCAACCTGAGCACGCCCGCTCTCGTCTGATCTCGGAAGCCAAGCAGGGTCGGGCCTGGTTAGTACTTGGATGGGAGACCGC......................       25
TGCTTABGGCCATACCACCCTGAGCACGCCCATCTCGTCCGATCTCGGAAGCTAAGCAGGGTCGGGCCTGGTTAGTACTTGAATGGGAGACCGC......................  26
       10        20        30        40        50        60        70        80        90       100       110
```

Figure 1.   The sequences of the spacer (a) and structural region (b) of oocyte 5S rRNA genes and cDNA (c) of oocyte 5S rRNA of *Misgurnus fossilis*.

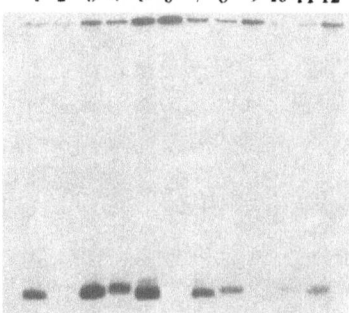

Figure 2.   The transcription in oocyte nuclei of pMf series of clones of
             5S RNA gene copies: 202 (lane 1), 206 (lane 3), 207 (lane 4),
             208 (lane 5), 209 (lane 7), 210 (lane 8), 309 (lane 10), 314
             (lane 11), 317 (lane 12); plasmid pBR322 as a control (lanes
             2, 6, 9). Autoradiographs of polyacrylamide gel analysis of
             transcriptional products.

coincides with the specific localization of base substitutions in the box C
of ICR interacting with TFIII A. We concluded that the oocyte 5S rRNA genes
are represented in the *Misgurnus* genome as multigene family with a heterogeneity in the primary structure. This was confirmed by primary structure
analysis of cloned cDNA of *Misgurnus* oocyte 5S rRNA (Figure 1c). The
regulation of transcriptional activity of different gene variants is
realized through the differences in their primary structure. Some of the
gene copies (no less than 20%) appear to be pseudogenes with very low
transcriptional activity, due to substitutions in box C of ICR. Earlier we
have shown (Shostak et al., 1984) that 5S rRNA genes are organized in the
Misgurnus genome as in *Xenopus* and many other animals and in plants, into a
cluster of repeats. We propose that the pseudogenes are interspersed with
the normal genes in the composition of cluster and may play some role in
regulation of the expression of normal 5S rRNA gene copies. These pseudogenes may form a complex with transcriptional factors and thus support the
active state of chromatin in all 5S rRNA gene clusters. We consider that
the active state of chromatin is necessary to support the high level of 5S
rRNA transcription during oogenesis.

     We proposed earlier that the spacer region ajacent to the trans-
criptional start point of the oocyte 5S rRNA gene can be involved in the
transcription modulation. It can be seen from Figure 1a that the base
substitutions are localized in a spacer close to the 5'-end of the structural
region. An upstream spacer signal, that influences the transcriptional
activity, has been described for 5S rRNA genes of *Bombyx mori* (Morton and
Sprague, 1984), *Dropophila* (Garcia et al., 1987), *Neurospora* (Selker et al.,
1986) and *Xenopus* somatic 5S rRNA genes (Reynolds and Azer, 1988). In order
to investigate our proposition on the role of the upstream spacer in
expression of 5S rRNA genes, we obtained mutants with deletion from the 5'-
end of the spacer to nt -34, -18, -8 and deletion in structural region to
nt +2, +4, +6, +9, +47, +61. The transcriptional activity of these mutants
was compared with the wild type gene in an oocyte microinjection assay. It
can be seen from Figure 3 that the deletion to nt -18 did not change the
transcriptional activity. The deletion to nt -8 and then to nt +2, +4, +6,
+9, drastically decreases and the deletion to nt +47 and +61 eliminates the
activity. In *Xenopus* oocytes similar results were obtained. The difference
is that the deletion to -8 hardly decreases transcriptional activity. In
the control experiments, oocytes were coinjected with deletion mutant or
wild type gene 5S DNA and DNA of tRNA gene. It was shown that the synthesis

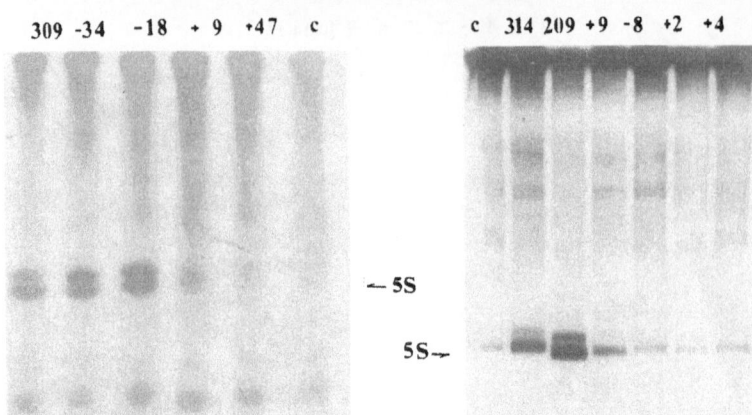

Figure 3.   The comparison of transcriptional activity of wild genes (pMf 209, 309, 314) and deletion mutants (-34, -18, -8, +2, +4, +9, +47). Autoradiographs of polyacrylamide gel analysis of transcriptional products.

of tRNA takes place in all cases, but the 5S rRNA synthesis strongly decreases in mutants +2 and +9, and is absent in mutant +61, due to damage of the gene ICR.

A sequence in upstream spacer of oocyte 5S rRNA genes therefore, have a signal that influences the transcriptional activity. The signal is located from nt -18 downstream to the structural region. The discovery of cis-acting signal in the spstream spacer of somatic (Reynolds and Azer, 1988) and oocyte 5S rRNA genes led us to conclude that these signal sequences could be involved in the developmental control of 5S rRNA gene expression in amphibians and fish. It would be logical to suppose that the expression of oocyte and somatic 5S rRNA genes is regulated by two types specific trans-acting factors. One of them, an oocyte transcriptional factor, interacts with the spacer of oocyte 5S rRNA genes. We propose that this factor is compartmentalized in oocyte nuclei and so could be designated as the nuclear oocyte transcriptional factor (NTFO). We further propose, that in maturing oocytes NTFO is inactivated after breakdown of the nucleus-germinal vesicle, when the nucleoplasm interacts with the cytoplasm of the oocytes.

REFERENCES

Garcia A. D., O'Connel, A. M. and Sharp S. J. 1987, Formation of an active transcription complex in the *Drosophila melanogaster* 5S rRNA gene is depentant on an upstream region, Mol. Cell. Biol., 7:2046.
Mashkova T. D., Serenkova T. I., Mazo A. M., Avodina T. A., Timofeeva M. J. and Kisselev L. L. 1981, The primary structure of oocyte and somatic 5S RNAs from the loach *Misgurnus fossilis,* Nucleic Acids Res., 9:2141.
Morton D. G. and Sprague K. V. 1984, In vitro transcription of a silkworm 5S RNA gene required an upstream signal, Proc. Natl. Acad. Sci. USA, 81:5519.
Reynolds W. F., and Azer K. 1988, Sequence differences upstream of the promoters are involved in the differential expression of the *Xenopus* somatic and oocyte 5S RNA genes, Nucleic Acids Res. 16:3391.
Selker E. U., Morzycka-Wroblewska E., Stevens J. N. and Metzenberg R. L. 1986, An upstream signal is required for *in vitro* transcription of *Neurospora* 5S RNA genes, Mol. Gen. Genet., 205:189.

Serenkova T. I., Mazo A. M., Mashkova T. D., Toots I., Nigul A., Timofeeva M. J. and Kisselev L. L. 1984, The secondary structure of oocyte and somatic 5S ribosomal RNAs of the fish *Misgurnus fossilis L.* from nuclease hydrolyses and chemical modification data, <u>Nucl. Acids Res.</u>, 12:5385.

Shostak N. G., Kupriyanova N. S., Serenkova T. I., Timofeeva M. J. and Bayev A. A. 1984, Organisation of genes coding 5S rRNA in the loach *(Misgurnus fossilis L.)*, <u>Molek. Biol. (Russ.)</u> 18:1352.

NUCLEOLAR DNA DISTRIBUTION AND ITS CONSEQUENCES FOR THE INTERPRETATION OF NUCLEOLAR COMPONENTS

E. Gwyn Jordan and David J. Rawlins*

Department of Biophysics
Cell and Molecular Biology
Division of Biomolecular Sciences
King's College London
Campden Hill, London   W8 7AH
United Kingdom

* Institute of Plant Science Research
John Innes Institute
Colney Lane, Norwich   NR4   7UH
United Kingdom

The most immediate question that must be answered to further our understanding of nucleoli is; "In which component are the actively transcribing genes?" The conclusion reached from the early studies employing the spreading technique and also autoradiography was that the active genes were in the dense fibrillar component. This interpretation has been seriously challenged by the newer approaches of EM immuno-gold localization (Scheer and Rose, 1984; Thiry et al. 1988). These later studies have shown the presence of polymerase I and DNA in the fibrillar centres but have failed to show them in the dense fibrillar component. It has been argued from this that there is no DNA or transcription in the dense fibrillar component.

A more recent study by Raska et al (1989) using the same type of EM immuno-gold localization has confirmed the presence of high levels of DNA and polymerase I in fibrillar centres. However this paper also reported, in very active cells, low levels of DNA and polymerase I in the dense fibrillar component. Derenzini et al. (1987) who employed the EM feulgen like osmiumammine reaction also report finding DNA in both these components of nucleoli.

In view of the uncertainties raised by some of the gold-labelling and cytochemistry it seems that other techniques should be explored and the limits of detection and sensitivity of the EM immuno-gold and other recent cytochemical procedures examined.

The most convincing evidence may yet come from work like Lamb and Daneholt's (1979) on the genes of the *Chironomus* Balbiani rings where the position of active transcription units was satisfactorily demonstrated in thin sections. Such pictures have eluded nucleolar workers so far and we are left to draw conclusions from less direct approaches. One of these is the use of the light microscope with the benefits of the recent image processing techniques.

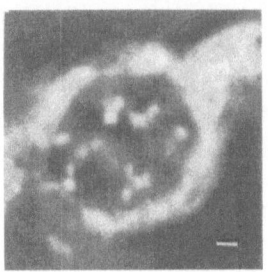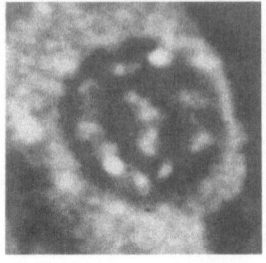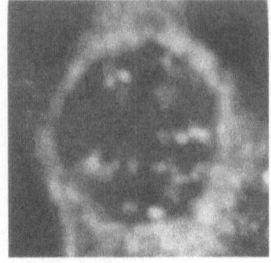

Figure 1.   Optical sections after computer de-blurring from three
             different nuclei of *Spirogyra grevilleana* with DAPI
             fluorescence showing intranucleolar DNA.   Scale 1μm.

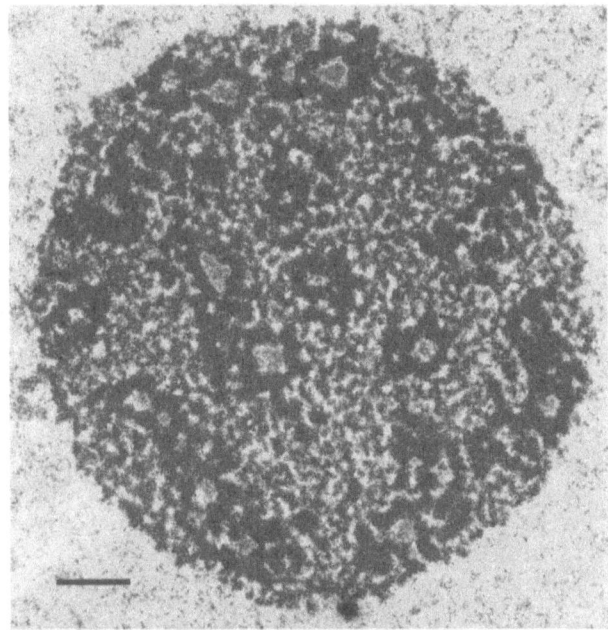

Figure 2.   Conventional 0.1μm section through a nucleolus of *Spirogyra
             grevilleana*.   The most lightly stained regions are fibrillar
             centres and their surrounding dense sheaths, the dense
             fibrillar component.   The rest of this nucleolus is accounted
             for by granular component.   The granular component has a
             spongy, reticulate or nucleolenemal form with many
             interstices.   Scale 1μm.

     We have investigated the 3D arrangement of nucleolar DNA in *Spirogyra*
using these techniques.   DNA was stained with DAPI and 0.5μm optical
sections collected using an ISIT video camera coupled to a fluorescence
microscope and a computer-controlled framestore (Rawlins and Shaw, 1988).
The nucleolus appears in these 0.5μm optical sections as a dark circle
within the bright fluorescence of the nuclei, Figure 1.   Within the
nucleoli small areas of fluorescing DNA can be seen as spots or small rows
of spots like beads on a string.   By comparison with electron micrographs
of the same species *(Spirogyra grevilleana)* it becomes clear that the
fluorescence arises principally from the fibrillar centres of the nucleolus,

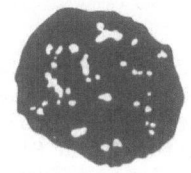

Figure 3.   A compilation of the images of fibrillar centres from five
             0.1μm serial electron micrographs prepared by tracing their
             outlines and blacking in all other components of the nucle-
             olus.   Scale 1μm.

Figure 2.   To obtain an image of fibrillar centres from electron micrographs that could be compared effectively with the light microscope optical
sections (thin sections are only 0.1μm while the deblurred optical sections
are 0.5μm thick) we added together the images of fibrillar centres from
five serial electron micrographs, Figure 3.   These are shown as white
regions in the figure and correspond well with the fluorescent images.   The
DNA structures look larger and less well defined in the fluorescent images
than in the electron micrographs but this may simply reflect the
differences inherent between the resolutions of the two procedures.   A final
decision on the precise extent of the fluorescence, especially how much
should be attributable to the dense fibrillar component that closely
ensheathes the fibrillar centre must await careful measurements on cells
with more uniform fibrillar centres but provisionally it seems that the
bright spots correspond fairly well with fibrillar centres.

      As well as the bright spots, we have also seen a faint fluorescence
arising from regions between and around the fibrillar centres (Figure 1).
But it is not possible to conclude with certainty that this weak fluorescence in regions that correspond to the dense fibrillar components accounts
for much of the nucleolar DNA.

      When it is recalled that DAPI staining can show up the dispersed DNA
of cell organelles as bright fluorescent spots, as we have seen in the
chloroplasts of the same cells, it becomes unnecessary to interpret the
bright fluorescent spots of the nucleoli as transcriptionally inert semi-
condensed chromatin and makes it more likely that the DNA of the fibrillar
centres is the transcriptionally active DNA.   A further argument
strengthens this interpretation.   The total amount of DNA revealed when the
number of fibrillar centres present in a whole nucleolus is taken into
consideration seems quite sufficient to account for the transcriptional
activity of a nucleolus (Jordan, 1987).

      This work fits more closely with the view that all transcription occurs
within fibrillar centres or at their surfaces than with earlier models for
the nucleolus.   Unless we are able to demonstrate higher levels of DNA in
the dense fibrillar component we have to conclude that we have either
failed to reveal all the DNA that resides there or that there is only
sufficient in these regions to account for the low level of DNA that must
be required to join the transcription sites together.   Finally, we interpret
our data as confirmation of the conclusions of Scheer and Rose (1984) and
Thiry et al. (1988) that the active genes reside within fibrillar centres
and that the evidence for DNA within the dense fibrillar component is
insufficient for us to conclude that it accounts for a major part of
nucleolar transcription.   It is clearly important to quantify the amount of
DNA present within the fibrillar centres as a confirmation of these
conclusions.

ACKNOWLEDGEMENT

    We thank Drs. D. A. Agard and J. W. Sedat (University of California, San Francisco) and P. J. Shaw (J.I.I.) for generously providing us with their computer programs for deblurring and projection of optical sections. D. R. was supported by the Agricultural and Food Research Council via a grant-in-aid to the John Innes Institute. Support was also received from the Gatsby Foundation.

REFERENCES

Derenzini M., Hernandez-Verdun D., Farabegoli F., Pession A. and Novello F. (1987). Structure of ribosomal genes of mammalian cells in situ. Chromosoma 95:63.
Jordan E. G. (1987). Nucleolar organizers in plants. In Chromosomes Today V. 9. Eds., Stahl. A., Luciani J. M., Vagner-Capadano A. M., Allen and Unwin, London. 272-283.
Lamb M. M. and Daneholt B. (1979). Characterization of active transcription units in Balbiani rings of Chironomus tentans. Cell 17:835.
Raska I., Reimer G., Jarnik M., Kostrouch Z. and Raska K. (1989). Does the synthesis of ribosomal RNA take place within nucleolar fibrillar centres or the dense fibrillar components? Biology of the Cell 65:79.
Rawlins D. J. and Shaw P. J. (1988). Three-dimensional organization of chromosomes of Crepis capillaris by optical tomography. J. Cell Sci. 91:401.
Thiry M., Scheer U. and Goessens G. (1988). Immunoelectron microscopic study of nucleolar DNA during mitosis in Ehrlich .tumor cells. E. J. Cell Biol. 47:346.
Scheer U. and Rose K. (1984). Localization of RNA polymerase I in interphase cells and mitotic chromosomes by light and electron microscopic immunocytochemistry. Proc. Nat. Acad. Sci. USA. 81:1431.

EFFECTS OF TOPOISOMERASE I INHIBITION ON NUCLEOLAR STRUCTURE AND FUNCTION

F. Novello, F. Farabegoli, M. Govoni and M. Derenzini

Dipartimento di Patologia Sperimentale
Via S. Giacomo 14
Bologna
I-40126
Italy

INTRODUCTION

Using the Feulgen-like osmium-ammine reaction as a selective staining method for DNA in thin sections, bi-dimensional and stereo-pair micrographs have shown that compact chromatin is composed of roundish units, which represent the nucleosomal organization of chromatin *in situ*. Nucleosome-like particles have never been visualized in ribosomal chromatin of interphasic NORs, which appears to be composed of numerous, interwoven, long filaments with a thickness of about 3nm. Ribosomal chromatin, present in the fibrillar components of the nucleoli was always in an extended non-nucleosomal configuration, independent of the transcriptional activity and the phase of the cell cycle (Derenzini et al., 1981; 1983; 1987; Hernandez-Verdun and Derenzini, 1983). Recently, there has been considerable interest in a group of enzymes, the DNA topoisomerases, which have the ability to convert one topological isomer of DNA to another. *In vivo* a possible role of topoisomerase I in ribosomal gene transcription has been suggested by studies of inhibition of rRNA synthesis by camptothecin, an inhibitor of topisomerase I activity (Zhang et al., 1988) and by the observation that topoisomerase I and RNA polymerase I have been immuno-cytochemically localized *in-situ* on the fibrillar components of nucleoli (Rose et al., 1988; Raska et al., 1989). These results led us to investigate the effects of camptothecin (National Cancer Institute, U.S.A.), a specific inhibitor of topoisomerase I, on the structural and functional organization of ribosomal chromatin in TG cells, a human tumor cell line.

RESULTS AND DISCUSSION

The biochemical consequences of camptothecin administration are summarized in Table I: it caused a reduction in the topoisomerase I activity as measured in cell extracts and a marked increase in breaks in DNA strands in treated cells compared with controls. Data on *in vivo* RNA synthesis (Table II) showed a marked reduction of labelled uridine into total RNA, while the inhibition was about 90% when ribosomal RNA synthesis was analyzed 180 min after camptothecin treatment. In thin sections of control TG cells stained with uranium and lead (Figure 1), small, roundish fibrillar centers were present (arrows). The dense fibrillar component appeared as a rim

Table I.   Effect of 25μM Camptothecin on the Activity of
            Topoisomerase I and the Percentage of Double
            Stranded DNA of TG Cells.

	TOPOISOMERASE I ACTIVITY*		% DOUBLE STRANDED DNA†
CONTROL	2.58	--	55
CAMPTOTHECIN 90'	1.70	66	5
CAMPTOTHECIN 180'	1.10	42	17

*Preparation of cells extracts and determination of enzyme activity
have been performed according to Tricoli et al.(1985).
†Percent of double strand DNA remaining 15 min. after partial alkali
treatment in fluorimetric analysis of DNA unwinding (F.A.D.U.)
according to Birnboim and Jevcak (1981).

Table II.   Effect of 25μM Camptothecin on RNA Synthesis of
            TG Cells.

	NUCLEAR RNA		50° RNA FRACTION*	
	3H Uridine incorporation	% control	3H Uridine incorporation	% control
CONTROL	0.13	--	0.29	--
CAMPTOTHECIN 90'	0.09	70	0.15	52
CAMPTOTHECIN 180'	0.05	39	0.03	10

*^{3}H Uridine (5 μC/ml) was added for the last 15' of culture. Nuclear
RNA was fractionated with phenol at different temperatures. The RNA
fraction, designed 50°C RNA,corresponded to nucleolar RNA (Derenzini et
al. 1983). Results are expressed DPM/ng RNA/Acid soluble fraction
(DPM/ng DNA).

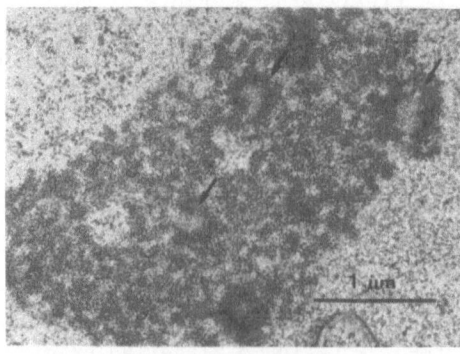

Figure 1.   Control TG cell. Nucleolus with three fibrillar centers
            (arrows) surrounded by the dense fibrillar component and
            granules. Paraformaldehyde-1% OsO₄ fixation. Uranium and
            lead staining.

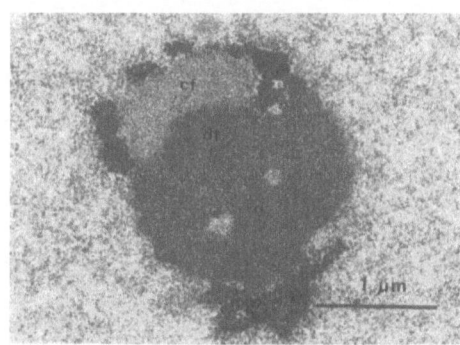

Figure 2.   TG cell 180min after 25µM camptothecin treatment. The
             nucleolar components are segregated into three well defined
             portions: clear fibrillar (CF), dense fibrillar (DF) and
             granular (G). 4% paraformaldehyde-1% OsO₄ fixation. Uranium
             and lead staining.

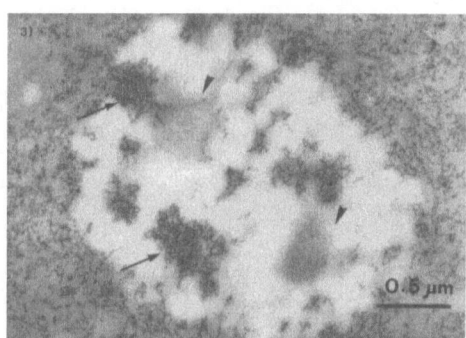 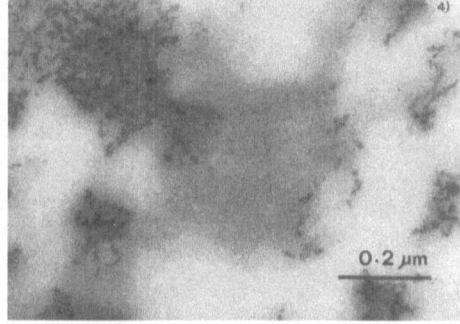

Figures 3   TG cell stained with the Feulgen-like osmium-ammine reaction
and 4.   for DNA 90min after 25µM camptothecin treatment. 4%
             paraformaldehyde fixation. Figure 3: the intranucleolar
             chromatin is composed of condensed (arrows) and loosened
             (arrowheads) agglomerates. Figure 4: at higher magnification
             thin filaments of extended DNA can be clearly seen.

around the fibrillar centers and was surrounded by the granular component.
After 90 or 180min treatment with camptothecin the morphological pattern of
the nucleoli of TG cells was profoundly changed (Figure 2): a segregation
of the ribonucleoprotein components occured similar to that observed after
actinomycin D treatment. The nucleolar components were visible in three
well defined zones consisting of granular (G), dense fibrillar (DF) and
clear fibrillar (CF) portion. In Figure 3 a TG nucleolus 90min after
camptothecin treatment is shown, in a thin section selectively stained for
DNA. Nucleolar chromatin appeared as highly compact clumps (arrows) or as
loosened agglomerates of very thin structures (arrowheads). These agglomerates are composed of very thin, extended, non-nucleosomal filaments and
never give rise to nucleosome-like structure as shown at higher magnification (Figure 4). These filamentous structures are perfectly super-
imposable on the loosened agglomerates of control cells. The reported
biochemical and morphological data suggested that the *in vivo* inhibition of
topoisomerase I activity by camptothecin did not cause a structural
modification towards nucleosome-like structures, for the extended non-
nucleosomal filaments of putative rDNA.

ACKNOWLEDGEMENTS

This work was supported by Progetto finalizzato 'Oncologia' 87/1378 and 88/781.

REFERENCES

Birnboim H. C., Jevcak J. J. 1981, Fluorimetric method for rapid detection of DNA strand breaks in human white blood cells produced by low doses of radiation. Cancer Res., 41:1889.

Derenzini M., Hernandez-Verdun D., Bouteille M. 1981, Subunit configuration of rat hepatocyte chromatin fixed in situ, as visualized in thin sections selectively stained for DNA. Biol. Cell., 41:161.

Derenzini M., Hernandez-Verdun D., Farabegoli F., Pession A., Novello F. 1987, Structure of ribosomal genes of mammalian cells in situ. Chromosoma, 95:63.

Hernandez-Verdun D., Derenzini M. 1983, Non nucleosomal configuration of chromatin in nucleolar organizer regions of metaphase chromosome in situ, Eur. J. Cell Biol., 31:360.

Raska I., Reimer G., Jarnik M., Kostrouch Z., Raska K. 1989, Does the synthesis of ribosomal RNA take place within the nucleolar fibrillar centers or dense fibrillar components? Biol Cell 65:79.

Rose K. M., Szopa J., Han F., Chen Y., Richter A., Scheer U. 1988, Association of DNA topisomerase I and RNA polymerase I: a possible role for topoisomerase I in ribosomal gene transcription. Chromosoma, 96:411.

Zhang H., Wang J. C., Liu L. F. 1988, Involvement of DNA topoisomerase I in transcription of human ribosomal RNA genes, Proc. Nat. Acad. Sci. USA 85:1060.

LOCALIZATION OF DNA AND CHARACTERIZATION OF ARGYROPHILIC STRUCTURES WITHIN

NUCLEOLAR COMPONENTS DURING INTERPHASE AND MITOSIS IN LEUKEMIA CELL LINES

D. Ploton, M. Menager, P. Jeannesson, A. Beorchia and
J. J. Adnet

Unité INSERM U314
Laboratoire d'Histologie – U.F.R. de Médecine and
    Laboratoire de Biochimie – U.F.R. de Pharmacie – Reims
France

INTRODUCTION

In previous work we demonstrated at the E.M. level the peculiar localization of Ag stained NORs around metaphasic chromosomes and also presence of numerous spherical argyrophilic components during telophase (D. Ploton et al., 1987). Several questions arose from these findings:

1) What are the spatial relationships of the argyrophilic components during all the phases of mitosis? 2) Do all the argyrophilic components contain very decondensed DNA, a supposed characteristic of rDNA? 3) What is the fine structure of argyrophilic components, particularly during metaphase and telophase?

In order to answer these questions we performed the studies presented in this paper: a) Observation of argyrophilic components with reflected light and also within thick sections observed with H.V.E.M. b) Localization of decondensed DNA with osmium ammine complex staining specific for DNA. c) Study of RNP components after EDTA regressive staining.

MATERIAL AND METHODS

Four leukemic cell lines were studied: K562, HL60, L1210, and Friend cells. Ag-NOR proteins were stained as previously described and observed with reflected light (D. Ploton et al., 1987). DNA was stained with osmium ammine complex (M. Derenzini et al., 1982). RNP were revealed by EDTA regressive staining (Wassef et al., 1979). 3D studies were performed by tilting thick-sections (0.5 to 2μm thick) in a Philips 300kV CM30.

RESULTS

1) Localization of Ag-NOR proteins has been achieved with reflected light (Figure 1). By using this very resolutive technique we showed principally the presence of two kinds of metaphasic NORs (Figure 1c): the first one constitutes a couple of dots and second are large contorted ribbon-like structures. Numerous tiny spots are seen within telophasic nuclei (Figure

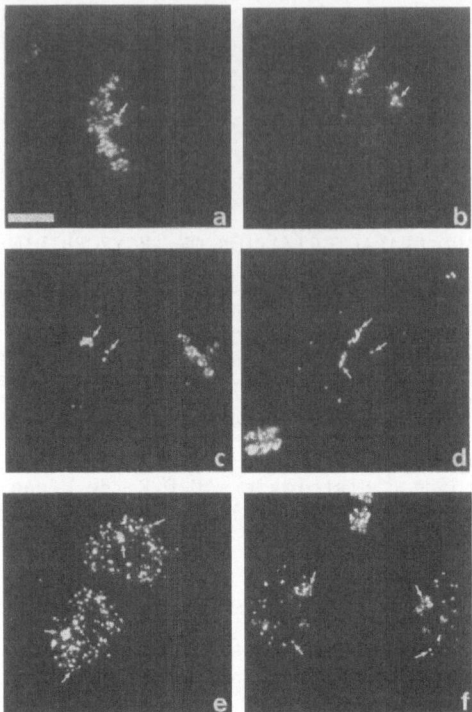

Figure 1.   Ag-NOR proteins localized with reflected light during:
             a) interphase, b) prophase, c) metaphase, d) anaphase,
             e) early telophase, f) late telophase. Arrows pointed to
             representative argyrophilic components. Bar = 5μm.

1e), besides less numerous bigger ones. During late telophase (Figure 1f)
the number of smaller dots decreases, whereas the volume of the bigger ones
increases suggesting a fusion of the former into the later.

2)   Localization of DNA (Figure 2). This study has focused on the
visualization of highly decondensed DNA, a characteristic of rDNA. Such DNA
was found in roundish structures, during interphase (Figure 2a), representing the nucleolar fibrillar centres and dense fibrillar component. During
metaphase (Figure 2b) decondensed DNA is seen within indentations at the
periphery of chromosomes. During telophase (Figure 2c and d) decondensed
DNA is found within structures similar to the larger argyrophilic components, but is absent from the smaller ones.

3)   Localization of RNP (Figure 3). RNPs are found as a sheath around
metaphasic chromosomes and also as elongated structures similar to metaphasic NORs (Figure 3a). Two sets of RNP structures similar to the two sets
of argyrophilic components are well defined during telophase (Figure 3b).

4)   3D study of argyrophilic components during telophase (Figure 3c). In
this thick section of two daughter cells, the two sets of argyrophilic
components and their spatial relationships appear clearly.

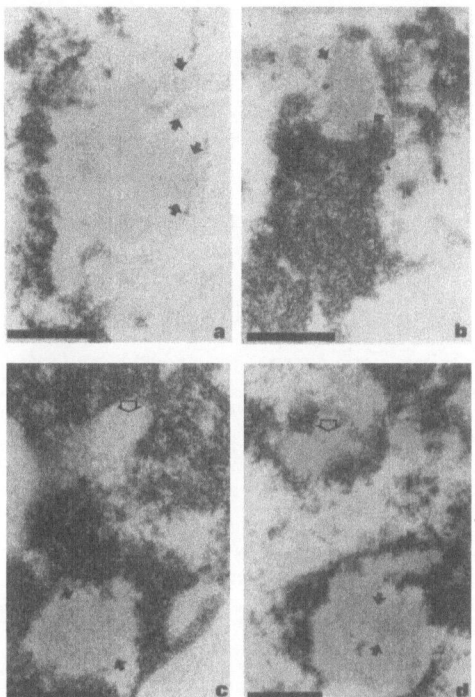

Figure 2.   Localization of decondensed DNA with osmium ammine complex
             within a) interphasic nuclwolus, b) metaphasic chromosome,
             c) early telophase and d) late telophase. Arrows pointed to
             decondensed DNA (➔) and to components without DNA (⬦).
             Bar = 0.5µm.

DISCUSSION AND CONCLUSION

     In this work we confirmed the frequent presence of contorted
metaphasic NORs besides the classical doublet of single spots, and the
presence of numerous spots, during early telophase, which fuse around NORs
to constitute new nucleoli. Localization of DNA also confirmed presence of
two sets of telophasic nucleolar components with decondensed DNA (larger
ones) and without decondensed DNA (smaller ones). 3D repartition of these
argyrophilic components demonstrated their regular dispersion within the
whole volume of telophasic nuclei. In conclusion, this work demonstrated
that argyrophilic components are not dispersed at random during prophase and
that two sets of these components may be observed; first, ones containing
rDNA: the NORs; second, ones probably dispersed around chromosomes and
fusing around NORs to constitute new telophasic nucleoli.

ACKNOWLEDGEMENTS

     We thank Pr. M. Derenzini for staining of DNA with osmium ammine
complex.

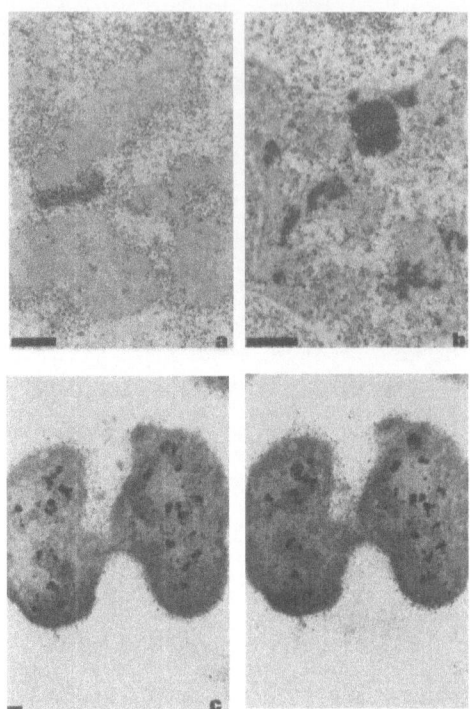

Figure 3.   Localization of RNP complex during a) metaphase and b)
             telophase. Bar = 0.5μm. c) stereo-pair of two telophasic
             daughter cells showing the spatial distribution of argyro-
             philic components. Bar = 1μm.

REFERENCES

Derenzini M., Viron A. and Puvion-Dutilleul F. 1982, The Feulgen-like
    osmium-ammine reaction as a tool to investigate chromatin structure in
    thin sections. J. Ultr. Res. 80:133.
Ploton D., Thiry M., Menager M., Lepoint A., Adnet J. J. and Goessens G.
    1987, Behaviour of nucleolus during mitosis. A comparative ultra-
    structural study of various cancerous cell-lines using Ag-NOR staining
    procedure. Chromosoma 95:95.
Wassef M., Burglen J. and Bernhard W. 1979, A new method for visualization
    of preribosomal granules in the nucleolus after acetylation. Biol.
    Cell 34:153.

RELATIONSHIP BETWEEN THE Ag-NOR PROTEINS AND THE FUNCTIONAL CHANGES IN
NUCLEOLI OF THE RAT HEPATOCYTES STIMULATED BY CORTISOL AND BY PARTIAL
HEPATECTOMY

A. Pession, D. Trerè, F. Farabegoli, F. Novello,
T. Romagnoli and M. Derenzini

Dipartimento di Patologia Sperimentale
Università di Bologna
Bologna, Italy

INTRODUCTION

In the interphase cell nucleus, the ribosomal genes are located in the
fibrillar centers and in the associated dense fibrillar component of the
nucleolus (Hernandez-Verdun, 1983; 1986; Goessens, 1984). Some acidic
proteins are associated with the fibrillar components and in particular with
the highly decondensed chromatin of the nucleolus. These proteins are
selectively stained by silver using the method already described for
visualizing proteins of metaphase Nucleolar Organizer Regions; for this
reason they are called Ag-NOR proteins. These proteins are involved in
transcription (Howell, 1982) suggesting they have a role in the structural-
functional organization of the nucleolar components. Nucleolin, the major
component of Ag-NOR proteins, might be involved in rRNA maturation (Herrera
and Olson, 1986) and it has been shown to induce chromatin decondensation
by binding to histone H1 (Erard et al., 1988).

Up to now, there appear to be no data available on the possible role of
Ag-NOR proteins in the structural nucleolar changes which are exclusively
related to the proliferating state of the cell. We have studied the
relationship between the Ag-NOR proteins and the structural-functional
organization of the nucleolus in rat hepatocytes stimulated to proliferate
by partial hepatectomy before DNA synthesis begins. Changes of Ag-NOR
protein quantity, evaluated by an automated image analyser, were correlated
with the preribosomal RNA synthesis in regenerating rat hepatocytes and in
8-hour cortisol-stimulated hepatocytes. The effect of cycloheximide, at a
dose which did not hinder rRNA synthesis, was also studied on Ag-NOR protein
quantity in regenerating rat hepatocytes.

The results indicated that the increase of Ag-NOR proteins in regenerating rat hepatocytes is not correlated to rRNA synthesis.

MATERIALS AND METHODS

Male Wistar rats were partially hepatectomized as described by Higgins
and Anderson (1931). Cycloheximide was injected into 7 hour and 12 hour
hepactectomized rats, at the dose of 0.025mg/100g b.w. Cortisol 20mg/100g
b.w. was administered to normal rats, 8 hours before killing. Control rats

Table I. Effect of Cortisol, Hepatectomy and Cycloheximide on Ag-NOR Protein Quantity and on Protein and RNA Synthesis.

HEPATECTOMY	CYCLOHEXIMIDE (0.025mg/100g bw)	CORTISOL (20mg/100g bw)	Ag-NOR PROTEIN AREA (um^2)	PROTEIN SYNTHESIS (dpm/mg protein)	RNA SYNTHESIS* nuclear	50° fraction
-	-	-	3.58 $\pm$ 0.67	6171 $\pm$ 122	100	100
7 hours	-	-	4.03 $\pm$ 1.11	9110 $\pm$ 665	127	160
12 hours	-	-	7.52 $\pm$ 2.46	10732 $\pm$ 939	127	160
12 hours	5 hours	-	3.62 $\pm$ 0.60	1584 $\pm$ 96	127	160
-	-	8 hours	4.28 $\pm$ 0.77	-	139	172

* Values are expressed as percentage of ^{3}H-orotic acid incorporation in control group: 584 $\pm$ 32 dpm/μg RNA and 718 $\pm$ 44 dpm/μg RNA for total nuclear and 50°C fraction RNA respectively.

received saline. RNA synthesis was measured in liver after 15 min ^{3}H-orotic acid injection; RNA fractions were isolated by stepwise extraction with phenol at 50°C (Dabeva et al., 1978; Hadjiolov et al., 1974). Protein synthesis was evaluated by ^{3}H L-leucine incorporation according to Verbin et al. (1969). Quantitative analysis of Ag-NOR proteins was carried out on frozen section of the same livers used for the biochemical evaluation, stained with the method described by Ploton et al. (1986). The analysis was performed using an automated image analyzer (Derenzini et al., 1989).

RESULTS

The quantity of Ag-NOR proteins measured using an automated image analysis of silver stained livers, did not change during the first 7 hours of regeneration (3.58 $\pm$ 0.67 μm^2 of control resting hepatocytes versus 4.03 $\pm$ 1.11 μm^2 of 7-hour regenerating hepatocytes); at 12 hours of regeneration we observed a marked increase of Ag-NOR proteins (7.52 $\pm$ 2.46 μm^2). The administration of cycloheximide at 7 hours regeneration, at a dosage which did not affect rRNA synthesis but inhibited protein synthesis (Table I), caused a reduction of Ag-NOR protein quantity. In order to study the relationship between the Ag-NOR protein quantity and ribosomal rRNA synthesis we compared the incorporation of ^{3}H-orotic acid in nucleolar RNA fractions in control, 12-hour regenerating and 8-hour cortisol treated rat livers. As shown in Table I cortisol treatment caused an increase in rRNA synthesis whereas the quantity of Ag-NOR proteins did not significantly change with respect to normal resting liver (4.28 $\pm$ 0.77 μm^2, 8-hour cortisol stimulated liver versus 3.58 $\pm$ 0.67 μm^2 normal liver).

To elucidate the role of Ag-NOR proteins in cells committed to duplicate we studied the changes of ribosomal chromatin *in situ* using the osmium-ammine DNA staining in 12-hour hepatectomized control and 5 hours cycloheximide treated rat livers. At 12 hours after hepatectomy there was an increase in the quantity of intranucleolar chromatin: throughout the enlarged nucleolar body many loose agglomerates of extended DNA filaments were present. After 5 hours cycloheximide treatment the nucleolar bodies appeared to be reduced in size, also the quantity of intranucleolar chromatin structures was reduced, together with the agglomerate of extended DNA filaments.

CONCLUSIONS

Our data have shown that the increase in Ag-NOR protein quantity occurs between 7 and 12 hours of regeneration. These proteins are newly synthesized; protein synthesis inhibition obtained by cycloheximide injection caused a marked reduction in Ag-NOR protein quantity in 12-hour regenerating livers. This increase occured before the begining of DNA synthesis and it was not related to rRNA synthesis. In fact, the reduction of Ag-NOR proteins by cycloheximide treatment did not affect the extent of pre-rRNA synthesis and in 8-hour cortisol stimulated hepatocytes the synthesis of rRNA was at the same level as 12-hour regenerating hepatocytes, whereas the quantity of ag-NOR proteins was very similar to control resting cells.

Ag-NOR proteins are known to be constantly associated with completely extended ribosomal chromatin: our results have demonstrated that in regenerating rat hepatocytes the quantity of extended ribosomal chromatin increases before DNA synthesis begins. Therefore, the increase of this type of chromatin is not due to the DNA duplication. Cycloheximide treatment, inhibiting protein synthesis prevents chromatin decondensation in regenerating hepatocytes.

These data indicate that Ag-NOR proteins have a structural role: they are involved in the decondensation of the portion of ribosomal genes which are in a condensed form. A decondensation of all ribosomal chromatin might be necessary for the duplication of ribosomal genes.

ACKNOWLEDGEMENTS

This work was supported by Progetto finalizzato 'Oncologia' 87/1378 and 88/781.

REFERENCES

Dabeva M. D., Dudov K. P., Hadjiolov H. H., and Stoyokova A. A. 1978. Quantitative analysis of rat liver nucleolar and nucleoplasmic ribosomal ribonucleic acid. Biochem. J. 121:367.

Derenzini M., Pession A., Farabegoli F., Trerè D., Badiali M. and Dehan P. 1989. Relationship between interphasic Nucleolar Organizer Regions and growth rate in two neuroblastoma cell lines. Am. J. Pathol. 134:925.

Erard M. S., Belenguer P. Caizergues-Ferrer M. Pantaloni A. and Amalric F. 1988. A major nucleolar protein, nucleolin, induces chromatin decondensation by binding to histone H1. Eur. J. Biochem. 175:525.

Goessens G. 1984. Nucleolar structure. Int. Rev. Cytol. 84:107.

Hadjiolov H. H., Dabeva M. D., and Mackendosky W. 1974. The action of α-amanitin in vivo on the synthesis and maturation of mouse liver ribonucleic acid. Biochem. J. 138:321.

Hernandez-Verdun D. 1983. The nucleolar organizer regions. Biol. Cell. 49:191.

Hernandez-Verdun D. 1986. Structural organization of the nucleolus in mammalian cells. Meth. Arch. Exp. Pathol. 12:26.

Herrera A. H., and Olson M. O. J. 1986. Association of protein C23 with rapidly labelled nucleolar RNA. Biochemistry 25:6258.

Higgins G. M., and Anderson R. M. 1931. Experimental pathology of liver. 1. Restoration of liver of white rat following partial surgical removal. Arch. Path. 12:186.

Howell W. M. 1982. Selective staining of Nucleolus Organizer Regions (NORs). In The Cell Nucleus, Vol. XI, Busch H., and Rothblum L. editors, Academic Press, New York. p. 89.

Ploton D., Menager M., Jeannesson P., Himber G., Pigeon F., and Adnet J. J. 1986. Improvement in the staining and in the visualization of the argyrophilic proteins of the nucleolar organizer region at the optical level. Histochem. J. 18:5.

Verbin R. S., Goldblatt P. J. and Farber E. 1969. The biochemical pathology of inhibition of protein synthesis in vivo. The effect of cycloheximide on hepatic parenchymal cell ultrastructure. Lab. Invest. 20:528.

ORIGIN AND ULTRASTRUCTURAL CYTOCHEMISTRY OF THE EXTRANUCLEOLAR BODIES AT

THE PREOVULATORY FOLLICLE STAGE AND DURING RAT OOCYTE MATURATION

N. Antoine and G. Goessens

Laboratoire de Biologie cellulaire et tissulaire
Université de Liège
20, rue de Pitteurs
B-4020 Liège
Belgium.

It is now well accepted that ordered changes occur in the fine structural organization of the nucleolus during follicular growth in mammalian
species. These morphological changes have been studied by many authors by
both light and electron microscopy in mouse oocytes (Chouinard, 1971; 1975),
in pig and bovine oocytes (Crozet et al., 1981; 1986) and in human oocytes
(Zybina et al., 1984). However, the majority of these authors have been
interested in the early stages of oocyte meiosis and further evolution of
the nucleolus of the mammalian oocytes has been less thoroughly described.
A systematic study of ultrastructural modifications which appear in the rat
oocyte nucleolus during follicular growth has therefore been performed
(Antoine et al., 1987).

The oocyte nucleolus in the primordial and primary follicles consists
of strands of dense fibrillar component and aggregates of granules. At the
secondary follicle stage, the morphological changes essentially result in
the appearance of several compact masses in the reticulated nucleoli. These
masses seem to fuse during antral formation and the nucleolus becomes
entirely compact being made up of an homogeneous mass at the preovulatory
follicle stage preceeding meiotic resumption (Takeuchi, 1984; Antoine et
al., 1987; 1988).

In order to define the nature of this homogeneous mass, microanalytical
X-rays analysis and cytochemical methods allowing detection of nucleic
acids, proteins and lipids were performed at the light microscopic and
ultrastructural levels. According to the results obtained, we postulated
that the nucleolar mass is composed of proteins rich in thiol groups and of
a small amount of RNA (Antoine et al., 1988). The function of this
proteinic nucleolar component at the end of folliculogenesis remains
unclear.

The nucleolar compaction seems to be a general feature in mammalian
oocytes but does not occur at the same stages in all species (Azevedo et
al., 1984). However, in all cases, the ultrastructural modifications are
correlated with a decrease in RNA synthesis during follicular growth
(Crozet et al., 1981; 1986). In fact, at the antral follicle stage, the
rate of transcriptional activity is very low in the rat oocyte nucleolus,
which appears to consist entirely of a compact mass. At this stage, some

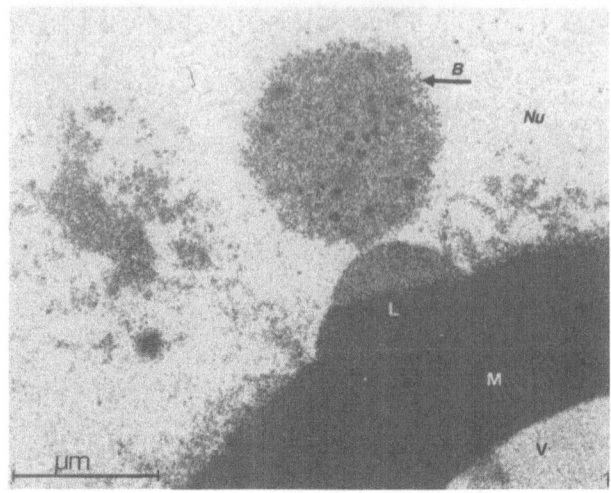

Figure 1.   Extanucleolar body formation. A lenticle (L) at the periphery
            of the vacuolated (V) nucleolar mass (M) gives rise to an
            extranucleolar body (B) in the nucleoplam (Nu).

authors (Baeckelant et al., 1986; Crozet et al., 1986) described a
vacuolisation of the nucleolus. This nucleolar vacuolisation in antral
follicles is not correlated with follicular atresia but could be associated
with the transport of previously synthesized material from the nucleolus to
the cytoplasm.

Thus, we have tried to follow nucleolar modifications appearing in the
nucleolus and in the germinal vesicle after inducing oocyte maturation by
gonadotropin hormone (LH). After hormonal treatment, all the nucleoli
observed are vacuolated and extranucleolar bodies appear in the germinal
vesicle.

The aim of the present paper is to supply new information concerning
the nucleolar morphological changes appearing in the germinal vesicle at
the preovulatory follicle stage in rat oocytes before the germinal vesicle
breakdown and finally the resumption of meiosis visualized by chromosome
condensation in the oocytes.

After hormonal stimulation, we initially observed that most of the
compact vacuolated nucleoli seem to bud nucleolar lenticles, giving rise to
extranucleolar bodies close to the nucleolar mass (Figure 1). These extranucleolar bodies are composed of several nucleolar components, which are
difficult to distinguish by usual ultrastructural methods. However, some
cytochemical methods led us to define the nature of the various components
in the extranucleolar bodies (Antoine et al., 1989). They originate from
the nucleolus at least in part from the nucleolar lenticles situated around
the nucleolar mass and are composed of argyrophilic proteins but also
contain a small amount of RNA.

Just before the germinal vesicle breakdown, the rat oocyte nucleolus
exhibits a rearrangement of the components. Vacuoles and budding of
fibrillar lenticles reflect the extrusion of nucleolar material leading to
the formation of extranucleolar bodies. This material is essentially made
up of ribonucleoproteins and could persist in the ooplasm at the germinal
vesicle breakdown. However, further evolution of the extranucleolar bodies

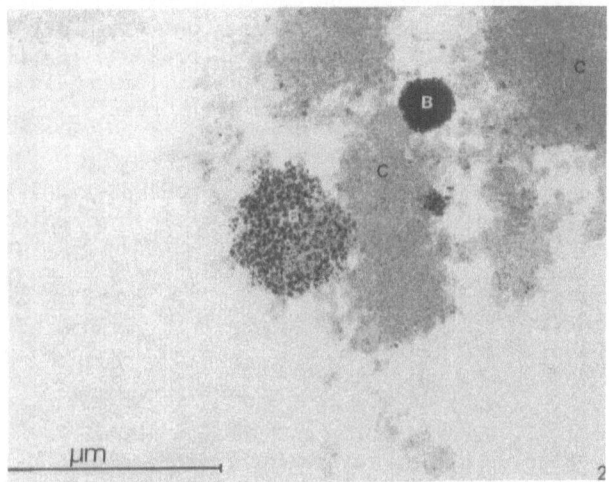

Figure 2.   Small argyrophilic extranucleolar bodies (B) are visualized
near condensed chromosomes (C).

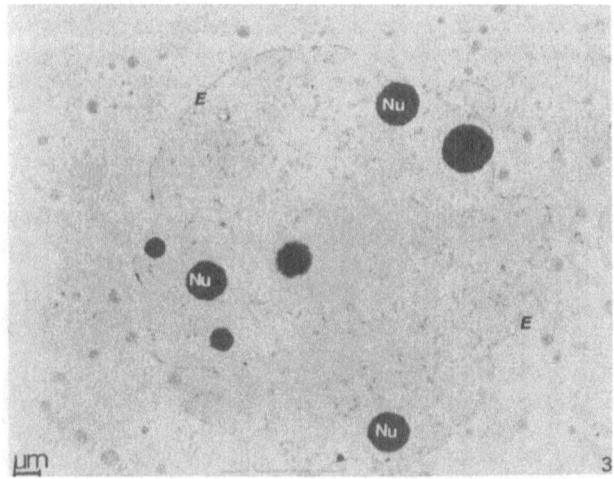

Figure 3.   Zygote nucleus.   Various 'small nucleoli' (Nu) are present in
one-cell embryo near the nuclear envelope (E).

has been little investigated.   Thus, we have tried to follow the resumption
of meiosis and the behaviour of nucleolar components at the early stage of
embryogenesis by silver staining method at the ultrastructural level.

4h30 after LH stimulation, the germinal vesicle breakdown occurs.
Condensed chromosomes are visualized in the ooplasm and the compact
nucleolar mass persist in the form of small heterogeneous bodies completely
covered by silver grains and localized near the chromosomes (Figure 2).
After fertilization, one cell embryo collected in the ovduct exhibit a
germinal vescile containing an important amount of small compact bodies.
These bodies are usually located at the nuclear periphery near the nuclear

envelope (Figure 3). This observation could suggest that the small bodies are the result of extranucleolar body association in the germinal vesicle just after oocyte fertilization. Sometimes decondensed fibrillar material seems to bud from these small bodies. This fibrillar budding could be the first sign of nucleolar decondensation before the resumption of transcriptional activity.

In summary, during follicular growth, the nucleolus develops, becomes compact then it disperses in the form of extranucleolar bodies during the resumption of meiosis and finally reappears in a condensed form at the first stage of embryogenesis. All these transformations are correlated with variations in nucleolar activity.

REFERENCES

Antoine N., Lepoint A., Baeckeland E. and Goessens G. 1987, Evolution of the rat oocyte nucleolus during follicular growth, Biol. Cell, 59:107
Antoine N., Lepoint A., Baeckeland E., and Goessens G. 1988, Ultrastructural cytochemistry of the nucleolus in rat oocytes at the end of folliculogenesis, Histochem., 89:221.
Antoine N., Thiry M., and Goessens G. 1989, Ultrastructural and cytochemical studies on extranucleolar bodies in rat oocytes at the preovulatory follicle stage, Biol. Cell, 65:61.
Azevedo C., Castilho F., and Coimbra A. 1984, Fine structure and cytochemistry of the oocyte nucleolus in the mollusk Helcion pellucidus (Prosobranchia), J. Ultrastruct. Res., 89:1.
Baeckeland E., Antoine N., Lepoint A., and Goessens G. 1986, Etude des vacuoles nucléolaires observées dans les ovocytes de rat, Bull. Assoc. Anat., 70:5.
Chouinard L. A., 1971, A light and electron microscope study of the nucleolus during growth of the oocyte in the prepubertal mouse, J. Cell Sci., 9:663.
Chouinard L. A., 1975, A light and electron microscope study of the oocyte nucleus during development of the antral follicle in the prepubertal mouse, J. Cell Sci., 17:589.
Crozet N., Motlik J., and Szollosi D. 1981, Nucleolar fine structure and RNA synthesis in porcine oocytes during the early stages of antrum formation, Biol. Cell, 41:35.
Crozet N., Kanka J., Motlik J. and Fulka J. 1986, Nucleolar fine structure and RNA synthesis in bovine oocytes from antral follicles, Gam. Res., 14:65.
Takeuchi J. K. 1984, Electron microscopic study of silver staining of nucleoli in growing oocytes of rat ovaries, Cell Tiss. Res., 236:249.
Zybina E. U., Grishchenko T. A., and Semenov V. M. 1984, Ultrastructure of the fibrillar center in oocyte nuclei at diplotene stage in golden hamster, Tsitologia, 26:1246.

SPATIAL DISTRIBUTION OF DNA AND HISTONES WITHIN EHRLICH TUMOR CELL NUCLEI

BY IMMUNOELECTRON MICROSCOPY

Marc Thiry

Laboratoire de Biologie cellulaire et tissulaire
Institut A. Swaen
20 rue de Pitteurs
B-4020 Liege
Belgique

It is now established that histones are essentially structural
components of chromatin. In particular, histones H2A, H2B, H3 and H4
spontaneously form octamers, around which a DNA molecule can wind. Such
histone octamer-DNA complex constitues a nucleosome (for review, see Kornberg, 1977). This structural organization seems to be characteristic for
packaging DNA in the inactive non-transcribed from of chromatin, whereas
active genes are apparently compacted in an altered nucleosome structure
(for review, see Reeves, 1984). Further, in the case of extremely active
genes, such as the ribosomal genes, the presence of histones even remains a
matter of much discussion.

After precise detection of DNA within Ehrlich tumor cell nucleoli
(Thiry, 1988; Thiry et al., 1988), the *in situ* spatial distribution of
histones within nucleoli has been investigated in the present work. In
order to simultaneously compare the spatial distribution of DNA and histones,
a double immunogold staining procedure has been developed on ultrathin
sections of Lowicryl-embedded cells. This technique involved the sequential
application of two distinct antibody-immunoglobulin gold procedures. For
the localization of DNA, a first indirect immunolabelling method using a
monoclonal anti-DNA antibody and an immunoglobulin coupled with gold
particles of 10nm diameter was applied by floating the ultrathin sections.
Once the first labelling was performed, the sections were mounted on
collodion-coated gold grids and dried. For the localization of histones,
the second face of the sections was then labelled using antisera raised
against purified histones of chicken erythrocytes and an immunoglobulin gold
complex formed with gold particles of 5nm diameter. This double immunogold
labelling procedure allows us to avoid co-labelling due to artefactual
cross-reactions between reagents since the labelling of each of the two
faces of the sections was independently performed. Further, in this
procedure, the first reaction cannot hinder the second one to any extent.
This is particularly advantageous when the two studied antigens occur close
together, as suspected in the present work.

Under these conditions, both small and large gold particles are
discretely visualized on the dense chromatin associated with the nuclear
envelope and with the nucleolus (Figure 1). A co-localization is also
revealed at all the nucleolar DNA positive sites (Figure 2). Specifically,

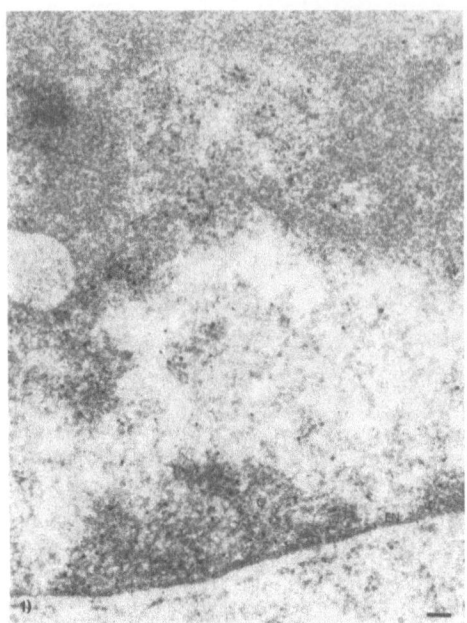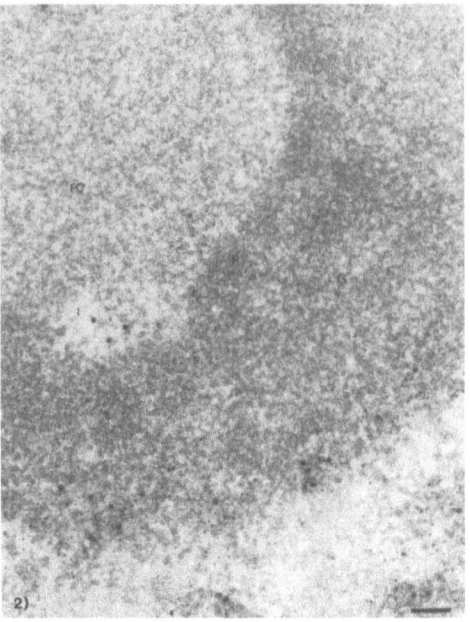

Figures 1   Simultaneous localization of DNA (large arrowheads) and
and 2.   histone H4 (small arrowheads) on ultrathin sections of 4%
              formaldehyde-fixed and Lowicryl-embedded Ehrlich tumor cells.
Figure 1.   Dense chromatin (C) associated with the nuclear envelope (NE)
              and with the nucleolus.
Figure 2.   Detail of a nucleolus.
              F: dense fibrillar component; FC: fibrillar center; G:
              granular component; I: nucleolar interstice. Bar = 0.1µm.

gold particles of two different sizes are found in the interstices coming
into contact with the fibrillar centers and in the peripheral regions of the
fibrillar centers. Interestingly, these two nucleolar sites have been
recently demonstrated to contain rDNA after *in situ* hybridization at the
electron microscope level. In contrast, the dense fibrillar component
appears to contain neither DNA nor histones.

    These results seem to indicate that all the DNA detected within the
nucleolus of Ehrlich tumor cells would be associated with histones. This
finding suggests that the ribosomal DNA including transcriptionally active
genes is bound to histones.

ACKNOWLEDGEMENTS

    The author is grateful to Dr. Muller who provided anti-histone antisera,
to Miss Skivée for the technical assistance, to the 'Fonds de la Recherche
Scientifique Médicale' (grant no 3.4512.86) and to the 'Action concertée'
(grant no 85/90-80 for their financial support. M.T. is 'Chargé de
Recherches auprès du F.N.R.S.'

REFERENCES

Kornberg R. 1977, Structure of chromatin, <u>Annu. Rev. Biochem.</u>, 46:931.

Reeves R. 1984, Transcriptionally active chromatin, <u>Biochim. Biophys. Acta</u>, 782:343.

Thiry M. 1988, Immunoelectron microscope localization of bromodeoxyuridine incorporated into DNA of Ehrlich tumor cell nucleoli, <u>Exp. Cell Res.</u>, 179:204.

Thiry M., Scheer U., and Goessens G., Localization of DNA within Ehrlich tumour cell nucleoli by immunoelectron microscopy, <u>Biol. Cell</u>, 63:27.

ELECTRON MICROSCOPE LOCALIZATION OF RIBOSOMAL RNA AND DNA AFTER *IN SITU*

HYBRIDIZATION

Marc Thiry and Lydia Thiry-Blaise

Laboratoire de Biologie cellulaire et tissulaire
Institut A. Swaen
20 rue de Pitteurs
B-4020 Liege
Belgique

It is now established that synthesis and processing of the cytoplasmic rRNA precursors (pre-rRNAs) as well as their assembly with specific proteins take place in the nucleolus (for review: see Hadjiolov, 1985). However, although various stages of pre-ribosome formation have been mapped to morphologically distinct nucleolar components, the precise intranucleolar location of the transcriptionally active rRNA genes is still a matter of debate. Immunocytochemical approaches have located RNA polymerase I (Scheer and Rose, 1984; Scheer and Raska, 1987) and DNA (Scheer et al., 1987; Thiry, 1988; Thiry et al., 1988) in the fibrillar centers, indicating that the rDNA transcription occurs within this nucleolar component. In contrast, it has been suggested from earlier autoradiographic studies based on the distribution of short-term labelled RNA that transcription takes place in or very near the dense fibrillar component (for reviews see Goessens, 1984; Fakan, 1986).

In order to clarify this problem, the spatial distribution of rDNA and rRNA within the nucleoli of Ehrlich tumor cells has been investigated by means of an *in situ* hybridization technique at the electron microscope level.

The probe used to localize the rRNA consists of an EcoRI-Sall fragment of 1.9kb containing 1.45kb 18S rDNA (Grummt and Gross, 1980) inserted into the pBR322 plasmid. This probe biotinylated by nick translation was hybridized to Lowicryl thin sections of cells. The hybrids were then detected by an anti-biotine antibody in consort with a secondary antibody coupled to colloidal gold.

Under these conditions, the gold particles are essentially found over the ribosome-rich regions of cytoplam and over the nucleolus of interphase cells. In the nucleolus (Figure 1), the labelling is particularly concentrated over the granular component and over the dense fibrillar component. In addition, small clusters of gold particles are frequently observed in the peripheral regions of the fibrillar centers, often in the vicinity of nucleolar interstices. By contrast, if the rRNA probe is replaced by either the pBR322 probe or only the hybridization buffer, there is no labelling. The labelling is also completely abolished when the RNA is specifically removed from the sections by treatment with RNase. Likewise, when the

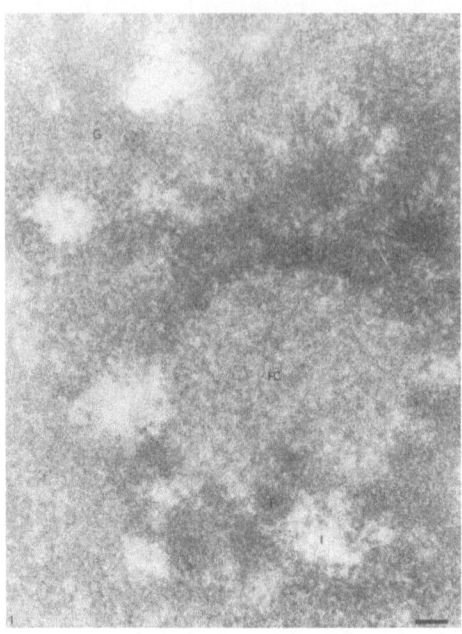

Figure 1.   Nucleolus after rDNA/rRNA hybridization on Lowicryl sections
of 0.2% glutaraldehyde-fixed Ehrlich tumor cells. F: dense
fibrillar component; FC: fibrillar center; G: granular
component; I: nucleolar interstice. Bar = 0.1µm.

biotinylated hybrids are detected only by the secondary antibody coupled to
colloidal gold, the sections are free of gold particles.

For the *in situ* detection of rDNA, it is indispensable to denature DNA
before attempting to hybridize it. However, the denaturation is a delicate
step because it directly depends upon the fixation performed. The choice of
the fixative is therefore of prime importance since it has, on the one hand,
to be effective independently of the aggressivity of the denaturation agents
and, on the other hand, to allow the denaturation. In order to research the
ideal fixation and DNA denaturation conditions, an immunocytochemical
approach was employed which uses an antibody reacting mainly with single-
stranded DNA. The results show that it is necessary to determine the best
denaturation conditions for each fixation used. For instance, when
formaldehyde is employed, the denaturation by heat gives an appreciable
labelling with the anti-single-stranded DNA antibody, whereas there is no
labelling after the action of protease and NaOH. Further, the biotinylated
probe used for the detection of rDNA corresponded to the only sense-strand
of a mouse rDNA fragment, hence avoiding any cross hybridization to rRNA.

Under these conditions, the labelling is exclusively found over
nucleoli, preferentially over some of its intranucleolar components.
Specifically, small clusters of gold particles are visualized in the
fibrillar centers, especially in their peripheral regions at the proximity
of the dense fibrils and the nucleolar interstices as well as within the
latter (Figure 2). In addition gold particles are detected over some clumps
of dense perinucleolar chromatin. If the rDNA probe is replaced by the
pBR322 or if the probe or the denaturation step is omitted, the labelling
is completely abolished.

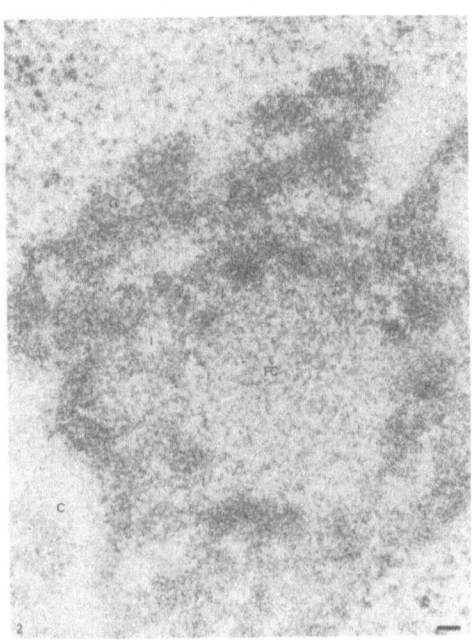

Figure 2.   Nucleolus after rDNA/rDNA hybridization on ultrathin sections
of Ehrlich tumor cells. 0.2% glutaraldehyde/Lowicryl K4M/
Protease-NaOH. C: dense chromatin; F: dense fibrillar
component; FC: fibrillar center; G: granular component;
I: nucleolar interstice. Bar = 0.1µm.

These results obtained after *in situ* hybridization seem to indicate
that the only sites where there is a co-localization of rDNA and rRNA within
the nucleolus of Ehrich tumor cells are the fibrillar center areas close to
the boundary regions of the dense fibrillar component. These findings,
together with the fact that RNA polymerase I is exclusively found in the
fibrillar centers (Scheer and Rose, 1984; Scheer and Raska, 1987), suggest
that the transcriptionally active rRNA genes are essentially located in the
confines of the fibrillar centers and the dense fibrillar component of the
nucleolus while others situated in the dense nucleolus-associated chromatin
are kept inactive.

ACKNOWLEDGEMENTS

The authors are grateful to Prof. I. Grummt for providing rDNA
fragment. This work was supported by grants from the 'Action concertée no
85/90-80' and from the 'Fonds de la Recherche Scientifique Médicale no
3.4512.86'. M.T. is 'Chargé de Recherches auprès du F.N.R.S.'

REFERENCES

Fakan S. 1986, Structural support for RNA synthesis in the cell nucleus,
    Meth. Achiev. Exp. Path., 12:105.
Goessens G. 1984, Nucleolar structure, Int. Rev. Cyt., 87:107.
Grummt I., and Groos H. 1980, Structural organization of mouse rDNA:
    comparison of transcribed and non-transcribed regions, Molec. Gen.
    Genet., 177:223.

Hadjiolov A. 1985, The nucleolus and ribosome biogenesis, <u>Cell Biol. Monographs</u>, 12:1.

Scheer U., Messner R., Hazan R., Raska I., Hansmann P., Falk H., Spiess E., and Franke W. W. 1987, High sensitivity immunolocalization of double and single-stranded DNA by monoclonal antibody, <u>Europ. J. Cell Biol.</u>, 43:358.

Scheer U., and Raska I. 1988, Immunocytochemical localization of RNA polymerase I in the fibrillar centres of nucleoli, <u>in</u> 'Chromosomes Today', A. Stahl, J. Luciani, A. Vagner-Capodano, eds., Allen and Unwin, London.

Scheer U., and Rose, K. 1984, Localization of RNA polymerase I in interphase cells and mitotic chromosomes by light and electron microscopic immunocytochemistry, <u>Proc. Natl. Acad. Sci. USA</u>, 81:143.

Thiry M. 1988, Immunoelectron microscope localization of bromodeoxyuridine incorporated into DNA of Ehrlich tumor cell nucleoli, <u>Exp. Cell Res.</u>, 179:204.

Thiry M., Scheer U., and Goessens G. 1988, Localization of DNA within Ehrlich tumour cell nucleoli by immunoelectron microscopy, <u>Biol. Cell</u>, 63:27.

DNA METHYLATION AND EXPERIMENTALLY INDUCED NUCLEOLAR DOMINANCE

C. De la Torre and A. González-Fernández

Centro de Investigaciones Biológicas
CSIC
Velázquez, 144
28006-Madrid
Spain

Our work shows that nucleolar dominance may be induced by hypomethylation of the NOR-chromosome. Hypomethylation of one of the two chromatids of both NOR-bearing chromosomes in a diploid cell was brought about in *Allium cepa L.* meristems by allowing one replication period in the presence of 10^{-6}M 5-azacytidine (5AZA), while the replication period of the subsequent cell cycle took place with no 5AZA at all. When chromosomal segregation occurred at a second mitosis, each chromosome was randomly split in two non-equivalent halves: one with native DNA, the other having incorporated 5AZA. Such incorporation will transform this chromatid in the hypomethylated state.

By analyzing the behaviour of both chromatids from telophase onwards, i.e. the behaviour of the telophasic chromosomes derived of each single NOR-bearing chromosome in metaphase, three facts were observed, namely:

1) Nucleologenesis was always faster in one of the two daughter chromosomes.

2) One of these two NORs was either permanently unable to organize any nucleolus on it, or the new nucleolus, which would be formed at later times, would be much smaller than that organized by its sister chromosome.

3) There was no co-segregation of the two strong and two weak NOR-bearing chromatids of the pair of homologous chromosomes.

The study was accomplished in synchronous cells labelled as binucleate by inhibiting the formation of the plate of cytokinesis, by a one-hour treatment with 5mM caffeine.

PECULIARITIES OF SILVER-STAINING OF NUCLEOLAR ORGANIZER REGIONS IN
CHROMOSOMES OF HUMAN PERMANENT CELL LINES AT THE LIGHT MICROSCOPICAL AND
ULTRASTRUCTURAL LEVEL

S. E. Mamaeva, L. G. Savelyeva and N. I. Komorova

Institute of Cytology of the USSR Academy of Sciences
Leningrad
USSR

Constant human and animal cell lines are widely used by scientists in
the study of such essential processes as proliferation, differentiation and
malignant transformation. It was shown that the activity of ribosomal
genes, determined by Ag-staining, depends on cell proliferation activity
(Hall et al., 1988), differentiation level (Reeves et al., 1984) and the
degree of malignancy (Crocker et al., 1987). It is essential to study the
activity of ribosomal genes in cell lines of different origins by silver
staining of metaphase chromosomes and interphase cells by light and electron
microscopy. In 30 cell lines (5 normal, 9 leukemic, 3 lymphoma and 13
tumor), the true number of acrocentric chromosomes, including rearranged
ones, the number of Ag^+ chromosomes, and the number and localization of
Ag-NORs on chromosomes were examined. Silver staining of NORs in interphase
cells from 5 cell lines (HSL-4, HL-60, U-937, A-431 and 293), which are distinguished by the character of silver staining of metaphase chromosomes, at
ultrastructural level was also carried out.

A karyotypic analysis of the cell lines showed that the true number of
acrocentric chromosomes, as a rule, ranges from 10 to 19-20 per cell (Table
I). To obtain this information it is necessary to identify all marker
chromosomes, the number of which, in some cases, exceeds 20-30 per cell.
In general, the acrocentric chromosomes are actively involved in structural
rearrangements (Table I). The major types of aberrations are deletions and
translocations, involving both long and short arms of acrocentric chromosomes. The number of Ag^+ chromosomes and silver staining patterns are
highly stable and important characteristics of the cell lines. A-431 is
characterized by essential intracellular heterogeneity, where NORs are
amplified, that evidently leads to the instability of the cell line. The
modal number of Ag^+ chromosomes varies from 3 in U-937 (AMMoL) to 18 in
MOLT-3 (T-ALL) (Table I). The cell lines are distinguished by the number
of Ag^+ chromosomes in D and G groups, the marker chromosomes as well as by
the intensity of silver staining, even in cases where the number of Ag^+
chromosomes was the same. In polyploid cells the number of Ag^+ chromosomes
was, as a rule, double if compared to that in diploid clones. In the
majority of cell lines, Ag-NORs are localized on the short arms of
acrocentric chromosomes, but in some lines silver staining revealed markers
with terminal localization of Ag-NORs on long arms of chromosomes as well
as markers with intercalated localization of NORs (K-562, RPMI-6410,
MG-63, HOsTE-85, A-431). In the majority of cell lines the number of
Ag-NORs and that of Ag^+ chromosomes was the same. However, in other lines

Table I.   Silver Staining of Metaphase Chromosomes in Cell Lines.

Lines	True number of NO chromosomes/ in markers	Modal number of Ag$^+$ chromosomes	Modal number of Ag-NORs	Lines	True number of NO chromosomes/ in markers	Modal number of Ag$^+$ chromosomes	Modal number of Ag-NORs
lymphoblastoid				tumor			
HSL-4	10/1	9	9	A-431	12/3	14	24
RPMI-1788	10/0	9	9	293	19/6	14	15
CCRF-SB	10/0	8	8	IMR-32	10/0	10	10
NC-37	10/1	8	8	HeLa-229	19/7	9	9
IM-9	10/1	8	8	HeLa TK$^-$	14/4	7	7
leukemia and lymphoma				HeLa Ohio	16/8	6	6
MOLT 3	20/0	18	18	M-HeLa	10/2	5	5
K-562	16/7	12	14	PA-1	10/1	8	8
RPMI-8226	12/5	12	12	Mg-63	11/7	8	10
Raji	10/1	10	10	Hos (TE-85)	10/4	8	10
RC-2A	10/2	9	9	COLO 320HSR	11/6	5	5
Namalwa	11/3	8	8	RPMI-2650	10/1	6	6
P3HR	11/2	8	8	SW-837	9/4	4	4
Jurkat	10/0	8	8				
RPMI-6410	10/2	8	8				
HL-60	10/0	8	8				
THP-1	10/1	5	5				
U-937	16/7	3	4				

(A-431, 293, U-937, HOsTE-85) number of Ag-NORs exceeds that of Ag$^+$ chromosomes (Figure 1a, b, c). By silver staining Ag-NOR amplification in cell line A-431 was found. This line contains 4-5 marker chromosomes with Ag-NOR number ranging from 3 to 6 (Figure 1d). The most objective criterion characterizing cell lines is the ratio of the Ag-NORs number to the total number of acrocentric chromosomes. The latter index varied from 25% to 200%. It was the highest in A-431 (200%), in Raji and JMR-32 (100%), the lowest in U-937 (25%). No direct correlation between the number of acrocentric chromosomes and that of Ag-NORs was detected.

Ultrastructural analysis of the pattern of silver staining of inter-phase nuclei showed two nucleolar structures to be stained: fibrillar centers (Fc) and dense fibrillar component (DFC). Each line has its own 'nucleolar phenotype' - number of Fc, size of Ag-positive structures as well as intensity of nucleolar staining. According to these criteria, cell line A-431 (Figure 2a), containing Fc up to 14 per nucleolus, was stained most intensively. In contrast, the least intensive staining was in HSL-4 (Figure 2b), which was characterized with poorly stained DFC and Fc, ranging from 1 to 3. There is no direct correlation between the number of AgNORs on chromosomes and the number of Fc in nucleoli. More detailed morphometric studies of Ag-stained interphase cells at light microscopy are necessary.

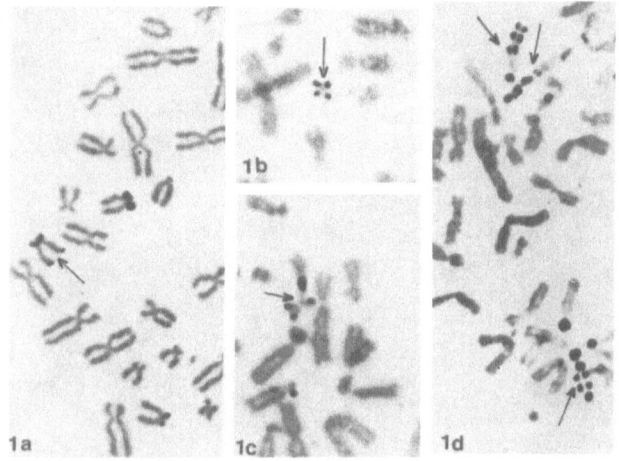

Figure 1.   Ag-NOR activity in acrocentric and marker (arrows) chromosomes
(a-d).   in 293(a), U-937(b), Hos (TE-85)(c) and A-431(d) cell lines.

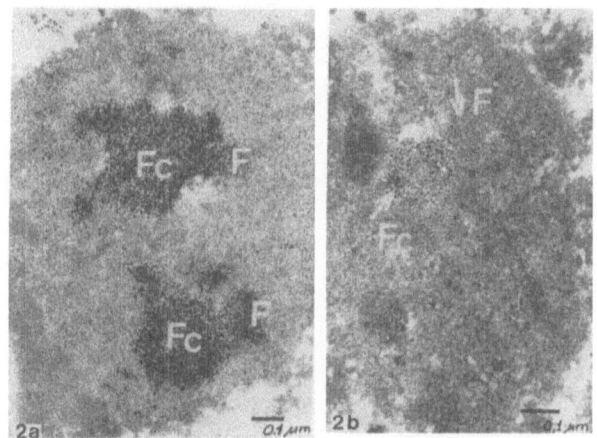

Figure 2.   Nucleoli in A-431(a) and HSL-4(b) cells after Ag-NOR staining.
(a, b).   Fibrillar centres (Fc) are surrounded by a layer of dense
              fibrillar component (F).

REFERENCES

Crocker J., and Nar P. 1987, Nucleolus organizer regions in lymphomas,
     J. Pathol., 151:111.
Hall P. A., Crocker J., Watts A., and Stansfeld A. G. 1988, A comparison of
     nucleolar organizer region staining and Ki-67 immunostaining in non-
     Hodgkin's lymphoma, Histopathol., 12:373.

MORPHOFUNCTIONAL CHARACTERISTICS OF SILVER-STAINED NUCLEOLI IN HUMAN NORMAL

AND PATHOLOGICAL CELLS

N. N. Mamaev

Department of Clinical Cytogenetics
I. P. Pavlov Medical Institute
Leningrad
USSR

Selective staining of nucleoli with silver nitrate has recently been shown to enhance the procedure of quantitative evaluation of the cells during ribosomal RNA (rRNA) and protein synthesis (Scheer et al., 1984). The experience from the use of Ag-stained human normal and pathological cells proves the method to be advantageous for the investigation of a wide range of problems in cell biology, including proliferation, differentiation and polyploidization. The aim of the present communication is to summarize the accumulated data and to consider some new trends for further research.

Bone marrow (BM), blood cells, cardiomyocytes (CM), normal and tumour cells from the oesophagus, stomach, colon and thyroid gland were studied. Blood and BM smears as well as smear-prints were made by conventional procedures. CM were studied in 10μm sections prepared from cardiobiopsy. Preparation of the specimens and their silver impregnation were carried out according to the previously published technique (Mamaev et al., 1985; 1989). The preparations were evaluated for the total content of nucleoli per 100 to 200 cells, for their shape, size and for the mean total quantity of Ag-grains in all nucleoli of a nucleus.

Our analysis demonstrated the number of nucleoli in silver-stained cells to be different in various types of tissues (Table I). It is minimal in CM and lymphocytes of the peripheral circulation in healthy subjects. In thyrocytes, mucosa cells from the gastrointestinal tract, erythroid and granulocytic elements of the BM, their number is greater, whereas in mega-karyocytes (MG) its content is the highest. The size of the nucleoli was greater in less differentiated cells than in mature ones. Large nucleoli were characteristic of myeloma cells. They were not uncommon in non-Hodgkin's and Hodgkin's lymphomas with, some prevalence in the latter (Berezovsky-Sternberg cells). Polymorphous nucleoli of a giant size were often evident in cancers from various localizations.

The number of Ag-grains in the nucleoli is the most important index of cell activity, with respect to rRNA synthesis and ribosome formation. Our findings show that during hemopoiesis the minimal count of Ag-material is seen in mature and maturing neutrophils, orthochromatic normoblasts and in circulating lymphocytes and monocytes. The level increases in the nucleoli of promyelocytes, basophilic and polychromatic normoblasts, pronormoblasts and blasts. The highest NOR activity is observed in MG, in spite of their

Table I. Mean Number of Nucleoli and NOR Activity in Various Types of Human Cells (Controls).

Cell type	Number of observations	Mean nucleoli number		Mean number of Ag-grains per a nucleus
		Range	M ± m	M ± m
Cardiomyocytes	4[+]	1.06 - 1.12	1.08±0.01	8.0± 0.4
Blood lymphocytes	6	1.02 - 1.25	1.18±0.2	*
Thyrocytes	5	1.5 - 2.08	1.8 ± 0.2	5.8±0.4
Gastric mucosa	8	1.5 - 3.2	2.0 ± 0.4	11.9±2.1
Colon mucosa	15	1.6 - 2.8	2.3 ± 0.3	12.3±3.2
Oesophagus mucosa	4	2.4 - 2.8	2.5 ± 0.2	9.8±1.4
Myelocytes	5	2.0 - 3.2	2.8 ± 0.3	5.8±0.8
Normoblasts (polychromatic)	5	2.9 - 3.6	3.3 ± 0.2	7.4±0.9
Pronormoblasts + basophilic normoblasts	5	2.6 - 3.7	3.2 ± 0.2	26.4±3.7
Promyelocytes	5	3.2 - 3.7	3.4 ± 0.1	16.7±2.0
Blasts	5	1.9 - 3.2	2.6 ± 0.4	28.4±4.3
Megakaryocytes	8	16.6 - 33.7	21.8 ± 4.3	76.4±11.5

Notes: [+] autopsy material of suddenly deceased subjects with no previous heart pathology is used.

* the figures are not given as the grains in most of lymphocytes were fused in one block

maturity. But CM, with identical nucleus ploidy, do not display similar activity of rRNA and protein synthesis. The NOR activity in mucosa cells from the gastrointestinal tract was almost the same throughout the length of the tract.

Proliferative potential of a cell may influence its NOR activity, as there is a strong correlation between the character of silver staining of the cells under study and the level of their staining with Ki-67 antibody (Hale et al., 1988). The observation is confirmed by our findings on the differential count of Ag-grains in the nucleoli of actively proliferating and non-proliferating blasts (of a large and small diameter respectively) in patients with acute leukemias (Mamaev et al., 1988). On the other hand, the fall in the activity of ribosome cistrons (RC) in the period of maturation of blood forming elements is common in the cells of both healthy subjects and leukemia patients (Mamaev et al., 1985). The same relationship is seen when comparing silver-stained cells from tumours of the same etiology, but at different stages of malignancy (Mamaev et al., 1986; Crocker and Nar, 1987).

As mentioned above, the highly polyploid human cells such as CM and MG possess a rather high level of rRNA and protein synthesis, in spite of their being mature and unable to proliferate. According to our data NOR activity of MG rises with the increase of their nuclear segmentation. The highest activity was observed in all types of cells in patients with immune thromboyctopenias, where supplementary hyperploidization of the nucleus is common. In contrast, the decrease in the activity of MG as compared to the controls, was observed in patients with chronic myelocytic leukemia, in which the presence of hypolobular MG with lowered ploidization of the nucleus is a peculiar feature of hemopoiesis. In line with MGs, the changes in the functional status of CM directly correlate to the changes in the activity of their NOR's. For example, the level of Ag-grains in CM nucleoli in patients with non-renal blood hypertension is closely dependent upon the mass of the myocardium ($r=1.0$, $p<0.001$). In addition, patients with various pathologies demonstrate a significant difference in the count of Ag-grains in CM nucleoli (Mamaev et al., 1989).

REFERENCES

Crocker J., and Nar P. 1987, Nucleolus organizer regions in lymphomas, J. Pathol., 151:111.

Hall P. A., Crocker J., Watts A., and Stansfeld A. G. 1988, A comparison of nucleolar organizer regions staining and Ki-67 immunostaining in non-Hodgkin's lymphoma, Histopathol., 12:373.

Mamaev N., Goodkova A., Amineva Ch., Kovalyeva O., and Almazov V. 1989. A method for evaluation of protein-synthesizing function in human cardio-myocytes, Arkh. Anat., 96, n5:69.

Mamaev N., Mamaeva S., Liburkina I., Kozlova T., Medvedeva N., and Makarkina G. 1985, The activity of nucleolar organizer regions of human bone marrow cells studied with silver staining. 1. Chronic myelocytic leukemia, Cancer Genet. Cytogenet, 16:31.

Mamaev N., Martinez U., Morozova E., and Ivanova N. 1988, Characteristics of Ag-stained nucleoli in blasts of various diameters from patients with acute leukemias, Tsitologiya, 30:1478.

Mamaev N., Mikhailov A., Irzhanov S., and Bykhovets I. 1986, Nucleolar organizer activity in epithelial tumour cells of the human colon, Bull. Exp. Med. (Engl. Transl.), 102:1774.

Scheer U., Hugle B., Hazan R., and Rose K. 1984, Drug-induced dispersal of transcribed rRNA-genes and transcriptional products: immunolocalization and silver-staining of different nucleolar components in rat cells treated with 5,6 - Dichloro-B-D-Ribofuranosylbenzimidazole, J. Cell Biol., 99:672.

NUCLEOLOGENESIS IN COW EMBRYO: RELATION BETWEEN ONSET OF TRANSCRIPTION AND

PENETRATION OF DNA INTO NUCLEOLAR PRECURSOR BODY

V. Kopecný

Research Institute of Animal Production
CS-104 00 Prague 10
Uhríneves
Czechoslovakia

INTRODUCTION

During the early period of the postfertilization development in the cow embryo, as well as in embryos of other mammalian species, functional nucleoli are absent. They develop only by the process of embryonal nucleologenesis, linking conspicuous morphological features with the onset of nucleolar functional activity. In the cow embryo, this process differs from the embryonic nucleologenesis of common laboratory mammals by its specific morphology and the relatively late localization of transcription onset to the 8-cell stage (Kopecný et al., 1985; Camous et al., 1986; Kopecný et al., 1989a). In addition to the actual practical breeding motivations for the study of nucleolar differentiation in the cow embryo (King et al., 1989), the nucleologenesis in this species also provides an alternative and attractive model for the study of temporal and spatial evolution of nucleolar components.

RESULTS

Ultrastructural Autoradiographic Evidence of Morphology and Transcription Onset in the Course of Embryonal Nucleologenesis in the Cow

The well-defined morphological features of nucleolus differentiation in the cow embryo allow the classification of four clear-cut stages of nucleologenesis, correlating with the autoradiographic detection of the steps of transcription re-establishment (Kopecný et al., 1989a), as demonstrated in Figure 1.

Stage 1. After fertilization, a nucleolus precursor body (NPB) appears, persisting until the early 8-cell cleavage stage. The NPB is of roundish shape and is composed from a dense network of uniform fibrils.

Stage 2. In the course of the 8-cell stage, i.e. in the 4th cell cycle, a big central lucid area of nucleoplasmic appearance develops in the NPB. Following recent nucleolar terminology, we named it the 'NPB-vacuole'. A contemporary feature was the association of prominent patches of condensed chromatin with the vacuolated NPB, but in relatively few and limited areas. There was no nucleolar incorporation of $(5-^3H)$ uridine during the first two stages of nucleologenesis.

Nuclear Structure and Function, Edited by J. R. Harris and
I. B. Zbarsky, Plenum Press, New York, 1990

187

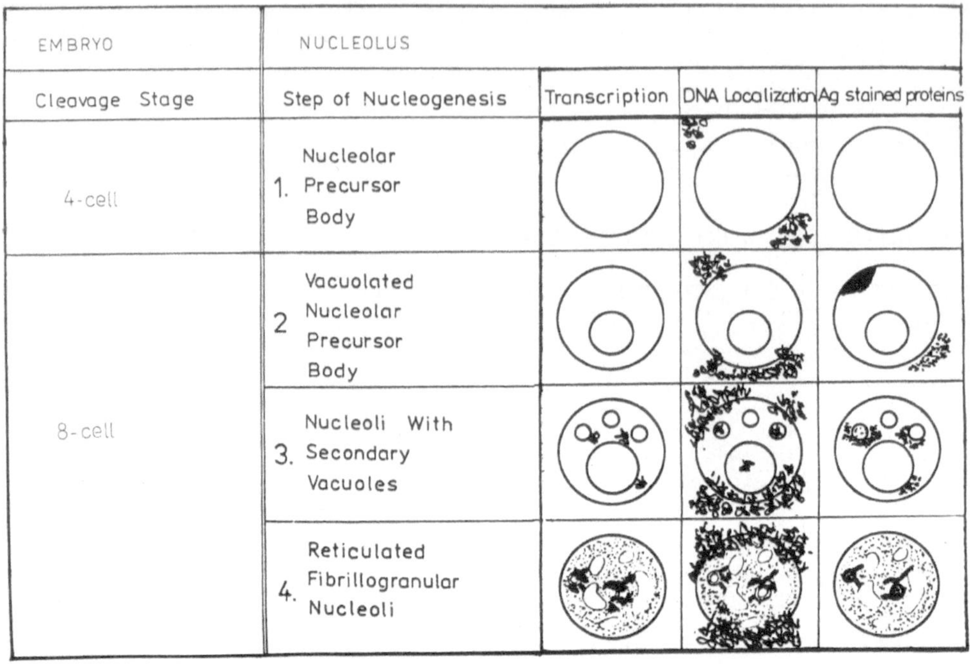

EMBRYO	NUCLEOLUS			
Cleavage Stage	Step of Nucleogenesis	Transcription	DNA Localization	Ag stained proteins
4-cell	1. Nucleolar Precursor Body			
8-cell	2. Vacuolated Nucleolar Precursor Body			
	3. Nucleoli With Secondary Vacuoles			
	4. Reticulated Fibrillogranular Nucleoli			

Figure 1. A schematic representation of morphological and molecular events during nucleologenesis in preimplantation cow embryo, showing transcription onset (fine-structure uridine-^{3}H autoradiography - Kopecný et al., 1989a), replicated-DNA localization (thymidine-^{3}H autoradiography - Kopecný et al., 1989b, for step 4 of nucleologenesis extrapolation from data of Tesarík et al., 1987) and localization of argyrophilic proteins (Antalíková et al., 1989).

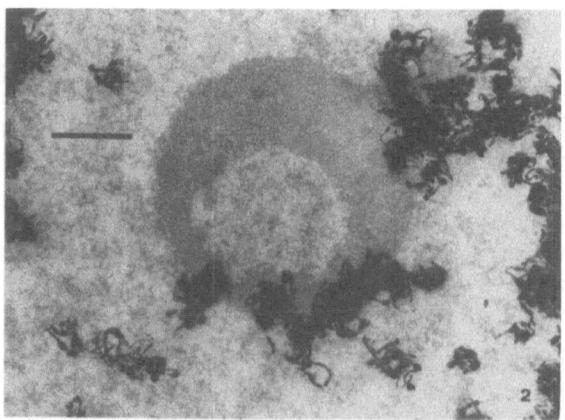

Figure 2. Fine-structure autoradiogram showing DNA-containing sites during the 2nd stage of nucleologenesis. Note the internalization of labelled perinucleolar chromatin into NPB in a limited number of association points and lack of labelling in the big central vacuole in NPB. Scale bar = 1.5µm.

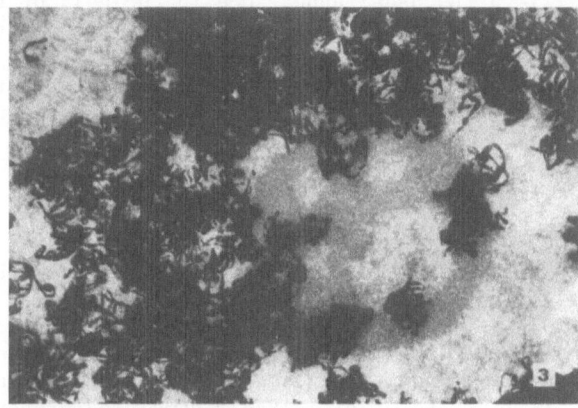

Figure 3.   During the 3rd stage of nucleologenesis similar label is seen
            encircling now almost completely the differentiating
            nucleolus and entering deeply into it, seemingly also in the
            numerous vacuoles.

    Stage 3. During further differentiation towards the typical nucleolar
structure secondary vacuoles appear, the primary vacuole loses its sharply
delineated boundary and a clump of chromatin often appears inside it. The
perinucleolar chromatin is internalized at that time and the first signs of
$(5-^3H)$ uridine incorporation in the dense fibrillar component are seen, as
well as the first granules of the granular component.

    Stage 4. In the rapid development that follows to the end of the
8-cell stage, typical fibrillogranular nucleoli appear showing fibrillar
centres, surrounded by labelled dense fibrillar component and an increasing
amoung of the granular nucleolar component. At the same time, the ability
to discern the primary embryonic fibrillar component of the original NPB is
lost (Kopecný et al., 1989a).

Replicated DNA-Containing Sites During Cow Embryonic Nucleologenesis

    At the stage of the NPB (= Stage 1 of nucleologenesis) replicated
embryonic DNA, visualized by thymidine-^{3}H labelling, was seen only outside
the NPB. During NPB vacuolation (Stage 2) the association with and
penetration into the NPB was seen. Labelled perinucleolar chromatin,
however, remained associated with the NPB superficially, but in a few
points, and no labelled DNA was seen in the large vacuole (Figure 2). A
massive penetration of DNA into NPB was seen at Stage 3, with the perinucleolar chromatin completely encircling the differentiating nucleolus.
Frequent sites of labelling were the vacuoles, where the label was
associated with their coarse fibrillar content (Figure 3) (Kopecný et al.,
1989b).

Localization of Silver-Staining Nuclear Proteins During Cow Embryonic
Nucleologenesis

    Silver-stained proteins were first detected in 8-cell NPBs as a
peripherally located convex-shaped area. As soon as the dense fibrillar
nucleolar component appeared (Stage 3), its staining by Ag was evident.
After formation of fibrillogranular nucleoli, the staining pattern

corresponded closely to the usual appearance in nucleoli of functionally active somatic cells. A distinct staining was seen also in areas closely adjacent to the NPB from the Stage 1 of nucleologenesis and, as far as the morphological background may have been interpreted, it was attributed either to the perinucleolar chromatin or to the nucleolar bodies (Antalíková et al., 1989).

DISCUSSION

Up to the beginning of the 8-cell cleavage stage the NPB of the cow embryo is probably composed, as is the case in other mamlian embryos (see Tesarík et al., 1987) of proteinaceous fibrils only. Since no silver-staining was ever detected in 4-cell cow NPB (Antalíková et al., 1989) its composition may parallel that of the compact nucleolar body occuring at the end of rat oogenesis (Antoine et al., 1988), namely by presence of acidic proteins but a total lack of argyrophilic proteins, as well as of nucleic acids.

The results reviewed in this paper suggest a clear-cut relation between DNA penetration into the cow proteinaceous NPB and the onset of RNA synthesis, leading to temporal and spatial differentiation of the embryonic nucleolus. Since synthesis of rRNA plays an essential role in nucleolo-genesis in somatic cells (reviewed by Jiménez-García et al., 1989), a similar relation is to be expected in the embryonic nucleologenesis in pre-implantation cow conceptuses. In mitotic nucleologenesis in somatic cells the nucleolus is reconstituted rapidly around specific loci on chromosomes containing rRNA genes (review Goessens 1984) in contrast to a lengthy gradual nucleologenesis in mammalian embryo where rRNA genes are probably not transcribed until associated with the NPB (Tesarík et al., 1987). Recent studies, however, show that prenucleolar bodies appearing during telophase of somatic cells mitosis are involved in a complex interaction among nucleolar proteins and rDNA as well as rRNA (see Jiménez-García et al., 1989). So similar methods and concepts as currently applied to the study of somatic cell nucleologenesis should be productive in the equivalent situation occurring in mammalian embryos.

REFERENCES

Antalíkova L., Kopecný V., and Fulka J. Jr. 1989, Ultrastructural localiz-ation of silver-staining nuclear proteins in early bovine embryos (Abstr.), Histochem. J., In Press.

Antoine N., Lepoint A., Baeckland E., and Goessens G., 1988, Ultrastructural cytochemistry of the nucleolus in rat oocytes at the end of the folliculogenesis, Histochemistry, 89:221.

Camous S., Kopecný V., and Fléchon J. E. 1986, Autoradiographic detection of the earliest stage of ^{3}H-uridine incorporation into the cow embryo, Biol. Cell., 58:195.

Goessens G. 1984, Nucleolar structure, Int. Rev. Cytol. 87:107.

Jiménez-Garcia L. F., Rothblum L. I., Busch H., and Ochs R. L. 1989, Nucleologenesis: use of non-isotopic in situ hybridization and immuno-cytochemistry to compare the localization of rDNA and nucleolar proteins during mitosis, Biol. Cell., 65:239.

King W. A., Chartrain I., Kopecný V., Betteridge K. J., and Bergeron H. 1989, Nucleolus organizer regions and nucleoli in mammalian embryos, J. Reprod. Fert., In Press.

Kopecný V., Fléchon J. E., Tománek M., Camous S., and Kanka J. 1985, Ultrastructural analysis of ^{3}H-uridine incorporation in early embryos of pig and cow (Abstr.), ECBO 9th Nucleolar Workshop, Krakov, p. 31.

Kopecný V., Fléchon J. E., Camous S. and Fulka J. Jr. 1989a, Nucleologenesis and the onset of transcription in the 8-cell bovine embryo: Fine-structural autoradiographic study, Molec. Reprod. Develop. 1:79.

Kopecný V., Fulka J. Jr., Pivko J., and Petr J. 1989b, Localization of replicated DNA-containing sites in preimplantation bovine embryo in relation to the onset of RNA synthesis, Biol. Cell., 65:231.

Tesarík J., Kopecný V., Plachot M., and Mandelbaum J. 1987, High-resolution autoradiographic localization of DNA containing sites and RNA synthesis in developing nucleoli of human preimplantation embryos: A new concept of embryonic nucleologenesis. Development 101:777.

STRUCTURAL TRANSITIONS IN MAMMALIAN CELL NUCLEOLI CONCOMITANT TO RIBOSOMAL

GENE ACTIVATION AND INACTIVATION

O. V. Zatsepina and Yu. S. Chentsov

A. N. Belozersky Laboratory of Molecular Biology and
    Bioorganic Chemistry
Moscow State University
Moscow 119899
USSR

The intranucleolar localization of transcriptionally active rRNA genes and the role of the fibrillar centers (FCs) and dense fibrillar component (DFC) in ribosome biogenesis is still a matter of debate (for reviews see Hadjiolov, 1985; Scheer and Rose, 1984). A possible approach to the understanding of the precise role of various nucleolar components in the transcription of rRNA genes is their ultrastructural analysis, in the nucleoli with different functional activity.

In this work we describe the structural and quantitative parametres of nucleoli, FCs and DFC observed in the process of natural as well as artificial modulation of ribosome gene (r-genes) activity.

As a test model, cultured pig kidney cells (PK-cells), containing only two nucleolus-organizing (NO) chromosomes (or two nucleolus-organizing regions, NORs) per haploid set, were chosen (Hozak et al., 1986; Zatsepina et al., 1988a). To observe the natural alterations in the nucleolar activity and architecture, PK cells at varying stages of the cell cycle (G_0, G_2 periods and mitosis) were used. It is well known that rRNA synthesis is completely absent in mitosis, is reduced in the G_0 period and proceeds vigorously in the G_2 period. Individual cells in the G_0 and G_2 periods were identified with the aid of double label autoradiography.

The cells were cut into serial ultrathin sections and were photographed under constant electron optical magnification (x 15,000). A Hewlett-Packard computer was used to measure the area of nucleoli, FCs and the DFC, as well as the perimeter of FCs. Based on this accumulated data, the surface area and volume of the nucleolar structures were calculated (Zatsepina et al., 1988a).

Analysis of the serial ultra-thin sections of the normal PK cells in the G_0 and G_2 periods shows that the interphase FCs and the mitotic NORs are practically the same in their structure: they are formed by loose filaments about 10nm in diameter. At metaphase each FC corresponds to one NOR of the NO-chromosome. It should be noted that the metaphase FCs are devoid of the DFC. The results of morphometric analysis of various nucleolar components of normal PK cells in the G_0, G_2 periods and at mitosis are given in Table I. This shows that the number of FCs is proportional to the degree of

Table I. Quantitative Characteristics of Nucleoli (Nu),
Fibrillar Centers (FC), and Dense Fibrillar
Component (DFC) in PK Cells in the G_0 Period,
the G_2 Period and Mitosis*).

Period, Cell ploidy	Total vol. of Nu (μm^3)	No of FC	Mean vol. of indiv. FC (μm^3)	Total vol. of FC (μm^3)	Total surface area of FC (μm^2)	Total vol. of DFC (μm^3)
G_0 period, 2C	8.1	7	0.033±0.005	0.212	3.54	0.405
G_2 period, 4C	23.4	33.7	0.014±0.001	0.430	8.27	1.115
Mitosis, 4C	–	6-8	0.025±0.002	0.15-0.20	2.85	–

*) 3 cells in the G_0 , 3 cells in the G_2 period and 5
cells in mitosis were analyzed.

nucleolar activity (or the overall nucleolar volume): in the G_2 period the
nucleoli contain four times more FCs than in G_0. At the same time, the
activity of the nucleolus correlates inversely with the mean volume of
individual FCs: the less active the nucleolus, the larger the volume of
individual FCs. Yet, despite the larger dimensions of individual FCs, the
total volume of FCs per cell in the G_0 period is half of that in the G_2.
However, in terms of a haploid set of chromosomes, the total volume of FCs
in these periods of interphase is about the same ($0.105\mu m^3$ in the G_0 and
$0.107\mu m^3$ in the G_2 period; see also Zatsepina et al., 1988b). Accordingly,
the change in the number of FCs in interphase is not accompanied by a
dramatic change in their total volume and may be due to the disassociation
or fusion of individual FCs.

Two experimental approaches were used to testify this supposition. As
shown previously, UV microbeam irradiation of one of the two mature
nucleoli within the same interphase nucleus causes significant dimimution
and inactivation of the irradiated nucleolus and compensatory growth and
activation on non-irradiated one (Sakharov and Voronkova, 1976). In this
work we concentrated our attention upon the quantitative parametres of FCs
and the DFC in the degraded (irradiated) and hypertrophied (non-irradiated)
nucleoli of PK cells in the G_1 and G_2 periods. The cell cycle stage was
identified on the basis of the behaviour of the centrioles in interphase PK
cells (Zatsepina et al., 1989).

Analysis of the serial ultra-thin sections of nucleoli shows that the
activation of additional r-genes within the hypertrophied nucleoli is
accompanied by an accumulation of the RNP-granular component, by an increase
in the number of FCs and a decrease in their size. Within the degraded
nucleoli the reverse processes, i.e. a loss of RNP-granules, an expansion of
individual FCs and a decrease in their number, was revealed. Furthermore,
the loss of the DFC occurs through the inactivation of r-genes, while in
the hypertrophied nucleolus a layer of DFC grows significantly. Rough
estimations show that the total volume of FCs per nucleolus in the degraded,
normal and hypertrophied states are practically the same. Figures 1 and 2

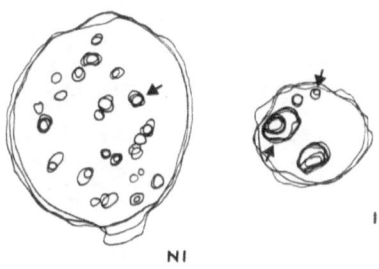

NI   I

Figure 1.   Ultrastructural reconstruction of FCs (arrows) within the
             degraded (irradiated, I) and the hypertrophied (non-irradi-
             ated, NI) nucleoli in a PK cell in the G_1 period.

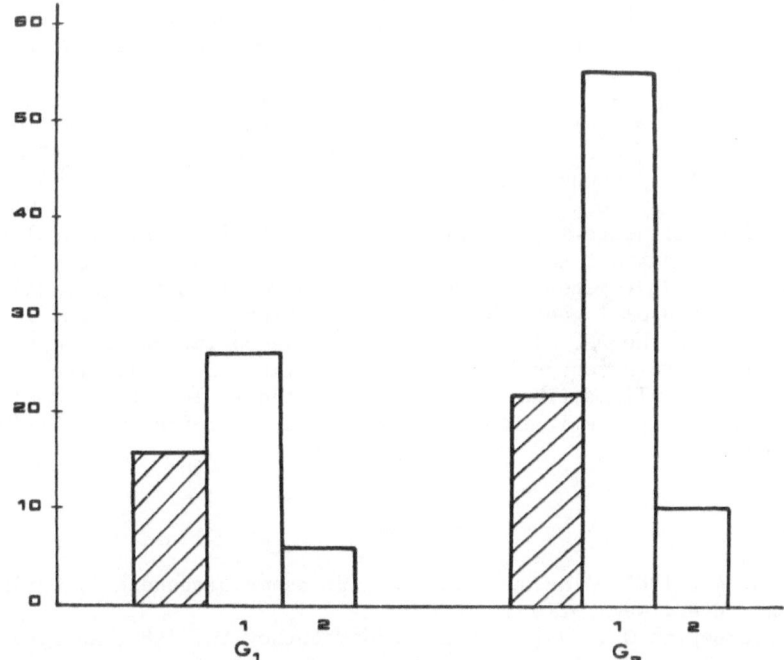

Figure 2.   A histogram illustrating the mean number of FCs in control
             (shaded), non-irradiated (hypertrophied, 1), and irradiated
             (degraded, 2) nucleoli in PK cells in the G_1 and G_2 periods.
             Ordinate, the FCs number.

sum up the results of the quantitative analysis of FCs in irradiated cells
in the G_1 and G_2 periods.

      These findings are in agreement with the data obtained following the
activation of nucleoli by treatment of the cells with 5-aza-C, an inhibitor
of enzymatic methylation of DNA. PK cells grown in the presence of 20μM
5-aza-C for 12-24h were transferred to the drug-free culture medium for
48-72h. Subsequent electron microscopy and the Ag-staining of nucleoli
revealed a 1.5-3 fold increase in the number of FCs and a decrease in their
linear size, compared with the control nucleoli.

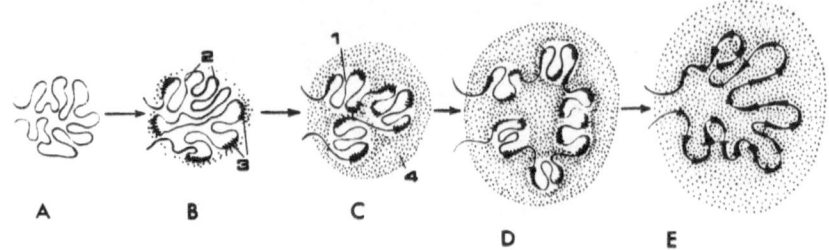

Figure 3.   The scheme illustrating the structural changes in the nucleolus through the increase of ribosomal gene activity.
1 - FC; 2 - silent r-gene; 3 - transcribing r-gene; 4 - RNP-
granules. For more details see the text.

Hence, alterations in nucleolar architecture obtained under natural or
artificial changes of the level of rDNA transcription, prompt us to
conclude that the activation of additional ribosomal genes is accompanied by
FC disassociation, while the NOR inactivation is accompanied by (or leads
to) FC fusion. As a result, a significant increase (decrease) of the
surface area of FCs on which pre-rRNA may be synthetized, occurs. On the
whole, our findings support the view, that the transcription of rRNA genes
takes place at the periphery of FCs.

The scheme illustrating the proposal concept of the structural transition of FCs concomitant to NOR activation is shown in Figure 3. When rDNA
is not transcribed, r-genes are located within FCs (NORs; Figure 3a). The
activation of ribosomal genes starts at the periphery of FCs, and leads to
the appearance of the DFC as well as to the disassociation of comparatively
large FCs into more numerous and small ones (Figure 3, b-d). The activation
of all the potentially active r-genes is expected to lead to a complete
disappearance of FCs, as it is in the amplified amphibian nucleoli (Miller,
1981, Figure 3e).

REFERENCES

Hadjiolov A. A., 1985, 'The nucleolus and ribosome biogenesis', Springer-
    Verlag, Wien, New York.
Hozak P., Zatsepina O., Vasilyeva I., and Chentsov Yu. 1986, An electron
    microscopic study of nucleolus organizing regions at some stages of the
    cell cycle (G_0 period, G_2 period, mitosis), Biol. Cell, 57:197.
Miller O. L. 1981. The nucleolus, chromosomes and visualization of genetic
    activity, J. Cell Biol., 91:15 s.
Sakharov V. N., and Voroncova L. N. 1976, Nucleolus degradation and growth
    induced by UV-microbeam irradiation of interphase cells growth in
    culture, J. Cell Biol., 71:963.
Scheer U., and Rose K. M. 1984. Localization of RNA polymerase I in inter-
    phase cells and mitotic chromosomes by light and electron microscopic
    immunocytochemistry, Proc. Natl. Acad. Sci. USA, 81:1431.
Zatsepina O. V., Hozak P., Babedjanyan D., and Chentsov Yu. 1988a,
    Quantitative ultrastructural study of nucleolus-organizing regions at
    some stages of the cell cycle (G_0 period, mitosis), Biol. Cell, 62:211.
Zatsepina O. V., Chelidze P. V., and Chentsov Yu. S. 1988b, Changes in the
    number and volume of fibrillar centres with the inactivation of
    nucleoli at erythropoiesis, J. Cell Sci., 91:439.
Zatsepina O. V., Voronkova L. N., Sakharov V. N., and Chentsov Yu. S. 1989,
    Ultrastructural changes in nucleoli and fibrillar centers under the
    effect of local ultraviolet microbeam irradiation of interphase
    culture cells, Exp. Cell Res., 181:94.

NUCLEOLAR CHANGES IN HYPOTONIC TREATMENT AND THE POSSIBILITY OF ISOLATION

OF NUCLEOLAR FIBRILLAR COMPLEXES

P. Hozák[1]*, O. V. Zatsepina[2] and Z. Likovský[1]

[1] Institute of Experimental Medicine
Czechoslovak Academy of Sciences
Lidových milicí 61
120 00 Prague 2
Czechoslovakia

[2] A. N. Belozersky Laboratory of Molecular Biology and
    Bioorganic Chemistry
Moscow State University
Moscow 119899
USSR

INTRODUCTION

Nucleoli regularly contain the following components: the dense fibrillar component, fibrillar centers the granular component and the nucleolar
chromatin (Schwarzacher and Wachtler, 1983; Goessens, 1984; Fakan and
Hernandez-Verdun, 1986). Morphological studies suggest a parallel
occurrence of the first two components: the fibrillar centers were never
found in the active nucleoli without adjacent dense fibrillar component, and
the fibrillar centers together with the dense fibrillar component (i.e.
fibrillar complexes) seem to correspond to individual nucleolar organizers
(Mirre and Stahl, 1981; Mirre and Kniebiehler, 1982; Hozák et al., 1986;
Cataldo et al., 1988; Hozák et al., 1989).

The present study describes the reaction of nucleolar organizers to a
simple hypotonic shock. This method enables a separation of fibrillar
complexes and can be effectively used for a detailed study of fibrillar
centers and the dense fibrillar component.

MATERIAL AND METHODS

Cells. We used lymphocytes obtained from peripheral blood of the pig,
as well as mouse Ehrlich tumor cells. The lymphocytes were separated by
centrifugation of defibrinated blood on a layer of 1.086g/ml Verografin
(Spofa). The cells were then triple-washed in 0.1M Sörensen phosphate
buffer (pH 7.3). The ICR C57 mice were intraperitoneally transplanted with
ascitic Ehrlich tumor. After seven days, the ascitic fluid with cells was
taken by intraperitoneal puncture and the cells centrifuged and triple-
washed in 0.1M Sörensen buffer.

Hypotonic treatment. Suspensions of cells in 0.1M Sörensen buffer were

---

* Correspondence.

Nuclear Structure and Function, Edited by J. R. Harris and
I. B. Zbarsky, Plenum Press, New York, 1990

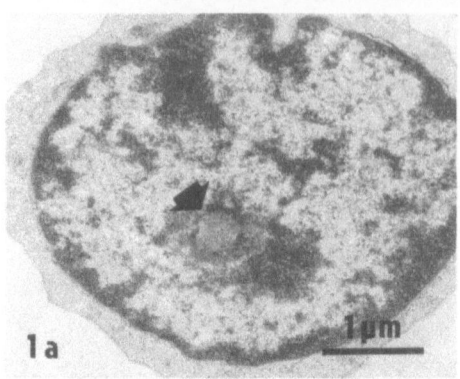

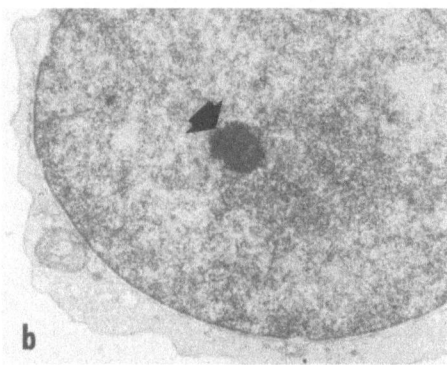
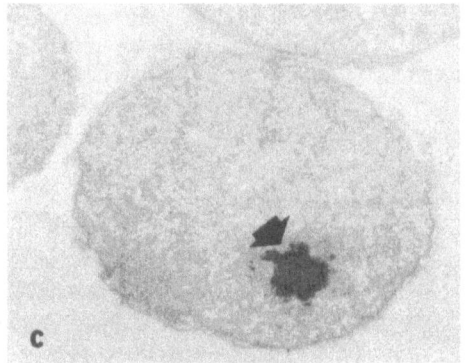

Figure 1.   a-c: Lymphocytes of the peripheral blood of the pig.
a: A lymphocyte following incubation in isotonic 0.1M
phosphate buffer: the arrow points to the nucleolar fibrillar
complex with a single large fibrillar center.
b: A lymphocyte following incubation in hypotonic 0.05M
phosphate buffer. The arrow points to the separated
fibrillar complex.
c: Silver-staining of lymphocyte nuclei isolated in hypotonic
environment. The fibrillar complex shows an intensive
argentophile reaction (arrow).

placed in Eppendorf test-tubes. After 2-minute centrifugation, the cells
were resuspended in the hypotonic phosphate Sörensen buffer (pH 7.3). The
buffer concentrations were graded in 0.01M increments from 0.01M to 0.08M.
As controls, the cells in the isotonic environment of 0.1M phosphate buffer
were used. All the groups were incubated at $37^{o}C$ for the duration of 10
minutes.

Isolation of cell nuclei. The isolation of cell nuclei was undertaken
in the environment of hypotonic phosphate buffer (pH=7.3) with a minimum
concentration of Nonidet P-40, at $0-2^{o}C$. The homogenization solution for
pig lymphocytes consisted of 0.085M phosphate buffer at 0.025% NP-40, 0.05mM
$MgCl_2$, 1mM phenylmethylsulfonyl fluoride, 10µg/ml of leupeptin and 1µM
pepstantin and for Ehrlich tumor cells it consisted of 0.04M phosphate

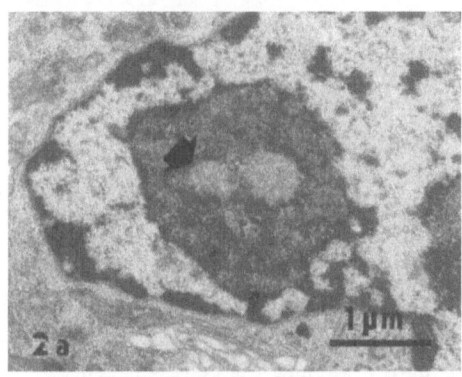

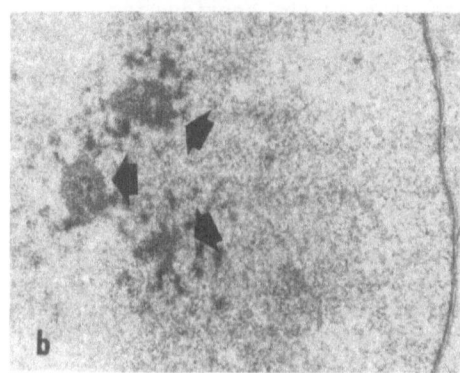
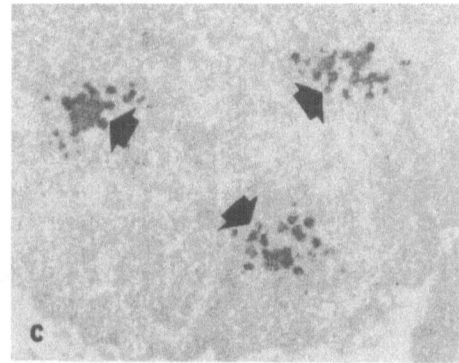

Figure 2.   a-c: Ehrlich mouse ascitic tumor cells.
a: Following incubation in 0.1M phosphate buffer; the arrow points to the fibrillar complex with two fibrillar centers, which is immersed in the nucleolus.
b: Cell section following incubation in hypotonic 0.035M phosphate buffer. A separation of fibrillar complexes took place (arrows).
c: Silver-staining of nuclei isolated in hypotonic environment. Argentophilic structures were provided by fibrillar complexes (arrows).

buffer with 0.2% NP-40, 0.05mM $MgCl_2$ and inhibitors of proteases. After 12 to 15 passes of a teflon pestle, the nuclei were centrifuged for 5 minutes at (100g) and resuspended in 0.25M saccharose solution in 0.07 or 0.04M phosphate buffer with 0.05mM $MgCl_2$. Following underlaying with 0.88M sucrose solution of the same concentration of buffer and $MgCl_2$, the nuclei were centrifuged once more for 7 minutes at 1,000g. The whole isolation procedure was monitored by means of electron microscopy.

Electron microscopy. The cells and the isolated structures were separated by a short centrifugation and immediately resuspended in the fixation solution (2.5% glutaraldehyde in buffers of corresponding concentration). Following 1-hour fixation at 4°C and washing in corresponding buffers, the cells were postfixed in 1% OsO_4 for 2 hours at room temperature and further dehydrated by ethanol, saturated with propylenoxide and cast in the Epon-Durcupan resin mixture. The ultrathin sections were contrasted in a standard manner by uranylacetate and lead nitrate according to Reynolds and observed under the OPTON EM 109 Turbo electron microscope at 50kV.

Silver staining. Prior to staining, the cells and the isolated structures were fixed for ten minutes in the 1.6% solution of glutaraldehyde in phosphate buffer of appropriate concentration, washed for 10 minutes in the identical buffer, and postfixed in the mixture of ethanol and acetic acid at the volume ratio 3 to 1 for the duration of 10 minutes. Following rehydration and washing in distilled water, the cells were silver stained in the mixture of 1 volume part of 50% solution of $AgNO_3$ and 1 part of stabilized formaldehyde with acetate buffer (Likovsky and Smetana, 1981) for four minutes at 60°C. The cells were thoroughly washed in redistilled water, omitting uranyl acetate, dehydrated and cast in the above manner. To achieve superior identification of cell structures, the sections were briefly contrasted with uranyl acetate.

RESULTS AND DISCUSSION

Comparison of the ultrastructure of control cells showed the distribution of nucleolar components corresponding to individual cell types: a large central fibrillar center with less developed other components in lymphocytes of low metabolic activity and compact nucleoli with significant granular component and larger number of fibrillar centers surrounded by the zones of the dense fibrillar component in active Ehrlich ascitic tumor cells (Figures 1a and 2a).

The reactions of nucleolar components were identical in active and resting cells, although the different cell types showed varying degrees of sensitivity to hypotonic treatment. At appropriately chosen low ionic strength, a large scale dispersal of the granular component took place and all the nuclear chromatin was decondensed into fibrils of about 25nm thickness. Where nucleoli were originally found, there are structurally unchanged fibrillar centers and the dense fibrillar component which, similarly to intact nucleoli, covers a part of the surface of fibrillar centers and forms bridges connecting individual fibrillar centers (Figures 1b and 2b).

The simultaneous presence of fibrillar centers and the dense fibrillar component is well-known from a number of morphological studies (Mirre and Stahl, 1981; Mirre and Kniebiehler, 1982; Goessens, 1984; Hozak et al., 1986; Cataldo et al., 1988; Hozak et al., 1989); it seem that the fibrillar complexes correspond to individual nucleolar organizers (Hozak et al., 1986; Hozak et al., 1989). As to the function of the two parts of fibrillar complexes, immunocytochemical methods recently revealed the presence of RNA-polymerase I and DNA in the peripheral parts of fibrillar centers, but not in the dense fibrillar component (Scheer and Rose, 1984; Derenzini et al., 1987; Scheer et al., 1987; Thiry et al., 1988a, b). Although a precise definition of these two nucleolar component has not yet been achieved, the fibrillar complexes are probably morphological-functional units of nucleoli. The above-described finding of the survival of fibrillar complexes following hypotonic shock is another indication in support of this opinion.

The above values of buffer concentrations, homogenization detergent and Mg^{2+} ions were found to be appropriate for the given cell types, enabling us to obtain the fraction of cell nuclei in which fibrillar complexes were the only significant structures that survived. The silver-staining reaction showed that fibrillar centers and the dense fibrillar component were

200

significantly stained, while the other structures in interchromatin areas were stained to a lesser degree (Figures 1c and 2c). The localization of argentophilic reaction product was impaired neither by the hypotonic treatment nor by isolation procedures. It seems that the protein composition of fibrillar complexes (or, at least, the presence of argentophilic proteins) was not affected by the above procedures.

The separation of fibrillar complexes by means of hypotonic media may become a useful approach, not only in the study of the internal organization of nucleoli, but also by providing detailed biochemical and immunocytochemical characteristics of the fibrillar complexes following their isolation from cell nuclei.

ACKNOWLEDGEMENTS

We thank Prof. K. Smetana, Institute of Hematology and Blood Transfusion, Prague, for helpful discussion throughout the study and Mrs A. Jelinková, Mrs V. Chobotová and Mrs A. Jirasková for their perfect technical assistance.

REFERENCES

Cataldo C., Souchier C., and Stahl A. 1988, Three-dimensional ultrastructure and quantitative analysis of the human Sertoli cells nucleolus, Biol. Cell, 63:277.

Derenzini M., Farabegoli F., Pession A., and Novello F. 1987, Spatial redistribution of ribosomal chromatin in the fibrillar centers of human circulating lymphocytes after stimulation of transcription, Exp. Cell Res., 170:31.

Fakan S., and Hernandez-Verdun D. 1986, The nucleolus and the nucleolar organizer regions, Biol. Cell, 56:189.

Goessens G., 1984, Nucleolar structure, in: 'International review of cytology', Academic Press, New York.

Hozák P., Novák J. T., and Smetana K. 1989, Three-dimensional reconstructions of nucleolus-organizing regions in PHA-stimulated human lymphocytes, Biol. Cell, 66, in press.

Hozák P., Zatsepina O., Vasilyeva I., and Chentsov Yu. 1986, An electron microscopic study of nucleolus - organizing regions at some stages of the cell cycle (G_0 period, G_2 period, mitosis), Biol. Cell, 57:197.

Likovský Z., and Smetana K. 1972, Further studies on the cytochemistry of the standardized silver staining of interphase nucleoli in smear preparations of Yoshida ascitic sarcoma cells in rats, Histochemistry, 72:301.

Mirre C., and Kniebiehler B. 1982, A re-evaluation of the relationships between the fibrillar centers and the NORs in reticulated nucleoli: ultrastructural organization, number and distribution of the fibrillar centers in the nucleolus of mouse Sertoli cell, J. Cell Sci., 55:247.

Mirre C. and Stahl A. 1981, Ultrastructural organization, sites of transcription and distribution of fibrillar centers in the nucleolus of the mouse oocyte, J. Cell Sci., 48:105.

Scheer U., Messner K., Hazan R., Raska I., Hansmann P., Falk H., Spiess E., and Franke W. 1987, High sensitivity immunolocalization of double-end single-stranded DNA by a monoclonal antibody, Eur. J. Cell Biol., 43:358.

Scheer U., and Rose K. M. 1984, Localization of RNA polymerase I in interphase cells and mitotic chromosome by light and electron microscopic immunocytochemistry, Proc. Natl. Acad. Sci. USA, 81:1431.

Schwarzacher H. G., and Wachtler F. 1983, Nucleolus organizer regions and nucleoli, Hum. Gen., 63:89.

Thiry M., Scheer U., and Goessens G. 1988a, Immunoelectron microscopic
    study of nucleolar DNA during mitosis in Ehrlich tumor cells, <u>Eur. J.
    Cell Biol.</u>, 47:346.
Thiry M., Scheer U., and Goessens G. 1988b, Localization of DNA within
    Ehrlich tumor cell nucleoli by immunoelectron microscopy, <u>Biol. Cell</u>,
    63:27.

TRANSCRIPTIONALLY ACTIVE AMPHIBIAN OOCYTE NUCLEOLAR CHROMATIN IS ORGANIZED

IN HIGHER ORDER STRUCTURE

A. G. Tsvetkov and V. N. Parfenov

Institute of Cytology of the USSR Academy of Science
Leningrad
USSR

The structure of transcriptionally active chromatin of amplified oocyte nucleoli of amphibians has been studied in detail (Miller, Beatty, 1969; Scheer, Zentgraf, 1982). It has been shown that tandem arrays of transcription units (TUs) are the basic structural elements transcribed in chromatin of nucleoli. However, light microscope data on the structure of nucleoli during oogenesis give some grounds to assume that transcriptionally active chromatin of nucleoli is organized into a higher order structure, as compared to the linearly arranged TUs. For instance, in vitellogenic amphibian oocytes, the spherical nucleoli acquire the appearance of bead-string rings (Callan, 1966). Ring structures with beads along the axis have been found directly in the body of spherical nucleoli during examination of thin section oocyte nuclei (Parfenov, 1983). The question of the correspondence of structures observed by different methods of the investigation is still to be answered. The goal of this study was to examine the amplified nucleoli by light and electron microscopy and to describe the organization of their transcriptionally active chromatin.

Figure 1 shows a nucleus, isolated from an oocyte of early vitellogenesis (size 0.6mm). The vacuolized nucleoli ranging from 6 to 10mkm are located under nuclear envelope, whereas lampbrush chromosomes and spheres (2mkm) around them are in the centre of the nucleus. The examination of thin section of such nucleoli (Figure 2) revealed the presence of two zones; dense fibrillar aggregate and granular material. With ionic strength decreasing to 25mM the nucleoli lose spherical shape and are transformed into beadstring structures of different size. The dissolution of the cortex of nucleoli results in the appearance of these structures (Figure 3). Ring-like pattern can be easily distinguished from lampbrush chromosomes (Figures 4 and 5). The beadstring ring are presented as series of 'fur-trimmed- beads of about the same size 2-2.5μm, the diameter of their dense core ranges from 0.6 to 0.8μm. It should be pointed out that the lose material surrounding the beads is similar to RNP-matrix of lateral loops of lampbrush chromosomes.

Using a gentle nucleolar dispersal procedure it has been demonstrated that the general view of nucleoli seen in electron microscope does not differ from ring-like structures seen in light microscope (Figures 6 and 7). The dispersed nucleoli represent a chain of rosette-like structures, where rosettes correspond to beads of the ring-like structures. Evidently under such experimental conditions the transcriptionally active nucleolar

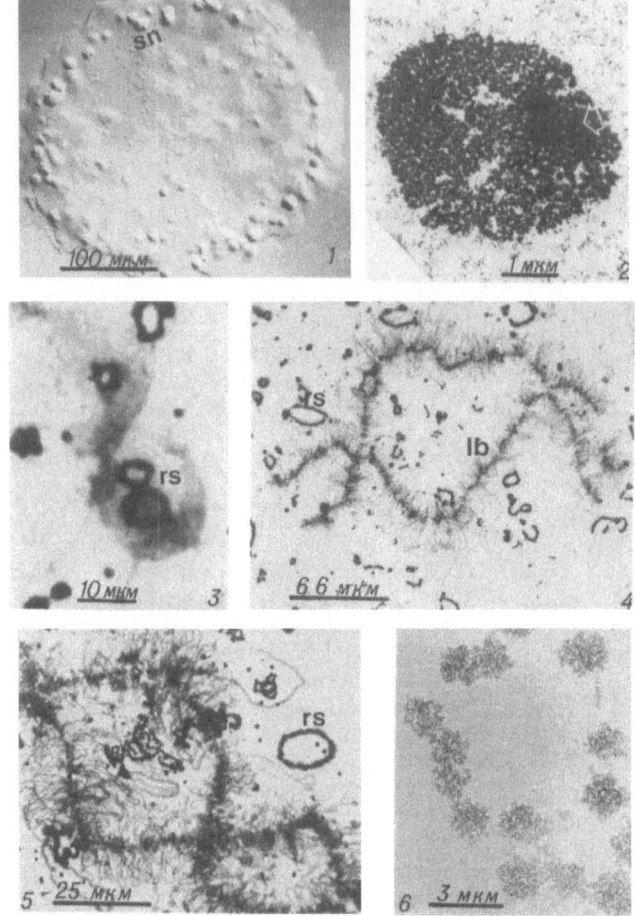

Figure 1.   General view of the nucleus, isolated manually from
            vitellogenic oocyte. At the periphery of the nucleus
            numerous spherical nucleoli (SN) are seen.
Figure 2.   Thin section of the nucleolus. Two zones can be distin-
            guished: dense fibrillar aggregate (arrow) and granular
            material.
Figure 3.   Dissolution of the nucleoli cortex and ring-like structure
            (RS) emergence.
Figures 4   Nuclear structures after dispersion. Ring-like structures
and 5.   (RS) are well distinguished from lampbrush chromosomes (LB).
Figure 6.   Morphology of the nucleolar chromatin in electron
            microscopic spread preparation.

chromatin has the least changed structure. Adjacent rosettes are connected
to each other either by regions of non-nucleosomal chromatin or by TUs of
ribosomal genes. The investigation of the structure of rosettes reveals
that the loops represent regions of transcriptionally active chromatin.
The number of loops attached to the rosette is from 6 to 10, the size of
dense core is about 1μm and the rosette diameter with loops ranges from
2 to 2.5μm (Figure 8), corresponds to parameters of beads in ring-like
structures seen in light microscope. The structure of the well dispersed
rosette is difficult to identify: only RNP-transcripts surrounding

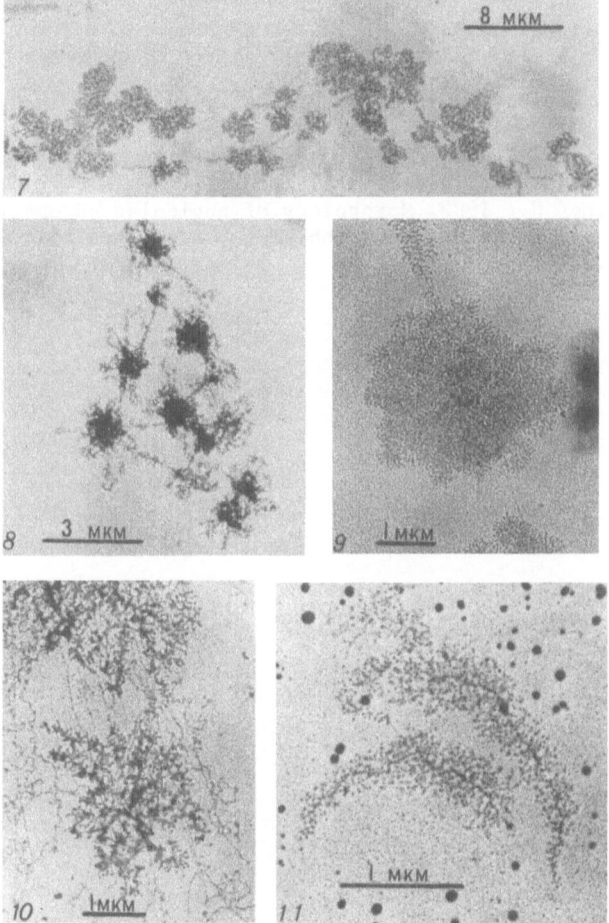

Figure 7.   Rosette-like structures.
Figure 8.   The appearance of rosette after uranyl acetate staining.
Figure 9.   The rosettes connected with adjacent structures by the means
             of nonbeaded chromatin as well as the means of transcription
             units.
Figure 10.   The loops of the rosettes presenting ribosomal transcription
             units.
Figure 11.   The structure of transcriptionally active nucleolar chromatin
             after complete dispersion. Only single ribosomal trans-
             cription units are seen.

rosettes are discernible (Figure 9). Figure 10 shows the rosette with
loops, which represent TUs of ribosomal genes. Complete dispersal of
nucleoli results in the damage of their initial structure that they are
presented by a set of TUs (Figure 11). Thus, transcriptionally active
chromatin of amplified nucleoli represents a regularly organized structure,
whose basic elements are discretely arranged rosettes rather than linear
tandem arrayed TUs, as has been previously considered.

REFERENCES

Callan H. G., 1966, Chromosomes and nucleoli of the axolotl, Ambystoma
    mexicanum, J. Cell Sci., 1:85.
Miller O. L., Beatty B. R., 1969, Visualization of nucleolar genes, Science,
    164:955.
Parfenov V. N., 1983, Nuclear structures of oocytes from atretic follicules
    in Rana ridibunds Pallas, Monitore zool. ital. (N.S.), 17:247.
Scheer U., Zentgraf H., 1982, Morphology of nucleolar chromatin, in 'The
    Cell Nucleus', Busch H. and Rothblum L., eds., Academic Press, New
    York, 11:143.

NUCLEOLAR ORGANIZING REGION (NOR) AND ARGENTOPHILIC NUCLEAR PROTEINS IN

FERTILIZATION AND INDUCED PARTHENOGENESIS

A. P. Dyban, E. L. Severova, E. M. Noniashvili,
O. V. Zatsepina* and Yu. S. Chentsov*

Department of Embryology
Institute for Experimental Medicine
Academy of Medical Science
Leningrad 197022
USSR

* Belozersky Laboratory of Molecular Biology
Moscow University
Moscow 117334
USSR

INTRODUCTION

Silver staining (Goodpasture and Bloom, 1975; Howell and Black, 1980) is widely used for visualizing in eukaryotic cells the transcriptionally active nucleolar organizing regions (NOR) (for review see Howell, 1982; Hubbell, 1985). It is well known that at very early stages of mammalian embryonic development chromosomes are genetically inactive, and ribosomal genes, like any others, are not transcribed (for review see Magnuson and Epstein, 1981; Johnson, 1981; Dyban and Baranov, 1987). The silver stained NORs (Ag-NORs) in metaphase chromosomes were detected in the 2 cell mouse embryos and at later stages (Engel, Zenzes, Schmid, 1977; Hansmann et al., 1978; Patkin and Sorokin, 1983). There is also data that parthenogenetically activated mouse ova have no Ag-NORs (Hansmann, Gebauer and Grimm, 1978). All these observations are very preliminary and need further study.

In our work, silver staining according to the method of Howell and Balck (1980) was used in light and electron microscopic studies for detection the localization of argentophilic nuclear proteins in fertilized and parthenogenetically activated mouse ova and cleaving embryos.

SILVER STAINED NORs IN METAPHASE CHROMOSOMES OF CLEAVING FERTILIZED OVA

Ag-NORs (pairs of dark black dots) were not detected in the metaphase chromosomes of a single 1-cell embryo. Unlike the 1-cell embryos, some 2-cell embryos had chromosomes with, and some without Ag-NORs. When the embryos were incubated in colcemide for 2-3h and the mitosis was blocked at the metaphase stage, no metaphase chromosomes with Ag-NORs were detected. When no colcemide was used, all prometaphases, but only some of the metaphases (23 out of 98) were found to have chromosomes with Ag-NORs. Their number per metaphase varied from 1 to 6.

In the third mitotic cleavage the metaphase plates always possessed chromosomes with clearly seen Ag-NORs. Their number per metaphase varied from 3 to 8 and did not depend on duration of the incubation in colcemide.

In the fourth and fifth mitotic cleavage all metaphases under study had chromosomes bearing Ag-NORs, their number varying from 4 to 8.

SILVER STAINED NORs IN METAPHASE CHROMOSOMES IN PARTHENOGENETICALLY ACTIVATED OVA

At the 1-cell stage in diploid as well as haploid ova, no Ag-NORs were detected in metaphase chromosomes. They were found beginning from the 2-cell stage onward. The average number of Ag-NORs per metaphase in diploid parthenogenic embryos was the same as in fertilized cleaving ova, reaching at the earlier stages (4-8 cell embryos) 6-8, and at the later stages 8-10. In the metaphase of haploid parthenogenes this number was 5.

ARGENTOPHILIC PROTEINS IN THE 1-CELL STAGE EMBRYOS

The localization of argentophilic proteins (Ag-proteins) was studied from fertilization or parthenogenetic activation until entering into the first mitotic cleavage. The distribution of Ag-proteins in parthenogenic and fertilized embryos was the same.

Immediately after a sperm penetrated the zone pellucida and entered the egg cytoplasm its head had no Ag-granules. When the sperm head started swelling, a large number of Ag-granules were found on the decondensing sperm chromatin. The metaphase and anaphase chromosomes of the second meiotic division possessed no Ag-NORs. Unlike this, a group of telophase maternal chromosomes was always covered with numerous dark black Ag-granules. This resembled the picture when a paternal pronucleus was being formed from the decondensing sperm chromatin. With the growth of both pronuclei, large Ag-granules surrounded by lighter zones were clearly seen in each of them.

In the fully formed pronuclei Ag-proteins were detected in the form of black granules located in pronucleoli. If the pronucleus had one pronucleolus, it was the site of all argentophilic granules, in case of several pronucleoli, silver grains accumulated in each.

In the electron microscopic study of embryos at these stages of development the pronucleoli were seen as homogenous electron dense bodies which had neither reticular, nor granular components. After silver staining the argentophilic regions made up of fine granular material were clearly visualized on the surface of pronucleoli in the form of caps. They were also detected inside the nucleoli. In some slides argentophilic 'caps' were less tight in the periphery, being composed of small loose granules detached from the pronucleoli entering the karyoplasm.

In prophase of the first mitotic cleavage most of the chromosomes were associated with pronucleoli which had already started to dissolve, and Ag-granules could be seen on the prophase chromosomes. In prometaphase Ag-granules could be detected, all this material being associated with the remnants of pronucleoli and prometaphase chromosomes. In telophase all chromosomes were covered with dark black (argentophilic) granules.

CONCLUSION AND SPECULATIONS

It should be pointed out that the method of Howell and Black (1980) detects a complex of acidic nonhistone Ag-proteins bound with ribosomal genes (Howell, 1982; Goessens, 1984). At least three proteins (C23, B23, and one designated as 'AgNOR protein') have been stained by silver nitrate. Protein C23 (nucleoline) is assumed to be a major silver staining nucleolus organizing region protein (Ochs et al., 1983; Ochs and Busch, 1984). Some of the other Ag-proteins correspond probably to a large subunit of RNA polymerase I (195000MW doublet) (Williams et al., 1982; Scheer and Rose, 1982; Goessens, 1984). This enzyme has a key role in the synthesis of rRNA, and the silver staining method used to detect NOR in the karyotypes of several organisms, including mice, is widely accepted as a method of visualizing the ribosomal genes which were transcribed during the previous cell cycle (Miller et al., 1976).

It was shown in our study that the metaphase chromosomes of the first cleavage division contain no Ag-NORs, the latter were found from the beginning of the 2-cell stage on. This is in good agreement with the available data, not only on the developmental stages when in mouse embryos Ag-NORs were detected (Engel et al., 1977; Hansmann et al., 1978; Patkin and Sorokin, 1983), but also with the finding that in the mouse embryos the synthesis of 18S and 28S RNA starts from the 2-cell stage on (Clegg and Piko, 1982).

According to our data, the synthesis of ribosomal RNA in partheno-genetically activated mouse ova starts at the same stage (2-cell) as in fertilized ova. If to judge by the number of Ag-NORs detected in parthen-ogenic embryos, the conclusion can be made that in haploid, compared to diploid fertilized embryos more ribosomal cistrons are transcriptionally active. We found Ag-proteins in the nucleoli of the 1-cell embryos, i.e. at a stage of development when the ribosomal genes are not transcribed, and in the metaphase chromosomes no Ag-NORs could be detected. Since Ag-proteins were detected 40-60 minutes after fertilization, it is reasonable to suppose that they were already present in the oocyte cytoplasm. Our observations allow a suggestion that after fertilization, Ag-proteins migrated from the cytoplasm into pronuclei, accumulated in pronucleoli and during mitosis entered into cytoplasm. In the beginning of the second cell cycle these granules were again located in the nuclei, accumulating in the nucleoli. There were reported cases of Ag-proteins being visualized in the nucleoli of somatic cells when RNA synthesis was inhibited by actinomycin D (Hubbell et al., 1980; Daskal et al., 1980), D galactosamine (Dimova, et al., 1982; Hadjiolova et al., 1986), and at the terminal stages of differentiation of blood cells (Smetana, 1980), i.e. when the ribosomal RNA is not synthetized (Hadjiolov, 1985). There is also evidence that Ag-proteins in nucleoli remain bound to the nontranscribed ribosomal genes (Ochs et al., 1983; Scheer and Rose, 1984; Escande-Gerand et al., 1985; Hadjiolova et al., 1986). Another example is the 1-cell mouse embryo.

As far as the role of argentophilic proteins in the early mouse embryos is concerned, now it can be only suggested that these proteins are used as matrix in the formation of definitive nucleoli and are some kind of precursor of Ag-NOR proteins. In the later stages of embryogenesis they go through some modifications to be able to bind with chromosome NORs, in order to be involved in the regulation of ribosomal gene transcription. It cannot be excluded, however, that pronucleoli of the 1-cell mouse embryo accumulate Ag-proteins, which have very little or no relation to the proteins involved in ribosomal gene transcription. It seems possible that Ag-proteins in the 1-cell stage mouse embryo belong to 'cyclins' - proteins involved in the cell cycle regulation (Maro et al., 1986; Dyban, 1988), or to proteins which participate in some other cell function. Further studies will show which of these assumptions is true.

REFERENCES

Clegg K. B., Piko L. 1982, RNA synthesis and cytoplasmic polyadenylation in the one-cell mouse embryo, Nature, 295:342.
Daskal Y., Smetana K., Busch H. 1980, Evidence from studies on segregated nucleoli that the nucleolar silver staining proteins C23 and B23 are in the fibrillar components, Exp. Cell Res., 129:285.
Dimova R. N., Markov D. V., Gajdardjeva K. C., Dabeva M. D., Hadjiolov A. A. 1982, Electron microscopic localization of silver staining NOR-proteins in rat liver nucleoli upon D-galactosamine block of transcription, Europ. J. Cell Biol., 28:272.
Dyban A. P., 1988, 'Early development of mammals' (In Russian), Nauka Publ., Leningrad.
Dyban A. P., Baranov V. S., 1987, 'Cytogenetics of mammalian embryonic development', Clarendon Press, Oxford.
Engel W., Zenzes M. T., Schmid M. 1977, Activation of mouse ribosomal RNA genes at the 2-cell stage, Hum. Genet., 38:57.
Goessens G. 1984, Nucleolar structure, Int. Rev. Cytol., 87:107.
Goodpasture C., Bloom S. E. 1975, Visualization of nucleolar organizer regions in mammalian chromosomes using silver staining, Chromosoma, 53:37.
Hadjiolov A. A. 1985, The nucleolus and ribosome biogenesis, Cell Biol. Monogr. (Springer-Verlag, Wien, New York), 12:1.
Hadjiolova R., Rose K. M., Scheer U. 1986, Immunolocalization of nucleolar proteins after D-galactosamine-induced inhibition of transcription in rat hepatocytes, Exp. Cell Res., 165:481.
Hansmann I., Gebauer J., Grimm T. 1978, Impaired gene activity for 18S and 28S rRNA in early embryonic development of mouse parthenogenones, Nature, 272:377.
Hansmann I., Gebauer J., Bihl L., Grimm T. 1978, Onset of nucleolus organizer activity in early mouse embryogenesis and evidence for its regulation, Exp. Cell Res., 114:263.
Howell W. M., 1982, Selective staining of nucleolus organizer regions (NORs) in 'The Cell Nucleus', H. Busch, L. Rothblum, eds., Acad. Press, New York.
Howell W. M., Black D. A. 1980, Controlled silver staining of nucleolus organizer regions with a protractive colloidal developer: a 1-step method, Experientia, 36:1014.
Hubbell H. R. 1985, Silver staining as an indicator of active ribosomal genes, Stain Technol., 60:285.
Hubbell H. R., Yan-Fai Lau, Brown R. L., Hsu T. C. 1980, Cell cycle analysis and drug inhibition studies of silver staining in synchronous HeLa cells, Exp. Cell Res., 129:139.
Johnson M. H. 1981, The molecular and cellular basis of preimplantation mouse development, Biol. Rev., 56:463.
Magnuson T., Epstein C. J. 1981, Genetic control of very early mammalian development, Biol. Rev., 56:369.
Maro B., Howlett S., Johnson M., 1986, Cellular and molecular interpretation of mouse early development: the first cell cycle in: 'Gametogenesis and Early Embryo', Alan R. Liss, New York.
Miller D. A., Dev V. G., Trantravahi R., Miller O. J. 1976, Suppression of human nucleolus organizer activity in mouse-human somatic hybrid cells, Exp. Cell Res., 101:235.
Ochs R. L., Busch H. 1984, Further evidence that phosphoprotein C23 (110kD/p I 5,1) is the nucleolar silver staining protein, Exp. Cell Res., 152:260.
Patkin E. L., Sorokin A. V. 1983, Nucleolar organizing regions of chromosomes in early mouse embryogenesis, Bull. Exp. Biol. Med. (In Russian), 96:92.
Williams M. A., Kleinschmidt T. A., Krohne G., Franke W. W. 1982, Argyrophilic nuclear and nucleolar proteins of Xenopus laevis oocytes identified by gel electronphoresis, Exp. Cell Res., 137:341.

THE COMPETITIVE REGULATION OF NUCLEOLAR ACTIVITY IN AN INTERPHASE NUCLEUS

REVEALED BY UV-MICROIRRADIATION OF THE NUCLEOLI

V. N. Sakharov and L. N. Voronkova

Institute of Chemical Physics Academy of Sciences of USSR
117334, Kosygin Str.
4, Moscow, V-334
USSR

The UV-microirradiation technique can be employed to selectively affect one of several nucleolar organizer regions (NORs) in an interphase nucleus. Such a local action may perturb the whole NORs system in the nucleus. This experimental model provides a unique opportunity for studying the inter-chromosomal aspects that deal with the regulation of the expression of these genes in a mature nucleus in the interphase state.

We carried out such studies on mammalian cells of the PK line (pig embryo kidney cells) in a monolayer culture. In so doing, we employed the methods of UV-microirradiation, time-lapse cinematography, autoradiography, three-dimensional ultrastructural analysis and silver staining.

We have already reported (Sakharov and Voronkova, 1976, 1979; Sakharov et al., 1978) that local UV-microirradiation can induce a rapid and profound size diminution of an irradiated mature nucleolus. However, this diminution or degradation took place only in the presence of another, unirradiated nucleolus in the same nucleus. Only in this case does the irradiated nucleolus shrink. This loss in nucleolar volume is, however, nearly completely compensated for by the growth of the unirradiated nucleolus. As a result, the volumes of individual nucleoli suffer a dramatic change, but the total nucleolar volume remains nearly constant. It was also shown that such UV-induced changes in the nucleolar volumes are accompanied by the corresponding alteration in the level of the synthesis of nucleolar rRNA at the NORs.

Therefore, not only the shrinkage in the volume of the nucleoli, but also the degree of suppression of RNA synthesis at the irradiated NOR is mediated by the presence of unirradiated NORs in the same nucleus.

The effect turned out to be reversible. The UV-induced changes of the nucleolar volumes reversed after the previously unirradiated NOR of the nucleus had been subjected to an additional UV-microirradiation (Figure 1). Such an action restored all the nucleoli to nearly their original sizes. It is natural that the original volumes of the nucleoli did not considerably change, after all the NORs of a nucleus had been irradiated simultaneously.

All this experimental data provides a strong indication that all NORs in an interphase nucleus exist in intimate interdependence. This inter-

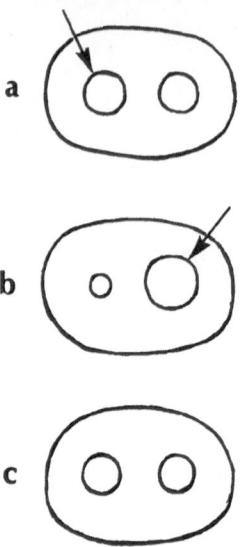

Figure 1.   The competitive regulation of the nucleolar activity in an
interphase nucleus revealed by UV-microirradiation of the
nucleoli.

dependence is revealed when an NOR of a nucleus is subjected to a partial
UV-microirradiation. These results permit us to assume the competition
between the NORs of the interphase nucleus with the predominance of one
NOR at the expense of suppression of the others.

     We have already suggested a possible mechanism for the interdependence
of the NORs in an interphase nucleus (Sakharov, Voronkova, 1976; 1979;
Sakharov et al., 1978). The postulated mechanism makes the following
assumptions:
1.   The active NORs in an interphase nucleus have a reserve for hyperactivation of the rRNA gene expression.
2.   The level of this expression is determined by a limited pool of
regulatory factors which are distributed in the NORs at dynamic equilibrium,
can migrate in the nucleus and redistribute between the NORs.
3.   These factors are distributed among the competitive NORs in proportion
to the number of potentially active rRNA genes at the NORs.

     The postulated mechanism enables us to explain all the features of
nucleolar changes that occur when they are subjected to UV-microirradiation.
It is known that UV-irradiation may inactivate some of the irradiated genes.
In such a case, the initial ratio of the number of potentially active genes
at the NORs changes in favour of the unirradiated NORs, when one of several
NORs is subjected to UV-irradiation. Such a change will lead to a new
equilibrium redistribution and steady state of the regulatory factors among
the NORs and to a subsequent change in the nucleolar sizes and activities.
The effect is reversible, if the additional UV-microirradiation leads to
restoration of the initial proportion of the number of potentially active
genes at the NORs of a nucleus. It is clear that, if all the NORs of a
nucleus are simultaneously irradiated, this will not result in violation of
the equilibrium among the NORs. In such a case, the partial loss of the
number of potentially active genes will be compensated for in each NOR in
the same way.

     Now we can report the experimental results of the ultrastructural
basis of the UV-induced nucleolar changes in an interphase nucleus

Table I.   Changes in the Nucleoli after UV-Microirradiation
One of Two Nucleoli of an Interphase PK Cell.

	Irradiated nucleolus (decrease, in %, from the initial value)	Unirradiated nucleolus (Increase, in %, from the initial value)
Area of nucleolus	53 ± 8	57 ± 9
The level of nucleolar RNA synthesis	78 ± 5	53 ± 9
Number of silver beads after silver staining	64 ± 10	54 ± 18
Number of fibrillar centers	~60	~100

(Zatsepina et al., 1989). The changes in the nucleoli were examined by means of complete series of ultrathin sections obtained from UV-micro-irradiated PK cells. It was shown that in the degraded nucleoli the number of fibrillar centers (FCs) decreases, but their dimensions increase compared with the control cells. On the other hand, hypertrophy of the unirradiated nucleoli is accompanied by a substantial increase in the number of FCs and by a decrease in their dimensions. One can assume that such ultrastructural changes reflect the activation of latent genes and formation of new additional sites for the active expression of rRNA genes in the unirradiated nucleolus. Despite the dramatic change in the number of FCs in irradiated and unirradiated nucleoli, the total number of such centers in a nucleus remains nearly constant. The total volume of the FCs in each nucleolus does not change either. It is probable that their number in the nucleoli increases and decreases by way of 'fragmentation' and 'fusion'.

There is now a general agreement that silver stainability may be considered a marker of the active and potentially active rRNA genes. We employed standardized silver staining for localizing Ag-positive proteins in PK cells throughout the cell cycle and also upon mature nucleoli modifications induced by UV microirradiation (Sakharov et al., 1988). In control mature interphase nucleoli this component is revealed in the form of silver bead chains. In a nucleus with two mature nucleoli each nucleolus contains, on average, nearly 28 such beads. In cells with UV-induced changes of the nucleoli, these beads are redistributed. In an irradiated degraded nucleolus their number greatly decreases, but greatly increases in the unirradiated one. Thus, the total number of silver beads per nucleus remains almost unchanged.

Table I summarizes the main quantitative experimental data. It is clear that the UV-induced changes in the nucleolar volumes, in the levels of the synthesis of nucleolar RNA, in the number of silver stained sites, and the number of FCs are strictly inverse in irradiated and unirradiated nucleoli and are highly correlated in each nucleolus. This permits us to assume that such changes are interconnected and interdependent. In line with the model suggested above, one can make the assumption that the observed changes in the volumes and ultrastructure of the nucleoli and in the expression of nucleolar RNA genes are associated with migration and redistribution of the limited pool of regulatory factors among the NORs of the interphase nucleus. Argyrophilic proteins may be the participants of such a regulation.

REFERENCES

Sakharov V. N., Voronkova L. N. 1976, Nucleolus degradation and growth induced by UV-microbeam irradiation of interphase cell grown in culture, J. Cell Biol., 71:963.

Sakharov V. N., Voronkova L. N., Pyrusyan L. A. 1978, Modification of nucleolar genes expression by the local microirradiation, Dokl. Acad. Nauk SSSR, 243:1048.

Sakharov V. N., Voronkova L. N. 1979, Modification of nucleolar expression by the UV-microbeam, in: Abstr. VI Europ. Nucleolar Workshop. Weimar, GDR.

Sakharov V. N., Voronkova L. N., Valova T. M., Maximova E. V. 1988, Silver staining of nucleoli in PK cells upon the UV microirradiation induced hyperactivation and inhibition of nucleolar RNA synthesis, Tsitologia (USSR), 30:949.

Zatsepina O. V., Voronkova L. N., Sakharov V. N., Cjentsov Yu. S. 1989, Ultrastructural changes in nucleoli and fibrillar centers under the effect of local ultraviolet microbeam irradiation of interphase culture cells, Exp. Cell Res., 181:94.

Ag-POSITIVE NUCLEOLAR ORGANIZER REGIONS IN INTERPHASE BLAST CELLS IN ACUTE

LEUKEMIAS

V. M. Pogorelov*, G. I. Kozinets and K. Smetana[+]

* Scientific Haematological Center
Moscow

[+] Institute of Haematology and Blood Transfusion
Prague

INTRODUCTION

The technique of silver impregnation of acrocentric chromosome
nucleolar organizer regions (Ag-NORs) (Goodpasture and Bloom, 1975) enables
cytological detection of ribosomal gene (rDNA) activities in human normal
or tumor cells to be performed. The reaction is based on a specific linkage
of metal ions by nucleolar proteins. Silver grains are localized on the
sites of transcription and their distribution pattern depends on the rDNA
condensation level; it is different in metaphase chromosomes and interphase
nucleoli.

This paper reports the results of interphase Ag-NORs studies in bone
marrow cells of healthy subjects as well as in patients with acute
lymphoblastic (ALL) and nonlymphoblastic (ANLL) leukemias.

MATERIAL AND METHODS

Bone marrows were taken from 47 untreated patients: 9 ANLL (median age
33 years), 18 ALL (21 years), 11 patients from the risk group (22 years) and
9 healthy bone marrow donors (24 years). Peripheral blood lymphocytes from
one donor were also studied after 48 hours in culture with PHA. A MOP-
Video-plan (Reichert, Austria) image analyzer was used for computer morph-
ometry of Giemsa stained cells at magnification 850x at GRT. The following
parameters were measured: nuclear diameter (mkm), nuclear perimeter (mkm),
area of cell and area of nucleus (both in sq. mkm), relative nucleolar
excentricity (arb. units), cytoplasminuclear ratio and intergrated nucleolar
area in cell (SNLY, sq. mkm): all these parameters were found to discrimin-
ate blasts of 2 classes, namely L1 and L2 in ALL. The mean number of
nucleoli in a cell (NNLY) was also calculated. The type of ALL was
established using the percentage of automatically classified L1 and L2
blasts. The smears were then impregnated by a 50% solution of $AgNO_3$
(Kotelnikov et al., 1988). As control for leukemic blasts, corresponding
normal counterparts were used, *viz.* donor myeloblasts for ANLL etc. In each
smear 100 Ag+cells were studied. We counted the number of silver grains
(NAg) and relative size of each grain (K2): the size of chromosomal NOR is
greatest - we consider it as 1. The interphase Ag-NORs are smaller, we
divide them into 3 groups identified as 0.1, 0.5, 0.9. The two-sided

Table I. Nucleolar Parameters of Bone Marrow Blast Cells in Acute Leukemia (N=27).

Case number	ANLL				ALL L1				ALL L2			
	NK	SNLY	NAg	K2	NK	SNLY	NAg	K2	NK	SNLY	NAg	K2
1.	1.81	8.64	4.99	0.44	1.25	3.40	6.45	0.35	2.92	4.95	9.05	0.21
2.	2.25	5.97	6.45	0.60	2.25	3.72	10.55	0.30	2.30	6.02	11.57	0.24
3.	2.92	5.08	4.92	0.27	1.85	1.85	4.70	0.48	2.36	3.26	6.60	0.29
4.	2.44	22.28	5.16	0.32	2.68	1.71	4.68	0.49	1.98	4.50	6.96	0.28
5.	1.86	5.57	5.10	0.35	2.02	2.53	5.50	0.32	1.91	3.62	9.40	0.22
6.	2.05	20.91	5.00	0.74	1.85	3.49	5.50	0.32	2.18	4.40	7.84	0.24
7.	1.95	9.59	3.24	0.49	2.10	3.40	7.28	0.41	1.76	6.11	13.10	0.23
8.	2.37	16.28	5.36	0.51	2.10	3.99	5.95	0.36	2.05	9.37	7.45	0.24
9.	1.60	19.36	3.11	0.43	2.08	3.37	5.90	0.42	1.55	3.74	8.45	0.23
Mean	2.14	12.63	4.81	0.46	2.02	3.05	6.29	0.38	2.11	5.11	8.94	0.24
SD	0.40	7.04	1.04	0.14	0.38	0.82	1.80	0.07	0.40	1.88	2.16	0.03
V, %	18.69	55.74	21.62	30.43	18.81	26.88	28.62	18.42	18.96	36.79	24.16	12.50
Median	2.05	9.59	5.00	0.44	2.08	3.40	5.90	0.36	2.05	4.50	8.45	0.24

Abbreviatin see text

1. For 48 hrs PHA-cells of donor (N=1) : NK=2.0, SNLY=18.0, NAg=9.0, K2=0.21
For normal bone marrow myeloblasts (N=9) : NK=3.0, SNLY= 6.69,NAg=9.7, K2=0.22
For "risk" bone marrow lymphocytes (N=11): NK=1.0,No data, NAg=3.7, K2=0.36

2. The only reason for high variability of nucleolar area inside the group ANLL
(V=55.7%) are the patients with M4 type of acute leukemia.

Wilcoxon's U-test and correlation analysis were used for statistical interpretation of the data.

RESULTS AND DISCUSSION

Bone marrow blast cells of ALL patients were easily discriminated by morphometric parameters (error was 12%), which permitted an objective classification of ALL subtype. The group ANLL was analysed without sub-classification. Our technique impregnation demonstrated silver grains almost in all smears. Table I presents all the results. The wide range of values of the parameters may be explained by non-uniformity of biosynthetic activities and proliferation of blast cells.

Interphase donor myeloblasts exhibit a large number of small granules, relatively high nucleolar coefficient and small nucleolar area. This data may indicate that myeloblasts, as well as their progenitors, are in a state of intensive division.

Donor lymphocytes, after 48 hours stimulation with PHA, are mainly in the prereplicative phase G1. Ribosomal cistron activities in these cells are elevated as well as in myeloblasts (a large number of small silver grains), but the nucleolar coefficient was lower ($p < 0.01$), while nucleolar area values in PHA-stimulated lymphoblasts are significantly higher ($p < 0.0001$). It apparently depends on the high rate of acrocentric chromosomes satellite associations after the long period G0, which is character-istic for peripheral blood lymphocytes with the ability of their nucleoli to undergo fusion soon after application of the mitogen (G0-G1 transition).

Bone marrow lymphocytes of 'risk' patients seem to be in the 'activated' (G1?) state: they have at least one nucleolus and an ability to transcribe rRNA (small numbers of large Ag-NOR). There is a strong negative correlation between K2 and NAg in this group ($N=11$, $r=-0.77$; $p<0.01$). All patients but one after examination were found to have no haematological disturbances. In one case, a Ph'-CML was diagnosed. In lymphocytes of this patient we found numerous small silver grains in large nucleoli.

There is no difference in the low rate of blast cell proliferation in different types of acute leukemia. But, there is strong evidence that ALL and ANLL are characterized by different argentophility of both metaphase and interphase cells. In ALL the intensity of Ag-NOR is much higher than in ANLL and normal bone marrow.

In our studies (Table I) ANLL blast cells were characterized by lower values of nucleolar coefficient ($p<0.01$) and NAg ($p<0.0001$) than normal myeloblasts, but higher values of the nucleolar area ($p<0.0001$) and the size of Ag-NORs ($p<0.0001$).

In lymphoblasts of ALL patients the nucleolar coefficient and NAg were higher ($p<0.0001$ for both parameters) but the size of Ag grains was lower ($p<0.05$), compared to the bone marrow lymphocytes of 'risk' patients. These differences were especially evident in the L2 type of ALL: the K2 was the same for L1 blasts and 'risk' lymphocytes, but significantly lower in L2 ($p<0.001$). The PHA-stimulated lymphocytes have significant differences compared to L1 blasts, in such parameters as nucleolar area ($p<0.0001$), NAg ($p<0.01$) and the size of Ag grains ($p<0.002$), but only in the nucleolar area ($p<0.001$), compared to L2 blasts.

There were statistically significant differences between L1 and L2 blasts in the area of nucleoli ($p<0.001$), NAg ($p<0.001$) and the size of grains (p<0.0001). The results of correlation analysis of nucleolar indices in ALL (for SNLY and K2: N=18, r=-0.64, p<0.05; for NAg and K2: N=18, r=-0.70, p<0.01) confirm the validity of morphometric classification of ALL subtypes.

Blasts from patients with ANLL have higher values of nucleolar area (p<0.0001) and size of silver grains (p<0.01) but the number of Ag-NORs is lower (p<0.001) than in ALL blasts.

These comparisons demonstrate that apparent differences with regard to Ag stainability among different types of acute leukemia depend rather on the rDNA packing (activity) in G1 phase of the cell cycle. If so, in ANLL (Table I) a relatively low value of nucleolar coefficient, large nucleoli, but a few large Ag-NORs, suggest the possibility of G1-late to G0 transition at least for a part of blast cell population. ALL L1 blasts which are apparently in the early (Postmitotic) G1, as evidenced by small nucleoli and 'metaphasic' appearance of Ag grains, while L2 blasts are in late G1 (a large number of small Ag-NORs). This hypothesis is also supported by the technique of premature chromosome condensation (Hittelman et al., 1979) when applied to acute leukemia blast cells.

REFERENCES

Goodpasture C., and Bloom S. E., 1975, Visualization of nucleolar organizer regions in mammalian chromosomes using silver staining, Chromosoma, 53:37.
Hittelman W. N., Broussard L. C., and McCredie K., 1979, Premature chromosome condensation studies in human leukemia. Pretreatment characteristics, Blood, 54:1001.
Kotelnikov V. M., Pogorelov V. M., Berger J., and Kozinets G. I. 1988, Cyclophosphamide induced generation of giant hypersegmented granulocytes in rat bone marrow: cell cycle distribution and silver nucleolar staining, Folia Haematol. 115:737.

POLYMORPHISM AND THE REGULARITIES IN THE INHERITANCE OF HUMAN NUCLEOLUS

ORGANIZER CHROMOSOMES

Natalia A. Liapunova, Natalia A. Egolina and
Victor V. Victorov

Institute of Medical Genetics Academy of Medical Science
115478 Moscow
USSR

The silver staining of nucleolus organizer regions (NORs) has made it possible to study ribosomal gene expression directly on metaphase chromosomes. It has been shown that the Ag-philic substance is phosphoprotein C23, which is involved in rRNA transcription and remains, together with RNA-polymerase I, bound on the rDNA in mitotic chromosomes (Bloom and Goodpasture, 1976; Spector et al., 1984). Under standard conditions of cell culturing, the Ag-stainability of NORs is constant within any individual and inheritable (Markovic et al., 1978; Zakharov et al., 1982). For the number of Ag-positive NORs and their staining intensity, a polymorphism has been discovered between various chromosomes, between homologous NO-chromosomes, between individuals and populations.

This work represents an attempt to assess, by the complex of Ag-stained NORs, the relative quantity of active ribosomal gene copies in the human genome cytogenetically we had to: (1) evaluate the quantitativeness of Ag-staining and (2) show the Ag-staining intensity to correlate with the relative number of rRNA gene copies.

Ag-staining of NORs was performed according to Howell and Black (1980) with slight modification. The intensity of Ag-staining in individual NOR was evaluated in grades 0, 1, 2 and 3 (Liapunova et al., 1988). Visual estimation of Ag-staining grades was compared to the cytophotometric estimation of Ag-grain sizes. 48 metaphases were sorted out from 7 individuals, that is, 480 chromosomes were present in the analysis. The results are given in Table I. The data obtained allow us to draw some conclusions. (1) Grades 2 and 3 represent more heterogenic variants than grade 1. (2) Mean values of light absorption for grades 1, 2, 3, correlate well with visual estimation, and each successive grade differs from the previous by some 8-9 conditional absorption units. (3) The visual estimation of Ag-NORs in grades 0 to 3 has every reason to serve as the semi-quantitative characteristic of individual Ag-NOR variants of human chromosomes.

Earlier, we studied the correlation between Ag-staining intensity of a given NO-chromosome and the relative copy number of rRNA genes it contained. For all 10 NO-chromosomes of one and the same individual we compared (1) the content of rRNA gene copies defined by the number of autoradiographic grains after *in situ* hybridization of ^{3}H-rDNA on metaphase chromosomes, (2) NOR's

Nuclear Structure and Function, Edited by J. R. Harris and
I. B. Zbarsky, Plenum Press, New York, 1990

Table I. Mean Values of Cytophotometricaly Estimated
Sizes of Ag-NORs Previously Evaluated
Visually in Grades 0, 1, 2 and 3.

Grades	n	T^a S.E.	δ^2	Mean Sizes of Ag-NORs (T_0-T)
0	65	$94,22\pm0,29^b$	5,59	0
1	72	$84,31\pm0,45$	17,13	9,91
2	196	$76,78\pm0,39$	29,07	17,44
3	147	$68,46\pm0,45$	30,12	25,76

[a]Transmission (T) was measured in arbitary units using
constant optical plug; on the empty field $T_{e.f.}$=100.

[b]Light absorption is due to Giemsa-stained Ag-negative
NORs (short arms of acrocentric chromosomes).

transcription activity detected by Ag-staining and (3) the satellite stalk
length. A high enough positive correlation was established between the
traits studied. It means that one can define with a certain precision a
comparative ribosomal gene copy number in a given NOR, by Ag-staining
intensity (Mkhitarova et al., 1988).

The criterium Σ (+) of 10 Ag-NORs (the sum of grades of 10 silver
stained NORs) has been proposed as the measure of the relative number of
rDNA active copies in the genome of a given individual. Figure 1 shows a
frequency distribution of individuals (n=159) with various values of Σ (+).
For comparison, a theoretical distribution, expected in case of equal
probability of grades 0-3 in each of all 10 NORs, is presented. It can be
seen that this distribution essentially differs from normal, ranging from
0 to 30 grades, the mean value being 15.5. In a real population the values
Σ (+) of Ag-NORs are within the range of 16 to 23 grades, the mean value
being about 19.0. Since the values under 16 and over 23 grades do not
exist in a real population, a stabilizing selection was suggested to exist,
directed at maintaining the 'normal reaction' of this quantitative,
biologically important trait of the genome at a level necessary for cell
viability. This selection must work at the stage of zygotes or in early
embryogenesis. To test this assumption, we performed a series of computer-
assisted imitation crossings. Real genotypes for which Ag-NOR variants of
5 pairs of identified chromosomes had been estimated in grades, were
taken in the crossing. In the course of meiosis, each crossing partner
forms 32 gamete variants on the basis of NO-chromosome combinations. The
fertilization results in one out of 1024 zygote variants being realized.
The results of several imitation crossings are given in Figure 2. From 10
to 45% of zygotes are shown to be subject to elimination in different
married couples.

The selective value of Ag-variants of different pairs of chromosomes
has been studied on the material of 34 individuals, using Ag- and G-
staining (Table II). It turned out that for all chromosomes variants '0'
and '1' occurred less than expected (25%), and variants '2' and '3' were
more frequent. The ratio K=(0;1)/(2;3), that is, low-copy NOR variants to
high-copy ones for all chromosomes (K_{13-22}), make up 0.35. By this index,

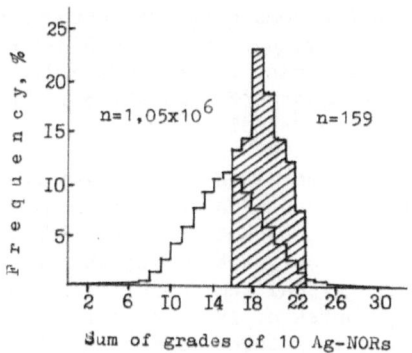

Figure 1.   Theoretical and empirical (shaded) distribution of individuals with different sum of grades of Ag-NORs.

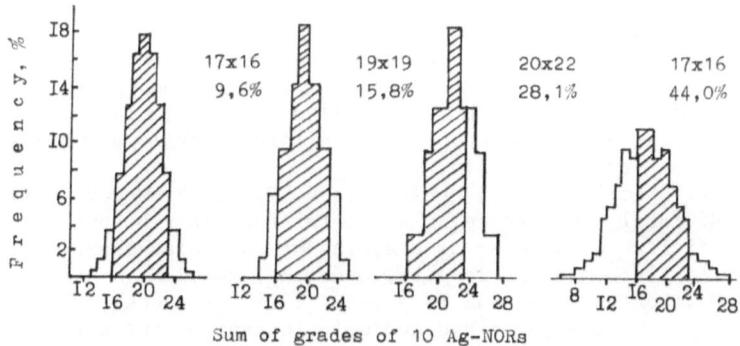

Figure 2.   Frequency distribution of zygotes with different sum of grades of 10 NORs obtained in imitation crossings of 4 couples. In the top right Σ (+) Ag-NORs of partner in crossing and the percentage of zygote variants subject to elimination (unshaded part of the histogram) are indicated for each particular case.

Table II.   Frequency of Grades (0 to 3) of Silver-Staining In The Different Pairs of NO-Chromosomes (%)[a].

No chr. Grades	13	14	15	21	22	Mean
0	11,19	19,70	17,22	10,83	17,73	15,33
1	14,39	7,85	10,03	7,41	12,50	10,44
2	39,10	41,79	38,30	38,30	46,08	40,71
3	35,32	30,67	34,45	43,46	23,69	33,52

[a]The data have been obtained on the basis of the analysis of Ag-staining of 64 chromosomes of each pair.

in the studied sample of NOR chromosomes, the chromosomes 13, 14 and 15 turned out to be close to the mean value (K_{13}= 0.34; K_{13}=0.38; K_{15}=0.37); high-copy variants were predominant in chromosome 21, (K_{21}=0.22), and low-copy were predominant in chromosome 22 (K_{22}=0.43). These criteria can be used for comparative population studies.

On the basis of the data obtained, we suggest that the main mechanism of maintaining a certain level of rRNA genes in man, is the combination of NO-chromosomes carrying different numbers of the ribosome gene's copies and the stabilizing selection on the basis of the sum of active gene copies in all 10 NORs. The visual semi-quantitative method of defining the relative number of active copies of rRNA genes in the genome by the intensity of Ag-staining, allows to establish inter-individual differences of the trait. This method is simple and reliable. Large groups of people can be tested by this method, which is important in studying population polymorphism of the trait.

REFERENCES

Bloom S. E. and Goodpasture C. 1976, An improved technique for selective silver staining of nucleolar organizer regions, Hum. Genet. 34:199.
Howell W. M. and Black D. A. 1980, Controlled silver-staining of nucleolus organizer regions with a propective colloidal developer: a 1-step method, Experientia, 36:1014.
Liapunova N. A., Egolina N. A., Mkhitarova E. V. and Victorov V. V. 1988, Interindividual and intercellular differences in ribosomal genes total activity detected by Ag-staining of nucleolar-organizing regions of acrocentric human chromosomes, Genetika, 24:1282.
Markovic V. D., Worton R. G. and Berg J. M. 1978, Evidence for the inheritance of silver-stained nucleolus organizer regions, Hum. Genet., 41:181.
Mkhitarova E. V., Egolina N. A., Garkavtsev I. V., Liapunova N. A. and Zakharov A. F. 1988, Correlation between transcription intensity and the content of rRNA genes in individual nucleolar-organizing regions of human chromosomes, Bull. Exp. Biol. Med., 105:63.
Spector D. L., Ochs R. L. and Busch H. 1984, Silver staining, immunofluorescence and immunoelectron microscopic localization of nucleolar phosphoproteins B23 and C23, Chromosoma (Berl.), 90:139.
Zakharov A. F., Davudov A. Z., Beniush V. A. and Egolina N. A. 1982, Genetic determination of NOR activity in human lymphocytes from twins, Hum. Genet., 60:334.

SATELLITE NUCLEOLI IN BLASTIC CELLS OF HUMAN ACUTE LEUKEMIAS

K. Smetana, R. Ochs and H. Busch

Institute of Hematology and Blood Transfusion
Prague
Czechoslovakia
Baylor College of Medicine
Houston
USA

INTRODUCTION

The visualization of satellite nucleoli (SN) was facilitated by the introduction of methods for the demonstration of characteristic nucleolar components such as protein B23, C23 and fibrillarin (Figure 1). Previous studies indicated that SN apparently represent solitary NORs which did not fuse and participate in the formation of characteristic nucleoli after mitosis (Smetana et al., 1987). Concerning blood cells, SN were observed only in human peripheral lymphocytes in a relatively small but constant percentage of cells (Smetana et al., 1986). Therefore, the present study was undertaken to provide more information on SN in blood cells and especially in early developmental stages of lymphocytes, granulocytes and monocytes. These cells are present in the peripheral blood of chemotherapeutically untreated patients suffering of acute leukemias and are easily accessible for specific immunoreactions.

MATERIAL AND METHODS

Mononuclear blastic cells were investigated in the peripheral blood of 5 patients with acute lymphoid, 6 patients with acute myeloblastic and 5 patients with acute myelomonocytic leukemia. All the patients investigated were not treated with cytostatic therapy at the time of taking samples for laboratory tests during the present study.

SN were visualized by specific immunostaining of protein B23 and fibrillarin. Fresh blood smears were fixed in 4% formaldehyde in phosphate buffered saline (pH 7.2) at room temp. for 10 minutes, and permeabilized with absolute acetone for 3-5 minutes at $-20^{\circ}C$. These proteins were then visualized by indirect immunofluorescence (see Smetana et al., 1986) using monoclonal mouse antibody to protein B23 and human autoantibodies to fibrillarin. Specific antibodies bound to nucleoli were detected with FTC conjugated goat anti-mouse or goat anti-human IgG (Cappel). Monoclonal antibody to protein B23 was prepared in the Department of Pharmacology, Baylor College of Medicine, Houston, USA and human autoantibodies to fibrillarin were prepared by Sigma, St. Louis, USA. SN were evaluated in

Table I.   SN in Leukemic Blast Cells[1].

Acute leukemia	SN per cell	% of cells with SN	Nucleolar coefficient
Lymphoid (lymphoblastic	$1.01(0.03)^2$	15.0(1.9)	1.56(0.01)
Myeloblastic	1.31(0.30)	15.4(3.3)	2.34(0.10)
Myelomonocytic	1.13(0.05)	15.3(1.2)	2.18(0.10)

1 - At least 200 cells were investigated in 5 patients with lymphoblastic, 6 patients with myeloblastic and 5 patients with myelomonocytic acute leukemia

2 - Mean and standard error of mean

at least 200 cells for each investigated patient, in smears stained for fibrillarin.

RESULTS

Disregarding procedures used for the visualization and the type of blastic cell such as lymphoblast, myeloblast or monoblast, SN appear as small but distinct bodies within the nucleolus in addition to being characteristic of certain nucleoli. The percentage of cells containing SN was small and constant without substantial differences in the type of acute leukemia investigated (Table I). The number of SN per cell ranged between 1 and 2. Most cells possessed only one SN disregarding the number of nucleoli and type of blastic cell (Table I). In contrast, the number of nucleoli per cell expressed by the values of the nucleolar coefficient (number of nucleoli divided by the number of evaluated cells) was dependent on the type of blastic cell (Table I). Largest values of the nucleolar coefficient were noted in myeloblasts of acute myeloblastic leukemias.

DISCUSSION AND CONCLUSIONS

The results clearly demonstrate that the number of blasts containing SN is small but constant, disregarding the type of acute leukemia. If SN are remnants of pronucleoli, prenucleolar bodies after the mitosis, this observation might indicate the presence of a constant postmitotic population of blastic cells in the peripheral blood of leukemic patients. This phenomenon, however is not specific for leukemias since the percentage of normal lymphocytes with SN in the peripheral blood of normal persons was approximately the same (Smetana et al., 1986).

The constant number of 1 - 2 SN per cell in all types of blast cells, i.e. lymphoblasts, myeloblasts and monoblasts also seems to be interesting. Moreover, it does not depend on the number of nucleoli in the investigated cells. Since the presence of 1-2 SN was noted, not only in the present but also in previous studies on completely different cell types (Smetana et al., 1987), all these observations might represent a general phenomenon which deserves further investigation.

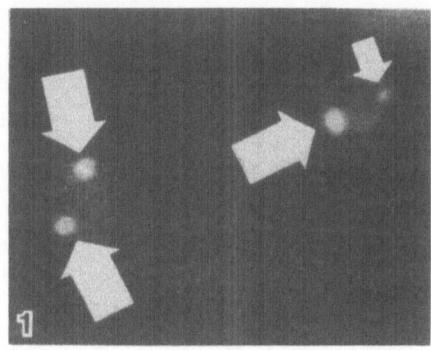

Figure 1.   Immature leukemic lymphocytes stained for protein C23.
Characteristic nucleoli - large arrows, satellite nucleoli
(SN) - small arrow. x approximately 1,700.

    The results of the present study also demonstrated that SN contain
fibrillarin, in addition to protein B23 and C23 (Smetana et al., 1987).
Since these proteins are characteristic for nucleolar regions containing
dense fibrillar RNP components and fibrillar centers (fibrillarin,
protein C23; Ochs et al., 1985) granular RNP components (protein B23;
Spector et al., 1984), a possibility exists that SN may contain all these
nucleolar components. Such information seems to be useful, because it is
difficult to characterize the structural organization of SN in ultrathin
sections by electron microscopy. SN in ultrathin sections might be
confused with peripheral portions of other nucleolar types and in addition,
the probability to find SN in ultrathin sections is also very low.

REFERENCES

Ochs R. L., Lischwe M. A., Spohn W. J., and Busch H. 1985, Fibrillarin a
    new protein of nucleolus identified by autoimmune sera, Biol. cell,
    54:123.
Smetana K., Likovsky Z., Ochs R., Novák J., and Busch H. 1986, Studies on
    satellite nucleoli in normal and leukemic lymphocytes, Virchows arch.
    (cell pathol.) 51:155.
Smetana K., Ochs R. L., Busch R. K., Lyeoman L. C., and Busch H. 1987,
    Studies on satellite nucleoli in rat hepatocytes and Novikoff hepatoma
    cells, Cell tissue res. 245:235.
Spector D. L., Ochs R. L., and Busch H. 1984, Silver staining, immunofluor-
    escence and immunoelectron microscopic localization of nucleolar
    phosphoproteins B23 and C23, Chromosoma (Berl.) 90:193.

STUDIES ON THE DISTRIBUTION AND CYTOCHEMISTRY OF NUCLEOLAR SILVER STAINED

PROTEINS IN RING SHAPED NUCLEOLI

K. Smetana and I. Jirásková

Institute of Hematology and Blood Transfusion
Prague
Czechoslovakia

INTRODUCTION

Silver stained proteins (SSPs) characteristic of NORs are present either in fibrillar centers (FCs) and/or in nucleolar regions containing RNP dense fibrillar components (Hernandez-Verdun et al., 1978; Daskal et al., 1980). FCs are considered to represent inactive parts of NORs (Goessens and Lepoint, 1979). Active portions of NORs are believed to be at FCs and in nucleolar regions which possess RNP dense fibrillar components (DFCs) containing the newly transcribed RNA (Bouteille et al., 1974; Smetana and Busch, 1974; Goessens and Lepoint, 1979; Raska et al., 1989). The present study was undertaken to provide more information on the distribution of SSPs in nucleolar regions adjacent to FCs. Ring shaped nucleoli of mature leukemic lymphocytes are convenient for such studies, since they contain one large fibrillar center surrounded by DFCs.

MATERIAL AND METHODS

Lymphocytes of the peripheral blood were investigated in therapeutically untreated patients with chronic lymphocytic leukemia (B-type). The silver reaction was performed before embedding as described previously (Daskal et al., 1980). Beside fixation in glutaraldehyde, specimens were fixed in 4% formaldehyde in distilled water (Lillie, 1954) for 10 minutes at room temperature. Fixed specimens were post-fixed in methanol-glacial acetic acid (3:1) for 5 minutes and rehydrated in 50%, 40% and 30% ethanol. After washing in cold redistilled water, specimens were stained with silver nitrate for 20 minutes at 60°C, then they were washed in ice-cold redistilled water, dehydrated in ethanol and embedded in Durcupan-Epon mixture. Silver staining solution was prepared by mixing silver nitrate 1g/1ml of redistilled water with 40% formaldehyde 1:1 before use. In cytochemical experiments formaldehyde fixed specimens were digested before post-fixation with ribonuclease (2mg/1ml of distilled water), deoxyribonuclease I (1mg/1ml of 0.3mM Mg acetate), or pepsin (1mg/1ml of 0.1N HCl) for 2 hours at 37°C. Aliquots were also treated in incubation media but without the presence of enzyme.

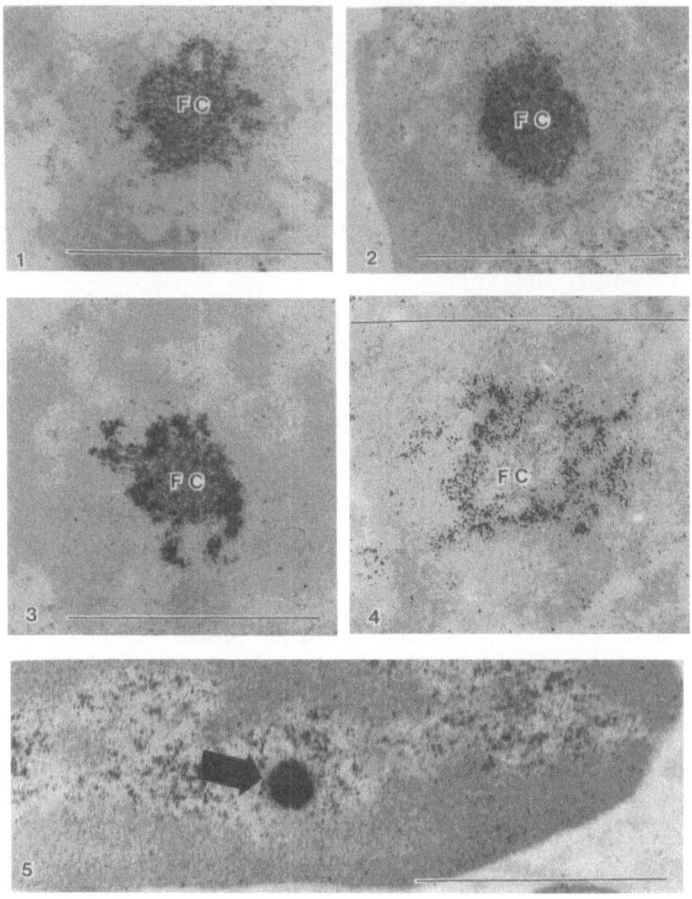

Figure 1.   Note protrusion like structures (PLSs) at FC.
Figure 2.   PLSs at FC are almost absent.
Figure 3.   PLSs at FC exhibit a alrger density.
Figure 4.   The density of FC was substantially reduced.
Figure 5.   A micronucleolus (arrow) in a mature granulocyte.
             Scale bars indicate 1μm.

RESULTS

     Similar to previous studies (Smetana and Busch, 1974) ring shaped
nucleoli were found to consist of one large light area (FC) surrounded by
DFCs. Dense granular RNP components were present in the peripheral
nucleolar region. DNA-containing components were represented by fine
filaments or a few chromatin clusters, in or at the FCs, which also contained protein filaments. In silver stained specimens SSPs were found in
FCs and the surrounding irregularly shaped regions including protrusion like
structures (PLSs) of the latter (Zacepina et al., 1984; Figure 1). However,
these PLSs were absent in some lymphocytes (Figure 2), as in micronucleoli
representing terminal nucleolar maturation stages in mature nucleolated
neutrophilic granulocytes (Figure 5).

     When silver staining was applied in specimens sucessfully digested
with ribonuclease, the density of nucleolar regions, including PLSs
adjacent to FCs was very high in comparison with that of the latter (Figure

3). The difference of the density between SSPs of FCs and adjacent nucleolar regions was also noted after digestion with pepsin or extraction with HCl. After such treatment, the intense silver reaction was preserved only in nucleolar regions adjacent to the periphery of FCs in which the silver positivity was prevented or substantially reduced (Figure 4).

DISCUSSION AND CONCLUSIONS

An interpretation of the varied distribution of SSPs in nucleolar regions adjacent to FCs is difficult at present. However, a possibility exists that the absence of PLSs adjacent to the FC is related to the cessation of nucleolar biosynthetic activity. This interpretation is supported by the similar absence of PLSs in micronucleoli of mature granulocytes. Such nucleoli represent terminal stages of the nucleolar development in the course of the cell maturation and are known to be inactive with respect to the nucleolar RNA synthesis (Smetana, 1980).

The presented study also provides a new information on the cytochemistry of nucleolar SSPs by showing differences in the sensitivity of the FC and adjacent areas, particularly to digestion with ribonuclease and extraction with HCl. SSPs of FCs were more sensitive to acidic extraction with HCl than those in adjacent regions. The distribution of the principal nucleolar SSPs does not explain this difference. Protein C23 and fibrillarin are present in both FCs and adjacent regions containing DFCs, and RNA polymerase I is present mainly in FCs (Spector et al., 1984; Ochs et al., 1985; Scheer and Raska, 1987). The increased density of nucleolar regions adjacent to FCs after digestion with ribonuclease might be produced by the removal of RNA containing DFCs which possibly mask SSPs in these regions.

REFERENCES

Bouteille M., Level M., and Depuy-Coin A. M. 1974, Localization of nuclear functions as revealed by ultrastructural autoradiography and cytochemistry in: The cell nucleus, vol. 1, p. 73, Busch, ed., Academic Press, New York.

Daskal Y., Smetana K., and Busch H., Evidence from studies on segregated nucleoli that nucleolar silver stained proteins C23 and B23 are in fibrillar components, Exp. Cell. Res. 127:285.

Goessens G., and Lepoint A. 1979, The nucleolus organizing regions (NORs): recent data and hypotheses, Biol. Cell 35:211.

Hernandez-Verdun D., Bourgeois C., and Bouteille M. 1978, Identification ultrastructurale de l'organization nucléolaire per la technique a l'argent, C. R. Acad. Sci. Paris 287:1421.

Lillie R. D. 1954, Histopathologic technic and practical histochemistry, Blakiston comp. New York.

Ochs R. L., Lischwe M. A., Spohn W. H., and Busch H. 1985, Fibrillarin: a new protein of the nucleolus identified by autoimmune sera, Biol. Cell 54:123.

Raska I., Reiner G., Jarník M., Kostrouch Z. and Raska K. jr. 1989, Does the synthesis of ribosomal RNA take place within fibrillar centers or dense fibrillar components. Biol. Cell. 65:79.

Scheer U., and Raska I. 1987, Immunocytochemical localization of RNA polymerase I in the fibrillar centers of nucleoli, Chromosomes today 9:284.

Smetana K. 1980, Nucleoli in maturing blood cells in: Topical reviews in hematology, Vol. 1, p. 115, S. Roath, ed. J. Wright, Bristol.

Smetana K., and Busch H. 1974, The nucleolus and nucleolar DNA in: The cell nucleus, Vol. 1, p. 73, H. Busch, ed., Academic Press, New York.

Spector D. I., Ochs R. L., and Busch H. 1984, Silver staining, immunofluorescence in immunoelectron microscopic localization of nucleolar phosphoproteins B23 and C23, Chromosama (Berl.) 90:139.

Zacepina O., Novák J., Smetana K., Schwarzacher H. G., and Wachtler F. 1984, Studies on nucleolar silver stained proteins in nucleoli of normal, stimulated and leukemic lymphocytes, Microscopie 41:278.

IMPLICATIONS FOR THE FUNCTION-STRUCTURE RELATIONSHIP IN THE NUCLEOLUS AFTER

IMMUNOLOCALIZATION OF DNA IN ONION CELLS (*)

Francisco Javier Medina, Marta Martin
and Susana Moreno Diaz de la Espina

Centro de Investigaciones Biologicas
C.S.I.C.
Velazques 144
28006 Madrid
Spain

THE NUCLEOLUS OF ACTIVE ONION CELLS

The architecture of the nucleolus in onion proliferating meristematic root cells is made up of the same structural components commonly described in other cell systems: fibrillar centers (FCs), dense fibrillar component (DFC), granular component (GC) and nucleolar matrix as basic and constant components, as well as two types of nucleolar vacuoles which appear occasionally, associated with certain stages of the cell cycle. However, the arrangement and morphometrical parameters of these components are some- what different from the most widely known nucleolar model, that of mammalian cells. The active nucleolus of onion cells has many small, discrete FCs immersed in an abundant and compact DFC which, in sections, may appear as several masses surrounded, in turn, by the GC. Furthermore, in low-active stages, FCs are larger and contain some inclusions of condensed chromatin; they are called heterogeneous FCs (see Risueño and Medina, 1986, for review). The main cause for these differences is the number of ribosomal genes from which the nucleolus is formed, around 35 times more in the onion than in mammals (see Hadjiolov, 1985). All these genes are capable of being transcribed, though some of them may be transitorily locked for transcription when the rate of nucleolar synthesis drops. These locked genes account for the condensed chromatin inclusions in heterogeneous FCs (Risueño and Medina, 1986). Thus, nucleolar chromatin can be found, in onion cells, in three different morphofunctional states: a) locked for transcription; b) unlocked, but non-transcribing; and c) active in pre-rRNA synthesis.

All these features make the onion cell nucleolus an advantageous system for the most accurate immunolocalization of DNA. This is a crucial step for the establishment of the molecular architecture of the nucleolus and for the assignment of known molecular processes to nucleolar structural components. Moreover, the recent apparent discrepancies between autoradiographic and immunocytochemical results in mammalian cells, concerning the site of nucleolar transcription (see Jordan, 1987; Medina, 1989) make it necessary

(*) Supported by CICYT (Spain) Grant No. PB88-037.

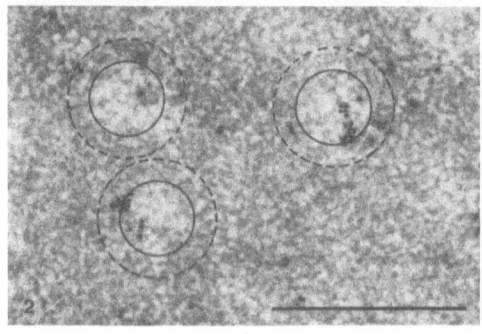

Figure 1.   Pattern of nuclear and nucleolar DNA immunolabelling in onion
             root proliferating cells. Extranucleolar condensed chromatin
             masses (chr) are heavily decorated by gold particles. The
             labelling on the nucleolus is much weaker and concentrates on
             fibrillar centers (arrows). However, gold particles, some-
             times clustered (thin arrow), also appear deep in the dense
             fibrillar component (F). The granular component (G) appears
             unlabelled. Fomaldehyde fixation, LR White embedding and 'on
             grid' indirect immunogold method. Uranyl counterstaining.
             Scale bar equals 1μm.

Figure 2.   Higher magnification of Figure 1. Clusters of gold particles
             appear covering FCs and the DFC immediately surrounding them.
             Circles have been drawn centered in the geometrical center of
             FCs. The radius of the inner one (continuous line) is the
             mean equivalent radius of FCs, calculated from their cross-
             sectional areas. The radius of the outer one (dotted line) is
             the mean plus the standard deviation of the distribution of FC
             equivalent radius. Scale bar equals 0.5μm.

to extend the studies to other cell systems. In particular, the features
and dimensions of the DFC in onion cells may facilitate the understanding of
this component with regard to the presence or absence of transcriptionally-
active rDNA (Goessens, 1984; Scheer and Raska, 1987; Thiry et al., 1988).

NUCLEOLAR STRUCTURES CONTAINING DNA

    Immunolocalization of DNA in onion root proliferating cells by means
of an anti-DNA monoclonal antibody (Scheer et al., 1987), shows the nuclear
condensed chromatin masses heavily decorated by gold particles, whereas the
labelling on the nucleolus is much weaker. This nucleolar labelling appears
mostly related to FCs, though some particles are located on the DFC,
without any apparent relationship to FCs. The granular component is

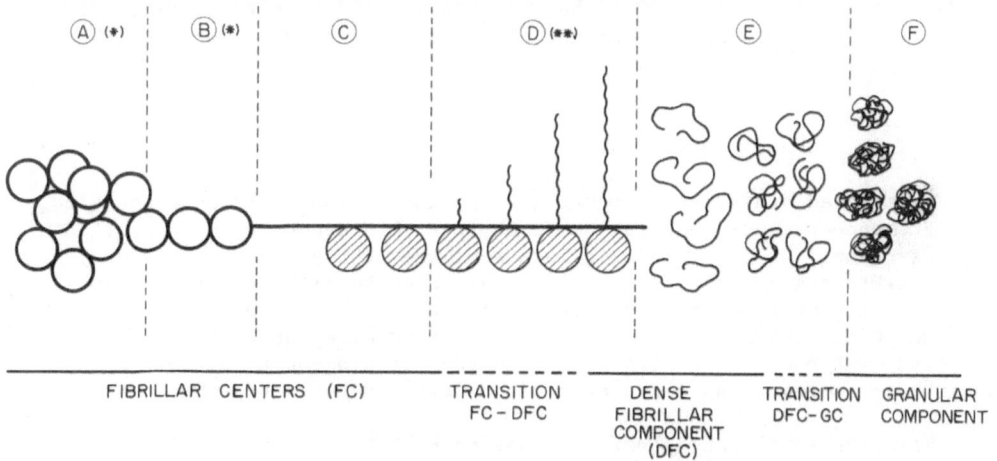

| FIBRILLAR CENTERS (FC) | TRANSITION FC - DFC | DENSE FIBRILLAR COMPONENT (DFC) | TRANSITION DFC-GC | GRANULAR COMPONENT |

Figure 3.   Schematic view of the function-structure relationship in the
             nucleolus. A: Condensed supranucleosomal chromatin. B:
             Nucleosomal chromatin. C: Binding of polymerases and factors
             to non-nucleosomal chromatin. D: Synthesis of pre-rRNA and
             earliest processing. E: Release of pre-rRNA from template and
             further processing. F: Pre-ribosomes. (*): Exclusive in cell
             systems with heterogeneous FCs (very high redundancy of rRNA
             genes), such as onion cells. (**): In onion cells, this
             portion comprises part of the DFC, due to the high number of
             rRNA genes.

unlabelled (Figure 1). At higher magnifications, clusters of gold particles
are seen covering, at the same time, portions of FCs and the adjacent DFC
(Figure 2).

     The accummulation of labelling inside and around FCs can be quantitat-
ively assessed; 36% of the gold particles are restricted to theoretical
circles whose radius is the mean equivalent radius (ER) of FCs (Figure 2),
occupying an area which is only 6% of the cross-sectional area of the
nucleolar fibrillar components. If the theoretical circle is enlarged to
the mean plus the standard deviation of the ER distribution (Figure 2), up
to 56% of the gold particles fall inside this area, which represents 16% of
the fibrillar components.

DNA IN THE DENSE FIBRILLAR COMPONENT

     The significance of these results is that, in the onion cell nucleolus,
the portion of DFC immediately surrounding FCs is a site of localization of
DNA in addition to the FCs themselves. Thus, in spite of its apparent
morphological uniformity the DFC appears to be made up of at least two
subcomponents with regard to the presence of DNA. Integrating these results
with the reported autoradiographic data, obtained after tritiated uridine
incorporation (Goessens, 1984; Risueño and Medina, 1986), we can conclude
that, in onion cells, the minor DFC-subcomponent nearest to FCs contains
transcriptionally active rDNA, while the major part of the DFC would contain
incompletely processed rRNP, released from the template. If we take into
account the number of rRNA genes and the particular dimensions of the DFC in
our cell system, these results are compatible with those obtained in
mammalian cells, in which DNA and active RNA polymerase I (RNA pol I) are

hardly detectable in the DFC (Scheer and Raska, 1987; Thiry et al., 1988; Raska et al., 1989), but the amount of DNA and the size of the DFC are considerably reduced.

'TRANSITION' VERSUS 'BORDERLINE'

A detailed analysis of Figure 2 shows that gold particles associated with FCs appear clustered, in such a way that, in each cluster one of the ends is clearly on the FC and the other on the DFC; however, it is very difficult to assign most intermediate particles to either of these components. Moreover, even in structural terms, it is not possible to establish definite borders between FCs and the surrounding DFC. These facts introduce the concept of 'transition FC-DFC' as playing a key role in defining the structures associated with nucleolar transcription.

The structural concept of 'transition FC-DFC' is the morphological expression of the molecular process of transcription of rRNA genes. Available data indicate that rDNA chromatin must be decondensed and activated before the first ribonucleotide is incorporated; in these previous steps, RNA pol I as well as other proteins and factors are involved (Hadjiolov, 1985; Jordan 1987). The resulting structures in the nucleolus would be FCs; gradually the DFC would be generated as pre-rRNA synthesis proceeds. This interpretation is in agreement with the localization of DNA (with non-nucleosomal structure) and active RNA pol I in FCs, preferentially at their periphery (Derenzini et al., 1982; Scheer and Raska, 1987; Thiry et al., 1988).

Figure 3 is an attempt to integrate our present knowledge on the function-structure relationship in the nucleolus in a schematic view that is intended to be of general value. Obviously, each cell system has its own pecularities, which produce differences in the arrangement and dimensions of the structural components; Figure 3 gives special attention to the particular features originated by the high number of rRNA genes on onion cells. However, most studies in other cell systems and the results presented here allow us to conclude that the bases for the function-structure relationship are the same, whatever the cell system.

REFERENCES

Derenzini M., Hernandez-Verdun D., Bouteille M., 1982. Visualization *in situ* of extended DNA filaments in nucleolar chromatin of rat hepatocytes. Exp. Cell Res. 141:463.
Goessens G. 1984. Nucleolar structure. Int. Rev. Cytol. 87:107.
Hadjiolov A. A., 1985. The nucleolus and ribosome biogenesis. Cell Biol. Monogr. vol. 12. Springer Verlag, 268pp.
Jordan E. G., 1987. At the heart of the nucleolus. Nature 329:489.
Medina F. J., 1989. The Nucleolus, in the soptlight. Eur. J. Cell Biol. 50: in press.
Raska I., Reimer G., Jarnik M., Kostrouch Z., Raska K., 1989. Does the synthesis of ribosomal RNA take place within nucleolar fibrillar centers or dense fibrillar component? Biol. Cell, 65:79.
Risueño M. C., Medina F. J., 1986. The nucleolar structure in plant cells. Cell Biology Reviews (RBC) vol. 7. Editorial Service University Basque Country, Spain-Springer International, 154 pp.
Scheer U., Raska I., 1987. Immunocytochemical localization of RNA polymerase I in the fibrillar centers of nucleoli. Chromosomes Today 9:284.

Scheer U., Messner K., Hazan R., Raska I., Hansmann P., Falk H., Spiees E.,
    Franke W. W., 1987. High sensitivity immunolocalization of double- and
    single-stranded DNA by a monoclonal antibody. Eur. J. Cell Biol.
    43:358.
Thiry M., Scheer U., Goessens G., 1988. Localization of DNA within Ehrlich
    tumor cell nucleoli by immunoelectron microscopy. Biol. Cell 63:27.

BIOTECHNOLOGY AND HUMAN TUMOR NUCLEOLAR ANTIGENS

Harris Busch, Rose K. Busch, James W. Freeman, Robert Larson,
Mohammed Haidar, Sissy Jhiang, Benigno Valdez and
Wei-Wei Zhang

Department of Pharmacology
Baylor College of Medicine
Houston
Texas   77030
U.S.A.

Cancer cells and most normal resting cells have been readily differentiated for a century by morphologic and biologic criteria but biochemical
and immunologic distinctions have not yet withstood the test of time (Busch,
1984).

EXPERIMENTAL STUDIES

Our working hypothesis has been that there are significant nuclear
macromolecular differences between cancer cells and cells of normal
resting tissues (Busch, 1984a; 1984b). Over a period of 25 years, we
investigated a variety of nuclear structural proteins, particularly histones; a variety of enzymes, including DNA and RNA polymerases; and a number
of other macromolecules in tumors and nontumor tissues. In these studies,
no consistent differences were detected by the separation, analytical, and
sequencing methods that were employed.

Approximately 12 years ago, we initiated immunologic approaches to
determine whether there were nuclear or nucleolar antigens that differentiated normal rat liver and rat tumors. Polyclonal antisera demonstrated
several differences by immunofluorescence, by Ouchterlony gels, and by
immunoelectrophoresis, particularly after the polyclonal antisera to the
liver and tumor nucleoli were exhaustively absorbed to remove potential
tumor or liver contaminants. Interestingly, on Ouchterlony gels, bands were
found that differed in mobility, and no lines of identity were found between
hepatoma and normal liver samples (Busch and Busch, 1977). This result was
surprising, for it suggested for the first time that there were nucleolar
antigens that differed in normal liver and in timor cells. Because the
nucleolus is known to be the site of rRNA synthesis and preribosomal RNP
assembly in cells, this type of distinction seemed improbable. It did not
fit the defined role of the nucleolus, and earlier studies showed no
differences in RNP products from normal liver and hepatomas.

INITIAL PROBLEMS WITH MONOCLONAL ANTIBODIES

Initially, whole tumor nucleolar preparations or specific fractions were used to immunize mice. During the initial studies on monoclonal antibodies to nucleolar proteins, it was found that there were equal reactivities of the antibodies with normal human liver and HeLa cells; at one point it appeared that differences found with polyclonal antibodies might not be reproduced.

With time, it became apparent that one of the nucleolar antigens, a protein coded as protein C23 (MW 110kD/pI 5.5) was highly immunodominant.

ANTITUMOR NUCLEOLAR MONOCLONAL ANTIBODIES

Mice immunized with antigen complexes freed from protein C23 developed antibody-producing clones that closely mimicked the behavior of the most satisfactory rabbit sera in distinguishing tumor from nontumor samples (Freeman et al., 1985; 1986).

MONOCLONAL ANTIBODY LIBRARIES TO NUCLEAR AND NUCLEOLAR ANTIGENS

With an increasing number of fusions from mice immunized with nuclear and nucleolar proteins, we have recognized a progressively larger number of PCNA antigens. Currently, more than 30 antigens are specifically identified by our monoclonal antibody library.

THE G_1 TIMETABLE

A most intersting development relates to the time of appearance of the antigens in the cell cycle. It is clear that a 'clock' operates for the events of the G_1 and S-phase of cell division. Initially, a timetable of 'onc' gene expression was demonstrated on the basis of the timing of gene expression and the appearance of 'onc' products. This strategy has proven useful in the analysis of the appearance of the various nucleolar and nuclear antigens we have been studying.

For example, the p120 antigen appears in nucleoli within 30 minutes after starved HeLa cells are fed, although it appears much later in PHA-stimulated lymphocytes that have been stimulated to grow and divide.

MAb PROTEIN P120

To develop the monoclonal antibody to the P120 nucleolar antigen, tumor nucleoli were treated with polyclonal antisera to normal human tissue nucleoli to block some determinant common to tumor and normal tissue nucleoli. Immunization of mice with these immune complexes resulted in the development of a monoclonal antibody (FB2) to a novel Mr 120,000 nucleolar proliferation-associated antigen. By indirect immunofluorescence, antibody FB2 produced bright nucleolar staining in a variety of malignant tumors, including cancers of the breast, liver, gastrointestinal tract, genito-urinary tract, blood, lymph system, lung, and brain. Although specific nucleolar immunofluorescence was not detectable in most normal tissues, it was detectable in some proliferating nonmalignant tissues including spermatogonia of the testes, ductal regions of hypertrophied prostates, and phytohemagglutinin-stimulated lymphocytes. The Mr 120,000 antigen was not detectable in 48h serum-deprived HeLa cells but was readily detectable (within 30 minutes) following serum refeeding. The Mr 120,000 antigen was not detected in retinoic acid-treated HL-60 cells following morphological differentiation but was detectable in 48h phytohemagglutinin-treated lymphocytes. These studies suggest that the Mr 120,000 antigen is a proliferation-associated antigen which plays a role in the early G_1 phase of the cell cycle (Freeman et al., 1988).

LOCALIZATION OF PROTEIN P120

The human proliferation-associated nucleolar antigen p120 was localized to substructures within HeLa cell nucleoli by immunofluorescence and immunoelectron microscopy of cells whose nucleoli were segregated by drug treatment or extracted with nucleases. By indirect immunofluorescence, protein p120 was localized diffusely throughout all interphase nucleoli. However, high resolution immunoelectron microscopy demonstrated that protein p120 staining delineated a network of 20-30-nm diameter beaded fibrils distributed throughout the nucleolus. This distribution was unique compared to that of the nucleolar proteins p145, RNA polymerase I, or B23 which were examined simultaneously. Drug-induced segregation of nucleoli by actinomycin D or dichlorobenzimidazole riboside, followed by immunoelectron microscopy, indicated that protein p120 was concentrated at the periphery of the granular region in segregated nucleoli. *In situ* nuclease digestion of cells with DNase I and/or RNase A did not release p120 from the nucleolus. Instead, p120 immunoreactivity was retained within phase-dense residual nucleoli. These results provide evidence that protein p120 is associated with and delineates a network of fibrils which is retained in the nucleolar residue fraction of proliferating cells (Ochs et al., 1988).

cDNA FOR PROTEIN P120

Because the 120kD proliferating-cell nucleolar antigen described by Freeman et al (1988) is the most cancer associated of the proliferation-associated nucleolar proteins identified thus far, the cDNA sequence and the corresponding amino acid sequence were determined:

AMINO ACID SEQUENCE: P120 PROTEIN

```
NH2-MGRKLDPTKE KRGPGRKARK QKGAETELVR FLPAVSDENS KRLSSRARKR  50
    AAKRRLGSVE APKTNKSPEA KPSPGKLPKG ISAGAVQTAG KKGPQSLFNA 100
    PRGKKRPAPG SDEEEEEDS  EEDGMVNHGD LWGSEDDADT VDDYGADSNS 150
    EDEEEGEALL PIERAARKQK AREAAAGIQW SEEETEDEEE EKEVTPESGP 200
    PKVEEADGGL QINVDEEPFV LPPAGEMEQD AQAPDLQRVH KRIQDIVGIL 250
    RDFGAQREEG RSRSEYLNRL KKDLAIYYSY GDFLLGKLMD LFPLSELVEF 300
    LEANEVPRPV TLRTNTLKTR RRDLAQALIN RGVNLDPLGK WSKTGLVVYD 350
    SSVPIGATPE YLAGHYMLQG ASSMLPVMAL APQEHERILD MCCAPGGKTS 400
    YMAQLMKNTG VILANDANAE RLKSVVGNLH RLGVTNTIIS HYDGRQFPKV 450
    VGGFDRVLLD APCSGTGVIS KDPAVKTNKD EKDILRCAHL QKELLLSAID 500
    SVNATSKTGG YLVYCTCSIT VEENEWVVDY ALKKRNVRLV PTGLDFGQEG 550
    FTRFRERRFH PSLRSTRRFY PHTHNMDGFF IAKFKKFSNS IPQSQTGNSE 600
    TATPTNVDLP QVIPKSENSS QPAKKAKGAA KTKQQLQKQQ HPKKASFQKL 650
    NGISKGADSE LSTVPSVTKT QASSSFQDSS QPAGKAEFIR EPKVTGKLKQ 700
    RSPKLQSSKK VAFLRQNAPP KGTDTQTPAV LSPSKTQATL KPKDHHQPLG 750
    RAKGVEKQQF AEQPFEKAAF QKQNDTPKGL SLPLCLPSVP AAPHQQRGRN 800
    LSPGATASCC YLRWLKTRRV AHCHCHQVGT LASVRMPSLL CIPMKFNTHF 850
    KTSGH-COOH
```

The underlined regions are the four domains of the protein, i.e. the basic, acidic, hydrophobic methionine-rich and the cysteine-protein rich regions.

mRNA LEVELS

To determine the relative messenger RNA (mRNA) level for protein p120, cellular mRNA was extracted, slot-blotted onto nitrocellulose filters, and hybridized to radioactive p120 cDNA fragments. Human tumor cells contained 15-60 times more p120 mRNA than human term placenta. The rat Novikoff hepatoma ascites cell mRNA hybridized to the p120 cDNA probes, but the p120 monoclonal antibody did not react with the Novikoff hepatoma proteins. Novikoff hepatoma mRNA contained 8 times as much p120 mRNA as normal rat liver. As a control, a cDNA was used for protein B23, an abundant nucleolar protein; there were 3.5, 29, and 14 times more B23 mRNA that p120 mRNA in normal rat liver, Novikoff hepatoma ascites cells, and HeLa cells, respectively. Whereas the increased levels of the mRNA for protein B23 reflect increased activity of the nucleolus for any increment of nucleolar function, the increased levels of p120 mRNA and the p120 protein reflect the activity of the G_1 phase of the cell cycle. The elevated level of p120 mRNA in tumors may reflect the heightened G_1 cascade in transformed cells (Hazlewood et al., 1989).

THE GENE FOR THE P120 PROTEIN

The exon sequence was identical to the sequence of the p120 cDNA. The human p120 gene was found in the DNA fragment 14.2kb long. The coding region of the gene comprises 15 exons. Exons 2-4, 5-8, 9-10, and 11-14 are clustered in regions containing fewer than 1kb. The 5' and 3' ends of the coding region are separated from the other clustered thirteen exons by introns greater than 1kb in size. The exons range in size from 767 bases (exon 15 at the 3' end) to 46 bases (exon 2).

A putative 'TATA' box is located at -82 and a 'CAAT' box at -669 with reference to the AUG. Primer extension and RNA protection assays have shown the 'cap' site to be located about 30 nucleotides downstream of the 'TATA' box. Two alu sequences are located at -971/-728 and -1297/-1025.

Two 'Sp1 binding sites' (CCGCCC or GGGCGG) are located -1082/-1077 (within the alu sequence) and -1507/1502. Other 'GC' rich areas are located at -422/-407 and -457/-440. An 'Sp1-like element' (CCCGCC) is located at -543/538. 'GC' boxes (GGGCGG or CCGCCC) are potential binding sites for the transcriptional factor Sp1, which activates the Simian virus 40 early promoter; Sp1 stimulates the transcription of several viral and cellular genes. Two NF-1 binding sites are at -2505/-2493 and at -2463/-2450. The TGGCA-binding protein sites are found nine times: -12, -257, -969, -1336, -1418, -2013, -2113, -2315, and -2454.

Other potential functional sites in the 5' flanking region are a 'heat shock' site at -2209/-2196, the palindromic sequence, ACCTGGGTTTCAGGT (-213/-199), a repeat of GTGCTGAGT at -112/104, -130/-122, and -144/-136, an AP2 binding site at -1231/-1224, an AP1 site (-2052/-2045), and a virus inducible site at -795/-789.

5' FLANKING SEQUENCE AND P120 GENE TRANSCRIPTION

To define the 5' flanking sequences necessary for efficient transcription of the P120 gene, we have constructed a hybrid gene in which the CAT gene was under the control of the P120 promoter (p-2532/+102; numbering is with reference to ATG). This hybrid construct was transfected into HeLa cells by the calcium phosphate co-precipitation method (ref.-ask Haidar). Transfactions were carried out using Rous Sarcoma virus long terminal repeat (pRSV CAT) as a positive control and ΔCAT (CAT coding sequence in

PUC vector without any P120 promoter region). CAT assays following 44 hours of incubation after transfection showed that the sequence - 2532/+102 is able to drive the transcription at about 25% of the efficiency of pRSV CAT.

Exonuclease III digestion was done to generate deletions from the 5' end (-2532) of the P120 promoter. Deletion mutants obtained ranged from -2192/+102 to -200/+102. Various deletion mutants of the p-2532/+102 CAT construct were transfected into HeLa cells to identify important cis-acting elements of the P120 promoter.

EFFECTS OF THE -535/-278 AND -1426/-1223 SEQUENCE ON HETEROLOGOUS TK PROMOTER

To analyze the elements -537/-278 and -1426/-1223 more precisely, these fragments were placed upstream of the pTK CAT construct in both orientations. These four constructs, along with the pTK CAT construct, were transfected into HeLa cells. Inclusion of the region -1426/-1223 in the sense orientation increased TK promoter activity about 50% over control (pTK CAT) but the antisense construct had no effect. The element, -537/-278, stimulated the TK promoter activity 2-3 fold in either orientation when placed upstream of the pTK CAT construct. This region may be an 'enhancer' (Haidar et al., in manuscript).

FUTURE STUDIES

These studies, along with additional information on the epitope region, provide a basis for assessing gene controls of P120 synthesis and utilization of structural information for drug development. The importance of such studies is obvious.

ACKNOWLEDGEMENTS

These studies were supported by the Cancer Research Grant CA-10893, P1, awarded by National Cancer Institute, Department of Health and Human Services; The DeBakey Medical Foundation; The Davidson Fund; The Pauline Sterne Wolff Memorial Foundation; H. Leland Kaplan Cancer Research Endowment; Linda and Ronny Finger Cancer Research Endowment Fund; and the William S. Farish Fund.

REFERENCES

Busch H. 1984a, Onc genes and other new targets for cancer chemotherapy. J. Cancer Res. Clin. Oncol., 107:1.

Busch H. 1984b, Molecular lesions in cancer. Mol. & Cell. Biochem., 61:111.

Busch R. K. and Busch H. 1977, Antigenic proteins of nucleolar chromatin of Novikoff hepatoma ascites cells. Tumori, 63:347.

Freeman J. W., Busch R. K., Gyorkey F., Gyorkey P., Ross B. E., and Busch H. 1988, Identification and characterization of a human proliferation-associated nucleolar antigen with a molecular weight of 120,000 expressed in early G_1 phase. Cancer Res., 48:1244.

Freeman J. W., Busch R. K., Ross B. E., and Busch H. 1985, Masking of nontumorous antigens for development of human tumor nucleolar antibodies with improved specificity. Cancer Res. 45:5637.

Freeman J. W., McRorie D. K., Busch R. K., Gyorkey F., Gyorkey P., Ross B. E., Spohn W. H., and Busch H. 1986, Identification and partial characterization of a nucleolar antigen with a molecular weight of 145,000 found in a broad range of human cancers. Cancer Res., 46:3593.

Haidar M. A., Henning D., and Busch H. The upstream sequence -537 to -278 is necessary for transcription of early G_1 proliferating cell nucleolar antigen P120 gene. In manuscript.

Hazlewood J., Fonagy A., Henning D., Freeman J. W., Busch R. K., and Busch H. 1989, mRNA levels for human nucleolar protein p120 in tumor and nontumor cells. Cancer Commun., 01:29.

Ochs R. L., Reilly M. T., Freeman J. W., and Busch H. 1988, Intranucleolar localization of human proliferating cell nucleolar antigen p120. Cancer Res., 48:6523.

ORIGIN OF ISOFORMS AND INTERACTION WITH NUCLEIC ACIDS

Mark O. J. Olson, Jin-Hong Chang and Tamba S. Dumbar

Department of Biochemistry
University of Mississippi Medical Center
Jackson
MS 39216-4504
U.S.A.

INTRODUCTION

Phosphoprotein B23 (M_r/pI = 38,000/5.1) is a RNA-associated protein located predominantly in the granular component of the nucleolus (Spector et al., 1984) proposed to be an assembly factor in the later stages of ribosome biogenesis. More recently, Borer et al. (1989) suggested that the protein is also a carrier of ribosomal proteins into the nucleus. Electrophoretic and immunoblot analyses in this and other laboratories (Chan et al., 1985) indicate the presence of at least two forms of the protein, differing slightly in molecular mass. Recently, a full-length cDNA for rat protein B23 was isolated and sequenced (Chang et al., 1988). During the course of screening a cDNA library a second cDNA clone was observed. The two cDNAs contained identical 5' regions but distinct 3' regions. This led to our first question: do the two isoforms arise from separate genes or from different expressed segments within a single gene?

The second question concerns the nucleic acid binding characteristics of protein B23. Although nucleic acid binding by nucleolin, the other major nucleolar protein, has been extensively investigated (Olson et al., 1983; Sapp et al., 1986; 1989) little is known about the interaction of protein B23 with RNA or DNA. The sequences of protein B23 and nucleolin (Lapeyre et al., 1987) have very little similarity suggesting that they may have different modes of nucleic acid binding. Therefore, the second part of this study was to elucidate the nature of this association in terms of thermodynamic parameters and effects on nucleic acid structure. Details on both topics are published elsewhere (Chang and Olson, 1989; Dumbar et al., 1989).

MATERIALS AND METHODS

Nucleoli were prepared from Novikoff hepatoma cells by the magnesium-sucrose sonication procedure (Rothblum et al., 1977). Protein B23 was purified from a nucleolar extract (Herrera and Olson, 1986) by chromatography on heparin-sepharose and DNA-cellulose. Association with nucleic acids was studied by circular dichroism (CD), gel retardation, competition binding and fluorescence methods. The cDNA clones were isolated from a rat brain cDNA library in λ gt11 screened by hybridization with synthetic

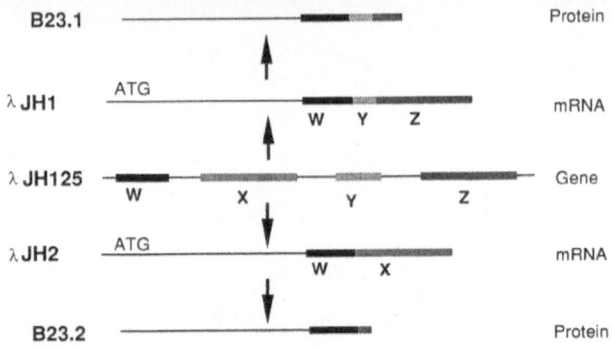

Figure 1.   Diagram of region of gene encoding 3' ends of mRNAs for
             proteins B23.1 and B23.2 whose sequences were derived from the
             cDNAs λ JH1 and λ JH2, respectively. Partial sequencing of
             genomic clone (λ JH125) revealed a 4kb fragment containing 4
             exons (designated W, X, Y and Z). Exons W & X code for the 3'
             end of the mRNA for B23.2 and exons W, Y and Z code for the 3'
             end of the mRNA for B23.1.

oligonucleotides (Chang et al., 1988). Genomic clones were isolated from
two rat liver genomic libraries in λ Charon 4A by similar methods.

RESULTS AND DISCUWSION

     Two cDNAs encoding two distinct forms of protein B23 were isolated and
sequenced. Isoforms B23.1 and B23.2 are polypeptides of 292 and 257 amino
acids, respectively (Figure 1). B23.1 and *Xenopus* protein NO38 (Schmidt-
Zachmann et al., 1987) share about 65% sequence identity, indicating they
represent mammalian and amphibian equivalents. Rat B23.1 and mouse B23
(Schmidt-Zachmann and Franke, 1988) differ by only two conservative
substitutions. The untranslated 5' regions of the two mRNAs and the N-
terminal 255 residues are identical in the two isoforms. However, the 3'
ultranslated regions of the mRNAs are completely different and the dipeptide
Gly-Gly in B23.1 (residues 256 and 257) is replaced by Ala-His in B23.2
indicating the former is not a precursor of the latter.

     Sequences from the genomic clones revealed the origin of the two
isoforms. One of the genomic clones contained a 6.5kb fragment encoding the
3' end of both cDNAs. This contains four exons designated W, X, Y and Z
(Figure 1). Exons W and X encode 36 amino acids at the C-terminus of B23.2,
whereas exons W, Y and Z encode the carboxyl terminal 71 amino acid residues
of B23.1. Exons X and Z each contain distinct 3' untranslated sequences in
which are found polyadenylation signals. Thus, it is probable that the two
different mRNAs are formed by alternative splicing of separate 3' segments.

     Alternative splicing at the mRNA level to generate various isoforms of
proteins is a widespread mechanism, especially in muscle and nervous tissues
and in immunological systems (for review see Andreadis et al., 1987). A
gene which resembles the B23 splicing model is the gene coding for calci-
tonin gene related product, where separate 3'-terminal exons, each with
distinct polyadenylation signals, are also spliced onto a 5' region to
produce two different mRNAs. Not only is the pattern of splicing the same,
but also the order of the 3 differentially spliced 3'exons is identical to
that found in the B23 gene.

     Among the advantages of alternative splicing are the ability to
generate families of proteins with variable and constant domains from a

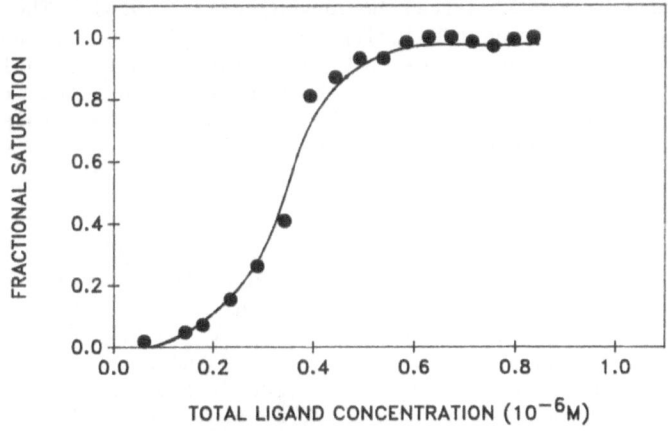

Figure 2.   Fluorescence enhancement of polyriboethenoadenylate upon
addition of protein B23. Typical curve when increasing
concentrations of protein were added to 5uM (nucleotides) of
the nucleic acid lattice in a binding buffer.

single gene and to provide for efficient and reversible switching of
proteins as the physiological requirements of the cell change (Andreadis et
al., 1987). In this case, the result of this mechanism is the deletion of
the C-terminal 35 residues and the substitution of two upstream residues to
transform B23.1 to B23.2. The removal of this C-terminal segment could
serve as a mechanism for modulating protein-protein interactions and/or
nucleic acid binding.

In the second area, preliminary studies suggested that protein B23 is
a single-stranded nucleic acid binding protein. This was confirmed in
competition binding assays with native or heat denatured linearized plasmid
pUC18 DNA where the protein showed a marked preference for the denatured
form. There was no apparent preference for single-stranded synthetic ribo-
versus deoxyribonucleotides. Equilibrium binding with polyriboetheno-
adenylic acid (Figure 2) indicated cooperative ligand binding with a protein
binding site size of 11 nucleotides and an apparent binding constant of
$5 \times 10^7 M^{-1}$ which includes an intrinsic binding constant (K) of $6.3 \times 10^4 M^{-1}$
and a cooperativity factor (ω) of 800 when data was analyzed according to
Kowalczykowski et al. (1986). In circular dichroism (CD) studies protein
B23, when combined with poly(rA), poly(rC) or poly(U), effected a decrease
in ellipticity and a shift toward higher wavelengths of the peak at 260-
270nm, indicating helix destablizing activity.

The most striking of the above findings is the relatively high level
of cooperativity exhibited by the B23-polyethenoadenylate interaction, with
an ω value of approximately 800. The only other eukaryotic RNA binding
protein for which these binding parameters have been determined is the
hnRNP A1 protein which exhibits a low level of cooperativity $\omega = 50$;
Cobianchi et al., 1988). We interpret the high level of cooperativity as an
indication that the protein can efficiently cover relatively long stretches
of single-stranded RNA at low protein concentrations, possibly for protection from nucleases or for exchange with ribosomal proteins (Borer et
al., 1989).

Another interesting property of protein B23 is its ability to alter
nucleic acid conformations. In CD studies the decrease in ellipticity at
275nm, as well as the shift of the maximum signal to the higher wavelengths
with poly (rA), is believed to be indicative of tilting and rotation of

bases when bound by proteins (van Amerongen et al., 1986). The precise biological role of helix destabilizing by protein B23 is not known; however, it is conceivable that the change in nucleic acid structure could facilitate attachment of ribosomal proteins to ribosomal RNA. Thus, the protein may have a 'shoehorn' effect in the ribosome assembly process. This possibility will have to await the development of *in vitro* ribosome assembly systems.

REFERENCES

Andreadis A., Gallego M. E., and Nadal-Genard B. 1987, Generation of protein isoform diversity by alternative splicing, Ann. Rev. Cell Biol., 3:207.

Borer R. A. L., Lehner M. H., Eppenberger C. F., and Nigg E. A. 1989, Major nucleolar proteins shuttle between nucleus and cytoplasm, Cell, 56:379.

Chan P. K., Aldrich M., Cook R. G., and Busch H. 1986, Amino acid sequence of protein B23 phosphorylation site, J. Biol. Chem., 261:1868.

Chang J. H., Dumbar T. S., and Olson M. O. J. 1988, cDNA and deduced primary structure of rat protein B23, a nucleolar protein containing highly conserved sequences, J. Biol. Chem. 263:12824.

Chang J.-H., and Olson M. O. J. 1989, A single gene codes for two forms of rat nucleolar protein B23 mRNA, J. Biol. Chem., 264:11723.

Cobianchi F., Karpel R. L., Williams K. R., Notario V., and Wilson S. H. 1988, Mammalian heterogeneous nuclear ribonucleoprotein complex protein A1, J. Biol. Chem. 263:1063.

Dumbar T. S., Gentry G. A., and Olson M. O. J. 1989, Interaction of nucleolar protein B23 with nucleic acids, Biochemistry, in press.

Herrera A. H., and Olson M. O. J. 1986, Association of protein C23 with rapidly labelled nucleolar RNA, Biochemistry, 25:6258.

Kowalczykowski S. C., Paul L. S., Lonberg N., Newport J. W., McSwiggen J. A., and von Hippel P. H. 1986, Cooperative and noncooperative binding of protein ligands to nucleic acid lattices: experimental approaches to the determination of thermodynamic parameters, Biochemistry 75:1226.

Lapeyre B., Bourbon H., and Amalric F. 1987, Nucleolin, the major nucleolar protein of growing eukaryotic cells: an unusual protein structure revealed by the nucleotide sequence, Proc. Natl. Acad. Sci. (USA), 84:1472.

Olson M. O. J., Rivers J. M., Thompson B. A., Kao W. K., and Case S. T. 1983, Interaction of nucleolar phosphoprotein C23 with cloned segments of rat ribosomal deoxyribonucleic acid, Biochemistry 22:3345.

Rothblum L. I., Mamrack M. D., Kunkle H. M., Olson M. O. J., and Busch H. 1977, Fractionation of nucleoli: enzymatic and two-dimensional polyacrylamide gel electrophoretic analysis, Biochemistry 16:4716.

Sapp M., Knippers R., and Richter A. 1986, DNA binding properties of a 110 kDa nucleolar protein, Nucl. Acids Res. 14:6803.

Sapp M., Richter A., Weisshart K., Caizergues-Ferrer M., Amalric F., Wallace M. O., Kirstein M. N., and Olson M. O. J. 1989, Characterization of 48-kDa nucleic acid binding fragment of nucleolin, Eur. J. Biochem., 179:548.

Schmidt-Zachmann M. S., and Franke W. W. 1988, DNA cloning and amino acid sequence determination of a major constituent protein of mammalian nucleoli, Chromosoma, 96:417.

Schmidt-Zachmann M. S., Hugle-Dorr B., and Franke W. W. 1987, A constitutive nucleolar protein identified as a member of the nucleoplasmin family, EMBO J., 6:1881

Spector D. L., Ochs R. L. and Busch H. 1984, Silver staining immunofluorescence and immunoelectron microscopic localization of nucleolar phosphoproteins B23 and C23, Chromosoma, 90:139.

van Amerongen H., van Grondelle R., and van der Vliet P. C. 1987, Interaction between adenovirus DNA-binding protein and single-stranded polynucleotides studied by circular dichroism and ultraviolet absorption, Biochemistry, 26:4646.

NUCLEOLIN MATURATION AND rRNA SYNTHESIS DURING *XENOPUS LAEVIS* DEVELOPMENT

Michèle Caizergues-Ferrer, Catherine Curie, Colette Mathieu,
and François Amalric

Centre de Biochimie et de Génétiques cellulaires du CNRS,
31062 Toulouse Cedex,
France.

INTRODUCTION

Nucleolin (also called C23 or 100kDa) is a major nucleolar phosphoprotein in exponentially growing cells and is believed to play a key role in
ribosome biogenesis (Jordan, 1987). It has been shown to be associated with
pre-rRNA (Bugler et al., 1987), preribosomes (Bugler et al., 1982; Herrera
and Olson, 1986) and chromatin (Olson and Thompson, 1983; Erard et al.,
1988). Several experimental approaches have shown a direct relationship
between the amount of nucleolin in the nucleolus, its phosphorylation by a
casein kinase NII and the rate of preribosomal RNA synthesis (Bouche et al.
1987, Belenguer et al., 1989). Correlations have been established between
phosphorylation of the protein and the process of its maturation into
defined subfragments (Bourbon et al., 1983a; Suzuki et al., 1985; Issenger
et al., 1988). By *in vitro* experiments, this maturation process was
suggested to directly control the transcription of ribosomal genes (Bouche
et al., 1984).

An *in vivo* approach to this process was carried out on the amphibian
Xenopus laevis. *Xenopus* is an useful model system due to the existence of
two developmental situations in which ribosome synthesis represents the
major effort of the cell: (1) following rDNA amplification in oogenesis,
when a huge quantity of ribosomes is produced and stored for use during
early development and (2) in embryogenesis, when the embryos begin to
synthesize new ribosomes. Furthermore, the relative timing of expression of
rRNA and nucleolin is well documented in *Xenopus* development (Davidson 1986,
Caizergues-Ferrer et al., 1989). In this report we present correlations
between nucleolin maturation and the level of rRNA synthesis during
oogenesis and embryogenesis.

RESULTS

Oogenesis

We have followed the expression of the nucleolin gene during oogenesis
after rDNA amplification when a large quantity of ribosomes accumulate
(Pierandrei-Amaldi et al., 1982). A cDNA encoding *Xenopus* nucleolin was
isolated and used to probe Northern blots. A quantification of several

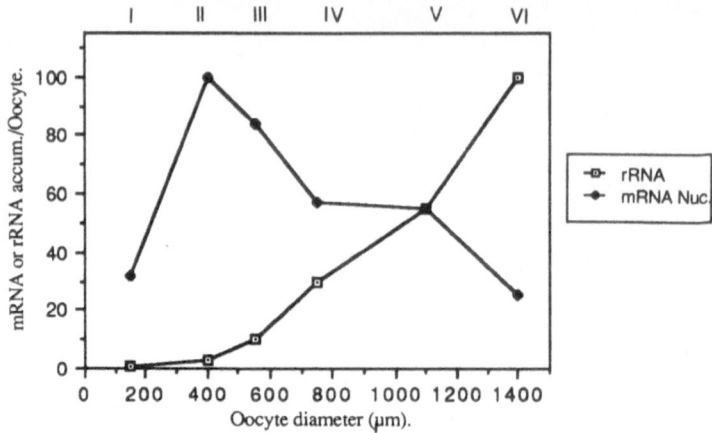

Figure 1.   Nucleolin mRNA accumulation during oogenesis.
             Quantitative data were obtained by densitometric scanning of
             four independent Northern blot experiments. Average values
             for each stage are expressed in percent of the value of the
             stage with the highest score. Numbers above the graph
             indicate the oocyte stage according to Dumont (1972).
             Abscissa: oocyte diameter in µm.

experiments is plotted in Figure 1. Nucleolin mRNA is already detectable in
the smallest oocytes (stage I), and keeps increasing up to the begining of
vitellogenesis (stage II) and then decreases.

     By immunofluorescence on frozen sections of *Xenopus* ovary using IgG
purified from a rabbit polyclonal antiserum raised against hamster nucleolin,
nucleolin and/or immunorelated peptides were detected in amplified nucleoli
in all stages (I to VI) (Figure 2).

     The same antiserum was used to characterize the corresponding proteins
on Western blots (Figure 3). When proteins from ovary were analyzed
(lane 0), the serum gave a weak reaction with two polypeptides of Mr 95,000
and 90,000 as in somatic cells, and a strong one with two smaller polypeptides Mr 59,000 and 48,000. In order to define at what stage nucleolin
maturation occurs, oocytes from stage I to VI were isolated and the
corresponding proteins were analysed on Western blot (lanes I to VI). The
polypeptide of Mr 59,000 could be detected from stage IV to VI, while polypeptides of Mr 95,000 and 90,000 were present in the first stages (I to III),
but in too low amount to be seen on the figure (one oocyte per lane). These
results establish that nucleolin is expressed as early as stage II and is
actively matured after stage III when rRNA synthesis is very high.

Embryogenesis

     Nucleolin mRNA accumulation has been studied during embryogenesis. We
have shown (Caizergues-Ferrer et al., 1989) that it increases very rapidly
between the mid blastula transition (MBT) stage 8 and the neurula stage 14-
16 after which it remains at a constant level up to stage 37 and decreases
slowly afterward. Quantitative analysis of Western blot shows that the two
polypeptides Mr 95,000 and 90,000 were detected at stage 16 and were present
until stage 42. The peptide Mr 59,000 could be detected at stage 24, it
becomes particularly prominent at stages 32 to 38 and then decreases.
Another polypeptide, Mr 35,000 that was not detected in oocytes was

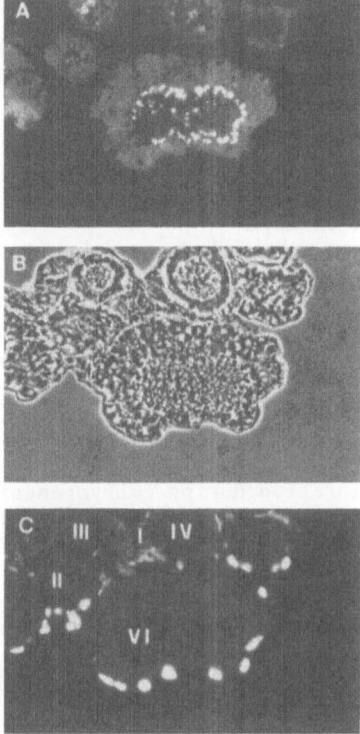

Figure 2.   Localisation of nucleolin in *Xenopus* oocytes.
Immunofluorescence staining of cryostat sections of *Xenopus laevis* ovaries using rabbit IgG against hamster nucleolin.
A: Reaction with anti nucleolin antibodies and FITC con-jugated anti-rabbit IgG.
B: Corresponding phase contrast.
C: Corresponding DAPI staining of DNA.
I to VI refer to the oocyte stage according to Dumont (1972).

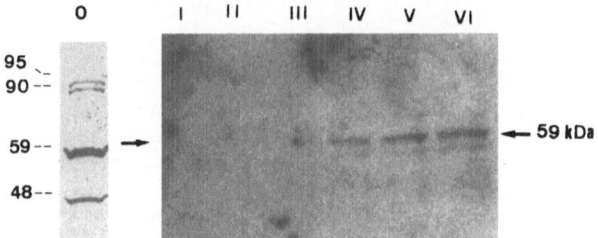

Figure 3.   Nucleolin maturation during oogenesis.
Protein extracts were separated on 10% SDS-PAGE, electro-blotted and probed using an antiserum raised against hamster nucleolin. Detection of the complexes was achieved either by a second antiserum coupled to alkaline phosphatase (lane 0) or by ^{125}I labelled protein A (lane I to VI).
Lane 0: Protein extracts corresponding to a piece of ovary.
Lanes I to VI: Protein extracts corresponding to one oocyte of each stage I to VI.

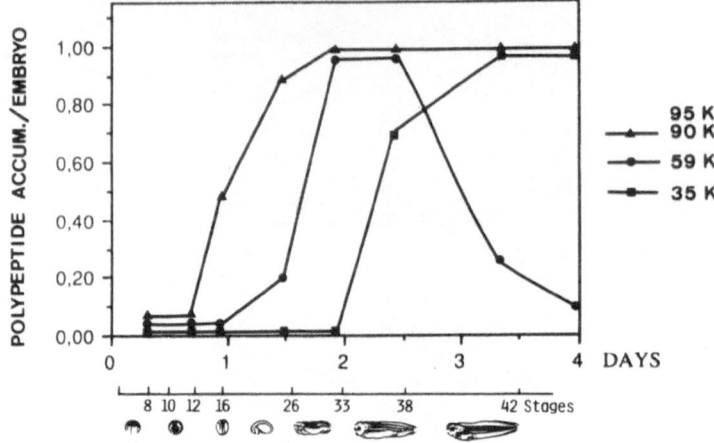

Figure 4.   Nucleolin maturation during embryogenesis.
            Quantitative data were obtained by densitometric scanning of
            three independent Western blot experiments. Average values
            for each stage are expressed in percent of the value of the
            stage with the highest score. Numbers below the graph indi-
            cate the embryo stage according to Nieuwkoop and Faber (1956).

accumulated from stage 38 (Figure 4). Maturation of nucleolin occurs after
stage 24 (tailbud stage) when ribosome accumulation is known to become very
active (Pierandrei-Amaldi et al., 1982). Furthermore, we have shown that,
at these same stages 24 and 32, nucleolin is highly phosphorylated.

DISCUSSION

    The general view which emerges from this study implies that, under
physiological conditions, a high level of rRNA synthesis is correlated with
nucleolin maturation. After rDNA amplification, in oogenesis when rRNA
synthesis is very high, the polypeptide Mr 59,000, which is immunorelated to
nucleolin, accumulates. This polypeptide is also detected after stage 26
when the embryo starts to produce its own ribosomes.

    Although the precise mechanism of this maturation is unclear, we have
shown that a thiol protease is implicated (Bourbon et al., 1983b) and a
consensus cleavage site (PAPAGKK) has been determined (Sapp et al., 1986 and
unpublished data). The presence of two prolines would introduce a high
degree of segmental flexibility in the polypeptide chain, and thereby expose
the adjacent basic cluster to proteolytic cleavage. This sequence is also
found in the insulin receptor precursor (Ebina et al., 1985) and in the
platelet-derived growth factor (PDGF) (Betsholtz et al., 1986). The
positions of these sequences in the nucleolin molecules so far characterized could explain the differences of length observed in the maturation
products in different organisms. Furthermore, it should be noted that
these sequences are found in the vicinity of the phosphorylation site
(Caizergues-Ferrer et al., 1987), a finding which is relevant to the
observed co-regulation between phosphorylation and maturation.

    The fate of the maturation products is open to experimental
investigation, but our opinion is that the carboxyl terminal part of the
molecule (Mr 59,000 and 48,000) which contains the RNA binding domain,

remains on the preribosomes and participates in pre rRNA processing and in nucleo-cytoplasmic exchange (Borer et al., 1989). On the other hand, the whole protein would be implicated in chromatin organization in the vicinity of rDNA (Erard et al., 1988). Antibodies obtained against the different subdomains of nucleolin could help answer this question. Furthermore the *Xenopus* oocyte provides an useful system to study the nucleolin maturation process by microinjection of exogenous protein that could be specifically followed.

ACKNOWLEDGEMENTS

We are grateful to F. Amaldi who introduced us in the *'Xenopus laevis'* world, to P. Mariottini, B. Lapeyre and G. Bouche for fruitful discussion, to N. Gas for immunofluorescence techniques to J. M. Rabanel and E. Barbey for the illustrations and to B. Stevens for comments and criticism of the manuscript.

This work was supported by grants from Centre National de la Recherche Scientifique, and Association pour le Développement de la Recherche sur le Cancer.

REFERENCES

Belenguer P., Baldin V., Mathieu C., Prats H., Bouche G. and Amalric F. 1989, Nucleic Acids Res. (in press).
Betsholtz B., Jonhson A., Heldin C. H., Westermark B., Lind P., Urdea M. S., Eddy R., Shows T. B., Philpott K., Mellor A. L., Knott T. J. and Scott J. 1986, Nature 320:695.
Borer R. A., Lehner C. F., Eppenberger H. M. and Nigg E. A. 1989, Cell 56:379.
Bouche G., Caizergues-Ferrer M., Bugler B. and Amalric F. 1984, Nucleic Acids Res. 12:3025.
Bouche G., Gas N., Prats V., Baldin J. P., Tauber J., Teissié J. and Amalric F. 1987, Proc. Natl. Acad. 84:6770.
Bourbon H. M., Bugler B., Caizergues-Ferrer M., Amalric F. and Zalta J. P. 1983a, Mol. Biol. Rep. 9:39.
Bourbon H. M., Bugler B., Caizergues-Ferrer M. and Amalric F. 1983b, FEBS Lett. 115:218.
Bugler H., Bourbon H. M., Lapeyre B., Wallace M. O., Chang J. H., Amalric F. and Olson M. O. J. 1987, J. Biol. Chem. 262:10922.
Caizergues-Ferrer M., Belenguer P., Lapeyre B., Amalric F., Wallace M. O. and Olson M. O. J. 1987, Biochemistry 26:7876.
Caizergues-Ferrer M., Mariottini P., Curie C., Lapeyre B., Gas N., Amalric F. and Amaldi F. 1989, Genes Dev. 3:324.
Davidson E. H. 1986, Gene activity in early development, 3rd edition, Academic Press, p 160-163.
Dumont J. N. 1972, J. Morphol., 136:153.
Ebina Y., Ellis L., Jarnagin K., Edery M., Graf L., Clauser E., Qu J., Masiarz F., Kan Y., Goldfine I. D., Roth R. A. and Rutter W. J. 1985, Cell 40:747.
Edard M., Belenguer P., Caizergues-Ferrer M., Pantaloni A. and Amalric F. 1988, Eur. J. Biochem. 175:525.
Herrera A. and Olson M. O. J. 1986, Biochemistry 25:6258.
Issenger O. G., Martin T., Richter W. W., Olson M. O. J. and Fujiki H. 1988, EMBO J. 7:1621.
Jordan G. 1987, Nature 329:489. Mol. Biol 97:611.
Nieuwkoop P. D. and Faber J. 1956, Normal table of *Xenopus laevis* (Daudin) North Holland, Amsterdam.

Pierandrei-Amaldi P., Campioni N., Beccari E., Bozzoni I. and Amaldi F. 1982, Cell 30:163.

Olson M. O. J. and Thomson B. A. 1983, Biochemistry 22:3187.

Sapp M., Richter A., Weissart K., Caizergues-Ferrer M., Amalric F., Wallace M. O., Kirstein M. and Olson M. O. J. 1988, Eur. J. Biochem. 179:541.

Suzuki M., Matsue H. and Hosoya T. 1985, J. Biol. Chem. 260:8050.

LOCALIZATION OF NON-HISTONE NUCLEAR PROTEINS BY IMMUNOCYTOCHEMISTRY IN

SOMATIC EMBRYOS AND POLLEN GRAINS

M. A. Sanchez-Pina, H. Kieft*, J. H. N. Schel*,
P. S. Testillano and M. C. Risueño

Centro de Investigaciones Biólogicas
C.S.I.C. Velázquez 144
28006 Madrid
Spain

* Department of Plant Cytology and Morphology
Agricultural University
Arboretumlaan 4
6703 BD Wageningen
The Netherlands

INTRODUCTION

Nuclear proteins, in general, as molecules involved in the regulation of differential gene expression play a main role in nuclear differentiation. They can either be structural factors, required for the attachment of specific DNA or RNA molecules to the nuclear scaffold (Verheijen et al., 1988) or specific DNA or RNA binding factors with a possible role in processing, transcription and/or replication (Berezney, 1984; Busch, 1984, Holoubek, 1984). In spite of their important role they are scarcely known, especially in plant cells; for this reason we have commenced this immuno-cytochemical study in two experimental systems, plant somatic embryogenesis and pollen grain development, where striking nuclear activity changes occur.

Somatic embryogenesis of *Daucaus carota* L. has a high efficiency of embryo induction together with a rather synchronized development (reviewed by Fujimura and Komamine, 1975; Sengupta and Raghavan, 1980; Raghavan, 1983). During this process some biochemical, and cytochemical probes have been used as markers of embryogenesis (Sung and Okimoto, 1981, 1983; Thomas and Wilde, 1987; de Vries et al., 1988; Wilde et al., 1988, Fransz et al., 1989). During pollen grain development, mainly in the postmeiotic interphase, nuclear structure changes strongly in order to be prepared for its gametophytic function (Risueño et al., 1988; Testillano and Risueño, 1988). Therefore, these developmental processes are ideal to study the presence of nuclear proteins during different nuclear activity periods.

In this paper we describe some preliminary data on the use of anti-bodies against three kinds of nuclear proteins: small nuclear ribonucleo-protein particles (snRNPs), nuclear matrix proteins and nucleolar proteins which were detected by immunofluorescence and immunogold labelling on cryosections from somatic embryos of *Daucus carota* L. and anthers of *Capsicum annuum* L. and *Scilla peruviana* L.

Nuclear Structure and Function, Edited by J. R. Harris and
I. B. Zbarsky, Plenum Press, New York, 1990

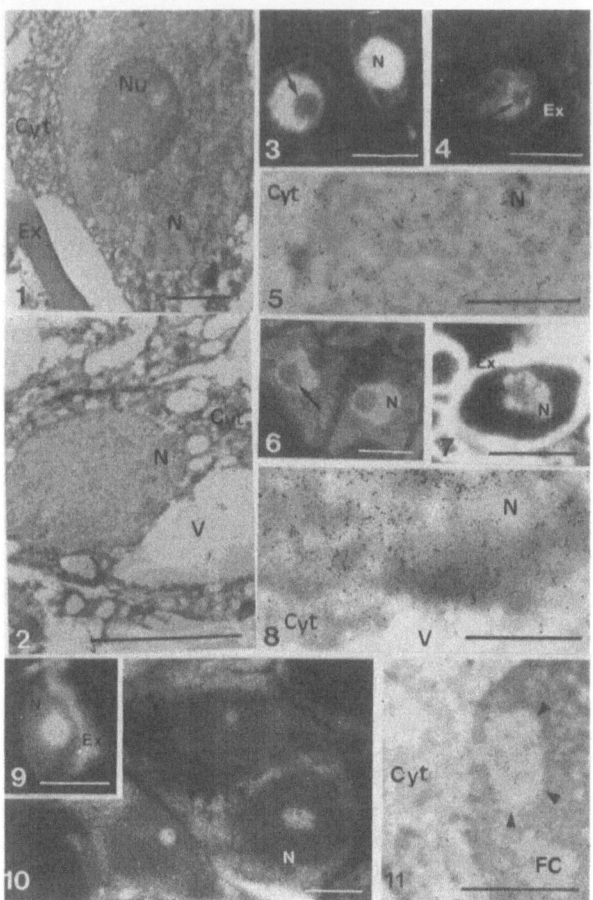

Figures 1 and 2.   Electron micrographs of a cryofixed and cryosectioned *Capsicum* pollen grain (Figure 1) and a 3 d.a.i. *Daucus* embryo cell (Figure 2). Nucleus of embryo cells shows a highly decondensed chromatin compared with that of the pollen grain. Note the good structural preservation of nucleus (N), nucleolus (Nu), cytoplasm (cyt), vacuoles (V) and cell walls, exine (Ex) in the pollen grain.

Figures 3, 4 and 5.   7.13 (anti-snRNPs) immunocytochemistry on cryosections of *Daucus* embryo cells (Figures 3 and 5) and *Capsicum* pollen grain (Figure 4). The nuclear fluorescence pattern is speckled (Figures 3 and 4) while nucleoli (arrows) are negative. The gold labelling is distributed irregularly, like a network in the nucleus (Figure 5).

Figures 6, 7 and 8.   RN-2 (anti-matrix protein) immunocytochemistry on cryosections of *Daucus* embryo cells (Figures 6 and 8) and *Capsicum* pollen grain (Figure 7). The nuclear fluorescence pattern is speckled (Figures 6 and 7) and more heterogeneous than after 7.13. The nucleolus (arrows) is negative. The gold particles are distributed in linear arrays over the condensed chromatin (Figure 8).

Figures 9, 10 and 11.   J26 (anti-nucleolar proteins) immunocytochemistry on cryosections from *Scilla* pollen grain and tapetal cells (Figures 9 and 10 respectively) and *Daucus* embryo cells (Figure 11). The nucleolar fluorescence is heterogeneous (Figures 9 and 10). The gold labelling is over the dense fibrillar and

MATERIALS AND METHODS

The materials used were embryogenic suspension cultures from *Daucus carota* L. and anthers from *Capsicum annuum* L. and *Scilla peruviana* L. Somatic embryogenesis was induced by adding to the culture medium B5 the hormone 2, 4 D (2, 4 dichlorophenoxyacetic acid) (Sanchez-Pina et al., 1989). Small cell aggregates were harvested at 0, 1, 2 and 3 days after the induction of embryogenesis (d.a.i.). Anthers were collected from plants grown in pots in a greenhouse with controlled temperature and photoperiod. Pollen grains studied were at different periods of the postmeiotic interphase.

Embryo clumps and anthers were cryofixed and cryosectioned as described in Sánchez-Pina et al. (1989).

The nuclear antibodies used were:
- 7.13 (anti-snRNPs) mouse monoclonal against protein D (16kD) of U1, U2, U4, U5 and U6 (Billings et al., 1982, 1985).
-RN-2 (anti-matrix protein) mouse monoclonal against a 250-300kD protein (Sánchez-Pina et al., 1989).
-J26 (anti-nucleolar proteins) human sera against 50kD and 86kD nucleolar proteins (Verheijen et al., 1986a).

The immunofluorescence and immunogold labelling was done essentially as reported in Sánchez-Pina et al. (1989); the only difference was that the second antibody for 7.13 and RN-2 was SwAM-FITC in embryos and GAM-FITC in anthers. Controls were made by omitting the first antibody incubation in both the light and electron microscopy experiments.

RESULTS AND DISCUSSION

Cryofixation of cells, followed by semithin or ultrathin cryosectioning, is a good method not only to study the nuclei *in situ*, but also to perform immunocytochemistry, because of the preservation of antigenicity, immobilization and accessibility of molecules to different reagents (for a survey, see Tokuyasu, 1986). As is shown in the Figures 1 and 2, the structural preservation of pollen grains and embryonic cells is good, because the main structural features of these cells can be clearly recognized.

We have studied three types of nuclear proteins, which are thoroughly immunocytochemically localized in animal and human cell systems, because of the regulatory functions they supposedly have. The results obtained using the monoclonal antibody 7.13 against snRNPs are shown in the Figures 3, 4 and 5. The fluorescence is localized specifically in the nucleus, in both plant systems, mostly in a speckled way; the nucleolus and the cytoplasm are negative. Cell walls and the large and numerous vacuoles in the embryo cells are completely negative (Figure 3), while the exine in the pollen grain shows some autofluorescence (Figure 4). After the immunogold labelling (Figure 5) most gold particles are found over the nucleoplasm, showing a rather irregular distribution like a reticular network. All controls, both for immunofluorescence and immunoelectron microscopy, were negative in

---

granular components, while fibrillar centres (arrow-heads) are devoid of them (Figure 11).
Fluorescence micrographs of pollen grains show the natural autofluorescence of the exine. Bars in light micrographs represent 10μm and in electron micrographs 1μm.

both plant systems. This localization pattern is similar to that found in other eukaryotic cells (Puvion et al., 1984, Verheijen et al., 1986b) which confirms the universal presence of the snRNP particles and the high level of conservation of the epitopes, which are recognized by this monoclonal antibody. Our data, in two very different plant systems, together with those obtained with animal and human cells, indicate the presence of the snRNPs in certain areas of the nucleus in most eukaryotic cells, including higher plants.

The nuclear matrix antibody RN-2 gives a specific nuclear labelling in both plant systems (Figures 6, 7, and 8). In all cases the nuclear fluorescence is speckled. The nucleolus and cytoplasm are negative. With 7.13 cell walls and vacuoles in the embryo cells are negative (Figure 6), but the pollen grain wall (exine) shows autofluorescence (Figure 7). In the electron microscopic immunodetection, nuclear labelling is preferentially located over the chromatin regions (Figure 8). The speckled pattern of this protein has also been found in rodent and human cell lines for other, non-lamin, nuclear matrix antigens (Chaly et al., 1984; Turner et al., 1985). The gold particle distribution seems to indicate that this could be a chromatin-bound protein. The positive reaction of this antibody in the plant systems studied reveals conservation of the antigenic sites of this protein.

The immunofluorescence labelling with the nucleolar antibody J26 is observed in the nucleolus in all the cells tested (Figures 9 and 10). This pattern is also found in the somatic tissues of the anthers, as is shown for tapetal cells in Figure 10. The autofluorescence of the exine is always observed (Figure 9). At the electron microscopic level gold particles are located over both dense fibrillar and granular components, while the fibrillar centres and nucleolar vacuoles appear practically devoid of labelling. The distribution of the gold particles can be correlated with the immunofluorescence pattern because this is also heterogeneous over the nucleolus (compare Figures 10 and 11, arrows). The localization of these antigens is similar to that obtained for C23/nucleolin and for B23 proteins that have been related to rRNA transcription, processing and preribosome assembly (Spector et al., 1984, see Reimer et al., 1987 for a review). The proteins immunodetected by us in both plant systems could also play a role in these nucleolar processes.

The immunolocalization of these antigens in such different plant systems as well as in animal and human cells indicates the high level of conservation of some epitopes in nuclear proteins throughout the evolution of eukaryots, as has been already demonstrated for other nuclear proteins such as B-36 (Risueño et al., 1988). This should be true for proteins with very basic and important roles in nuclear function.

ACKNOWLEDGEMENTS

We would like to thank: Dr S. C. de Vries (Department of Molecular Biology, Agricultural University) for supplying the embryo cultures; Dr F. Ramaekers (Department of Pathology, University of Nijmegen) for his generous gift of 7.13 and RN-2 antibodies and Dr W. Habets (Department of Biochemistry, University of Nijmegen) for his kind gift of the J26 antibody. This work was partially supported from the project DIGICYT/CSIC 88/91 PB 033201. Dr Sánchez-Pina was supported by a grant from the C.S.I.C. during her stay at the Agricultural University.

REFERENCES

Berezney R. 1984. Organization and functions of the nuclear matrix. In: 'Chromosomal non-histone proteins'; vol. IV. 'Structural associations', L. S. Hnilica, ed., CRC Press, Boca Raton, Florida.

Billings P. B., Allen R. W., Jensen F. C. and Hoch S. O., 1982. Anti-RNP monoclonal antibodies derived from a mouse strain with lupus-like autoimmunity. J. Immun. 128:1176.

Billings P. B., Barton J. R. and Hoch S. O., 1985. A murine monoclonal antibody recognizes the 13,000 molecular weight polypeptide of the Sm small nuclear ribonucleoprotein complex. J. Immun. 135:428.

Busch H., 1984. Nucleolar proteins: purification, isolation and functional analyses. In: 'Chromosomal non-histone proteins'; vol. IV. 'Structural associations', L. S. Hnilica, ed., CRC Press, Boca Raton, Florida.

Chaly N. T., Bladon G., Setterfield J. E., Kaplan J. G. and Brown D. L., 1984. Changes in distribution of nuclear matrix antigens during the mitotic cell cycle. J. Cell Biol. 99:661.

DeVries S. C., Booij H., Meyerink P., Huisman G., Wilde H. D., Thomas T. L. and van Kammen A., 1988. Acquisition of embryogenic potential in carrot cell suspension cultures. Planta 176:196.

Fransz P. F., de Ruijter N. C. A. and Schel J. H. N., 1989. Isozymes as biochemical and cytochemical markers in embryogenic callus cultures of maize (Zea mays L.). Plant Cell Reps. 8:67.

Fujimura T. and Momamine A., 1975. Effects of various growth factors regulators on the embryogenesis in a carrot cell suspension culture. Plant Sci. Lett. 5:359.

Holoubek V., 1984. Nuclear ribonucleoproteins containing heterogeneous RNA. In: 'Chromosomal non-histone proteins', Vol. IV. 'Structural associations'. L. S. Hnilica, CRC Press, Boca Raton, Florida.

Puvion E., Viron A., Assens C., Leduc E. H. and Jeanteur Ph., 1984. Immunocytochemical identification of nuclear structures containing snRNPs in isolated rat liver cells. J. Ultr. Res. 87:180.

Raghavan V., 1983. Biochemistry of somatic embryogenesis. In: 'Handbook of plant cell culture', vol. 1. D. A. Evans, W. R. Sharp, P. V. Ammirato and Y. Yanada, eds., McMillan Publ. Co., New York-London.

Reimer G., Raska I., Tan E. M. and Scheer U., 1987. Human autoantibodies: probes for nucleolus structure and function. Virchows Arch. B. 54:131.

Risueño M. C., López-Iglesias C., Testillano P. S., Olmedilla A. and Christensen M. E., 1988. Immunoelectron microscopy localization of the nucleolar protein B-36 in plant nucleoli. Inst. Phys. Conf. Serv. No 93, Vol. 3, Chapt. 3:69.

Risueño M. C., Testillano P. S. and Sánchez-Pina M. A., 1988. Variations of nucleolar ultrastructure in relation to transcriptional activity during G_1, S and G_2 periods of microspore interphase. In: 'Sexual Reproduction in Higher Plants'. M. Cresti, P. Gori and E. Pacini, eds. Springer-Verlag, Berlin.

Sánchez-Pina M. A., Kieft H. and Schel J. H. N., 1989. Immunocytochemical detection of non-histone nuclear antigens in cryosections of developing somatic embryos from Daucus carota L. J. Cell Sci. 93:615.

Sengupta C. and Raghavan V., 1980. Somatic embryogenesis in carrot cell suspension. I. Pattern of protein and nucleic acid synthesis. J. Exp. Bot. 31:247.

Spector D. L., Ochs R. L. and Busch H., 1984. Silver staining, immuno-fluorescence, and immunoelectron microscopic localization of nucleolar phosphoproteins B23 and C23. Chromosoma 90:139.

Sung Z. R. and Okimoto R., 1981. Embryogenic proteins in somatic embryos of carrot. Proc. Natl. Acad. Sci. USA, 78:3683.

Sung Z. R. and Okimoto R., 1983. Coordinated gene expression during somatic embryogenesis. Proc. Natl. Acad. Sci. USA, 80:2661.

Testillano P. S. and Risueño M. C., 1988. Evolution of nuclear interchromatin structures during microspore interphase periods. In: 'Sexual Reproduction in Higher Plants'. M. Cresti, P. Gori and E. Pacini, eds. Springer-Verlag, Berlin.

Thomas T. L. and Wilde D., 1987. Analysis of carrot somatic embryo gene expression programs. In: 'Plant tissue and cell culture', C. E. Green, D. A. Somers, W. P. Hackett and D. D. Biesboer, eds., Alan R. Liss Inc., New York.

Tokuyasu K. T., 1986. Applications of cryoultramicrotomy to immunochemistry, J. Microsc. 143:139.

Turner B. M., Davies S. and Whitfield W. G. F., 1985. Characterization of a family of nuclear and chromosomal proteins identified by a monoclonal antibody. Eur. J. Cell Biol. 38:344.

Verheijen R., Kuijpers H., Vooijs P., van Venrooij W. and Ramaekers F., 1986a. Protein composition of nuclear matrix preparations from HeLa cells: an immunochemical approach. J. Cell Sci. 80:103.

Verheijen R., Kuijpers H., Vooijs P., van Venrooij W. and Ramaekers F., 1986b. Distribution of the 70k U1 RNA-associated protein during interphase and mitosis. J. Cell Sci. 86:173.

Verheijen R., van Venrooij W. and Ramaekers F., 1988. The nuclear matrix: structure and composition. J. Cell Sci. 90:11.

Wilde H. D., Nelson W. S., Booij H., de Vries S. C. and Thomas T. L., 1988. Gene expression programes in embryogenic and non-embryogenic carrot cultures. Planta 176:205.

THE 'ARGENTAFFIN' REACTION OF THE NUCLEOLUS. SILVER REDUCING SITES AFTER

MERCURIC ACETATE AND COPPER TETRAMMINE TREATMENTS

M. C. Risueño, P. S. Testillano, M. A. Ollacarizqueta[1]
and C. J. Tandler[2]

Centro de Investigaciones Biologicas
Estructuras Celulares
[1]Servicio de Microscopía Electrónica Analítica
C.S.I.C.
Velázquez 144
28006 Madrid, Spain
[2] Institute de Biología Celular
CONICET, Fac. de Medicina
Buenos Aires
Argentine

INTRODUCTION

'Argentaffinity', as opposed to 'argyrophilia', denotes the reduction
of silver salts to metallic silver by tissue or cell component *per se*,
without addition of any extraneous reducing substance and in the dark.
Aldehydes, which are known to reduce diammine silver, are not needed for the
argentaffin reaction to take place; thus, staining after ethanol or Carnoy
fixation was entirely attributable to an intrinsic property of the cell
components (Tandler, 1954). A novel type of 'argentaffin' histochemical
staining reaction based on the reduction of diammine silver nitrate by
mercuric acetate postfixed tissues has been recently applied to striated
muscle cells (Tandler and Pellegrino de Iraldi, 1989; Tandler et al., 1989).

The 'argyrophilic' staining of the nucleoli has been reported after
using different silver staining methods in different tissues by many
cytologists and it was shown that the argyrophilic material was of protein
nature (for review see Hubbell, 1985 and Risueño and Medina, 1986). On
the contrary, the 'argentaffinity' of the nucleoli was scarcely noted,
Tandler and Das being the first authors to report it (Tandler 1954, 1955;
Das and Alfert, 1959, 1963 and Das, 1962). A large number of papers have
been published on the nucleolus in human, animal and plant cells making use
of silver staining methods under normal and experimental conditions (Stahl,
1982; Schwarzacher and Wachtler, 1983; for review see Risueño and Medina,
1986). This current interest has promoted us to study the 'argentaffin'
properties of the nucleolar material in plant cells.

This paper deals with the first application of the argentaffin
technique to examine the plant nucleolar components in interphase and
during mitosis at the light and electron microscope level (L. M. and E. M.),
using the Hg-Ag technique previously described (Tandler et al., 1989) and
the Cu-Ag technique, a modified procedure using ammoniacal copper salts

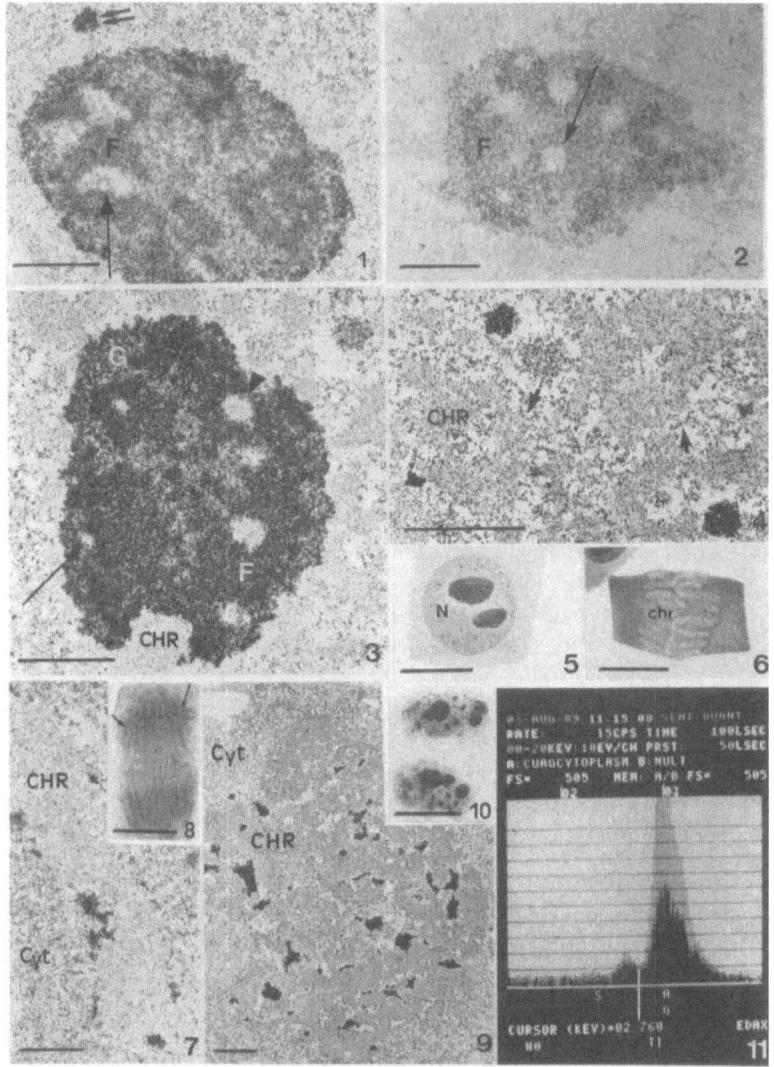

Figures 1   Hg–Ag staining of onion root meristematic cells after
and 2.   glutaraldehyde fixation (Figure 1) and formaldehyde fixation
              (Figure 2). Nucleolar dense fibrillar component (F) is
              preferentially stained. Fibrillar centres (arrows) are devoid
              of silver grains. Nuclear bodies (double arrows) are densely
              stained. Nucleoplasm staining is lower after formaldehyde
              fixation.

Figures 3   Cu–Ag staining of glutaraldehyde fixed onion root meristematic
to 10.   cells.

Figures 3, 4, Cells at the E.M. level stained with uranyl.
7 and 9.

Figures 5, 6, Squashed cells at the L.M. level.
8 and 10.

Figure 3.   Active interphase nucleolus showing the heavily stained DFC
              intermingled with the granular component (G). Heterogeneous
              (arrowheads) and homogeneous (arrows) fibrillar centres are
              devoid of silver grains.

Figure 4.   Interphasic nucleus showing a different distribution pattern
              of silver between condensed chromatin (chr) and interchromatin

instead of mercuric acetate. We also studied the correlation between silver deposits and protein concentration, by means of an X-ray microanalysis.

MATERIAL AND METHODS

The material used was root meristematic cells from *Allium cepa L*. Onion bulbs were grown under standard conditions at 15°C.

Root tips were fixed with 3% glutaraldehyde in H_2O for 2h at room temperature for both Hg-Ag and Cu-Ag techniques and with 4% formaldehyde in H_2O for 24h at 4°C in the case of Hg-Ag. Then, all the samples were washed in cold distilled water. After that samples were divided in two series: - Hg-Ag samples were, as in Tandler and Pellegrino de Iraldi (1989), immersed in 5% mercuric acetate in 1% acetic acid overnight, in darkness, at room temperature and then washed with 1% acetic acid and rinsed with distilled water. - Cu-Ag samples were immersed in an aqueous solution of copper tetrammine sulfate or acetate, for 12-24h at room temperature. The reagent was prepared by adding dropwise concentrated ammonium hydroxide (25-27% NH_3) to a 10% copper sulfate or acetate solution until the copper hydroxide precipitate redissolves. After that they were washed with 5% (v/v) aqueous ammonia, then with distilled water and with 5% (v/v) acetic acid until the blue colour of the tissue disappears. Finally, the acid was washed away with distilled water.

From now on, both series were processed in the same way and placed in the ammoniacal silver nitrate overnight, in darkness, at 43°C. Later, they were washed in 5-10% sodium sulfite, rinsed in distilled water, dehydrated in ethanol series and embedded in Epon. Ultrathin sections stained with uranyl acetate or unstained were observed in a Philips EM 300 or a Hitachi EM 7000. Squashed preparations were made on a drop of glycerin and examined at a Zeiss light microscope. The results were compared with samples processed by the conventional technique for E.M.

X-ray microanalysis was used to check the relationship between the sulfur quantity, as an indicator of proteins presence, and the silver

---

	region. Nuclear bodies showing different structures are densely stained as well as interchromatin-like fibres (small arrows).
Figure 5.	Interphasic cells. Besides the nucleolus, nuclear bodies are clearly distinguished after the Cu-Ag method.
Figure 6.	Metaphasic cell. Chromosomes appear unstained by silver.
Figure 7.	Anaphasic cell. The argentaffin material is surrounding the chromosomes. Note the cytoplasmic staining in this phase.
Figure 8.	Anaphasic chromosomes. Argentaffin nucleolar material densely stained forms a coat around the unstained chromosomes (arrows).
Figure 9.	Early telophase showing the argentaffin nucleolar masses in between the chromosomes.
Figure 10.	Nucleologenesis at late telophase. Prenucleolar bodies and newly formed nucleolus show argentaffin reaction.
Figure 11.	X-ray microanalysis superposed spectra from Cu-Ag stained cells. Black spectrum corresponds to cytoplasm and grey spectrum to nucleolar DFC. Note that the S and Ag peaks are much higher in the DFC than in the cytoplasm. Bars in electron micrographs represent 1μm. Bars in light micrographs represent 10μm.

precipitates. Ultrathin sections on carbon coated titanium grids were observed unstained in a Philips EM 420 fitted with an energy-dispersive X-ray spectrometer and an EDAX 9100 analyzer computer system.

RESULTS AND DISCUSSION

The use of Hg-Ag and Cu-Ag techniques gives a constant pattern of nuclear protein staining in all cells examined. Silver grains were preferentially and heavily distributed on the dense fibrillar component (DFC) of the nucleoli, while the granular component (GC) was less intensively stained; no staining was observed on the fibrillar centres (FCs) during interphase (Figures 1, 2 and 3) and chromosomal NOR during mitosis (Figures 6, 7 and 8). Silver precipitates were also found in the nucleoplasm (Figures 1, 2, 3 and 4). When formaldehyde was used as fixative in the Hg-Ag technique, the chromatin and interchromatin regions were mostly devoid of silver grains (compare Figures 1 and 2). Besides nucleoli many nuclear bodies showed a strong reactivity (Figures 1 and 4), being clearly distinguished even at the L.M. level (Figure 5).

Particularly, after the Cu-Ag technique a differential silver distribution between condensed chromatin and interchromatin region was obtained (Figure 4). Over the chromatin masses the pattern of less numerous silver grains was homogeneous while in the interchromatin region it was heterogeneous, forming a fibrillar-like network (Figures 3 and 4). These fibres were observed in relation to the nuclear bodies and/or dense chromatin (Figure 4). Nuclear bodies seem to be different types, as some were smaller and heavily stained and others were bigger with an uneven distribution of precipitate (Figure 4).

During mitosis metaphase chromosomes remain completely unstained (Figure 6), which corresponds with the poor stainability of the deoxy-ribonucleoproteins in interphasic condensed chromatin (Figure 4). In late anaphase the argentaffin material forms a thin coat around the unstained chromosomes (Figures 7 and 8). During telophase the masses of nucleolar material (Figure 9), as well as the prenucleolar bodies and the newly formed nucleolus (Figure 10), showed the argentaffin reaction as was firstly reported by Tandler (1959) in a L.M. study. The cytoplasm of metaphase-anaphase cells usually showed a greater stainability (Figures 7 and 8) suggesting the presence of dispersed argentaffin nuclear material.

Contrary to Ag-NOR argyrophilic techniques, the present argentaffin methods do not visualize the NOR in mitotic chromosomes. It has been shown that heavy metals with high affinity for sulphydryl groups, such as Cu and Hg, prevent the silver staining NOR reaction (Capoa et al., 1982). Our data clearly show that the argentaffin reaction of the other nucleolar components is not affected to any extent by Cu and Hg. These results indicate that the mechanisms involved in the argyrophillic techniques and in this argentaffin method must be different. We could state that the NOR region of chromosomes and FCs in interphase nucleoli are argyrophillic (Hernández-Verdun, 1983; Goessens, 1984) but not argentaffinic. On the other hand the DFC as well as the prenucleolar bodies are reactive with both types of techniques. Nevertheless, a relationship between argentaffinic proteins and NOR argyrophillic proteins seems evident, as judged by the reactivity of the DFC which contains the active part of the NOR interphase chromatin (Risueño et al., 1982; Thiry et al., 1985; Raska et al., 1989).

In an attempt to correlate protein concentration with the amount of metallic silver formed over the cellular components where the reaction takes place, using sulfur as representative of proteins, the Ag/S ratio was studied by X-ray microanalysis. This was only done on nucleoli and cytoplasm, because the sulfur content of histones in chromatin is very low. The Ag/S ratio was rather constant for the nucleolar components as well as for the cytoplasm, and the Ag and S quantities were much higher in the nucleolar DFC (Figure 11). This points to the conclusion that the preferential stainability of the DFC after the argentaffin reaction is mainly due to its higher concentration of argentaffinic proteins, but a tight spatial conformation at the molecular level should also be considered.

This novel silver technique has revealed that these argentaffinic proteins coexist with the argyrophilic ones, except for those of the NOR in mitosis. The argentaffinic proteins that have not been studied as intensively as the argyrophillic ones, and should now be explored, in order to determine their nucleolar role.

ACKNOWLEDGEMENTS

We wish to thank Dr Sánchez-Pina for her helpful comments and critical reading of the manuscript. Dr C. J. Tandler was supported by a grant from the Ministerio de Educación y Ciencia of Spain during his stay at the Centro de Investigaciones Biológicas. This work was supported by the project DIGICYT/CSIC 88/91 PB 033201.

REFERENCES

Capoa A., Ferraro M., Lavia P., Pellicia F. and Finazzi-Agro A., 1982, Silver staining of the nucleolus organizer regions (NOR) requires clusters of sulphydryl groups, J. Histochem. Cytochem., 30:908.
Das N. K., 1962, Demonstration of a non-RNA nucleolar fraction by silver staining, Exp. Cell Res., 26:428.
Das N. K. and Alfert M., 1959, Detection of a nucleolar component and its behaviour during the mitotic cycle, Anat. Rec., 134:548.
Das N. K. and Alfert M., 1963, Silver staining of a nucleolar fraction, its origine and fate during the mitotic cycle, Ann. Histochim., 8:109.
Goessens G., 1984, Nucleolar structure, Int. Rev. Cytol., 87:107.
Hernández-Verdun D., 1983, The nucleolar organizer regions, Biol. Cell. 49:191.
Hubbell H. R., 1985, Silver staining as an indicator of active ribosomal genes, Stain Tech., 60:285.
Raska I., Reimer G., Jarnik M., Kostrouch Z. and Raska K. 1989, Does the synthesis of ribosomal RNA take place within nucleolar fibrillar centres or dense fibrillar component?, Biol. Cell., 65:79.
Risueño M. C. and Medina F. J., 1986, 'The nucleolar structure in plant cells'. Rev. Biol. Cel., Serv. Edit. Univ. Pais Vasco. Bilbao, Spain.
Risueño M. C. Medina F. J. and Moreno-Diaz de la Espaina S., 1982, Nucleolar fibrillar centres in plant meristematic cells: ultra-structure, cytochemistry and autoradiography, J. Cell Sci., 58:313.
Schwarzacher H. G. and Wachtler F., 1983, Nucleolar organizer regions and nucleoli, Human Genet., 63:89.
Stahl A., 1982, The nucleolus and nucleolar chromosomes, in: 'The Nucleolus' E. G. Jordan and C. A. Cullis, eds. Cambridge University Press. Cambridge.
Tandler C. J., 1954, An argentaffin component of the nucleolus, J. Histochem. Cytochem., 1:165.
Tandler C. J., 1955, The reaction of nucleoli with ammoniacal silver nitrate in darkness; additional data, J. Histochem. Cytochem., 3:196.
Tandler C. J., 1959, The silver-reducing property of the nucleolus and the formation of prenucleolar material during mitosis, Exp. Cell Res., 17:560.
Tandler C. J., González D. A., Remorini P. G. and Pellegrino de Iraldi A.,

1989, A silver-reducing component in rat striated muscle. II.
Isolated sarcoplasmic reticulum vesicles, Histochemistry, 92:23.

Tandler C. J. and Pellegrino de Iraldi A., 1989, A silver-reducing component
in rat striated muscle. I. Selective localization at the level of the
terminal cister/transverse tuble system. Light and electron microscope
studies with a new hystochemical procedure, Histochemistry, 92:15.

Thiry M., Lepoint A. and Goessens G., 1985, Re-evaluation of the site of
transcription in Ehrlich tumour cell nucleoli, Biol. Cell, 54:57.

IMMUNOCHEMICAL IDENTIFICATION OF A NOVEL NUCLEAR RNP COMPLEX OF 70-110S FROM

RAT LIVER

V. Aidinis, A. Dangli, P. Markrinos, M. Patrinou-Georgoula,
C. E. Sekeris and A. Guialis

Biological Research Center
The National Hellenic Research Foundation
48 Vassileos Constantinou Avenue
Athens 11635
Greece

INTRODUCTION

Nuclear ribonucleoprotein particles (RNPs) are abundant structures, currently implicated as key components in pre-mRNA processing. They are considered dynamic structures, assembled into a variety of functionally related complexes. Of these the best so far characterized are the 40S hnRNPs or monoparticles and U-snRNPs (for recent reviews see Dreyfuss, 1986; Steitz et al., 1988).

The existence of a novel heterogeneous RNP particle of 70-110S in rat liver nuclei has been reported (Hatzoglou et al., 1985). Distinct biochemical features ascribed to this particle include resistance to high salt and mild RNase digestion, as well as unique protein components in the range of 90-40kD, with two polypeptides of 72/74kD as prominent species. Since specific antibody probes have been proven as valuable tools in deciphering structural/functional aspects of RNP, we attempted the production of antibodies against components of the 70-110S RNP and used these antibodies to characterize the antigenic polypeptides and the complex itself.

MATERIALS AND METHODS

The preparation of nuclear extracts from rat liver nuclei, the isolation of 70-110S RNP structures and the CsCl density gradient analysis have been described in Hatzoglou et al. (1985). The a-Sm antiserum was obtained from the Centers for Disease Control, Atlanta, Georgia, USA. Immunoprecipitation reactions were as Guialis et al. (1987), and the protocols for protein analysis, immunoblotting, affinity purification of antibodies, as well as indirect immunofluoresence, as in Dangli et al. (1988).

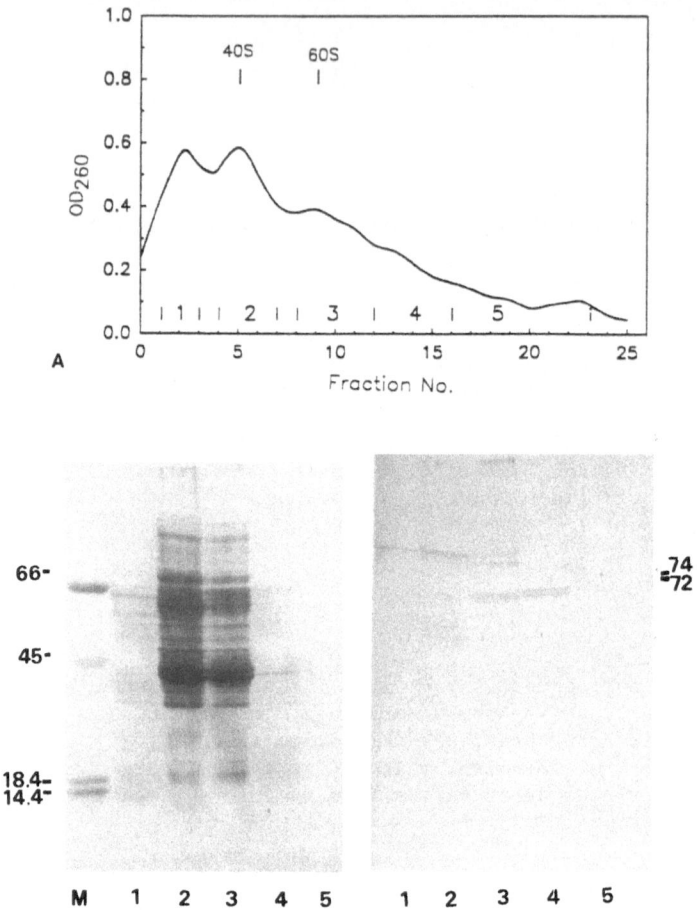

Figure 1.   Identification of the polypeptides recognized by the a-70SI
serum, and their distribution pattern along the sucrose-
glycerol gradient. A: Absorbance profile (OD 260) of nuclear
extracts fractionated on 15-30% sucrose-18% glycerol gradient.
B: Protein analysis of pooled gradient fractions (regions 1-5)
on a 8-15% SDS-Polyacrylamide gel. Ponceau-S staining following transfer to nitrocellulose. C: Immunodetection of the
a-70SI antigenic proteins on the same nitrocellulose filter.

RESULTS

Antibody Production and Characterization of Antigenic Peptides

We elicited antibodies in rabbits against polypeptide components of
the 70-110S RNP, making use of its salt-resistance property as a further
purification step. As pointed out before (Hatzoglou et al., 1985) the two
prominent 72/74kD polypeptides remain associated to the complex after salt-
treatment. The steps in the preparation of the immunogen were: isolation of
rat liver nuclei in spermine containing buffer at pH 5.2, extraction at
pH8 and 0.14M salt and fractionation of nuclear extracts on sucrose/glycerol
gradients. The 70-110S material was then adjusted to 0.7M NaCl and the
salt-resistant components, collected by high speed centrifugation, were used
for immunization.

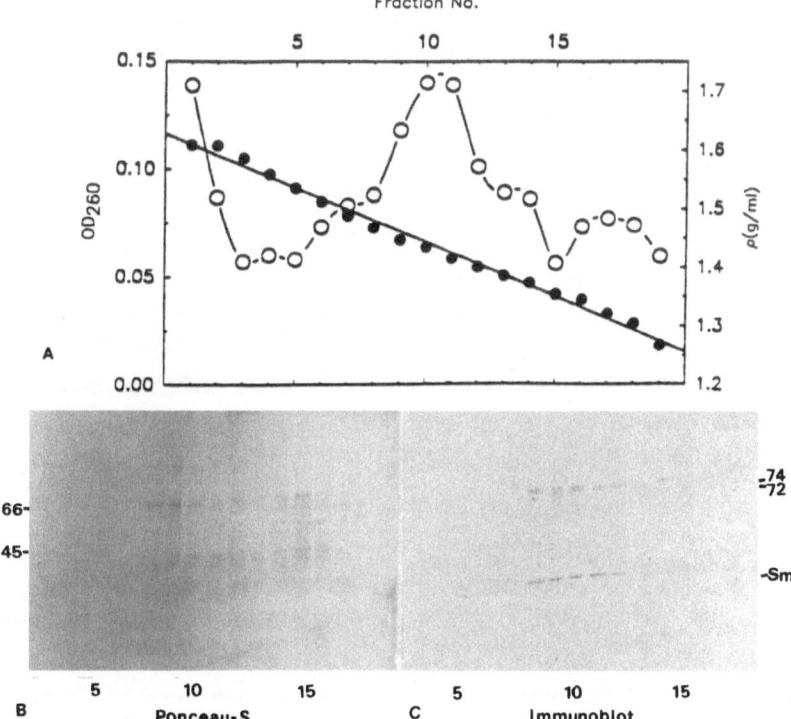

Figure 2.   CsCl gradient analysis of the unfixed 70-110S RNP. A:
             Absorbance (OD 260, •-•) and density (o-o) profile of CsCl
             gradient. B: Electrophoretic analysis of the proteins in each
             gradient fraction. Ponceau-S staining following transfer to
             nitrocellulose. C: The nitrocellulose filter shown in B
             probed with a--70SI and a-Sm sera.

     To characterize the polypeptides recognized by the serum (a-70SI), we
examined their sedimentation pattern following sucrose-glycerol gradient
fractionation of nuclear extracts. This was done by immunoblotting of the
proteins from pooled gradient fractions (Figure 1). The polypeptides
present in pooled gradient fractions (panel A regions 1-5) following
analysis by SDS-PAGE and transfer to nitrocellulose are shown in panel B.
The majority of the proteins were recovered in regions 2 and 3, which
corresponded to 40S monoparticles and related 60S RNP entities, respectively
(Hatzoglou et al., 1985; and our unpublished observations). When probing
with a-70SI the two proteins of 72/74kD were recognized as the major antigens in the 70-110S material (region 4 of the gradient), where they were
largely localized. Additional immunoreactive protein species (mainly a
96kD polypeptide) were also detected but were mainly found in fractions
lighter than 60S. It is likely that the latter were minor contaminants in
the salt-resistant 70-110S material used as immunogen. The a-70SI serum,
although polyspecific, can be used as an almost monospecific serum in the
analysis of material originating from the 70-110S region of the gradient.
This also applies to the components contained in the high speed pellet of
nuclear extracts treated with 0.7M NaCl (data not shown).

     The antibodies obtained were then used to characterize the antigenic
polypeptides. Immunoblotting of proteins from the 70-110S fractions that
had been analyzed by two-dimensional (NEPHGE/SDS-PAGE) electrophoresis

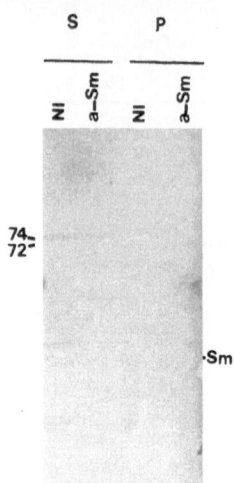

Figure 3.   Immunodetection of the 70-110S RNP fraction by a-Sm and non-
            immune (NI) sera. The proteins present in the immune pellets
            (P) or the supernatants (S) were separated by SDS-PAGE and
            probed by a-Sm and a-70SI sera. Protein bands corresponding
            to either the Sm or the 72/74kD antigens are indicated.

revealed that the 72/74kD immunoreactive polypeptides were basic proteins
(pI 8.2), exhibiting considerable charge heterogeneity (data not shown).
The ability of the a-70SI serum to recognize the same protein doublet in
three different cell lines rat (RV), mouse (MEL) and human (HeLa) cells
indicated that the 72/74kD proteins were evolutionary conserved, at least in
mammals (data not shown). Both the 72kD and 74kD polypeptides were
recognized by the same antibody population. This was shown by the finding
that on Western blots, both polypeptides reacted with antibodies affinity-
purified from either the 72kD or the 74kD protein band (data not shown).
These results indicated then that the 72kD and 74kD polypeptides were
structurally related species sharing common epitopes. The possibility that
the 72kD polypeptide was a proteolytic product of the 74kD protein could not
be excluded but seemed rather unlikely since both protein bands were immunodetected in extracts of different cell lines in the presence of protease
inhibitors.

Structural Features of the 70-110S RNP

     The a-70SI serum was further utilized for the structural character-
ization of the RNP complex itself. Although efficiently performing both in
immunoblots and indirect immunofluorescence (see below), the serum did not
so far immunoprecipitate more than approximately 10% of its antigens, an
efficiency not allowing direct analysis of the RNP components. We have,
nonetheless, attempted to probe the structure of the heterogeneous RNP
complex by combining immunodetection experiments with CsCl gradient
analysis. First, to verify that the 72/74kD polypeptides were indeed
authentic components of an RNP complex and not co-migrating protein
contaminants, we determined by slot-blot analysis the distribution of the
antigenic proteins following analysis of the formaldehyde-fixed 70-110S
material on a CsCl gradient. Since a specific immunoreaction was localized
in the region of the gradient with a density (1.43g/ml), characteristic of
an RNP complex, we concluded that the 72/74kD protein doublet was authentic
RNP component (data not shown). Given its notable salt-resistance, the
70-110S RNP complex was expected to withstand CsCl density analysis without

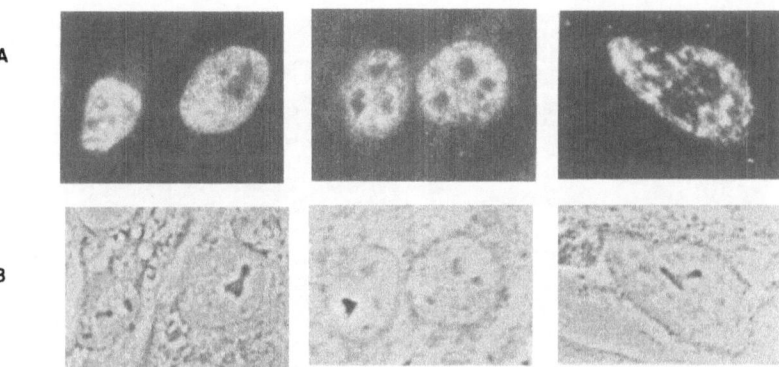

Figure 4. Indirect immunofluorescence on a rat cell line (RV) with
either the a-70SI serum (1), the affinity purified antibodies
reacting with the 72/74kD proteins (2), or the a-Sm autoantibodies (3). A: Phase contrast. B: indirect immunofluorescence.

prior aldehyde fixation. This property then would allow determination of
the protein composition of every fraction of the CsCl gradient by SDS-PAGE
and the subsequent recognition of the immunoreactive polypeptides on
Western blot. From the combined data presented in Figure 2, panels A, B and
C, it was clear that a large portion of the unfixed 70-110S material had
indeed a density of 1.43g/ml. In addition to the immunoreactive 72/74kD
polypeptides shown in panel C and in line to the previous findings
(Hatzoglou et al., 1985), Ponceau-S staining (panel B) identified a limited
number of polypeptides banding at the density of 1.43g/ml as putative
protein components of the RNP complex. Moreover, immunoblotting (panel C)
revealed the presence of the 72kD Sm-polypeptide which exhibited a similar
distribution to that of the 72/74kD protein doublet. This was not
unexpected, since U-snRNAs (U_1-U_6) have been detected in the 70-110S
material (Hatzoglou et al., 1985) and are most likely in the form of snRNP
complexes, as deduced from their ability to be Sm-precipitated (unpublished
observation). The property of the unfixed snRNP complexes to band on CsCl
gradients at a density of 1.43g/ml, has been previously reported (Brunel et
al., 1981). To determine then whether the U-snRNPs were co-fractionating or
directly associated with the 70-110S RNP on sucrose gradients, we immunoprecipitated the 70-110S material using anti-Sm and looked for the
concomitant presence of the 72/74kD polypeptides and the Sm antigen in the
immune pellet. As shown in Figure 3, the 72/74kD polypeptides were not
detected in the anti-Sm immunoprecipitate but remained in the supernatant of
the reaction. We concluded, therefore, that the 70-110S RNP and the
U-snRNPs simply co-purified on sucrose gradients. This was further
supported by the finding that upon native agarose gel electrophoresis of the
70-110S material the 72/74kD doublet and the Sm-antigen displayed different
gel mobilities (data not shown).

Cellular Localization of the a-70SI Antigens

The a-70SI serum was used in immunofluorescence experiments to determine the cellular localization of its antigenic polypeptides and the
70-110S RNP complex itself.

As shown in Figure 4(1), the antigenic protein species recognized by

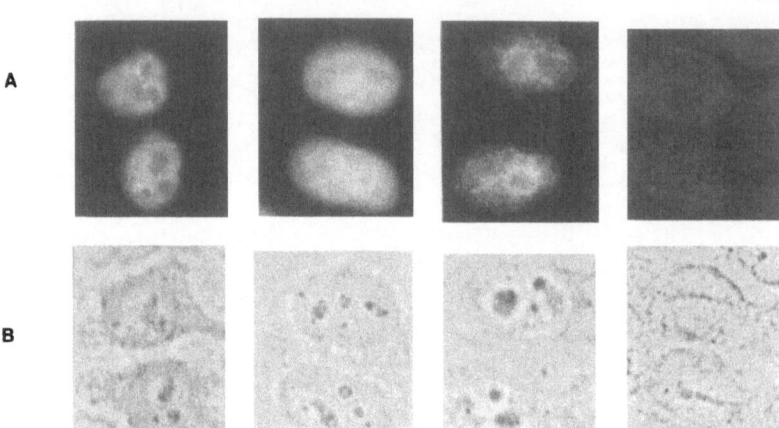

Figure 5.   Association of the a-70SI antigens with nuclear structures.
             Indirect immunofluorescence on HeLa cells untreated (1), or
             sequentially extracted with detergent (2), DNase/High salt
             (3), RNase/High salt (4). A: Phase contrast. B: Indirect
             immunofluorescence.

the a-70SI serum in a rat (RV) cell line were localized in the nucleoplasm
and not in the nucleolus. Moreover, as expected for RNP associated
components, they were not detected on mitotic chromosomes (data not shown).
The observed staining pattern of the a-70SI serum was speckled with a rather
homogeneous background. When the affinity purified a-72/74kD antibodies
were used instead, the staining pattern was more speckled resembling that of
a-Sm (Figure 4 (2 and 3)). Double labelling experiments using a-70SI and
a-SM sera to identify the 70-110S RNP and U-snRNPs, respectively, showed
that the two complexes did not co-localized in the nucleoplasm (data not
shown). These findings, are in line with the biochemical data discussed in
the previous section which did not support a direct association between
these two RNP populations. Concerning the association of the a-70SI
antigens to other nuclear structures, we investigated their possible
association to the nuclear matrix. To do this HeLa cells attached to cover-
slips were treated sequentially with detergent, DNase, high salt and RNase
followed by high salt, and the a-70SI antigens were identified by immunofluorescence. As seen in Figure 5 the a-70SI antigens were only removed
from residual nuclear structures by RNase/high salt treatment. Thus, the
antigenic protein species of the a-70SI serum were not components of the
RNA-depleted nuclear matrix and they required RNA integrity to remain
associated to it. This conclusion was further supported by the finding that
the 72/74kD doublet was not detected by immunoblotting amongst the protein
components of the nuclear matrix prepared by the analogous biochemical
protocol of Staufenbiel (1983).

DISCUSSION

    From the studies presented here two novel antigens of 72/74kD were
identified as potentially important nuclear constituents. Immunochemical
characterization defined these antigens as distinct, basic proteins with
shared homology and exclusive nucleoplasmic localization.

The 72/74kD antigenic polypeptides were not present in the RNA-depleted nuclear matrix and had a cellular localization characteristic of RNP structures. It is thus, of interest to identify the nuclear structures into which these antigens are organized. Considering the previous biochemical studies (Hatzoglou et al., 1985) and the present findings from CsCl gradient analysis, we believe that the 72/74kD polypeptides are authentic RNP components associated to the heterogeneous RNP complexes of 70-110S. Additional protein components of this complex should be represented amongst the 40-96kD polypeptides co-purifying with the 72-74kD doublet. With regard to the RNA composition, the co-fractionating snRNAs appear to belong to snRNP entities not directly associated with but rather co-purifying with the heterogeneous RNP complex. HnRNA molecules found in the 70-110S material could be RNA components of the complex, as suggested by the observed reduction in the RNP complex recovered from cells treated with α-amanitin (Hatzoglou et al., 1985). Nevertheless, the exact RNA protein composition of the 70-110S RNP is not yet established.

By the biochemical and immunochemical criteria thus far presented, the 70-110S RNPs are rather large, not abundant nuclear entities which are not related to 40S monoparticles or polyparticles. They are, also, distinct from the co-purifying snRNPs with which they exhibit similar salt-resistance properties. The 70-110S RNPs have some properties in common with the heterogeneous complexes described by Jacob and co-workers (1981). These include, in addition to size heterogeneity, RNase and salt resistance properties, similarities in the overall protein composition. However, the proteins of the heterogeneous complexes are found to be acidic proteins (pI 5-7), while the prominent 72/74kD polypeptides of the 70-110S RNP were basic polypeptides. The exact correlation of the 70-110S RNP to the nuclear matrix fibrils and to other nuclear structures remain to be established.

ACKNOWLEDGEMENTS

This work was partly supported by grant I/62 945 from the Stiftung Volkswagenwerk.

REFERENCES

Brunel C., Widada J. S., Lelay M. N., Jeanteur P. and Liautard J. P. 1981, Purification and characterization of a simple ribonucleoprotein particle containing small nucleoplasmic RNAs (snRNP) as a subset of RNP containing heterogeneous nuclear RNA (hnRNP) from HeLa cells, Nucl. Acids Res., 9:815.

Dangli A., Guialis A., Wretou E. and Sekeris C. E. 1988, Autoantibodies to the core proteins of HnRNPs, FEBS Lett., 231:118.

Dreyfuss G. 1986, Structure and function of nuclear and cytoplasmic ribonucleoprotein particles, Ann. Rev. Cell Biol., 2:459.

Guialis A., Dangli A. and Sekeris C. E. 1987, Distribution of snRNP complexes in rat liver nuclear extracts: Biochemical and immunochemical analysis, Mol. Cell Biochem., 76:167.

Hatzoglou M., Adamtziki E., Margaritis L. and Sekeris C. E. 1985, Isolation and characterization of nuclear particles containing rapidly labelled hnRNA and snRNA in combination with a distinct set of polypeptides of MW 74000 and 72000, Exp. Cell Res., 157:227.

Jacob M., Devilliers G., Fuchs J. P., Gallinaro H., Gattoni R., Judes C. and Stevenin J. 1981, Isolation and structure of the ribonucleoprotein fibrils containing heterogeneous nuclear RNA in: 'The Cell Nucleus' H. Busch, ed., Vol. 8, Academic Press, New York.

Staufenbiel M. and Deppert W. 1983, Nuclear matrix preparations from liver tissue and from cultured vertebrate cells: differences in major polypeptides, Eur. J. Cell Biol., 31:341.

Steitz J. A., Black D. L., Gerke V., Parker K. A., Kramer A., Frendwey D. and Keller W. 1988, Function of the abundant U-snRNPs in: 'Structure and Function of Major and Minor Small Nuclear Ribonucleoprotein Particles', M. L. Birnstiel, ed., Springer-Verlag, Berlin.

VISUALIZATION OF CHROMATIN ARRANGEMENT IN GIANT NUCLEI OF MOUSE TROPHOBLAST

BY VIDEOMICROSCOPY AND LASER SCANNING MICROSCOPY

Markus Montag, Angelika Wild, Herbert Spring and
Michael F. Trendelenburg

Institute of Experimental Pathology
German Cancer Research Center
Im Neuenheimer Feld 280
D-6900 Heidelberg
F.R.G.

INTRODUCTION

In the past years substantial progress has been made in the development of novel light microscopic (LM) instrumentation, which provided new insight in structure and dynamics of nuclear chromatin organization. (i) High resolution videomicroscopy allowed investigations on cellular structures which could previously only be demonstrated by electron microscopic (EM) techniques and therefore were largely inaccessible to light microscopy (Allen, 1985; Trendelenburg et al., 1986). (ii) More recently, laser scanning microscopy (LSM) has allowed us to get insight into the very complex organization of intact cells. Using the confocal scanning mode it is possible to record series of consecutive optical sections through thick specimens. These image stacks can then be further processed for 3-D visualization and 3-D reconstruction (Ploem, 1987; White et al., 1987; Montag et al., 1989a). Therefore, videomicroscopy and confocal laser scanning microscopy can be applied for the examination of large numbers of samples and yield a high amount of structural as well as spatial information in a relatively short time, as compared to the conventional approach using serial thin section electron microscopy.

A particularly attractive application of these techniques is the study of intranuclear chromatin organization. In addition to the analysis of the overall chromatin distribution, particularly the identification of defined chromatin regions within intact nuclei had been reported (Agard and Sedat, 1983; Arndt-Jovin et al., 1985; Brakenhoff et al., 1985; Spring et al., 1988; Montag et al., 1989b).

In the present study, we describe an application of videomicroscopy and CLSM to investigate the organization and spatial arrangement of nuclear chromatin of giant mouse trophoblast cells. During early placental differentiation, giant nuclei are known to reach a very high DNA content by endomitotic replication cycles. So far, analysis of different chromatin configurations was predominantly carried out on squashed and Feulgen-stained nuclei (Barlow and Sherman, 1974; Zybina et al., 1975). In order to compare these data with observations on non-squashed nuclei it appeared to be valuable to apply video- and laser scan microscopy.

Nuclear Structure and Function, Edited by J. R. Harris and
I. B. Zbarsky, Plenum Press, New York, 1990

Table I. Nuclear Volumes and DNA-Contents of Mouse Trophoblast Giant
Cells During Early Gestation.
Values for the maximum nuclear volumes (expressed in pico-
liters) and DNA contents (expressed in C-values) are dependent
on the day of gestation. For this study, isolated nuclei
were measured, fixed and subjected to Feulgen cytophotometry
according to standard techniques (see Bohrmann et al., 1986).

Day of gestation	6	7	8	9	10	11	12
Max. nuclear volume (pl.)	9	15	40	99	275	442	203
Max. ploidy level (C-Value)	16-32	32-64	64-128	128-256	256-512	512-1024	256-850

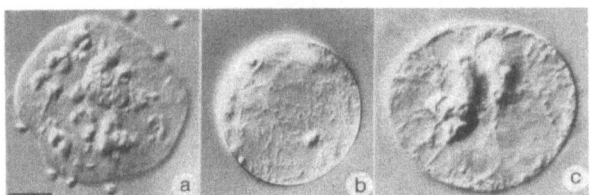

Figure 1.   Three nuclei representing the homogenous (a), intermediate (b)
and reticulate (c) nuclear chromatin morphology type are shown
by video-enhanced differential interference contrast (DIC).
Nuclei were manually isolated at day 10 of gestation and
transferred to an observation chamber for further microscopic
analysis (see Trendelenburg et al., 1986). Chromatin arrays
in defined focal planes through large intact giant nuclei can
be documented by video-enhanced Nomarski DIC. Bar indicates
20µm.

RESULTS AND DISCUSSION

     Giant trophoblast nuclei can be manually isolated from the mouse
trophoblast tissue from day 6 up to day 14 of gestation. This facilitates
the morphometric and cytophotometric analysis of individual nuclei, in
order to get a first impression of the most obvious changes of the nuclear
shape and the DNA content. Table I summarizes the results which we obtained
for the analysis of the nuclear volume and the ploidy level at consecutive
days of gestation.

     Up to day 11 the observed increase of maximum nuclear volumes is
accompanied by a progressive increase of the DNA content, as expressed in
C-values. From day 12 onwards degenerative processes occur in the giant
cell layer and finally giant nuclei can no longer be isolated in large
numbers at later days of gestation. These data imply the occurrence of a
presumable endomitotic replication cycle characterized by a mean cycle
length of 22-26 hours, which would (i) result in the exponential increase
of the DNA content and (ii) are likely to be accompanied by a time-dependent
change of the intranuclear chromatin organization.

     So far, using analysis of squashed and Feulgen-stained preparations
three different types of nuclei could be identified, showing either a
homogenous, intermediate or reticulate type respectively, are shown in

Table II. Quantitation of Nuclei With Typical Chromatin Configuration During the Cell Cycle.
Up to 100 nuclei were isolated at the indicated time points and chromatin configurations immediately determined using DIC microscopy. Only mice kept at a constant day/night rhythm were taken to assure a good correlation of the data.

Day of gestation / Chromatin distribution	Day 7		Day 11	
	9 a.m.	9 p.m.	9 a.m.	9 p.m.
Homogenous	8.9%	48.3%	75.8%	21%
Intermediate	33.9%	25.7%	14.5%	47.4%
Reticulate	57.2%	25.9%	9.7%	31.6%

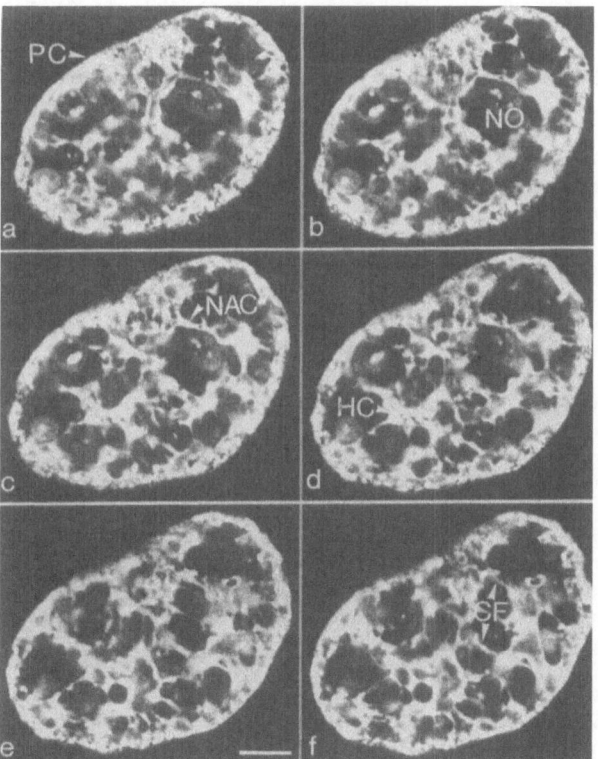

Figure 2.   CLSM analysis of a day 10 nucleus with an 'intermediate' to 'reticulate' chromatin morphology. For DNA-fluorescence a trophoblast tissue whole mount was prepared and stained by the Feulgen-technique as described (Montag et al., 1989a). The optical sections in a-f were recorded in 500nm steps in z-axis using an argon laser at 488nm. Every image was averaged over 16 frames and stored on hard disk. Using this technique we can demonstrate several intranuclear domains such as peripheral chromatin (PC), heterochromatin (HC), nucleolus (NO), nucleolus associated chromatin (NAC) and finally chromatin fibers (CF) and their arrangement in consecutive optical sections. Bar indicates 10µm.

video-enhanced differential interference contrast (DIC) microscopy. It has been previously shown that the combined application of videomicroscopy and DIC is a fast and simple method to record optical sections from intact and non-squashed giant nuclei (Montag et al., 1989b). Our results show that the different nuclear morphology types are mainly characterized by the pattern of lateral association of the intranuclear chromatin fibers.

To obtain further information about the dynamics of interconversion of the different chromatin arrangements, we have examined the percentage of isolated nuclei exhibiting one of the above described chromatin patterns at 9 a.m. and 9 p.m. at days 7 and 11 of gestation. The result is shown in Table II and can be interpreted in the following ways: (i) the differences in the distribution pattern between 9 a.m. at the days 7 and 11 clearly indicates that the cycle length is not precisely 24 hours. (ii) Obviously there is a change in chromatin organization within one day of gestation and therefore the classification of homogenous, intermediate and reticulate nuclei has to be extended with regard to chromatin morphologies which are likely to interconnect these major patterns.

In order to extend our knowledge on structural as well as functional aspects of the different chromatin organizations, CLSM was used for a precise 3-D analysis of the DNA-distribution within intact nuclei. In Figure 2, the CLSM analysis of a nucleus with a transitional chromatin arrangement between the 'intermediate' and the 'reticulate' nuclear morphology type is shown. The sample shown represents a nucleus located within a tissue whole mount of a day 10 layer of trophoblast giant cells which had been stained by the Feulgen-technique for DNA-fluorescence (for detailed methodological description see Montag et al., 1989a). By CLSM, 40 consecutive optical sections were recorded with a stepsize of 500nm in z-axis; only 6 characteristic optical sections of this image stack are shown. In the sequence of images shown, the paths of individual chromatin fibers which span through the whole nucleus can be clearly seen. At present we are combining this type of image analysis with *in situ* hybridization. This will finally enable us to detect the precise sequence of chromatin configuration changes in defined areas of the giant cell nucleus.

SUMMARY AND CONCLUSIONS

Using the high resolution videomicroscopy and laser scanning microscopy we have studied the chromatin structure of isolated nuclei of the polyploid mouse trophoblast giant cell layer during the early course of gestation. We found a correlation between cell-cycle dependent changes of the chromatin organization and the increase in nuclear volume and DNA-content. Furthermore, the novel LM techniques we applied, are particularly suited to investigate aspects of the spatial intranuclear chromatin distribution. Therefore, we can now study in detail the changes in the nuclear chromatin morphology which are thought to be mainly due to different levels of lateral association of intranuclear chromatin fibers.

Further studies are in progress to investigate the intranuclear distribution of chromosomal marker genes by *in situ* hybridization. This will help us to clarify the sequential changes of chromatin configuration during the endomitotic DNA replication cycle.

ACKNOWLEDGEMENTS

We thank Dr. Günter Kiefer (Institute of Pathology, University of Freiburg) for help in DNA measurements and our colleagues Ansgar Hofmann, Bärbel Meissner, Herbert Steinbeißer and Helmut Tröster for helpful

discussion as well as Roger Fischer, Daniela Frey and Angelika Frenznick
for expert technical assistance. We are also indebted to the Division of
Applied Microscopy, Carl Zeiss, Oberkochen, Fed. Rep. Germany. This study
was supported by the DFG, Bonn-Bad Godesberg (grant to M.F.T., Tr 147/6-3).

REFERENCES

Agard D., and Sedat J. 1983, Three-dimensional architecture of a polytene
  nucleus, Nature, 302:676.
Allen R. D. 1985, New observations on cell architecture and dynamics by
  video-enhanced contrast optical microscopy, Annu. Rev. Biophys. Chem.,
  14:265.
Arndt-Jovin D. J., Robert-Nicoud M., Baurschmidt P., and Jovin T. M. 1985,
  Immunofluorescence localization of Z-DNA in chromosomes: Quantitation
  by scanning microphotometry and computer-assisted image analysis, J.
  Cell Biol., 101:1422.
Barlow P. W., and Sherman M. I. 1974, Cytological studies on the organiz-
  ation of DNA in giant trophoblast nuclei of the mouse and rat,
  Chromosoma, 47:119.
Bohrmann J., Kiefer G., and Sander K. 1986, Inverse correlation between DNA
  content and cell number in nurse cell clusters of Drosophila,
  Chromosoma, 94:36.
Brakenhoff G. J., van der Voort H. T. M., van Spronsen E. A., Linnemans W.
  A. M., and Nanninga N. 1985, Three-dimensional chromatin distribution
  in neuroblastoma nuclei shown by confocal scanning laser microscopy,
  Nature, 317:748.
Montag M., Spring H., Trendelenburg M. F., and Kriete A. 1989a, Methodical
  aspects of 3-D reconstruction of chromatin architecture in mouse giant
  trophoblast nuclei, J. Microsc., in press.
Montag M., Spring H., and Trendelenburg M. F. 1989b, Light microscopic
  optical sectionning using fluorescence videomicroscopy, Eur. J. Cell
  Biol., Suppl. 25:47.
Ploem J. S. 1987, Laser scanning microscopy, Appl. Optics, 26:3-26.
Spring H., Trendelenburg M. F., and Montag M. 1988, DNA-fluorescence of
  mammalian intra-nucleolar chromatin detected by confocal laser
  scanning microscopy (CLSM), Biol. Cell, 64:371.
Trendelenburg M. F., Allen R. D., Gundlach H., Meissner B., Tröster H.,
  and Spring H. 1986, Recent improvements in microscopy towards analysis
  of transcriptionally active genes and translocation of RNP-complexes,
  in: 'Nucleocytoplasmic Transport', R. Peters, M. F. Trendelenburg, eds.
  Springer Verlag Berlin Heidelberg.
White J. G., Amos W. B., and Fordham M. 1987, An evaluation of confocal
  versus conventional imaging of biological structures by fluorescence
  light microscopy, J. Cell Biol., 105:41.
Zybina E. V., Kudryatseva M. V., and Kudryatseva B. N. 1975, Polyploidization and endomitosis in giant cells of rabbit trophoblast, Cell Tiss.
  Res., 160:525.

EFFECTS OF A DNA INTERCALATOR, ETHIDIUM BROMIDE ON CHROMATIN STRUCTURE OF

CHICKEN ERYTHROCYTE

Zhu Jing-de, Sun Xiao-ping, and Wang Fan

Shanghai Institute of Cell Biology
Shanghai Open Laboratory in Life Sciences
Academia Sinica
320 Yue-yang Road
Shanghai 200031
PR China

INTRODUCTION

DNase I hypersensitive sites (DHS) are the most loosely packed regions in chromatin, which are ultrasensitive, typically 100 times more than bulk chromatin, to DNase I as well as other DNA cleaving agents including methidiumpropy EDTA, a DNA intercalator (Elgin, 1988; Gruss and Garrard, 1988). DHS have been mapped to many functionally crucial regions, including promoter, enhancer, silencer and terminator of transcription, DNA replication origins, recombination elements as well as the centromere and telomere. Relevant to the transcription of genes, the occurrence of DHS can be constitutive or inducible, and developmentally or tissue-specifically regulated. DHS have been correllated with the chromatin region of nucleosome-free, or of atypical DNA structure, or associated with trans-acting protein or proteins, or both. Conformational polymorphism of the DNA component of genome, notably bending, unpairing, cruciform and Z-form, is both sequence and environment-dependant (Rich et al., 1984; McLean and Wells, 1988; Jaworski et al., 1987). DNA supercoiling (Tsao et al., 1989) and DNA binding proteins are two important environmental factors. The former may also play a role in formation and maintenance of DHS (Villeponteau, et al., 1984). DNA-protein interaction, especially the sequence-specific interaction (Pabo and Sauer, 1986; Dynan and Tjian, 1985), may also participate in formation of DHS *in vivo* (Emerson and Felsenfeld, 1984). DNA intercalators alter DNA conformation by inserting their planer aromatic rings between the base pairs of the neighbouring nucleotides, without impairing the structural integrity of DNA (Neidle and Abraham, 1984). Treatment of eukaryotic chromatin with DNA intercalators causes a selective release and displacement of nuclear protein along chromatin (Schoter et al., 1985; Wang and Zhu, 1989; McMurry and Van Hodle, 1986). The most accessible regions in chromatin for DNA-intercalators are probably DHS, in the sense of both DNA conformation and interaction of DNA-protein.

The chicken β^A-globin gene in the erythrocyte is a well characterized model system for the study of chromatin-structural involvement in cellular differentiation during development (Felsenfeld et al., 1986). In this report, we used the chicken β^A-globin gene in both mature chicken erythrocytes (RBC) and reticulocyte (Ret) as model systems to study the effects of

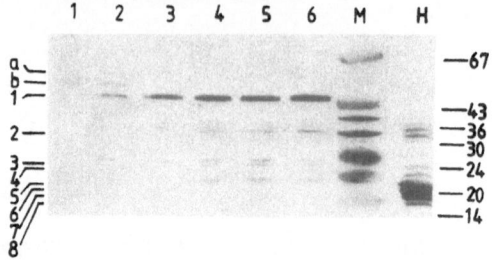

Figure 1.   Protein profile of the extracts from RBC nuclei with or
             without EB.   15mg/ml RBC nuclei were incubated in the nuclei
             preparation buffer consisting of 10mM Tris. HCl, pH 7.4, at
             $4^{\circ}C$, 10mM NaCl, 3mM $MgCl_2$, 20% glycerol, 0.1mM PMSF with
             various amounts of EB.   No-EB control (lane 1) was set in
             parallel.   20µl of supernatants from each extracts were
             analyzed in a 5-20% SDS-PAGE gel system and silver-stained.
             Lanes 2-6, extracts by EB at concentrations of 5, 10, 15, 20
             and 25mM respectively.   Calf thymus histones (Sigma, lane H)
             and protein size markers (Sigma, lane M) were used.   Their kD
             value are indicated on right side of the photograph.

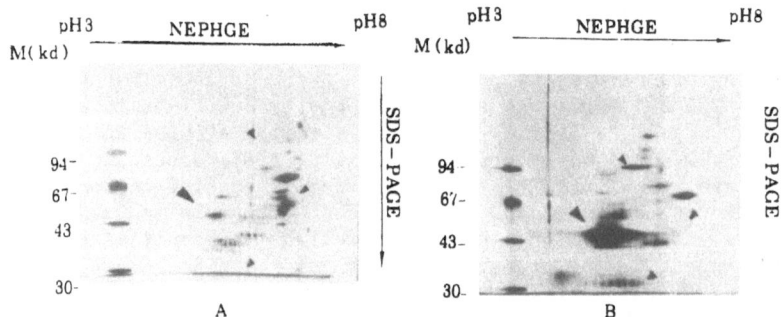

Figure 2.   2D-electrophoretic analysis of 2% TCA precipitates of 25mM EB
             extract and no-EB control of RBC nuclei.
             2% TCA precipitate of 50ul (1µg) of control (panel A) and 25mM
             EB (panel B) extracts of RBC nuclei were compared in a micro-
             2D-electrophoretic system.   The first dimension is a pH 3-8
             nonequilibrium polyacrylamide gel electrofocusing, and the
             second dimension is a 7% SDS-PAGE electrophoresis.   The
             proteins were revealed by silver-staining.   The filled
             triangles are used to pinpoint the groups of protein-dots
             which differ in density between two panels.

a DNA-intercalator, ethidium bromide (EB) on chromatin structure, by
assaying both the protein profile extracted by EB and effects on the DHS
pattern of β^A-globin gene from both nuclei.

RESULTS AND DISCUSSION

     Without EB, only a few proteins were extracted at low quantity (0.1mg/
ml) from RBC nuclei (15mg/ml).   EB (at concentrations of 5-20mM) altered
the pattern of released nuclear proteins from RBC nuclei (Figure 1).   Some
proteins became less extractable in the presence of EB (a and b) and
disappeared from the extract when EB was 10mM (lane 3).   On the other hand,

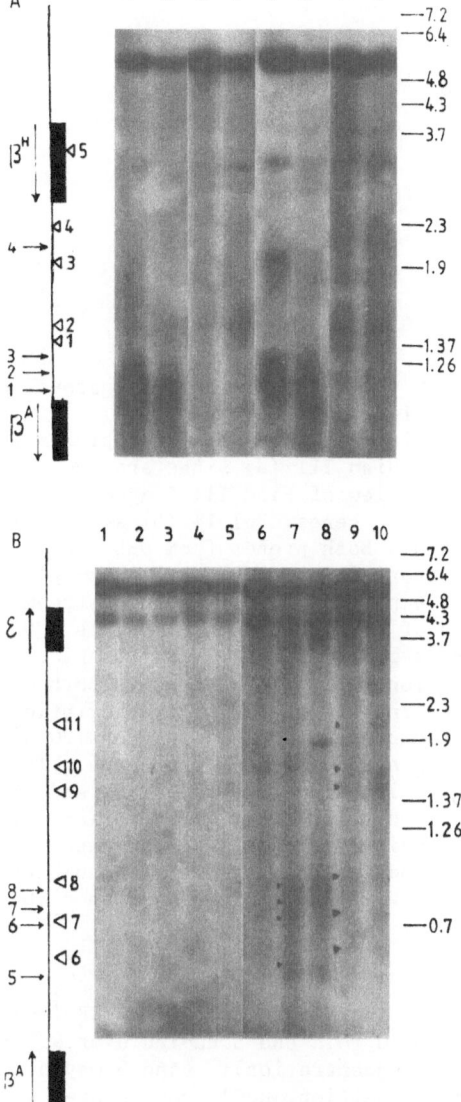

Figure 3.   DHS and DHS^{EB} at both upstream and downstream of β^A-globin
             gene in both RBC and Ret. nuclei.
             DNA purified from DNase I digested nuclei with EB and no-EB
             was restricted with Hind III and Southern-blotted and
             hybridized with either Bgl II-Hing III fragment (panel A) or
             Hind III-Sac I fragment (panel B). DNA of lanes 1 and 6 in
             panel B, were no-DNase I control for RBC and Ret nuclei,
             respectively.

Panel A, lanes   Panel B, lanes
RBC   1, 2,   2, 3,
RBC (EB)   3, 4,   4, 5,
Ret   5, 6,   7, 8,
Ret (EB)   7, 8,   9, 10.

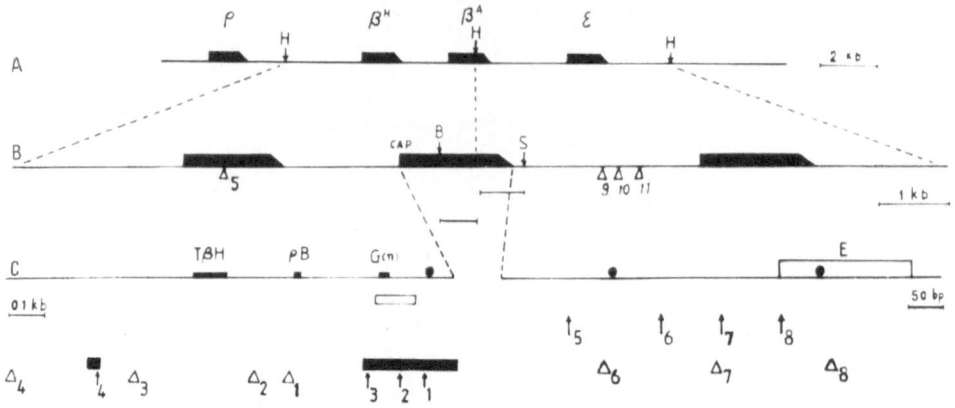

Figure 4.   A summary for both DHS and DHS   around of β -globin gene in
Ret nuclei.
Line drawings: A, presents the chicken β-like globin gene
cluster. Hind III (H) sites are shown with arrows. B, ia an
amplified view of Hind III fragments which contain β^H-, β^A-,
and ε-globin genes, Bgl II (B) and Sac I (S) sites are
indicated. Both probes from DHS mapping experiments are shown
by two short lines underneath of β^A-globin gene. C, is a
further amplified form of both upstream and downstream
regions of β^A-globin gene. TβH, is a 90bp inverted repeat,
and ρB is promoter for a short transcript in an opposite
orientation to β^H-globin genes. Both were suggested being
involved in termination of transcription of β^H-globin gene.
G(n) is a consecutive sequence of 18G at -175 to -193 on the
noncoding strand of β^A-blobin gene. The open bar is referred
to the nucleosome-free region. E and the open bar underneath
is referred to enhancer region. The frilled circles are the
cleavage-sites for DNA topoisomerase II cleavage *in vivo*.
DHS are indicated by arrows with numbers, and DHS^{EB} are
indicated with open triangles with numbers.

some proteins were preferentially extracted by EB (1-8). Protein 1 (Mr=45kD)
increased as much as 20-70 fold and occupied over 90% in quantity in EB
extracts (over 10mM EB concentration). Band 3 may be HMG14, and band 5 may
be HMG17 (N.B. in this condition, HMG17 co-migrated with H3). Bands 5-8
were core histones H3, H2B, H2A and H4, respectively. Core histones were
released by 15mM EB, but were not released with the further increase of EB
concentration. Detailed comparison of 2% trichloroacetic acid precipitates
of EB extracts with non-EB control (Figure 2), by a two dimensional PAGE
electrophoretic method, we observed that in addition to 45kD and 32kD
proteins, a 94kD protein was also EB extractable. They were all acidic
proteins with pI, 4-5 for the 45kD, 4-5.5 for the 32kD, and 5-6 for the
94kD protein. A few protein dots, Mr *ca* 45kD to 67kD with pI 7.0 in
panel A were absent in panel B, indicating that EB made them less extractable. In addition, there were some proteins with no response to EB
treatment, as indicated by their approximately equal density between panels
A and B, in Figure 2. In this aspect, there was no significant difference
between RBC and Ret (not shown).

The nuclei of both RBC and Ret after 10mM extraction were used to study
the effect of EB on the DHS pattern of the β^A-globin gene, in view of a

Table I. Occurrence of DHS and DHS Around β^A-Globin Gene in RBC and Reticulocyte Nuclei.

DHS[-bp]*	RBC No**	RBC EB	Ret. No	Ret. EB	DHS$_{EB}$[-bp]	RBC No	RBC EB	Ret. No	Ret. EB
1[-50]	+	-	+	-	1[-450]	-	+	-	+
2[-150]	+	-	+	-	2[-550]	-	+	-	+
3[-250]	+	-	+	-	3[-900]	-	+	-	+
4[-1000]	-	-	+	-	4[-1250]	-	+	-	+
5[+1600]	-	-	+	-	5[-2500]	-	+	-	+
6[+1720]	-	-	+	-	6[+1640]	-	-	-	+
7[+1800]	-	-	+	-	7[+1750]	-	-	-	+
8[+1890]	-	-	+	-	8[+1950]	-	-	-	+
					9[+2850]	-	-	-	+
					10[+3150]	-	-	-	+
					11[+3450]	-	-	-	+

* The numbers in [ ] are referred to the distance of each DHS and DHS$_{EB}$ from , either upstream[-] or downstream[+] of the cap site of β^A-globin gene.
** no-EB control.

sharp increase in the amount of protein extracted by EB at this concentration. Three effects of EB on chromatin structure of RBC and Ret nuclei were observed: 1. DNA in EB treated nuclei was 6-8 folds more resistant to DNase I digestion than DNA in no-EB control. 2. The electrophoretic pattern of DNA derived from DNase I digested non-EB treated RBC nuclei was a faint ladder with two-nucleosomal repeat over a smearing background, whereas, that for EB-treated RBC nuclei shows a ladder with the mononucleosome as its repetitive unit (not shown). 3. The DHS pattern, upstream and downstream, of the β^A-globin gene in both nuclei were altered by pretreatment of nuclei with 10mM EB (Figures 3 and 4, Table I), i.e., the original DHS disappeared and a new set (DHS^{EB}) occurred.

Our results suggested that: 1. EB could preferentially extract a set of nuclear proteins from nuclei, probably by its effects on the secondary structure of DNA, and these proteins were likely to be enriched at DHS. 2. EB could alter the DHS pattern of active genes in nuclei by the same mechanisms. Accompanying the disappearance of DHS, DHS^{EB} appearance at new sites could be a case similar to the hierarchial alteration of S1 sensitive sites in the promoter region of chicken β^A-globin. Namely, the deletion of the S1 site at the higher hierarchy induced a new S1 site which was at the lower hierarchy in negatively supercoiled molecules (Schon et al., 1983), or associated with displacement of the proteins responsible for formation of DHS from the original sites to the new sites. To further verify the above notion, it is desirable to characterize the binding affinity of EB released proteins to the regulatory regions of the β^A-globin gene, at both primary and secondary sequence levels, and evaluate their role in the formation of DHS *in vitro* with the nucleosome reconstitution assay (Emerson and Felsenfeld, 1984).

REFERENCES

Dynan W. S., and Tjian R., 1985, Control of eukaryotic messenger RNA synthesis by sequence-specific DNA binding proteins, Nature, 316:774.
Elgin S. C. R., 1988, The formation and function of DNase I hypersensitive sites in the process of gene activation, J. Biol. Chem., 263(36):19259.
Felsenfeld G., Emerson B. M., Jackson P. D., Hesse J. E., Liever M. R., and Nickol J. M., 1986, Chromatin structure near an active gene, In 'New

frontier in the study of gene function', Poste G., and Cook S. T., eds, Plenum, New York.

Gruss D. S., and Garrard W. T., 1988, Nuclease hypersensitive sites in chromatin, Ann. Rev. Biochem., 57:159.

Jaworski A., Hsieh W-T., Balho J. A., Larson J. E., and Wells R. D., 1987, Left-handed DNA in vivo, Science, 238:773.

McLean M., and Wells R. D., 1988, The role of sequences in the stabilization of left-handed DNA helices in vitro, BBA, 950:243.

McMurry C. T., and Van Holde K. E., 1986, Binding of ethidium bromide causes dissociation of nucleosome core particle, Proc. Natl. Acad. Sci. USA, 83:8472.

Neidle S., and Abraham Z., 1986, Structural and sequence-dependent aspects of drug interaction into nucleic acids, CRC Critical Reviews in Biochemistry, 17(1):73.

Pabo C., and Sauer R. T., 1986, Protein-DNA recognition, Ann. Rev. Biochem., 55:292.

Rich A., Nordeheim A., and Wang A. H. J., 1984, The chemistry and biology of left-handed Z-DNA, Ann. Rev. Biochem., 53:791.

Schon E., Evans T., Welsh J., and Efstratiadis A., 1983, Conformation of promoter DNA: fine mapping of S 1-hypersensitive sites, Cell, 35:837.

Schoter H., Maier G., Ponsting H., and Nordheim A., 1985, DNA intercalator induce specific release of HMG14, HMG17 and other DNA-binding proteins from chicken erythrocyte chromatin, The EMBO J., 4(13B):3867.

Tsao Y-P., Wu H-Y, Liu L. F., 1989, Transcription-driven supercoiling of DNA: direct biochemical evidence form in vitro studies, Cell, 56:111.

Villeponteau B., Lundell M., Martison H., 1984, Torsional stress promotes the DNase I sensitivity of active genes, Cell, 39:469.

Wang F., and Zhu J. D., 1989, The effects of DNA intercalators on chromatin of the chicken red blood cells———— Differential extraction of nonhistone proteins, Cell Research, 1:105.

THE FINE STRUCTURE OF NATIVE DNA STUDIED WITH THE SCANNING TUNNELING

MICROSCOPE

Li Min-qian, Zhu Jing-de*, Zhu Jie-qin, Hu Jun, Gu Ming-min,
Xu Yao-liang, Zhang Lan-ping and Huang Ze-qin

Shanghai Institute of Nuclear Research
Academia Sinica
P.O. Box 8204
Shanghai 201800
P. R. China

* Shanghai Institute of Cell Biology
Shanghai Open Laboratory of Life Sciences
Academia Sinica
320, Yue-yang Road
Shanghai 200031
P. R. China

The advantages of scanning tunneling microscopy (STM) for the study of
surface structure down to atomic level, such as simplicity of operation,
Ångstrom resolution, and imaging in a variety of environments, including air
or liquids (Binnig and Rohrer, 1986; Hansma and Tersoff, 1987), has greatly
encouraged scientists to analyze the structural details of biological
materials using STM. Objects of earlier researches in life sciences
included bacteriophage virus particles (Baro et al., 1987), metal-shadowed
recA-DNA complexes (Amerein et al., 1988) and native closed circular DNA.
The latest development in this field is the direct observation of native
DNA structure without metal shadowing (Beebe et al., 1988). In this work,
although the major groove and minor groove of DNA, as well as the pitch of
helix turn could be resolved, the pitch of the DNA helix turn varies greatly
and measurement of the width of grooves was not possible. We have tried to
modify our experimental conditions such as the size and concentration of DNA
as well as ionic strength of solvent with success.

A high resolution STM was built, based on a single piezotube three-
dimensional scanner (Li et al., 1989). Coarse positioning of the sample to
the tip is obtained by a fine-pitched head screw and a spring-lever system.
One of the most striking modifications is that the coarse positioning
system and sample-tip set up are entirely separated during tunneling.
Etched tips of tungsten wires were chosen and the image with high resolution
was collected with an elapsed interval after the tip scanning over the
sample surface. With this instrument, the atoms on the surface of highly
oriented pyrolytic graphite (HOPG) can be resolved at a resolution of less
than 1Å.

To achieve better results, we have reduced the size of DNA by an
intensive sonication down to an average size of 1kb, and reduced KCl
concentration from 10mM down to 0.05mM, as well as reduced DNA concentration

Figure 1.   STM image of native DNA in a top-view-three-dimensional format
             with gray scale. Original STM image was directly collected
             without any postacquisition-data-processing. Field size,
             203Å by 203Å. Height of vertical axis, 48Å. Sample bias,
             150mV. Current setpoint, 1.0nA. The tipscanning velocity,
             2000Å/Sec. This high resolution image shows that a DNA
             segment with two and half DNA double helix turns. The major
             and minor grooves are indicated by the thick and thin arrows
             respectively.

with triple-distilled water from 1mg/ml down to 5μg/ml (about 5 fold the
concentration of DNA required to form a unilayer on a surface after
evaporation[1]). The DNA left occupied an area of diameter 3-4mm, with
several concentric rings due to evaporation. In these experimental conditions, neither aggregates of DNA nor salt crystals could be detected.
Even in the center of the DNA area, the tunneling current was still stable.
All processes in this experiment were carried out at room temperature and
in air, the image was obtained in the conventional constant-current mode.
An image in a three-dimensional format in a color scale was directly
collected by scanning the DNA deposited on the graphite surface. The field
size was 203Å x 203Å (Figure 1). This image is representative of all the
images (from a total of 7 fields) obtained. In this STM image, the DNA
molecules with two and an half double helix turns appear in red and pink on
a pale blue background, and both major and minor grooves of DNA helix turn
can also be resolved (Figure 1). Figure 2A is a three dimensional contour
image from the raw image in Figure 1. Comparing the simplified model of a
Watson-Crick double helix of DNA, i.e. a right-handed B form DNA, in Figure
2B, we demonstrate: 1, the right-handiness of the DNA double helix, which
was schematic presented in Figure 2A; 2, the pitch of the helix turn,

---

[1] The average molecular weight of one base pair of nucleotides is 660
   Daltons, 1mg of DNA consists of 10^{18} base pairs, or 10^{17} helix turn of
   B-form DNA. If the occupied area of one turn of B-form DNA is approximately 680Å^2 (i.e. 34Å in length, times 20Å at diameter), 20μl of DNA is
   spotted and evaporated, the area is around 4mm in diameter. The amount
   of DNA required to form a unilayer for such an area should be 20ng/20μl.
   Hence, the concentration for 5 fold more concentrated DNA is 100ng/20μl.

Table I. Comparison of the Structural Features of DNA Double Helix (B-form) From STM Studies and the X-Ray Crystallographic Analysis.

```
==================================================================
Features      X-ray analysis*  Beebe`s study** This work
------------------------------------------------------------------
Handiness     Right-handed      Right-handed    Right-handed
of helix
------------------------------------------------------------------
Pitch per     33.8 Å           36 A[27-63 A]   34.4 ± 2.0 A***
helix turn                                      [32-36 Å]
------------------------------------------------------------------
Major groove                                    spacing:
              width:11.7 Å      not measured    22.6 Å
              ----------------------------------------------------
              depth:8.5 Å       not measured    7 A
------------------------------------------------------------------
Minor groove                                    spacing:
              width:5.7 Å       not measured    12 Å
              ----------------------------------------------------
              depth: 7.5 Å      not measured    2 Å
------------------------------------------------------------------
Ratio of major
/minor groove width:2.2                         spacing:1.88
==================================================================
```

* Saenger, 1984,
** Beebe, et al, 1989,
*** Totally 7 helical turns were measured and the figure given in table 1, is the mean ± 2 SD.

approximately 34.6Å, and 3, the spacing of the major and minor grooves of each turn, approximately 22.6Å and 12Å respectively. The depth of the major groove is approximately 7Å, whereas the depth of the monor groove is about 2Å. In view of the constancy of the above figures among all the imaging fields, the comparison of our data would be meaningful with the 'orthodox' features of DNA analyzed with other approaches, such as X-ray crystallography. Table I presents a comparison of our data with that of Beebe et al. (1989) and that obtained from X-ray crystallographic analysis. It is clear that our results are very close to the data from X-ray crystallography, although the latter was derived from analysis with crystallized DNA oligonucleotides, while ours was from the air-dried DNA molecules. Two major differences between our data and that from the X-ray crystallographic study were observed. The first is related to the width of the major and monor grooves. Since the STM tip scans over the surface of DNA molecules, the resolution of the surface image would be limited by the contouring space of the object in the sense of both spatial hindrance and local density of electronic state to allow the STM tip to scan. The image obtained (Figures 1 and 2) did not show the surface features in sufficient detail for us to demark the width of the major groove and minor groove, as was possible in the X-ray crystallographic studies (Saenger, 1984); rather we measured the spaces between the neighboring ridges, which are likely corresponding to the 'backbones' of DNA strands. To avoid the possible confusion, the term 'spacing' was used to define the distance between two neighboring ridges. It is apparent that the spacing ratio of the major groove to the minor groove is not very different from the width ratio. The former is approximately 1.88, and the latter is 2.2. The second disagreement is related to the figure of the depth of the minor groove. Studied by X-ray crystallography, the minor groove is almost as deep as the major groove, i.e. 7.5Å. But, in our analysis, the depth of the minor groove is not bigger than 2Å. The reasonable explanation for this is probably the spatial hindrance, namely, the width of the minor groove is too narrow to allow the STM tip to scan the side surfaces of the valley of the minor

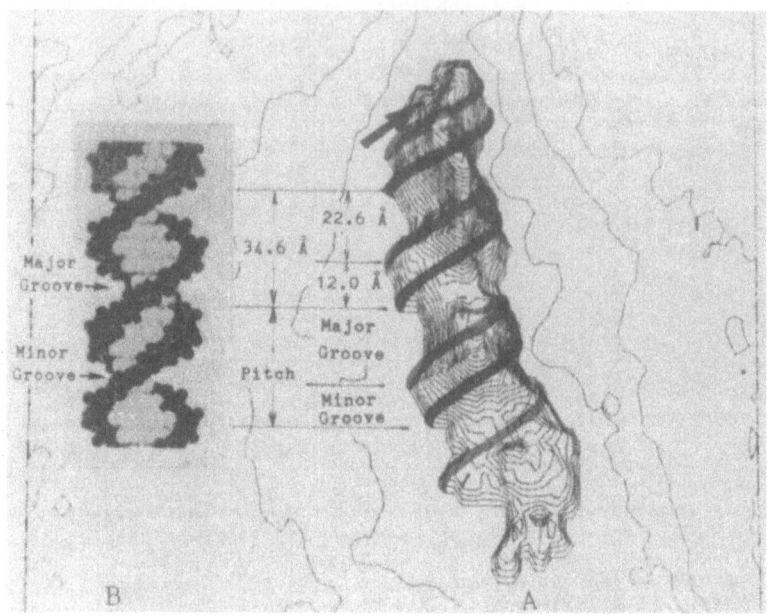

Figure 2.   Three dimensional contour image of the DNA segment in Figure
1, in comparison with the typical right-handed B-form DNA.
A. Three dimensional contour image was constructed from
Figure 1. Field size, 203Å by 203Å. Postacquisition image
processing consisted of assigning each scanning point a
height which is proportional to its corresponding contour
number and a 3 by 3 two-dimensional smoothing treatment and
the background corrugation suppressed, as well as viewing from
a perspective 85 degree above the plane. Two pitches of
double helical turns as well as the major and minor grooves
with their spacing are indicated. Two sugar-phosphate backbones are schematically presented with two lines and the
right-handness of DNA double helix is indicated by an arrow.
B. A spack-filling model of the DNA double helix (B-form)
with one and half turns.

groove. Alternatively, the local density of electronic state is not ideal
for the tunneling between the surface and the STM tip.

The conformational polymorphism of DNA has been known for decades and
created great enthusiasm in scientists searching for their functional
implication in biological processes of genome (Saenger, 1974; Rich et al.,
1984; Wells and Harvey S. C., 1988). With STM, it is possible to compare
the structural details of the conformational variants of DNA, notably, A,
C, D, and Z-form DNA, triple helix as well as cruciform structure, DNA
bending and kinking. The other potentially fruitful field is probably the
study of the effects of DNA reacting reagents, such as intercalating
chemicals, on DNA structure (Niedle and Abraham, 1984).

After we finished this work, Lee et al. (1989) presented their result
on the B-form DNA structure with STM. In comparison to their data, we have
added two pieces of information to our understanding of the structural
features of DNA by STM. First, we have given a quantitative measurement on
both 'spacing' and depth of both major and minor grooves of DNA (which
maintains the expected ratio between major and minor grooves, namely,

approximately 2). Although Lee et al. (1989) also measured the spacing between major and minor grooves, their figure is the same for both major and minor grooves. Furthermore, they did not touch upon the matter of the depth of grooves. Second, their assessment of the pitch of turn was much more constant than that of Beebe et al. (1989), with which we made a direct comparison earlier in this report. But it is much shorter than the 'orthodox' figure, even when they included both spermidine and hexaammine cobalt (III) in DNA solution, this fugure is still 7% shorter. In contrast, our measurement on the pitch of turn of DNA without any 'stabilizing factors' was very similar to the 'orthodox' figure derived from X-ray analysis.

ACKNOWLEDGEMENT

This work was supported by SINR Institutional Research Grant and Shanghai Branch of Academia Sinica. Thanks due to Wang Ya-hui and Yan Fu-jia for encouragement and support and K. Besocke for providing HOPG.

REFERENCES

Amerein M., Stasiak A., Gross H., Stoll E., and Travaglinin G., 1988, Scanning tunneling microscopy of rec-A-DNA complexes coated with a conducting film, Science, 240:514.

Baro A. M., Garcia N., Binnig G., Rohrer H., Gerber C. H., and Carrascosq J. L., 1985, Determination of surface of topography of biological specimens at high resolution by STM, Nature, 315:253.

Beebe T. P., Tray J. R., Wilson E., Ogletree D. F., Kats J. E., Balhorm R., Salmeron M. B., and Siekhous W. J., 1989, Direct observation of native DNA structure with scanning tunneling microscope, Science, 243:370.

Binning G., and Rohrer H., 1986, Scanning tunneling microscopy, IBM J. Res. Dev. 30:355.

Hansama P. K., and Tersoff J., 1987, Scanning tunneling microscope, J. Appl. Phys., 61:R1.

Lee G., Arscolt P. G. and Bloomfield V. A., 1989, Scanning Tunneling Microscopy of nucleic acids. Science 244:475.

Li M. Q., Zhu J. Q., Hu J., Gu M. M., Xu Y. L., Zhang L. P., and Huang Z. Q. 1989, A scanning tunneling microscope with Å level resolution, Nature (China) in press.

Rich A., Nordheim A., and Wang A. H. J., 1984, The chemistry and biology of left-handed Z-DNA, Ann. Rev. Biochem., 53:791.

Saenger W., 1984, Principles of nucleic acid structure, Chapter 9-14, 220-330, Springer-Verlag, New York.

Travaglinin G., Rohrer H., Amerein M., and Gross H., 1987, Scanning tunneling microscopy on biological matter, Surf. Sci., 181:380.

Neidle S., and Abraha Z., 1984, Structural and sequence-dependant aspects of drug intercaltion into nucleic acids, CRC Critical Rev. in Biochem., 17(1):73.

Wells R. D., and Harvey S. C., 1989, Unusual DNA structures, Springer-Verlag, New York.

CYTOCHEMISTRY AND IMMUNOLOGY OF SOME NUCLEAR PROTEINS IN A PRIMATIVE

DINOFLAGELLATE

L. Salamin Michel*, M.-O. Soyer-Gobillard**, M.-L. Géraud**
and A. Gautier*

*Center of Electron Microscopy
University of Lausanne
CH-1005 Lausanne
Switzerland

**Arago Laboratory
University of Paris VI
F-66500 Banyuls-sur Mer
France

INTRODUCTION

Dinoflagellates represent an outstanding experimental model for the study of nuclear structure-function relationships, due to the permanently condensed DNA in 'arch-shaped' chromosomes and the constant presence of nucleoli and nuclear membrane throughout the whole cell cycle. The fact that in the nucleoli the chromosomes remain distinct inside the Fibrillar Centers (FC), makes it easier to distinguish the proteins bound to DNA from those which are not. This allows a better interpretation of *in situ* labelling studies of nuclear components.

With the help of the Ag-staining reaction and immunolabelling, we have studied the ultrastructural distribution of argyrophilic proteins in the nucleus of the primitive dinoflagellate *Prorocentrum micans E*. We show that these proteins are distributed in every chromosome and in the nucleolar Dense Fibrillar Component (DFC). In the FC we did not find argyrophilic proteins, apart those located on the chromosomes included in this region.

The comparative study of immunoreactivity of nuclear proteins between dinoflagellates and higher eukaryotes shows that these proteins are closely related, in spite of the considerable phylogenic distance between these phyla.

MATERIALS AND METHODS

Biological materials. *Prorocentrum micans E.*, a marine, free-living dinoflagellate was chosen for this study. Cultivated mammalian cells (mastocytoma P815) were used as controls.

Ag-Reaction Procedure

Aldehyde fixation followed by various embedding procedures (Epon,

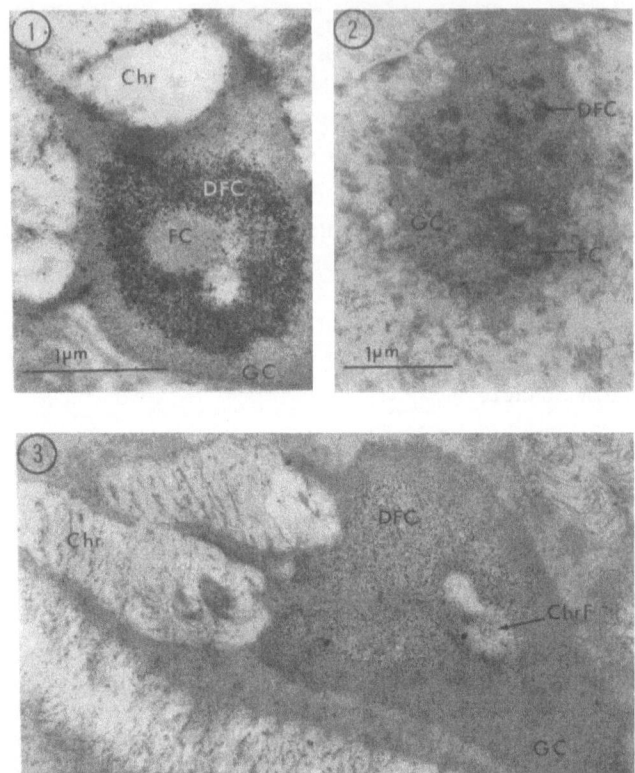

Figures 1   Silver-staining. Figure 1: Cryosection of *P. micans*/AgM,
to 3.   20°C. Figure 2: controls, Lowicryl section of P815/AgM, 20°C.
               Figure 3: Lowicryl section of *P. micans*/AgM, 31°C: the
               chromosomal labelling is reduced.

Lowicryl K₄M) were used (Gautier et al., 1987). Thin sections were stained
using the Ag-staining procedure of Ploton et al. (1982) as modified by
Moreno et al. (1985). Reaction temperatures were tested from 18 to 50°C.
(AgM reaction).

   Pretreatments before Ag-reaction. Thin sections were floated on
various protease solutions before Ag-staining (Salamin Michel et al., in
preparation).

Immunolabelling Experiments

   Preparation procedures. *P. Micans* were prefixed in aldehyde 1hr at
20°C, then either PLT embedded in Lowicryl K4M (Carlemalm et al., 1982) or
embedded in 5% gelatin, infused with 2.1M sucrose, frozen and cryosectioned
(Tokuyasu, 1986).

   Polyclonal antibodies. Anti-B23: this chicken serum, raised against
the purified 38kD protein B23, was a kind gift from S. Kaufmann (Fields et
al., 1986). Anti-nucleolin: this rabbit serum raised against purified 100kD
nucleolin, was a kind gift from F. Amalric (Bugler et al., 1982).

   Labelling. On K4M thin sections: incubation with the primary antibody,

292

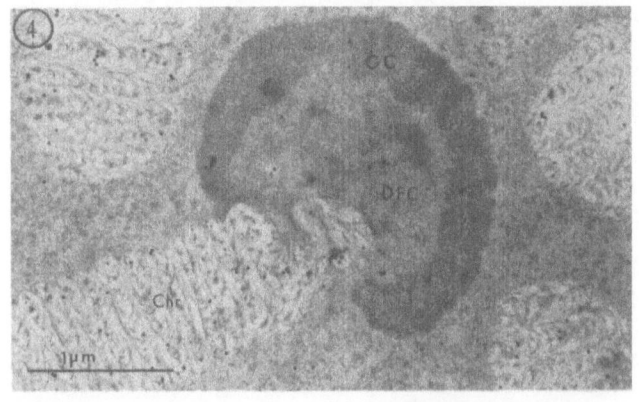

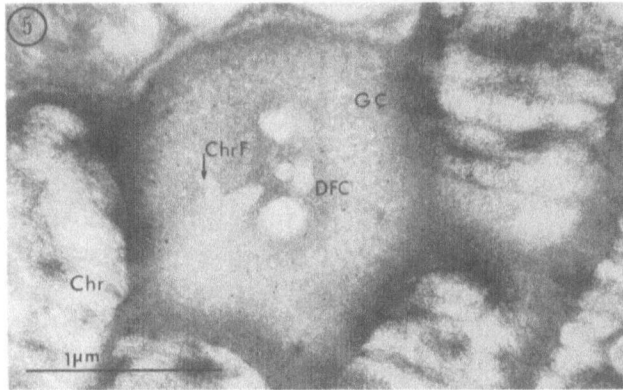

Figures 4 and 5.   Immunolabelling of *P. micans,* Figure 4: anti-B23 labelling on Lowicryl Sections. Figure 5: anti-nucleolin labelling on cryo-sections.

17hrs at 4°C; incubation with the secondary antibody (rabbit anti-chicken), 1hr at 20°C; incubation with protein A-gold complex, according to Roth (1982). On cryosections: incubation with the primary antibody, 1hr at 20°C; incubation with the secondary antibody 30 minutes at 20°C, if necessary; incubation with protein A-gold or with immunoglobulins (Janssen Life Sciences), coupled with 10nm gold probes.

RESULTS

On *P. micans* thin sections, numerous arch-shaped cylindrical chromosomes are observed. One or several chromosomes are observed included in the nucleoli with one of their extremities; this part of the chromosome is more or less decondensed (Figures 3 and 4). Depending on the orientation of the section, the penetrating chromosomes are frequently seen as fragments of chromosomes located inside the FC. We hereafter call them Intranucleolar Chromosomes Fragments (Figures 3, 5, 6 and 7: ChrF).

One or two nucleoli are generarly observed. They are composed of three main components, morphologically similar to those observed in interphasic cells of higher eukaryotes. The Fibrillar Centers (FC), which seem frequently to be totaly filled by the included chromosomes (Figures 3 and 4), are surrounded by the Dense Fibrillar Component (DFC), which itself is

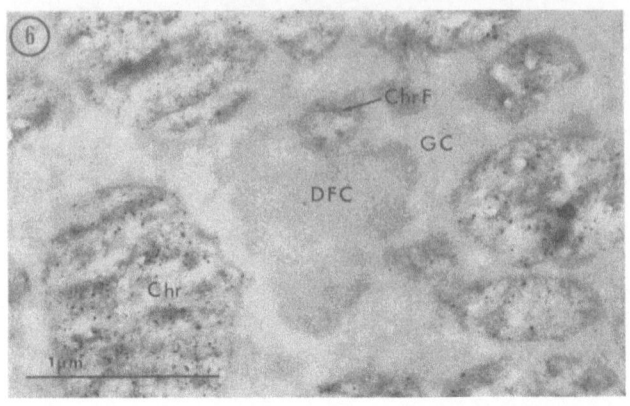

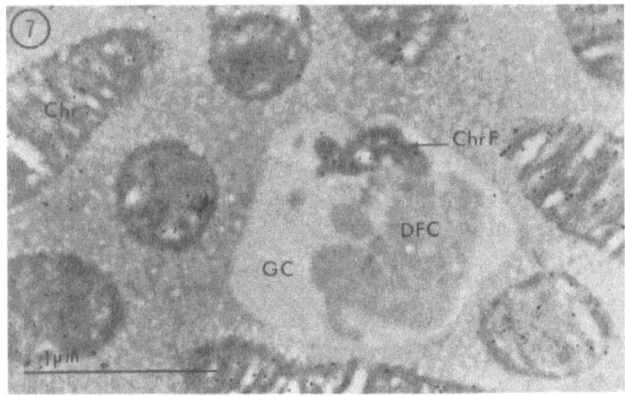

Figures 6   Protease pretreatments of Lowicryl sections of *P. micans*,
and 7.   Figure 6: AgM, 20°C, Figure 7: anti-B23. In both cases, the
              labelling disappears on nucleoli but remains on chromosomes.

surrounded by the Granular Component (GC) (Figure 1). It is to note that
the DFC in dinoflagellates appears usually less dense than the GC (Figure
4).

Ag-Reaction

     In the mammalian controls, the AgM-reaction only shows, as expected,
nucleolar labelling, dense in the DFC and lighter in the FC (Figure 2),
whereas in simultaneously treated *P. micans* thin sections of either
cryofixed or conventionaly embedded specimens, two reaction-sites were
always observed: a nucleolar and a chromosomal site.

     On the nucleolus, Ag-targets are localized on the DFC forming a large
ring around the FC (Figure 1) whereas the FC and the GC are devoid of
labelling.

     On the chromosomes, including the intranucleolar chromosomal fragments,
silver-grains are scattered, usually in close contact with chromosomal
fibrils. If the reaction temperature is increased, thereby accelerating,
the chromosomal labelling is decreased, but not the nucleolar one. (Figure
3).

     Proteolytic section pretreatments with pronase or proteinase K suppress any nucleolar silver-staining, but <u>not</u> the chromosomal staining (Figure 6).

Immunocytochemistry

The Ag-staining results were corroborated by preliminary experiments, using polyclonal antibodies against two mammalian argyrophilic nucleolar proteins, B23 and nucleolin (=C23), on Lowicryl K4M sections and/or on cryosections. With both antibodies, the labelling is found scattered on the DFC and the GC but not on the FC (Figures 4 and 5). This labelling is, however, less abundant in the DFC than after Ag-labelling.

Furthermore, with anti-B23, the labelling is observed not only on the nucleoli but also on the chromosomes and on the intranucleolar chromosomal fragments. This labelling seems to be as abundant as the silver-grains obtained on the chromosomes after AgM-reaction. Proteolytic section pretreatments eliminate the nucleolar labelling, but not the chromosomal one (Figure 7). With anti-nucleolin, no chromosomal labelling is observed (Figure 5).

DISCUSSION

In dinoflagellate nucleoli, some chromosomes are observed included with one of their extremities in the FC, so it is easy to observe that the argyrophilic proteins, either cytochemically or immunocytochemically detected in these FC, are bound to DNA and are not integral components of the nucleoli, contrary to the argyrophilic proteins detected in the DFC, which are genuine DFC components. The nucleolar and chromosomal silver targets present different cytochemical reactions with regard to proteolytic, acidic and alkaline pretreatments, and with regard to change in the reaction temperature, thus demonstrating that the corresponding argyrophilic proteins are not identical either in localization, or in reactivity.

Besides this, preliminary experiments with immunolabelling against two mammalian argyrophilic proteins, B23 and nucleolin, demonstrate that they are present in the DFC and in the GC of dinoflagellates. This is in agreement with the findings of Biggiogera et al. (1989) on mammalian cells, where anti-B23 and anti-C23 label only the DFC and the GC but not the FC. These observations suggest that B23 and nucleolin, or proteins with similar epitopes, are preserved through evolution from dinoflagellates to higher eukaryotes. This labelling is however much weaker and more scattered than the silver reaction products. We deduce from this observation that other argyrophilic proteins, besides B23 and nucleolin, are localized in the nucleolus. Furthermore, the labelling with anti-B23 serum is also specifically found on the chromosomes, where it seems to be in the same proportion and to react in the same manner to enzymatic pretreatments, as the Ag-labelling. We deduce therefore that the antigen recognized by this serum may correspond to the argyrophilic chromosomal target.

This work illustrates that dinoflagellates with their easily localized chromosomes are an useful model for the study of chromatin and its relationships with the other nuclear components in cells of higher eukaryotes.

REFERENCES

Biggiogera M., Fakan S., Kaufmann S. H., Black A., Shaper J. H. and Busch H., 1989, Simultaneous Immunoelectron Visualization of Protein B23 and C23 Distribution in HeLa Cell Nucleolus, <u>J. Histochem. Cytochem.</u>, 37:1371.

Bugler B., Caizergues-Ferrer M., Bouche G., Bourbon H., and Amalric F., 1982, Detection and Localization of a Class of Proteins Immunorelated to a 100kDa Nucleolar Protein, Eur. J. Biochem., 128:475.

Carlemalm E., Garavito R. M., and Billiger W., 1982, Resin Development for Electron Microscopy and Analysis of Embedding at Low Temperature, J. Microsc., 126:123.

Fields A. P., Kaufmann S. H., and Shaper J. H., 1986, Analysis of the internal Nuclear Matrix. Oligomers of a 38kD Nucleolar Polypeptide Stabilized by Disulfide Bonds, Exp. Cell Res., 164:139.

Gautier A., Michel-Salamin L., Tosi-Couture E., McDowall A. W., and Dubochet J., 1986, Electron Microscopy of the Chromosomes of Dinoflagellates *in situ*: Confirmation of Bouligand's Liquid Crystal Hypothesis, J. Ultrastruct. and Molec. Struct. Res., 97:10.

Moreno F. J., Hernandez-Verdun D., Masson C., and Bouteille M., 1982, Silver Staining of the Nucleolar Organizer Regions (NORs) on Lowicryl and Cryo-Ultrathin Sections, J. Histochem. Cytochem., 33:389.

Ploton D., Bobichon H., and Adnet J. J., 1982, Ultrastructural Localization of NOR in Nucleoli of Human Breast Cancer Tissues Using a One-Step Ag-NOR Staining Method, Biol. Cell., 43:229.

Roth J., 1982, The Protein A-Gold (pAg) Technique-A Qualitative and Quantitative Approach for Antigen Localization on Thin Sections, in: 'Techniques in Immunocytochemistry', Bullock, G. R., and Petrusz P., ed., Acad. Press, London, New York.

Tokuyasu K. T., 1986, Application of Cryoultramicrotomy to Immunocyto-chemistry, J. Microsc., 143:139.

HIGHER ORDER CHROMATIN STRUCTURE IN TRANSCRIPTIONALLY ACTIVE AND INACTIVE

CELLS

V. I. Vorob'ev, E. V. Karpova and T. N. Osipova*

Institute of Cytology of the Academy of Sciences of the USSR
Leningrad
194064
USSR

* Institute of Biological Research of the Leningrad State
  University
Leningrad
198904
USSR

INTRODUCTION

At present there is no general view on the organization of nucleosome chain packaging in the so called 30nm chromatin fiber. Comparative studies of chromatin differing in transcriptional activity and in linker DNA lengths represent a fruitful approach to the problem of higher order chromatin structure (Osipova et al., 1980; 1986; Zalenskaya et al., 1981; Thomas et al., 1986). This communication reports the analysis of the sedimentation data for chromatin from nuclei of pigeon brain cortical neurones, rat thymus and sea urchin sperm, characterized by linker DNA sizes of 20, 50, and 100 base pairs, respectively.

RESULTS AND DISCUSSION

The sedimentation behaviour of nucleosome oligomers in relation to the number of nucleosomes in the chain and the ionic strength in the range between 5mM to 85mM has been investigated. Figure 1 shows the dependency of the sedimentation coefficient ($s_{20,W}$) of oligonucleosomes on the size of the chain, for rat thymus and sea urchin sperm chromatin, at three different ionic strengths. Similar results were obtained also for chromatin from cortical neurones (Osipova et al., 1986). These data demonstrate that, in spite of considerable differences in linker DNA length in the investigated chromatin, their compaction processes are quite similar. It is seen from Figure 1 that a characteristic feature of the sedimentation behaviour of all the types of chromatin investigated is the sharp increase of $s_{20,W}$ for hexanucleosomes at ionic strengths higher than 25mM.

Earlier, we suggested the method for the analysis of the dependency of sedimentation coefficients on the number of nucleosomes in oligonucleosome chains. This approach, based on different hydrodynamic theories for the cylinder model, was described in detail in our previous papers (Osipova et al., 1980; 1986; Osipova, 1987). It enables us to calculate such structural

Nuclear Structure and Function, Edited by J. R. Harris and
I. B. Zbarsky, Plenum Press, New York, 1990

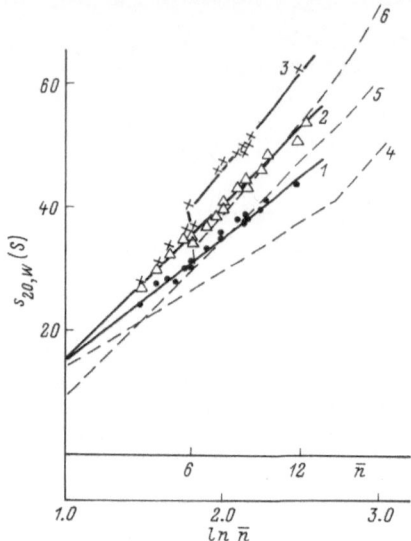

Figure 1. Dependence of sedimentation coefficients $s_{20,W}$ on the number of nucleosomes ($\bar{n}$) for oligonucleosomes from chromatin of sea urchin sperm (1, 2, 3) and rat thymus (4, 5, 6), in solutions of different ionic strength: 5mM (1, 4), 25mM (2, 5), 65mM (3, 6). $\bar{n}$ - weight-average number of nucleosomes in the oligonucleosomal chain.

Table I. Structural Parameters of Oligonucleosome Chromatin Fragments on the Dependence of the Ionic Strength of Solution Calculated from the Sedimentation Data on the Basis of the Cylinder Model*.

The source of chromatin	Linker DNA size (base pairs)	Ionic strength								
		5 mM			25 mM			65 mM		
		l_o nm	d nm	y	l_o nm	d nm	y	l_o nm	d nm	y
Cortical neurones	20	11.8	19	4.7	7.5	21	7.5	6.4	22	8.8
Rat thymus	50	11.6	20	5.8	7.1	22	9.3	6.1	22	11.0
Sperm of the sea urchin	100	10.6	22	8.0	8.6	21	9.8	6.7	20	13.0

* The range of mean-square errors was from 2% to 7% for l_0 and y, and from 2% to 15% for d.

298

parameters of the chain of nucleosomes as M_L - the average mass per unit length, l_0 - the length of the chain per nucleosome, d - the hydrodynamic diameter of the chain and y - the DNA packing ratio. All the parameters obtained are represented in the Table I. The similar diameter values (about 20nm) for all the types of chromatin investigated and at different ionic strengths indicate that the structure of oligonucleosomes differing in linker DNA sizes is in good agreement with the three-dimensional model of a zig-zag shaped chain, in which the mass per unit length and the DNA packing ratio grows with increasing linker DNA length.

We tried to construct various models for DNA packing in the nucleosome fiber, in an attempt to fit the calculated parameters for chromatin differing in linker DNA sizes. So, for short linker DNA chromatin from brain neurones at low ionic strength, it was necessary to assume that part of the linker DNA was unfolded in such a way that two full turns of DNA characteristic for the chromatosome could not be formed. This assumption is confirmed by corresponding circular dichroism data (Ramm et al., 1986). For chromatin with longer linker DNA, a part of the linker DNA (about 20 base pairs) should be folded around the histone octamer forming the chromatosome, even at low ionic strength. At this ionic condition the remaining linker DNA (30 base pairs) in thymus chromatin is extended in order to fit the value $l_0 = 12$nm. As to sea urchin sperm chromatin, the remaining part of the linker DNA (about 80 base pairs) should be partly folded, forming a kind of loop to achieve the value $l_0 = 10$nm.

With increase of ionic strength up to 25mM the neighbouring nucleosomes come closer (l_0 decreases) without significant changes in the value of the diameter of the chain (d = 20-22nm). This compaction process may include some additional folding of the loop of long linker DNA in sperm chromatin, bending or folding of linker DNA in rat thymus chromatin and coiling over the histone octamer some additional part of linker DNA in neurone chromatin. However, it is not enough to complete the second DNA turn on the histone octamer and to form a chromatosome stable subunit with two DNA turns.

Further increasing of ionic strength up to 65mM results in further condensation of all types of chromatin due to two processes: shortening the distance between the neighbouring nucleosomes inside zig-zag shaped structure and the formation of compact helical structure for oligonucleosomes containing more than 11-12 subunits, in agreement with our data demonstrating that the value showing dependence of $s_{20,W}$ on ionic strength increases from 0.09 for oligonucleosomes containing from 6 to 10 nucleosomes to 0.14 for longer nucleosomal chain. We suppose that long band of zig-zag nucleosomal chain consisting of two rows of nucleosomes is twisted into a compact double superhelix with the increase of ionic strength to physiological value. The condensation is accomplished by drawing adjacent nucleosomes nearer, both along the chain length and between the neighbouring turns of the double superhelix, the loops of linker DNA being inside superhelix. Similar to the model suggested by Woodcock et al. (1984), in our model the chromatin fiber is formed by a zig-zag chain of nucleosomes folded into the superhelix, but the difference is that in our model linker DNA is folded and the pitch of the double superhelix is independent of the length of linker DNA.

The suggested model seems to be reasonable for the explanation of our sedimentation data. The jump in the dependence of $s_{20,W}$ on $\ln \bar{n}$ for the fragments containing more than 5 nucleosomes (Figure 1) can be related to the fact, that beginning only from hexanucleosomes, can optimal conditions for cooperative interactions between nucleosome cores and histone H1 of different nucleosomes be achieved. The sharp increase in the degree of compaction of oligomers with more than 10-11 nucleosomes may indicate the formation of the first turn of the double superhelix.

REFERENCES

Osipova T. N., 1987, Structural parameters of short DNA fragments and oligonucleosomes on the basis of the cylinder model, <u>Studia Biophysica</u>, 120:171.

Osipova T. N., Karpova E. V., Svetlikova S. B., Kukushkin A. N. and Pospelov V. A., 1986, Structure of pigeon brain chromatin: sedimentation of analysis of oligonucleosome compaction, <u>Molekul. Biol.</u>, 20:78.

Osipova T. N., Pospelov V. A., Svetlikova S. B. and Vorob'ev V. I., 1980, The role of histone H1 in compaction of nucleosome, <u>Eur. J. Biochem.</u>, 113:183.

Ramm E. I., Ivanov G. S., Pospelov V. A., Svetlikova S. B. and Kukushkin A. N. 1986, Peculiarities of structural state of DNA in chromatin with short linker, <u>Biofizika</u>, 31:404.

Thomas J. O., Rees Ch. and Butler P. J. G., 1986, Salt-induced folding of sea urchin sperm chromatin. <u>Eur. J. Biochem.</u>, 154:343.

Woodcock C. L. F., Frado L.-L. Y. and Rattner J. B., 1984, The higher-order structure of chromatin: evidence for a helical ribbon arrangement, <u>J. Cell Biol.</u>, 99:42.

Zalenskaya I. A., Pospelov V. A., Zalensky A. O. and Vorob'ev V. I., 1981, Nucleosomal structure of sea urchin and starfish sperm chromatin. Histone H2B is possibly involved in determining the length of linker DNA, <u>Nucleic Acids Res.</u>, 9:473.

CHROMOMERE - THE STRUCTURAL UNIT OF THE CHROMATIN IN THE INTERPHASE NUCLEUS

O. V. Zatsepina, I. I. Kirevev, E. I. Frolova,
V. Yu. Polvakov and Yu. S. Chentsov

A. N. Belozerskv Laboratory of Molecular Biology and
  Bioorganic Chemistry
Moscow State University
Moscow 119899
USSR

The higher order structure of chromatin in the interphase nucleus of eukaryotic cells is currently under extensive study. According to the general concept, DNA is attached to an intranuclear skeleton termed the nuclear matrix. DNA loops can be observed after extraction of nuclear proteins using high salt solutions, but they have never been seen in nuclei fixed *in situ* (Hancock and Boulikas, 1982). That is why, in this work we turn our attention to study (1) the structural organisation of chromatin in animal nuclei fixed in situ, and (2) to the folding of DNP fibrils following experimental decondensation of chromatin by a decrease in Ca^{2+} ion concentration in low salt media. A noticeable loss of chromatin material does not occur through this treatment (Kiryanov et al., 1982).

1. STRUCTURAL 'DOMAINS' IN THE INTERPHASE NUCLEUS REVEALED UPON THE ARTIFICIAL DECONDENSATION OF CHROMATIN

Cultured Chinese hamster cells (clone 237) were consequently treated with a 'condensation buffer': 3mM $CaCl_2$; 1mM $MgCl_2$; 0.1% Triton X-100; 5mM Tris-HCL (pH 7.0); 0.1mM PMSF, with descending concentrations of $CaCl_2$ in 5mM Tris-HCL (pH 7.0). The cells were fixed in glutaraldehyde diluted to 2.5% by the solutions used for the cell treatment. After the standard procedure of postfixation, the cells were embedded in Epon (Zatsepina et al., 1983).

Figure 1 shows a region of the interphase nucleus of the Chinese hamster cell fixed *in situ*. DNP fibrils form numerous structures of a higher order, many of which are globular formations 80-120nm in diameter. They correspond in size to the chromomeres - the discrete chromatin structures, which have been previously described in mitotic chromosomes (Zatsepina et al., 1983).

An initial action of the 'condensation buffer' (3mM Ca^{2+}) causes a strong compaction of chromatin. The chromomeres aggregate into short strands and large chromatin clumps. As Ca^{2+} is gradually removed from the incubation medium, the chromomeres decondense. This process begins at the periphery of chromomeres, forming a structure containing numerous DNP fibrils radiating from an electron dense core (Figure 2). Fibril diameter

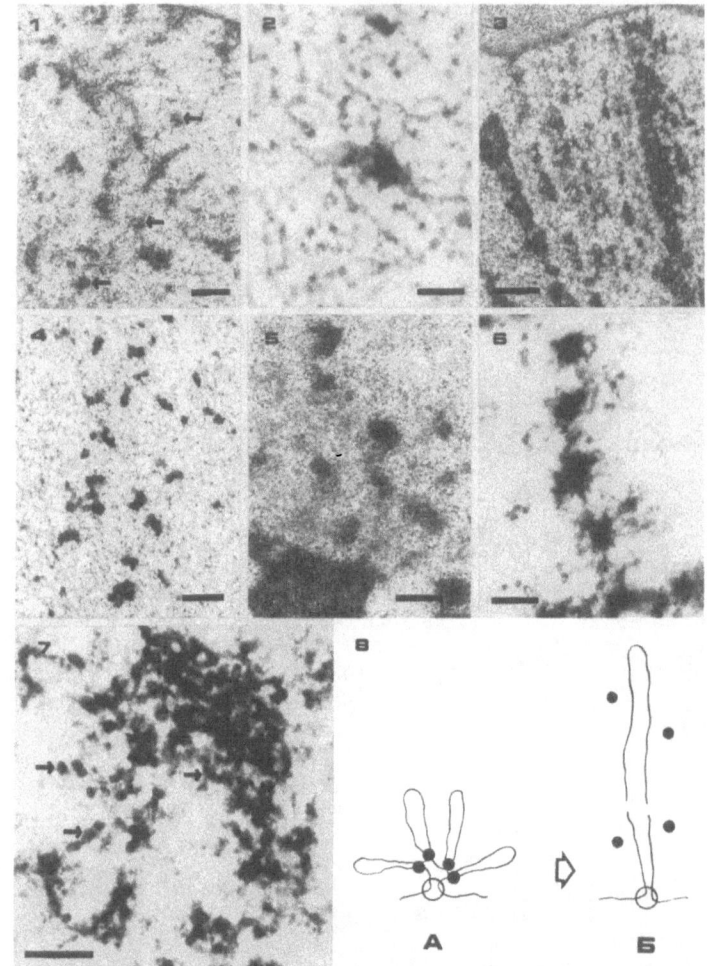

Figure 1.   Chromomeres (arrows) in the interphase nucleus of Chinese hamster cell fixed *in situ*. Bar, 0.5μm.

Figure 2.   A rosette-like chromomere in 0.1mM $CaCl_2$. Bar, 0.1μm.

Figure 3.   Compact chromomeres in a band of Drosophyla virilis polytene chromosomes. Bar, 0.5μm.

Figure 4.   Rosette-like chromomeres of polytene chromosomes observed upon artificial decondensation 0.1mM $CaCl_2$. Bar, 0.3μm.

Figure 5.   A group of compact chromomeres in the S PCCs fixed *in situ*. Bar, 0.3μm.

Figure 6.   The rosette-like chromomeres observed upon artificial decondensation of PCCs in the S period. Bar, 0.1μm.

Figure 7.   Ultrastructure of the 'reconstructed' chromatin; DNP globules are indicated by arrows. Bar, 0.3μm.

Figure 8.   Scheme, illustrating a possible transformation of a rosette-like chromomere to a loop-like one after removal of nuclear proteins.

ranges from 10 to 25nm. DNP fibrils form loops, so that a decondensing chromomere resembles a 'rosette'. In Ca^{2+}-free Tris-HCl buffer, chromomeres decondense to DNP fibrils 10-20nm in diameter. Keeping in mind the nuleomeric model of DNP fibrils (Kiryanov et al., 1982), the content of DNA in the chromomeres was estimated to be about 10-30mkm or 30-100Kbp.

2. THE STRUCTURAL TRANSITIONS OF CHROMOMERES AND BANDS IN *Drosophyla virilis* POLYTENE CHROMOSOMES UPON ARTIFICIAL DECONDENSATION

As already known one of the structural elements of bands in diptera polytene chromosomes is the chromomere (Sorsa, 1974). The size of the chromomeres of polytene chromosomes are comparable with those of chromomeres observed in the interphase chromatin (Figure 3). To determine the more precise correlation between these structures, changes were studied during artificial decondensation of polytene chromosomes under the condition of chromomeric existence in Chinese hamster cells.

Analysis of ultra-thin sections of isolated salivary glands of *Drosophyla virilis* shows that in 0.2-0.1mM $CaCl_2$, the bands are composed of loosely packed chromomeres, resembling the 'rosettes' (Figures 2 and 4). They have an electron dense core with DNP fibrils radiating from it (Figure 4).

3. CHROMOMERIC LEVEL OF DNP FOLDING PREMATURELY CONDENSED CHROMOSOMES (PCCs)

To study the chromomeres more, PCCs induced in Chinese hamster cells were used. The PCCs forming in the G_1, S and G_2 phases of interphase were analysed using serial ultra-thin sections. It was shown that chromatin globules, about 100nm in diameter - the chromomeres, can be observed within the S phase PCCs fixed *in situ* without any treatment (Figure 5). Chromomeres can be also viewed through the artificial decondensation of the G_1, S, and G_2 PCCs (Figure 6). The chromomeres thus obtained have a rosette-like structure as well (Figure 6).

4. STRUCTURAL 'RECONSTRUCTION' OF SOME LEVELS OF FOLDING OF DNA WITHIN THE INTERPHASE NUCLEUS

Chinese hamster cells were treated with a Ca^{2+}-free medium (5mM Tris-HCL or H_2O) containing 0.1% Triton X-100. This treatment causes the transition of thick (20nm) DNP fibrils to thin (10nm) ones. A subsequent increase in $CaCl_2$ concentration causes a cooperative transition of 10nm DNP fibrils to fibrils 20nm thick, which in turn form chromatin threads and globules about 40-50nm in diameter (Figure 7). The typical chromomeres (i.e. chromatin globules approximately 100nm in diameter) cannot be reconstructed. It is reasonable to believe that 'fasteners' preserving the integrity of chromomeres are destroyed in low ionic strength solution. It should be noted that interactions between the DNP fibrils and nuclear 'skeleton' have suffered in low salt (Rasin and Yarovaya, 1984).

As we and others have shown previously, the chromomeres, as discrete chromatin structures about 100nm in diameter, are found within mitotic chromosomes of animals (Zatsepina et al., 1983) as well as higher plants (Gornung et al., 1986). Hence, we assume that the chromomeres represent a common level of folding of DNP fibrils in interphase and mitosis. In the chromomere, DNA forms a rosette-like structure resembling 'rosettes' which are obtained under conditions of partial removal of chromatin proteins

(Prusov et al., 1983). An approximate estimation shows that the chromomeres and 'rosettes' correspond in DNA content (30-10Kbp) to the loops of DNA which are viewed within salt-extracted nuclei by electron microscopy (Hancock and Boulikas, 1982), as well as to 'domains' revealed using the biochemical approaches (Igo-Kemenes et al., 1982).

Therefore, we believe that the chromomeres, 'rosettes' and loops are the elements of the same chromatin structures which differ with respect to the degree of folding of their DNP fibrils (Figure 8).

REFERENCES

Hancock R., and Boulikas T. 1982, Functional organisation in the nucleus, Intern. Rev. Cytol., 79:165.

Igo-Kemenes T., Hoers W., and Zachau H. G. 1982, Chromatin, Ann. Rev. Biochem., 51:89.

Kiryanov G. I., Smirnova T. A., and Polyakov V. Yu. 1982, Nucleomeric organization of chromatin, Eur. J. Biochem., 124:331.

Gornung E. M., Polyakov V. Yu., and Chentsov Yu. S. 1986, Levels of DNA compactization in interphase and mitotic chromosomes of higher plants, Tsitologia (USSR), 28:911.

Prusov A. N., Polyakov V. Yu., Zatsepina O. V., Chentsov Yu. S., and Fais D. 1983, Rosette-like structures from nuclei with condensed (chromomeric) chromatin but not from nuclei with diffuse (nucleomeric) chromatin, Cell Biol. Int. Report, 7:849.

Rasin S. V., and Yarovaya O. V. 1986, Low ionic strength of nuclease treated nuclei destroys the apparent attachment of transcriptionally active DNA to the nuclear skeleton elements. Mol. Biol. (USSR), 20:646.

Sorsa V. 1974, Organisation of chromomeres, Cold Spring Harbor Symp. Quant. Biol., 38:602.

Zatsepina O. V., Polyakov V. Yu., and Chentsov Yu. S. 1983, Chromonema and chromomere. Structural units of mitotic and interphase chromomomes, Chromosoma, 88:91.

AN IMMUNOELECTRONMICROSCOPIC ANALYSIS OF THE ORGANIZATION LEVELS OF THE

METAPHASE CHROMOSOMES

E. V. Kiseleva, N. D. Belyaev*, A. G. Shilov, N. P. Kulyiba**
and A. V. Kozlov**

Institute of Cytology and Genetics and Institute of
  Bioorganic Chemistry*
Academy of Sciences
Siberian Department
Novosibirsk;
Institute of Biology of Leningrad University
Leningrad**
USSR

Analysis of the DNA packing modes and organization in interphase
chromatin and metaphase chromosomes, as well as the position of DNA and
histone proteins relative to each other in the compact chromatin fiber, is
helpful in clarifying gene expression and regulation. It was hoped that an
approach to understanding of DNA packing and histone topography would be
provided by combining ultrastructural immunochemistry with the Miller
spreading procedure. We used Miller's technique in this study of the DNA
packing levels in the chromosomes isolated from Chinese hamster cells and
mink fibroblasts. With the use of antibodies to H1 and H2A and protein
A-colloidal gold complex (pAcg), we estimated the extent of accessibility
of H1 and H2A to antibodies in the compact and partly decondensed metaphase chromosomes. Decondensation was achieved by 15-30 minutes incubation
in 0.1-0.5mM borate buffer (pH 9.0). The metaphase chromosomes were
incubated with the antibodies (Fakan et al., 1986), and the chromosomes
were incubated with the pAcg without the antibodies in the control
experiments.

In the Miller spreads, the metaphase chromosomes are seen as compact
X or U-shaped structures with closely packed 30nm fibres (Figure 1a).
After exposure to 0.1-0.5mM borate buffer, chromosomes or rosette-like
structures with DNA loops arranged with a radial orientation, nucleomeres
and nucleosomes (measuring 100nm, 30nm, 12nm in diameter, respectively)
(Figure 1b-d) appear. Noteworthy is the heterogeneity of the response of
the different regions along the chromosome length to the decondensing effect
of the borate buffer, this is in agreement with the previous observation on
the differential decondensation of the metaphase chromosomes (Zatsepina et
al., 1985). The heterogeneity may be due to the different strength of
protein-DNA interactions in the metaphase chromosome. In the more
decondensed regions of the metaphase chromosomes, thin fibers containing
regularly arranged nucleosomes not forming a zigzag pattern are observed.
Their absence suggests that either H1 has dissociated from these regions or
its concentration has much decreased. Immunoelectron microscopy demonstrates that the decondensed chromosome regions virtually cease binding

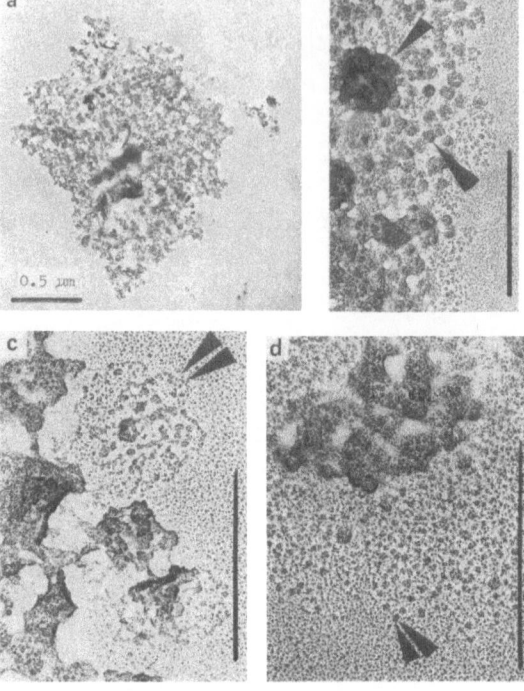

Figure 1.   Electron micrographs of metaphase chromosome spreads; (a) a
             compact intact chromosome, (b-d) fragments of partly decon-
             densed chromosomes with different levels of DNA organization.
             Arrows indicate: chromomere (single black); rosette-like
             (double black); nucleomere (single black); nucleosome (double
             black).   Scale bars indicate 0.5μm.

anti-H1 antibodies (Figure 2a), although labelling heavily with anti-H2A
antibodies. Figures 2b,c show that gold grains associated with DNP fibers
are distributed linearly like the nucleosome globules are. The compact
chromosomes do not label with anti-H2A antibodies and as heavily with
anti-H1 antibodies as observed in the metaphase spreads (Figure 2d) and in
the unspread samples (Figure 2e). Thus H1 in a 30nm fiber of the metaphase
chromosomes appears to be completely unaccessible to the antibodies. This
disagrees with the results reported for the location of the linker histones
in condensed chromatin, suggesting their internalization in the 30nm fiber
(Yorkin, 1987).

     In an attempt to reconcile this discrepancy, explanations are offered:
(1) the metaphase chromosomes and interphase chromatin may differ in the
structural organization of the 30nm fiber and the position of DNA and H1
relative to each other; this appears to be very unlikely; (2) the location
of H1 inside the 30nm fiber may be the same in chromatin and metaphase
chromosomes; however, because such a fiber in a chromosome twists and folds
during further packing, H1 may become accessible to the antibodies in these
regions; (3) the more plausible explanation would be in terms of the domain
organization of the H1 molecule; one of the H1 domains may presumably be
inside, and some other outside, the 30nm fiber of chromatin; if this were
so, one would expect different results with antibodies to different H1
domains. The evidence obtained, so far, does not allow us to give

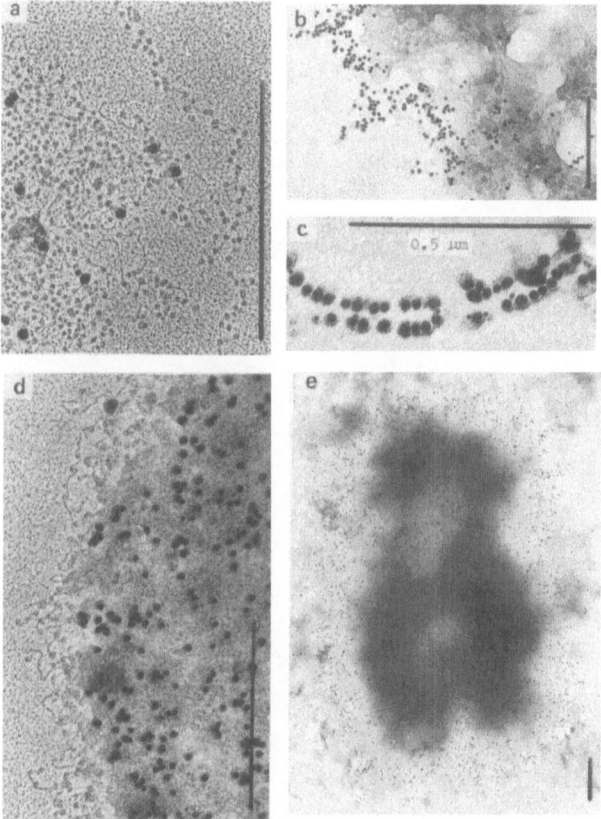

Figure 2.   Visualization of the antigen determinants of H1(a,d,e) and
              H2A (b,c) histones in the partly decondensed (a-c) and
              compact (d,e) metaphase chromosomes. Scale bars indicate
              5µm.

preference to any one of the models suggested for the organization of the
30nm chromatin fiber.

REFERENCES

Fakan S., Leser G., and Martin T. E. 1986, Immunoelectron microscope
    visualization of nuclear ribonucleoprotein antigens within spread
    transcription complexes, J. Cell Biol., 103:1153.
Yorkin A. M. 1987, Chromatin on the membrane: study of histone H5
    accessibility for antibodies in the supernucleosomal structure, Mol.
    Biol., 21:688 (In Russian).
Zatsepina O. V., Polyakov V. Yu., Chentsov Yu. S. 1985, Differences in the
    structural organization of G and R-bands recognized by the differ-
    entially decondensed chromosome technique, Tsitologia, 27:865 (In
    Russian).

THE EFFECT OF SULFHYDRYL-OXIDIZING AND SULFHYDRYL-BLOCKING REAGENTS ON

CHROMATIN PREPARATIONS

M. Gaczynski and L. Klyszejko-Stefanowicz

Department of Cytobiochemistry
Institute of Biochemistry
University of Lodz, ul. Banacha 12/16
92-327 Lodz
Poland

INTRODUCTION

Washing cell nuclei with EDTA-containing solutions and Tris buffer has been one of the most widely used non-enzymatic methods of obtaining a nucleoprotein complex termed as 'chromatin'. The effect of oxidizing or reducing conditions on the yield and quality of chromatin preparations was studied in our work.

METHODS

Cell nuclei were isolated by the sucrose method (Blobel and Potter, 1966) without the Triton X-100 washing step in the presence of 10mM iodo-acetamide (IAA) or 2mM sodium tetrathionate (NaTT) (Kaufmann and Shaper, 1984). Control experiments were performed without any additive.

Chromatin was obtained from cell nuclei by washing with NaCl-EDTA and Tris (Spelsberg and Hnilica, 1971) and cell nuclear lysates were prepared by demetalizing the nuclear suspension in 0.25mM EDTA with Chelex 100 (Makarov et al., 1983).

DNA and proteins were estimated by a spectrophotometric method based on absorption at 230 and 260nm (Kalb and Bernlohr, 1977).

Two-dimensional electrophoresis was performed according to the NEPHGE-SDS scheme (O'Farrell et al., 1977). Nonidet P-40 was replaced with CHAPS, in protein solubilizing solution and the first dimension gel (Rabilloud et al., 1986). The gels were stained with Coomassie Brilliant Blue R.

RESULTS

Cell nuclei were isolated without washing with Triton X-100 and not exposed to oxidative contaminants present in commercial batches of this detergent (Ashani and Catravas, 1980). Three kinds of cell nuclei were obtained: a) control, b) sulfhydryl-blocked (10mM IAA present throughout the preparative work) and c) sulfhydryl-oxidized (prepared in the presence of 2mM NaTT). Both 'washed' chromatin and total nuclear lysate were prepared from all types of nuclei.

Two general observations can be made: 1) the demetalized nuclear lysate was not influenced by the oxidizing/blocking conditions and the differences between chromatin from sulfhydryl-blocked or oxidized nuclei appear only after washing them with NaCl-EDTA and/or Tris and 2) control chromatin preparations were very similar to the IAA-treated ones.

Almost all the nuclear DNA (90%) was found in chromatin from oxidized nuclei, while only 35-45% appeared in chromatin prepared from control or IAA-treated ones. The missing DNA was found in the NaCl-EDTA and Tris washes.

The amount of protein released from nuclei was 26-29% during the washing with NaCl-EDTA and 4-6% in the Tris washes and did not depend on the oxidizing conditions. The electrophoretic patterns (not shown) of these proteins from the three kinds of nuclei resembled each other very closely, except for a few quantitative changes. There were some differences in the UV-spectra of the chromatin samples. The A320 to A260 ratio was higher for the control and sulfhydryl blocked chromatin than in oxidized preparations (0.20 versus 0.09).

DISCUSSION

Three explanations for the limited release of DNA from NaTT-treated nuclei are possible: a) oxidation of sulfhydryl groups inhibits endogenous nucleases, b) fewer digestion sites are exposed on DNA and susceptible to nucleases or c) digestion products are retained inside the nucleus. Careful analysis of the kinetics of chromatin digestion with endogenous nucleases and deoxyribonuclease I (Tas and Walford, 1982) or micrococcal nuclease (Tas et al., 1980) led to conclusions supporting the second possibility rather than the first.

In our experimental conditions there was no massive release of nuclear matrix proteins described by Kaufmann and Shaper (1984) after reducing -S-S- bonds and the released proteins were very similar in blocked and oxidized nuclei.

The A230 to A260 ratio is higher in sulfhydryl-blocked and control nuclei and indicates a higher degree of chromatin aggregation (Dixon and Burkholder, 1985), not found in demetalized nuclear lysates. It could be caused by 1) lowering the DNA to protein ratio in sulfhydryl blocked nuclei after washing them with NaCl-EDTA and Tris or 2) it needs two factors to appear: reduced -SH groups and the presence of trace amounts of divalent cations not removed during washing the nuclei.

In conclusion, oxido-reducing conditions during cell nuclear isolation do not change the protein composition of chromatin to a greater extent. However, they do influence its yield and quality. It does not seem possible to obtain satisfactory chromatin preparations by washing nuclei with NaCl-EDTA unless some -SH groups are artificially oxidized.

REFERENCES

Ashani Y., and Catravas G. N. 1980, Highly reactive impurities in Triton X-100 and Brij-35: partial characterization and removal, Anal. Biochem., 109:55.

Blobel G., and Potter V. R. 1966, Nuclei from rat liver: isolation method that combines purity with high yield, Science, 154:1662.

Dixon D. K., and Burkholder G. D. 1985, The effect of sodium and magnesium-ion interactions on chromatin structure and solubility, Eur. J. Cell Biol., 36:315.

Kalb V. F., and Bernlohr R. W. 1977, A new epectrophotometric assay for protein in cell extracts, Anal. Biochem., 82:362.

Kaufmann S. H., and Shaper J. H. 1984, A subset of non-histone nuclear proteins reversibly stabilized by the sulfhydryl cross-linking reagent tetrathionate. Polypeptides of the internal nuclear matrix, Exp. Cell Res., 155:477.

Makarov V. L., Dimitrov S. I., and Petrov P. T. 1983, Salt-induced contormational transitions in chromatin. A flow linear dichroism study, Eur. J. Biochem., 133:491.

O'Farrell P. Z., Goodman H. M., and O'Farrell P. H. 1977, High resolution two-dimensional electrophoresis of basic as well as acidic proteins, Cell, 12:1133.

Rabilloud T., Hubert M., and Tarroux P. 1986, Procedures for two-dimensional electrophoretic analysis of nuclear proteins, J. Chromatogr., 351:77

Spelsberg R. C., and Hnilica L. S. 1971, Proteins of chromatin in template restriction, Biochim. Biophys. Acta, 226:202.

Tas S., Tam C. F., and Walford R. L. 1980, Disulfide bonds and the structure of the chromatin complex in relation to aging, Mech. Aging Devel., 12:65.

Tas S., and Walford R. I. 1982, Influence of disulfide-reducing agents on fractionation of the chromatin complex by endogenous nuclease and deoxyribonuclease I in aging mice, J. Gerontol., 37:673.

FRACTIONATION AND ANALYSIS OF DNA MOLECULES DIFFERING IN THEIR ATTACHMENT

TO THE NUCLEAR INTERIOR

A. V. Lichtenstein, M. M. Zaboikin, and R. P. Alechina

Institute of Carcinogenesis
All-Union Cancer Research Centre
Moscow 115478
USSR

Some time ago, we elaborated a chromatographic method for fractionation of nucleic acids as constituents of nucleoprotein complexes (Lichtenstein et al., 1975). In this method, isolated nucleoproteins are adsorbed on Celite which binds proteins irreversibly but not nucleic acids. The latter are then eluted from the immobilized complexes with the use of agents capable of dissociating DNA (RNA)-protein interactions (salts, urea, and temperature). This procedure, termed nucleoprotein-Celite chromatography (NPC-chromatography), enabled us to separate the major classes of cellular nucleic acids (18S and 28S rRNA, heterogeneous nuclear RNA, and DNA) in a single chromatographic run, as well as to obtain an insight into the nature of the chemical bonds stabilizing the respective complexes (Lichtenstein et al., 1982; Sjakste et al., 1985; Zaboikin et al., 1988). NPC-chromatography appeared also effective when applied to the analysis of such complex structures as isolated nuclei or, even, whole cell lysates (Lichtenstein et al., 1982). In this case, the chromatographic patterns of DNA eluted from immobilized nuclei were found to be to some extent cell cycle-specific. Namely, the potential of cells to proliferate correlated with the portion of nuclear DNA most tightly attached to the nuclear skeleton. This fraction, termed DNA-II, eluted from nuclei at elevated temperature (70°-90°C) significantly differed, by structural and functional criteria, from the less tightly bound DNA-O and DNA-I fractions eluted by the salt and salt-urea gradients, respectively.

These earlier results raised a question of genome compartmentalization, i.e. co-existence within a single cell of covalently non-linked DNA molecules which differ from each other in their protein environment. In the present study, we have undertaken the analysis of DNA compartmentalization in more detail. Three successive gradients, namely of NaCl (0-3M at 2°C), LiCl-urea (0-4M LiCl, 8M urea at 2°C) and temperature (2°-96°C, 4M LiCl, 8M urea used as an eluent), were performed to elute from the immobilized nuclei the DNA-O, DNA-I, and DNA-II fractions, respectively. It was established previously that DNA-O originates from chromatin fibres having no contacts with the nucleoskeleton, while DNA-I and DNA-II are attached to the nucleoskeleton by weak and tight bonds, respectively (Sjakste et al., 1975; Zaboikin et al., 1988).

Figure 1 shows the chromatographic patterns of DNA eluted from Djungarian hamster fibroblasts transformed by SV40 (line 4/21) (A), from

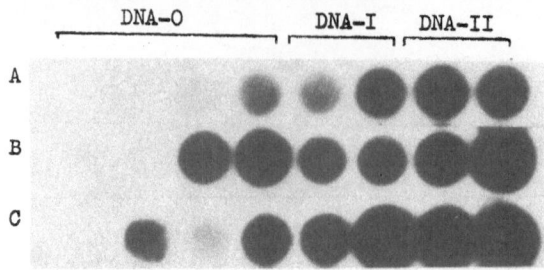

Figure 1.   NPC chromatography of DNA from 4/21 cells (A), Zajdela
             ascites hepatoma cells (B) and mouse kidney cells (C). DNA
             fractions were transferred to nylon filters and hybridized to
             nick-translated genome DNA of respective origin.

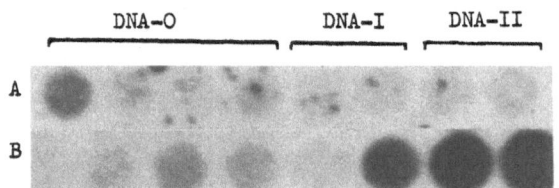

Figure 2.   Dot hybridization of fractionated DNA from 4/21 cells to
             v-Ha-ras probe (A) and total genome DNA (B).

Zajdela ascite hepatoma cells (B), and from mouse kidney cells (C). In all
these cases, approximately 10^6 cells were lysed in Triton X-100 and
immediately subjected to NPC-chromatography as described earlier (Zaboikin
et al., 1988). DNA fractions were concentrated by hydroxyapatite adsorption
and transferred to membrane filters (Lichtenstein et al., 1989) for dot
hybridization to nick-translated total genome DNA as a probe. We prefer
such indirect labelling for DNA detection in order to avoid DNA labelling
with radioactive precursors *in vivo,* which can bring about irradiation-
induced DNA fragmentation. A very heterogeneous distribution of DNA is
clearly seen upon chromatographic analysis of all the cell types studied.
We have explored the possibility that some artefacts, such as DNA mechanical
shearing or cleavage by some endogeneous nucleases may be responsible for
the DNA fragmentation revealed. The results (to be published elsewhere)
have shown that DNA shearing during the chromatography procedure does not
take place and that enzyme digestion, though not ruled out completely, is
unlikely to occur. In fact, isolation of 4/21 cell nuclei under the
conditions that either reduce endogenous nuclease activity (near-zero
temperature, EDTA, diethyl pyrocarbonate) or facilitate it (Mg^{2+},
incubation at 37°C for 30 minutes) did not significantly affect the DNA
patterns obtained.

    There is a strong body of evidence for nonrandom distribution of DNA
sequences with respect to the chromatographic DNA fractions. Figure 2A
shows the results of dot hybridization of DNA fractions from 4/21 cells with
nick-translated cloned BS-9 DNA which contains a 0.49kb fragment of the
v-Ha-ras oncogene (Ellis et al., 1981). After the removal of label, the
filter was reprobed with nick-translated total genome DNA (Figure 2B). It
is clearly seen that the c-Ha-ras gene sequences are differentially
represented in the DNA fractions.

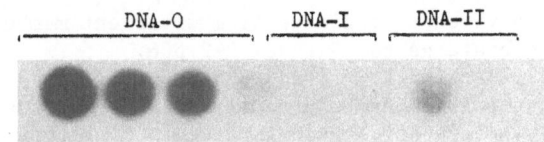

Figure 3.   Chromatographic pattern of DNA labelled by nick-translation
of isolated 4/21 cell nuclei *in situ*. Isolated nuclei were
labelled with the use of a nick-translation kit (Amersham)
according to manufacturer's specifications and then subjected
to NPC chromatography. DNA fractions were collected,
transferred to a nylon filter and autoradiographed.

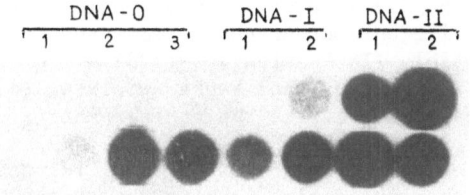

Figure 4.   Chromatographic patterns of DNA pulse labelled with ^{32}PdCTP
in permeable 4/21 cells (upper row) and chased for 50 min. in
an excess of cold dCTP (lower row). See text for details.

In the following experiment, we have attempted to detect a chromatographic pattern of DNA sequences selectively labelled by nick-translation
of isolated nuclei *in situ,* bearing in mind that the preferable targets of
this reaction are DNAase I-sensitive, i.e. transcriptionally competent,
chromatin regions. Figure 3 shows that the DNA fraction with the lowest
adherence to the nuclear interior is the most intensively labelled, thus
giving additional support to the idea of nonrandom distribution of
functionally significant sequences.

Finally, the experiments aimed at revealing whether there exists a
distinct compartment assigned for DNA synthesis were carried out.
Djungarian hamster 4/21 cells were made permeable in Triton X-100 (van der
Velden et al., 1984) and labelled with ^{32}PdCTP for 10 minutes at 37°C. One
portion of cells was used for NPC chromatography immediately (pulse), while
the other portion was additionally incubated with an excess of cold dCTP for
50 minutes at 37°C and chromatographed after that (chase). DNA fractions
obtained were transferred to a nylon filter and autoradiographed (Figure 4).
The results of this pulse-chase experiment favour the notion of a
precursor-product interrelationship between DNA-II and the rest of nuclear
DNA. Thus, the tightest DNA-nucleoskeleton complexes (type II) seem to be
the site of DNA synthesis.

We believe that the phenomenon of nuclear DNA compartmentalization
(nonrandom dissection of genome into pieces of a large size) may be of
significant interest with regard to its nature, possible biological role
and chromatin architecture.

REFERENCES

Ellis R. W., Do Feo D., Shih T. Y., Gonda M. A., Young H. A., Tsuchida N.,
    Lowy D. R., and Scolnick E. M., 1981, The p21 src genes of Harvey and

Kirsten sarcoma viruses originate from divergent members of a family of normal vertebrate genes, <u>Nature</u>, 292:506.

Lichtenstein A. V., Alechina R. P., and Shapot V. S., 1975, Ribosomal and informational ribonucleoprotein complexes of animal cells. Study on rat liver ribonucleic acids as constituents of ribonucleoprotein complexes by chromatography on nucleoprotein-Celite columns, <u>Biochem. J.</u>, 147:447.

Lichtenstein A. V., Sjakste N. I., Zaboikin M. M., and Shapot V. S., 1982, Rearrangement of DNA-protein interactions in animal cells coupled with cellular growth-quiescence transitions, <u>Nucl. Acids Res.</u>, 10:1127.

Lichtenstein A. V., Moiseev V. L., and Zaboikin M. M., 1989, Hydroxyapatite-mediated transfer of DNA from diluted solutions to nitrocellulose or nylon membranes, <u>J. Biochem. Biophys. Methods</u>, 18:77.

Sjakste N. I., Zaboikin M. M., Erenpreisa E. A., Lichtenstein A. V., and Shapot V. S., 1985, Analysis of cellular nucleoproteins by the nucleoprotein-Celite chromatography. Two types of DNA-matrix interactions, <u>Molek. Biol.</u> (In Russian), 19:1231.

Van der Velden H. M. W., Poot M., and Wanka F., 1984, *In vitro* DNA replication in association with the nuclear matrix of permeable mammalian cells, <u>Biochim. Biophys. Acta</u>, 782:429.

Zaboikin M. M., Moiseev V. L., Shapot V. S., and Lichtenstein A. V., 1988, Localization of the replication fork in the tight DNA-matrix complex (Type II), <u>Molek. Biol.</u> (In Russian), 22:1119.

PLANT NUCLEAR MATRIX EFFECTS OF DIFFERENT EXTRACTION PROCEDURES ON ITS

STRUCTURAL ORGANIZATION AND CHEMICAL COMPOSITION

M. A. Cerezuela and S. Moreno Diaz de la Espina

Centro Investigaciones Biologicas
C.S.I.C. Velazquez 144
28006 Madrid, Spain

The lengthy controversy about the existence of a nuclear substructure
in vivo has been resolved in the recent years by the identification of a
nuclease and high salt resistent nuclear matrix in cells ranging from lower
eukaryotes to insects and mammalian cells, and also by the increasing
evidence of the implications of this structure in primary nuclear functions
(Berezney, 1984; Verheijen et al., 1988; Razin, 1987). The isolated
nuclear matrix is built up almost entirely of proteins. Its better-known
components are the lamins, belonging to the intermediate filament proteins
(Krohne and Benavente, 1986) which have been immunologically localized in
the residual nuclear envelope. In contrast, the polypeptides of the internal
matrix are not so well characterized at present (Verheijen et al., 1988).
The recovery of a well-organized intranuclear matrix from the extraction
procedures appears to be dependent, not only on the cell type, but also on
the isolation procedure: conditions of nuclease digestions, use of
alkylating agents, divalent cations, protease inhibitors, etc (Kaufmann et
al., 1981; Verheijen et al., 1988).

In spite of the great advances gained in the study of the nuclear
matrix, very little information exists dealing with this structure in plant
cells (Barthelemy and Moreno Diaz de la Espina, 1984; Stolyarov, 1984; Ghosh
and Dey, 1986). To determine whether the internal matrix and residual
nucleolus in plant cells result from certain procedures of nuclear matrix
preparations, or are characteristics of the cell type, we undertook a
comparative study by electron microscopy (EM) and one-dimensional SDS-PAGE
of isolated nuclear matrices of onion cells prepared following 4 different
extraction procedures: 1. - Sequential extraction of nuclei with low and
high salt buffers, non-ionic detergents and nucleases. 2. - The same
procedure as before but performing RNase digestion before extraction in high
salt buffer; 3. - Inducing reversible disulfide bonds by NaTT in isolated
nuclei and 4. - reducing the disulfide bonds with DTT in 2M NaCl at the end
of the extraction procedure. The residual DNA in nuclear matrices has been
monitored by DAPI fluorescence and EM immunocytochemistry with an anti-DNA
monoclonal antibody. Residual RNA was detected by pyronine fluorescence, as
well as preferential EM staining.

ULTRASTRUCTURAL ORGANIZATION OF PLANT NUCLEAR MATRICES

Our data demonstrate the existence of a proteinaceous residual structure

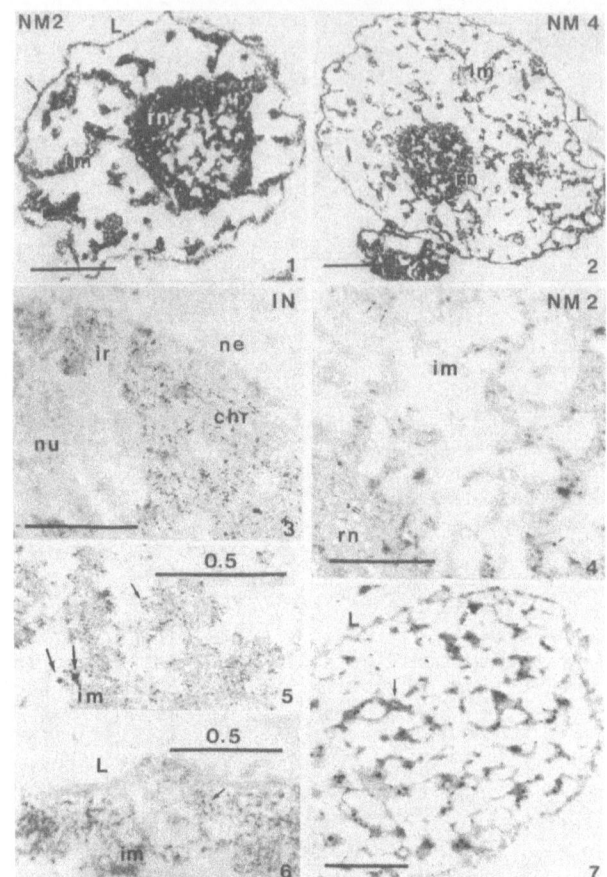

Figures 1 and 2, –EM images of nuclear matrices prepared from isolated
nuclei by procedures 2 and 4, respectively. In both cases
they reveal well-organized matrices with the three typical
components.

Figures 3 and 4, – Immunolocalization of DNA in isolated nuclei and NM2
respectively, using an anti DNA-Mab and gold coupled
secondary antibodies. Note that the im DNA (↑) does not
appear to be associated with the granules of the internal
matrix. Nucleolar-matrix associated DNA (↑↑).

Figure 5.   Cytochemical detection of RNPs by EDTA staining. Big
granules (↗) and also the 10-15nm granules (↑), but not the
fibrils of the internal matrix appear to be stained.

Figure 6.   The granules of the internal matrix (↗) are very conspicuous
when divalent cations are eliminated from the incubation
medium.

Figure 7.   Cytochemical detection of interchromatin granules and
phosphorylated proteins by bismuth oxinitrate staining. Note
that the stained granules of the matrix (↗) do not correspond
to the small ones detected by EDTA staining.

<u>Abbreviations</u>: L: lamina: <u>im</u>: internal matrix: <u>rn</u>: residual nucleolus: <u>nu</u>:
nucleolus: <u>ir</u>: interchromatin regions: <u>chr</u>: chromatin: <u>ne</u>:
nuclear envelope. When not indicated bars = 1μm.

in plant nuclei resistent to different procedures of extraction using low salt and high salt buffers, non-ionic detergents, digestion with nucleases, and alkylating agents, as has been reported in other eukaryotes (Berezney, 1984; Verheijen et al., 1988). Whatever the procedure used for the preparation of nuclear matrices, they showed a similar morphological organization at the EM level, with lamina, residual nucleoli and internal elements of the matrix (Figures 1 and 2) (Barthelemy and Moreno Diaz de la Espina, 1984). This pattern of organization is similar to that previously found in other cell types (Berezney, 1984; Verheijen et al., 1988). The lamina appears to be the most stable element of plant nuclear matrix, in contrast with a previous report on a plant nuclear matrix lacking a lamina (Ghosh and Dey, 1986). It appears in sections as a continuous network of 10nm fibres associated with nuclear pores (Figures 1 and 2). Residual nucleoli are, under the experimental conditions used, constant components of plant nuclear matrices. They are mainly composed of 10nm fibres, in contrast to the fibrillo-granular organization of the *in situ* nucleoli, and appear to be related to the elements of the internal matrix. Nevertheless, some authors claimed that the persistence of a residual nucleoli in nuclear matrices depends on the method of extraction. Especially following RNase digestion (Bouvier et al., 1982; Verheijen et al, 1988), we detected differences in the degree of localization of nucleolar components, after the different extraction procedures, but never a complete disorganization of the structure. The components of the internal matrix appear in sections to be formed by patches of 10nm fibrils and granules interconnected with each other, as well as with the lamina and residual nucleoli (Figures 1 to 7). The internal matrix is the most variable component regarding size and compactness of elements in the different extraction procedures. It is formed by 10nm fibrils and 10-25nm granules. The results of EDTA and Bismuth oxynitrate stainings reveal that some of the granules forming the internal matrix correspond to ribonucleoprotein particles and also contain highly phosphorylated proteins (Figures 5 and 7). These results corroborate previous work identifing the internal matrix with the residual RNP-network involved in hnRNA association and processing (Gallinaro et al., 1983).

POLYPEPTIDE COMPOSITION OF THE NUCLEAR MATRIX

The protein patterns of the nuclear matrices obtained by the four experimental procedures are complex, revealing many bands between 70 and 12 kD which vary according to the extraction procedure used. Matrices from procedures 3 and 4 present more bands than those obtained by the more extractive two first protocols. Some of the predominant bands are constant: the most conspicuous are those in the range of 60-40kD and 37kD. Histones also appear to be constant component of the matrices. These results agree with previous data in onion nuclear matrices prepared by a procedure similar to number 2 (Stolvarov, 1984).

DIFFERENT EFFECTS OF THE EXTRACTION PROCEDURES ON THE ORGANIZATION AND COMPOSITION OF NUCLEAR MATRICES

Although procedures 1 and 2 appear to extract more proteins from nuclei than the milder 3rd and 4th methods, the ultrastructural components of the internal matrix and residual nucleolus in nuclear matrix 1 and 2 appear a little collapsed at the EM level (Figure 1). We never found these components to be sensitive to RNase digestion, when digestion was performed before high salt extraction, as reported in fibroblasts and hepatocytes (Bourgeois et al., 1982; Kaufmann et al., 1981). This is in agreement with Stolyarov (1984), even when magnesium ions were eliminated from the nuclear incubation medium (Figure 10). Only by elimination of magnesium from all the extraction buffers does complete disruption of nuclear matrices occur

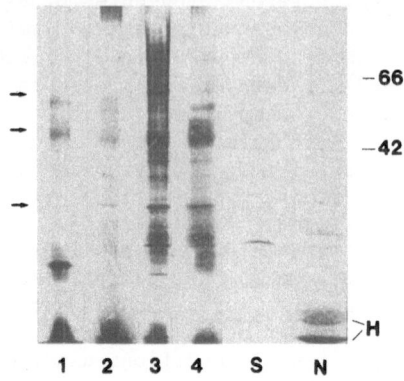

Figure 8.   Polypeptide composition of isolated nuclei (N) and nuclear
matrices obtained by the different procedures: Proc. 1. -
1st. - 0.25mM MgCl$_2$. 2nd. - 2M NaCl in 0.25mM MgCl$_2$. 3rd. -
0.5% Triton X-100 and 4th. - 200µg/ml DNase and RNase
respectively. Proc. 2. - 1st. - 0.5% Triton X-100. 2nd. -
200µg/ml DNase and RNase respectively. 3rd. - 0.25mM MgCl$_2$.
4th. - 2M NaCl in 0.25mM MgCl$_2$. Proc. 3. - 1st. - 2mM NaTT.
2nd. - 200µg/ml DNase and RNase respectively. 3rd. - 0.25mM
MgCl$_2$. 4th. - 2M NaCl in 0.25mM MgCl$_2$. Proc. 4. - 1st to
4th: the same as procedure 3, and 5th. - 2M NaCl. 0.25mM
MgCl$_2$ and 20mM DTT. as separated in 10% acrylamide gels. The
common bands are marked by arrows. S: Mw standards.

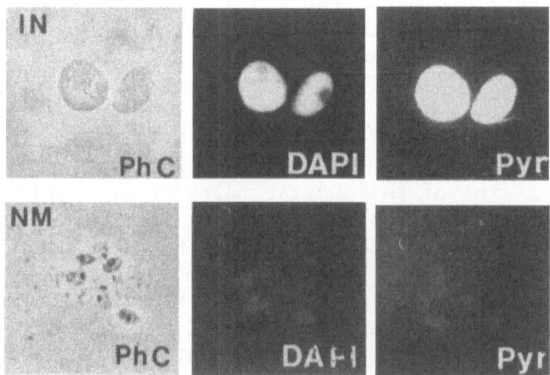

Figure 9.   Detection of DNA and RNA in isolated nuclei and nuclear
matrices by double staining with 10^{-5}M DAPI and 5.10^{-5}M
Pyronine; fluorescence at 365 v 546nm respectively. Micrographs from matrices are underdeveloped to show the structures.

(data not shown here). The formation of intermolecular disulfide bonds
induced by NaTT in isolated nuclei does not dramatically affect either
intranuclear organization or protein composition of plant nuclear matrices,
as compared with the DTT-reduced matrices, although the later do show a
looser ultrastructure and lack some protein bands (Figures 2 and 8). This
contrasts with the results obtained from rat hepatocytes (Kaufmann et al.,
1984), and implies that the internal elements of the matrix are not entirely
dependent on the formation of fortuituous disulfide bonds during extraction.

Although the residual DNA associated with the nuclear matrix (nmDNA) is not detectable by DAPI fluorescence, we were able to reveal it by using a much more sensitive approach: immunogold detection with an anti-DNA monoclonal antibody (Figures 3, 4 and 9), the amount of DNA detected in nuclear matrices is very much lower than that in isolated nuclei and DNA-rich nuclear matrices processed under the same conditions (Figures 3 and 4). The DNA appears to be associated with the internal matrix and, to a lesser extent with the residual nucleolus. This DNA could correspond to the specific nmDNA sequences which play a role in the topological organization and functioning of DNA (Razin, 1987), although a fortuituous trapping of DNA by nuclear proteins during the extraction cannot be completely discarded.

RNPs were found to be associated mainly with the internal matrix by preferential EDTA staining for ribonucleoproteins, but not by pyronine fluorescence after blocking with DAPI (Figures 5 and 9). The EDTA staining reveals a ribonucleoprotein network of small particles 10-20nm diameter which sometimes form beaded fibres (Figure 5). This network is also present in isolated nuclei as revealed by the same technique (Figure 7), and appears very conspicuous in isolated nuclear matrices when magnesium ions and diethylpirocarbonate are eliminated from the incubation medium of nuclei (Figure 6). This would correspond to the RNP network involved in hnRNA association and processing (Gallinaro et al., 1983).

Interchromatin granules (Medina et al., 1989) are also a part of the internal matrix as detected by bismuth oxynitrate staining, and as reported in animal cells (Krzyzowska Grucca et al., 1983). They show distribution within the matrices to be similar to that found in the interchromatin regions of isolated nuclei (Figure 7).

CONCLUSIONS

The present results confirm the existence of a residual nuclear matrix in plant cells, homologous to those previously described in other cell systems.

Whatever the extraction procedure used, the matrix has the same general organization with a lamina, residual nucleolus and internal matrix, indicating that all their elements are at least partially resistent to RNase digestion, and do not depend on the formation of fortuitous disulfide bonds during extraction, but they appear to be highly dependent on the concentration of magnesium ions during the extraction.

Plant nuclear matrices have a complex polypeptide composition which varies according to the procedure of extraction used, but a large majority of bands appear as constant components of the matrices with molecular weights 60-40kD and 37kD.

The internal matrix seems to be the site of DNA and RNA association, as revealed with anti-DNA monoclonal antibodies and EM-cytochemical detection of ribonucleoproteins. Interchromatin granules are also a part of the nuclear matrix.

REFERENCES

Barthelemy I., Moreno Diaz de la Espina S., 1984. Preliminary studies on the plant nuclear matrix ultrastructure. Cienc. Biol. 9:138.

Berezney R., 1984. Organization and functions of the nuclear matrix. in: 'Chromosomal non-histone proteins. Vol. IV'. L. S. Hnilica, ed CRD Press Boca Raton.

Bouvier D., Hubert J., Seve., Bouteille M., 1982. RNA is responsible for the three-dimensional organization of nuclear matrix proteins in Hela Cells. Biol. Cell. 43:143.

Gallinaro H., Puvion E., Kister L., Jacob M., 1983. Nuclear matrix and hnRNP share a common structural constituent associated with premessenger RNA. EMBO. J. 6:953.

Gosh S., Dey R., 1986. Nuclear matrix network in Allium cepa. Chromosoma 93:429.

Kaufmann S. H., Coffey D. S., Shaper J. A., 1981. Considerations in the isolation of rat liver nuclear matrix, nuclear envelope, and pore complex lamina. Exp. Cell. Res. 132:105.

Khrone G., Benavente R., 1986. The nuclear lamins. Exp. Cell Res. 162:1.

Kryzowska-Grucca S., Zborek A., Grucca S., 1983. Distribution of interchromatin granules in nuclear matrices obtained from nuclei exhibiting different degree of chromatin condensation. Cell Tissue Res. 231:427.

Medina M. A., Moreno Diaz de la Espina S., Martin M., Fernandez-Gomez M. E. 1989. Interchromatin granules in plant nuclei. Biol. Cell. 67:no 2.

Razin S. V., 1987. DNA interactions with the nuclear matrix and spatial organization of replication and transcription. Bio Essays. 6:19.

Stolyarov S. D., 1984. Isolation and characterization of the nuclear matrix of the onion Allium cepa. Cytologia. 26:874.

Verheijen R., Van Venrooij W., Ramaekers F., 1988. The nuclear matrix: structure and composition. J. Cell Sci. 90:11.

SYNTHESIS AND PHOSPHORYLATION OF THE 125 K NUCLEAR MATRIX PROTEIN

MITOTIN DURING THE CELL CYCLE

I. T. Todorov, N. Z. Zhelev, R. N. Philipova,
*V. Bibor-Hardy and A. A. Hadjiolov

Department of Molecular Genetics
Institute of Cel Biology and Morphology
Bulgarian Academy of Sciences
1113 Sofia, Bulgaria

* Institut du Cancer de Montreal
Hospital Notre Dame
Montreal, Quebec
Canada H21 4M1

The preparation of eukaryotic cells for entry into mitosis is related to a cascade of G2 phase phosphorylations of several nuclear proteins (Adlakha et al., 1985; Ottaviano and Gerace, 1985; Lee and Nurse, 1988) driven by mitosis specific protein kinases (Ha-leck et al., 1987; Dessev et al., 1988; Draetta and Beach, 1988).

Recently using a specific monoclonal antibody, we have identified in human cells a 125kD/pI6.5 protein associated with the intranuclear matrix that is present in proliferating, but not in quiescent cells (Philipova et al., 1987). This protein displays a speckled nucleoplasmic distribution throughout interphase and is markedly increased during late G2 and M pahses of the cell cycle and was designated as 'mitotin' (Todorov et al., 1988a). This antigen is also increased in meiotic cells during rat spermatogenesis (Hadjiolova et al., 1989). In the present work we studied the synthesis and turnover of mitotin in cell cycle synchronized human WISH cells. We studied further the *in vivo* phosphorylation of mitotin during S and G2 phases of the cell cycle.

The study of the immunocytochemical distribution of mitotin at different stages after release from the double thymidine block (Figure 1) shows that in S phase cells the antigen displays a speckled nucleoplasmic distribution (Figure 1A) and is markedly increased from the beginning and during mitosis (Figure 1C). At the ultrastructural level, using the immunogold technique in Lowicryl sections, we found the antigen to be located in small distinct centers (arrows) (Figure 2) corresponding to the speckles in the immunofluorescent microscopy. In S phase cells, these centers are localized in the nucleoplasm, associated probably with the interchromatin granules (Figure 2A). At the end of G2 phase the reaction is markedly intensified and a lot of granules could be found in the nucleolus (Figure 2B) suggesting a role of mitotin in nucleolar disintegration.

In a preliminary set of experiments we tried to quantitate the differences in the amounts of p125 in lysates from interphase and mitotic

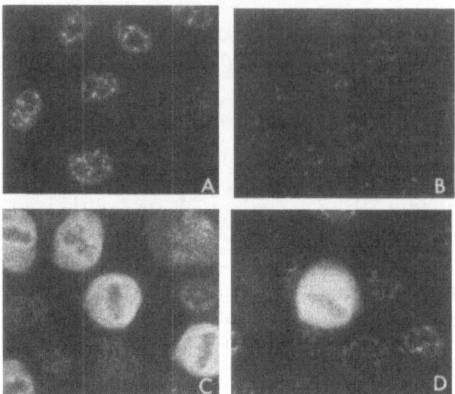

Figure 1.   Immunofluorescence analysis of synchronized WISH cells with
the anti-mitotin monoclonal antibody. A. Cells taken at 2h
after double thymidine block (corresponding to S phase of the
cell cycle); B. S phase cells treated with cycloheximide
(5µg/ml) for 2h; C. Cells taken at 9h after the block
(corresponding to G2/M and M phase); D. Cells in G2/M and M
phase treated with cycloheximide for 2h.

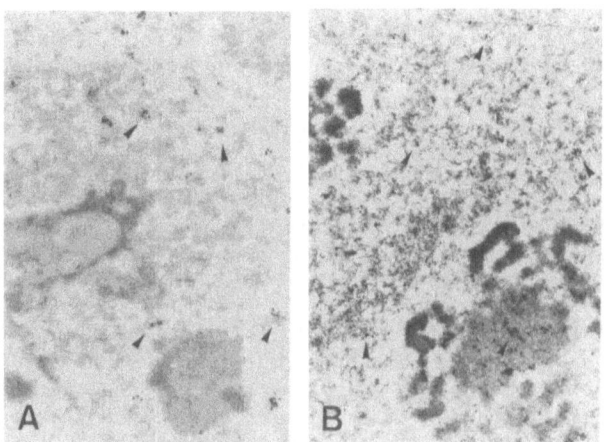

Figure 2.   Ultrastructural localization of mitotin in Lowicryl sections
of synchronized BHK cells using monoclonal antibody and
Protein A-Gold. A. S phase cells; B. G2/M cells.

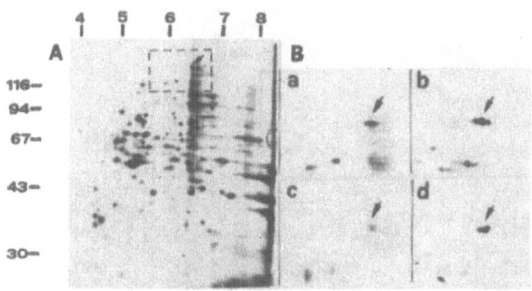

Figure 3.   Synthesis and turnover of mitotin during the cell cycle. A.
Autoradiography of a 2D acrylamide gel electropherogram of
total proteins from lysates of WISH cells labelled for 2h
with ^{35}S-methionine (1mCi/ml) in S phase (starting at 2h
after release from the double thymidine block); B. Magnifications corresponding to the boxed area from panel A:
(a) Cells labelled fro 2h in S phase (as in A); (b) Cells
labelled in S phase, as in (a), treated with cycloheximide
for 2h; (c) Cells labelled for 2h in G2 and G2/M phase
(starting at 8h after release from the block); (d) Cells
labelled in G2/M phase and treated with cycloheximide for
2h. Arrows indicate the position of mitotin - p125/6.5.

cells obtained as described earlier (Todorov et al. 1988a). Lysates from
equal numbers of cells were fractionated by 2D gel electrophoresis, stained
with Coomassie brilliant blue R250 and the absorbancy of the eluted mitotin
spots determined. The results show that the amount of p125 in mitotic cells
is about 7 fold higher than interphase cells. Unexpectedly, short-term
labelling with ^{35}S-methionine did not show such a dramatic increase, the
labelling of p125 in mitotic cells being only about 2-fold more intense than
in interphase cells. In order to clarify the reason for this discrepancy we
carried out experiments with WISH cells synchronized by a double thymidine
block (Figure 3). The results show that p125 is synthesized throughout the
cell cycle and the labelling in G2 phase is noticeably higher than in S
phase. These results are in line with experiments with a cDNA probe
showing that the amount of p125 mRNA is not more than 2 fold higher in G2,
as compared to S phase WISH and Raji cells (Todorov et al., 1988b).

We studied the stability of mitotin using cycloheximide block of the
protein synthesis. The treatment of synchronized WISH cells with
cycloheximide for 2h leads to a distinct decrease of the antigen (Figure
1B). However, this decrease is not so pronounced in the case of mitotic
cells (Figure 1D). Further, we studied the short-term turnover of mitotin
labelled with ^{35}S-methionine, upon 2h cycloheximide block of protein
synthesis. This turnover is very rapid in S phase cells and the protein
labelled for 2h is barely detectable after 2h cycloheximide. In contrast,
mitotin labelled during G2 phase decreases only slightly upon cycloheximide
treatment. These results demonstrate that mitotin synthesized during G2
and G2/M is metabolically stabilized, and the result of its continuing
synthesis is the recorded marked increase in the amount of the protein in
mitotic cells. Observation of the 2D gel electropherograms reveals that
mitotin labelled with ^{35}S-methionine during G2/M phase reproducibly
produced 3 spots instead of the single spot observed in S phase (see Figure
3B). The additional spots of p125 labelled during G2 phase are located at
slightly more acidic pI, suggesting the modification of this protein in premitotic and mitotic cells. The two more acidic spots appear to be
metabolically stabilized in cycloheximide-treated cells, thus suggesting the
possibility of mitotin phosphorylation during the G2 phase. We therefore

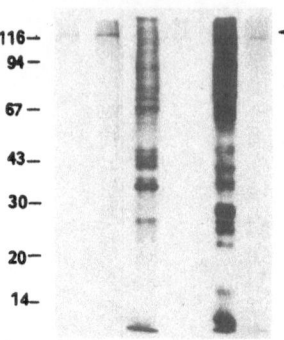

Figure 4.   SDS-polyacrylamide gel electropherograms of proteins
             immunoprecipitated with the anti-mitotin monocolonal antibody
             from lysates of WISH cells labelled during different phases
             of the cell cycle. A. Immunoprecipitated proteins from cells
             labelled with ^{35}S-methionine for 2h in S phase; B. Immuno-
             precipitated proteins from cells labelled with ^{35}S-methionine
             for 2h in G2/M phase; C. Total proteins from cells labelled
             with ^{32}P-phosphate (100μCi/ml) for 2h in S phase (starting
             at 2h after the block); D. Immunoprecipitated proteins from
             cells labelled with ^{32}P-phosphate in S phase (as in C); E.
             Total proteins from cells labelled with ^{32}P-phosphate for 2h
             in G2/M phase (starting at 8h after the block); F. Immuno-
             precipitated proteins from cells labelled with ^{32}P-phosphate
             in G2/M phase.

carried out experiments to study the phosphorylation state of mitotin at
different stages of the cell cycle.

    To obtain direct evidence for the phosphorylation of the protein in
mitotic cells we carried out comparative acrylamide gel electrophoresis
analysis of ^{35}S-methionine and ^{32}P-phosphate labelled proteins,
immunoprecipitated with the anti-mitotin monoclonal antibody (Philipova et
al., 1987). The results (Figure 4) show that after 2h labelling in S phase,
the antibody precipitates a ^{35}S-methionine labelled protein, but does not
precipitate any detectable phosphorylated product, while in G2 and M cells
a single phosphorylated polypeptide of about 125kD is observed. Upon 2D
polyacrylamide gel electrophoresis (Figure 5) the monoclonal antibody
immunoprecipitates three polypeptide products with an identical Mr of 125kD
and similar pI (Figure 5A), two of which are phosphorylated (Figure 5B).
It is noteworthy that the same polypeptides remain stable after cycloheximide block of the protein synthesis (see Figure 3b and 3d).

    The results obtained in this work demonstrate that in cultured human
cells, the nuclear matrix associated protein mitotin accumulates during late
G2 and G2/M phases of the cell cycle. Two phosphorylated froms of this
protein are present in mitotic cells and are not found in S phase cells.
These phosphorylated forms of mitotin are metabolically stabilized, thus
resulting in the marked accumulation of the protein in premitotic and
mitotic cells. The observed considerable (5 to 10 fold) phosphorylation-
related accumulation of mitotin in premitotic nuclei and mitotic cells
strongly suggests that this protein may be more directly involved in the
complex cascade of nuclear events preparing the cell for mitosis (Lee and
Nurse, 1988; Draetta and Beach, 1988).

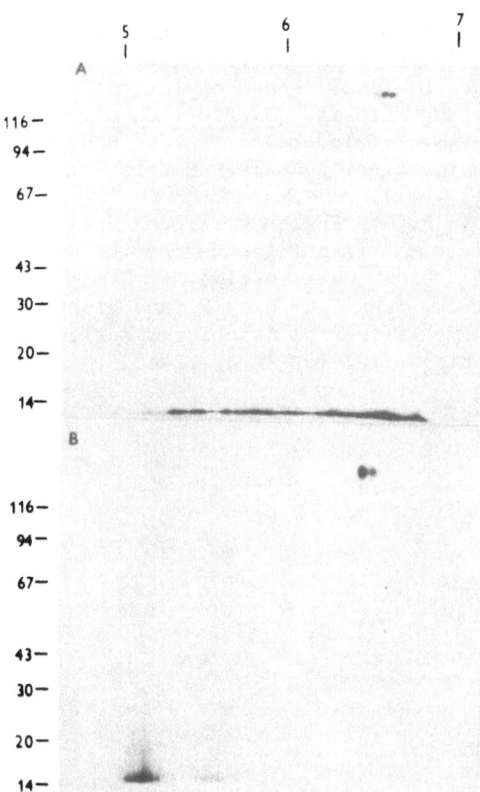

Figure 5.   2D acrylamide gel electrophoretic analysis of the proteins
             immunoprecipitated with anti-mitotin monoclonal antibody from
             lysates of WISH cells labelled as shown in Figure 4. A.
             Immunoprecipitated proteins from cells labelled with ^{35}S-
             methionine in G2/M phase for 2h; B. Immunoprecipitated
             proteins from cells labelled with ^{32}P-phosphate in G2/M phase.

REFERENCES

Adlakha R. C., Davis F. M. and Rao P. N. (1985) Role of phosphorylation of
    nonhistone proteins in the regulation of mitosis. In Control of
    animal cell proliferation (A. L. Boynton and H. L. Leifert, eds.) vol.
    1, pp. 488, Academic Press, N. Y.
Dessev G., Iovcheva C., Tasheva B. and Goldman R. (1988) Protein Kinase
    Activity Associated with the Nuclear Lamina. Proc. Natl. Acad. Sci.,
    USA, 85:2994.
Draetta G. and Beach D. (1988). Activitation of cdc 2 Protein Kinase during
    Mitosis in Human Cells: Cell Cycle-Dependent Phosphorylation and
    Subunit Rearrangement. Cell, 54:17.
Hadjiolova K. V., Martinova Y. S., Yankulov K. Y., Davidov V., Kancheva L.
    S. and Hadjiolov A. A. (1989) An Immunocytochemical Study of the
    Proliferating Cell Nuclear Matrix Antigen p125/6.5 during Rat
    Spermatogenesis. J. Cell Sci., 93:173.
Halleck M. S., Lumley-Sapanski K. and Schlegel R. A. (1987) Mitosis-specific
    Cytoplasmic Protein Kinases. In Molecular Regulation of Nuclear Events
    in Mitosis and Meiosis (Schlegel R. A., Halleck M. S., Rao P. N., eds),
    227, Academic Press, N. Y.

Lee M. G., and Nurse P. (1988) Cell Cycle Control Genes in Fission Yeast and Mammalian Cells. Trends in Genetics, 4:287.

Ottaviano Y., and Gerace L. (1985) Phosphorylation of the Nuclear Lamins during Interphase and Mitosis. J. Biol. Chem., 260:624.

Philipova R. N., Zhelev N. Z., Todorov I. T. and Hadjiolov A. A. (1987) Monoclonal Antibody Against a Nuclear Matrix Antigen in Proliferating Cells. Biol. Cell, 60:1.

Todorov I. T., Philipova R. N., Zhelev N. Z. and Hadjiolov A. A. (1988a) Changes in a Nuclear Matrix Antigen during the Cell Cycle: Interphase and Mitotic Cells. Biol. Cell, 62:105.

Todorov I. T., Lavigne J., Sakr F., Foisy S. and Bibor-Hardy V. (1988b) cDNA Encoding the Proliferation Associated Nuclear Matrix Protein Mitotin, J. Cell Biol., 107 (6, Pt 3), 744a.

IDENTIFICATION OF MYOSIN-LIKE PROTEINS IN CELL NUCLEI, THEIR INTERACTION

WITH CHROMATIN COMPONENTS

T. N. Priyatkina and O. R. Zarembskaya

Department of Biochemistry
Leningrad State University
199164 Leningrad, USSR

Contractile proteins, which were once considered to be specific
components of muscles, are now identified in different types of non-muscle
cells (Pollard, 1981). The variations in quantitative maintenance,
intracellular localization and organization of actomyosin structural
orientation in the cells of different origin and function indicate possible
diversity of the non-muscle functions of contractile proteins. The
evidence for components of the contractile complex being present in cell
nuclei is of great interest (Ohnishi et al., 1964; LeStourgeon et al.,
1975). The intranuclear localization of actin analogos has already been
substantially documented (Katsumaru, Fukui, 1982). The question of the
myosin component cannot, however, be considered completely settled. Due to
a number of difficulties in technique, myosins were not isolated and
studied as whole molecules in their native form. Thus their degree of
structural homology of muscle and non-muscle myosins is so far unknown. The
lack of sufficiently pure samples of the myosin-like protein is the main
obstacle in determining their intranuclear localization by means of
immunochemical techniques.

We developed procedures for purification of myosin-like protein (MP)
from rat liver chromatin, electrophoresis in PAG of semipreparative amounts
of the native myosin and also its two-dimensional electrophoresis. Using
these methods we identified a nuclear protein as a structural analog of
myosin and studied its content in chromatin at different levels of
transcription activity.

The myosin-like proteins were isolated from liver chromatin, dissolved
in non-ionic 0.25M sucrose by sonication (4 $w \cdot cm^{-2}$, 60 s, 22 kHz). The
following scheme of purification was used: the chromatin was precipitated by
means of adding $MgCl_2$ at a final concentration of 0.005M (or NaCl - 0.15M)
and the pellet was extracted by 0.3M NaCl, 0.005M $MgCl_2$, 0.01M PP_i, 0.01M
Tris-HCl buffer (pH 8.0). To remove actin and other ballast proteins,
ammonium sulphate was added at final concentration 20% (w/v). The myosin-
like proteins were precipitated from the supernatant by decreasing ionic
strength, during dialysis against bi-distilled water. This method produces
purification of the myosin-like protein by 30-40 times (determined by ATP-
ase activity) from its original level in nucleus (Figure 1). There was
about 2% of the protein and 50% of the K^+-EDTA-ATPase activity present in
the initial chromatin preparations. The specific K^+-EDTA-ATPase activity
was about 0.3-0.4mkM $PP_i \cdot mg^{-1}$ of protein$\cdot min^{-1}$. The component which

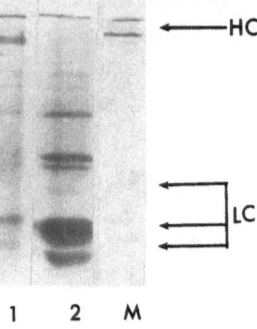

Figure 1.   Electrophoregrams of proteins from rat liver chromatin in
             dissociating conditions.
             1 - myosin-like protein from chromatin.
             2 - total soluble chromatin.
             M - rabbit skeletal muscle myosin.
             HC, LC - high and light chains, respectively.

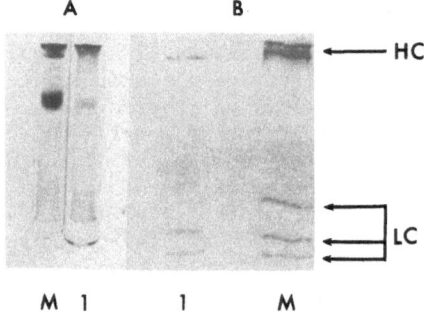

Figure 2.   Two-dimensional electrophoresis of myosin-like protein from
             rat liver chromatin.
             A - first direction - non-dissociating conditions. Staining
             of ATPase activity.
             The proteins were solubilized in electrophoretic buffer (0.05M
             Tris-glycine, 5mM ATP), pH 8.5 with 0.08M PP_i; PAG: T-4%,
             C-2.5%. For each sample a quantitiy of 40µg was applied;
             B - the second direction - in the presence of SDS. Protein
             staining. (The gel pieces containing proteins were cut out
             and then applied to the SDS-gel).
             1 - myosin-like protein from chromatin;
             M - rabbit skeletal muscle myosin;
             HC, LC - high and light chains, respectively.

increases during purification, simultaneously with the level of ATPase
activity, according to densitometry, corresponds by its electrophoretic
mobility (in the presence of SDS) to skeletal muscle myosin high chains (data
are not presented here). The electrophoretic data in non-dissociating
conditions (Figure 2A) and two-dimensional electrophoresis (Figure 2B)
revealed that ATPase activity of MP is caused by one component, which has the
same migration rate in gels as muscle myosin and dissociates in the presence
of SDS into heavy subunits (co-migrating with heavy chains of muscle myosin)
and light ones. The latter are of two types (Figure 2B).

Table I.   Some Properties of Chromatin from Nuclei with
            Transcribed and Nontranscribed DNA.

	Erythrocytes		Liver cells	
	hen*	frog	rat	mouse
Soluble chromatin (DNA %)	55,0	94,0±0,9	90,0±2,0	80,0±7,8
$\frac{Protein}{DNA}$	1,3	1,1±0,7	2,4±0,4	2,2±0,6
$\frac{RNA}{DNA}$	0,03	0,02±0,01	0,07±0,02	0,09±0,02
K^+-EDTA-ATPase activity ($mkM\ P_i \cdot min^{-1}$ per DNA)	0,032	0,021±0,008	0,065±0,026	0,083±0,008

* - mean data from three experiments are presented.

The soluble chromatin was obtained after sonication of nuclei in
non-ionic 0,25 M sucrose and removing the pellet by centrifugation
(5000 g, 20 min).

Thus the protein isolated from liver chromatin exhibits basic
properties of myosin analogs: high specific K^+-EDTA-ATPase activity, the
ability to be revealed in gels by means of Ca^{2+}-ATPase activity, the same
electrophoretic mobility in non-dissociating conditions and typical myosin
subunit structure.

The amount of MP associated with chromatin apparently depends on the
level of transcription. Thus the content of MP (according to K^+-EDTA-ATPase
activity per DNA) in liver chromatin was 2-3 times higher than that of inert
chromatin of erythrocytes. The K^+-EDTA-ATPase activity per RNA was
approximately constant and comprised about 1 $mkM \cdot min^{-1}$ (Table I). The
sonication of liver nuclei in non-ionic sucrose solution produces homogeneous
20-nucleosomes in length soluble chromatin fragments, with which more than
80% of nuclear MP is associated. Their further digestion by Ca^{2+}, Mg^{2+}-
endonuclease (Hewish, Burgoyne, 1973) into 1-6 nucleosomal fragments
(according to electrophoretic data, not presented here), leads to
precipitation of a DNP fraction in non-ionic solution (about 30% on DNA
scale). More than 70% of both the MP and RNA of the initial chromatin were
found in this fraction.

We assume that the distribution of MP along the nucleohistone fibrils
has a discontinuous pattern and coincides with the distribution of RNA. The
results obtained confirm once again the prescence of myosin-like components
in cell nuclei and their possible involvement in the intranuclear organization or transport of RNA. Further investigations are clearly required to
provide conclusive answers to this question.

REFERENCES

Hewish D. R., Burgoyne L. A. 1973, Chromatin substructure. The digestion
    of chromatin DNA at regularly spaced sites by a nuclear deoxyribo-
    nuclease, Bioch. Res. Commun., 52:504.

Katsumaru H. and Fukui Y. 1982, In vivo identification of Tetrahymena actin probed by DMSO induction of nuclear bundles, Exp. Cell Res., 137:353.

Lestourgeon W. M., Forer A., Yang Y., Bertram J. and Rusch H. 1975, Contractile proteins. Major components of nuclear and chromosome nonhistone proteins, Biochim. Biophys. Acta., 379:529.

Ohnishi T., Kawamura H. and Tanaka Y. 1964, Die Aktin und Myosin annlichen proteins im Kalbsthymus Zellkern, J. Biochem., 56:6.

Pollard T. D. 1981, Cytoplasmic contractile proteins, J. Cell Biol., 91:156s.

DNA SEQUENCES NONRANDOMLY DISTRIBUTED IN RESPECT TO RAT LIVER NUCLEAR

MATRIX

J. Rzeszowska-Wolny, J. Lanuszewska and J. Rogolinski

Institute of Oncology
Department of Tumour Biology
44 - 100 Gliwice
Poland

INTRODUCTION

Eukaryotic DNA is organized into loops which from functional domains. Two main types of loop configuration seem to exist: 1) compact, where DNA is packaged into higher order chromatin structures and interacts with the nuclear matrix only at loop bases; and 2) extended into internal nuclear matrix by dynamic DNA-protein interactions (i.e. 'matrix attached' configuration). It has been proposed that gene activation is directed by a multistage process begining with loop uncoiling and extension, then followed by action of specific transcription factors (Bodnar 1988). In this work we identified genes present in both types of configurations in rat liver cells.

MATERIALS AND METHODS

Nuclear matrix, bulk chromatin and total nuclear DNA fractions of rat liver cells were isolated according to Berezney et al. (1974) with modifications as described elsewhere (J. Rzeszowska-Wolny et al., 1988). DNA was isolated and labelled according to methods described by Maniatis et al. (1982). The distribution in total, matrix and bulk chromatin DNA fractions of a particular sequence was assayed on the basis of DNA-DNA dot-blot hybridization. The following plasmids were used as labelled radioactive probes: p-fos-1 (v-fos sequence), pM29 (v-myc gene), pM1.8 (hsp70 gene), pDHFR11 (DHFR gene), pMtk1 (TK gene), pKC1 (KC gene), pAEPst0.45 (v-erbA gene), pAeBam0.5 (v-erB gene), pHM2A (c-mos gene) and pSMTC (v-fms gene).

RESULTS AND DISCUSSION

Dot-blot analysis of DNA present in total, nuclear matrix and bulk chromatin fractions of rat liver cells is shown on Figure 1. DNA of these three fractions was tested for the presence of sequence coding for proto-oncogenes fos, myc, erbA, erbB, mos, fms, and dihydrofolate reductase (DHFR), thymidine kinase (TK), gene KC and heat shock protein 70 related (hsp 70-like). Six out of the ten genes under study (fos, myc, DHFR, TK, KC, and hsp 70-like) have been enriched in matrix attached fraction,

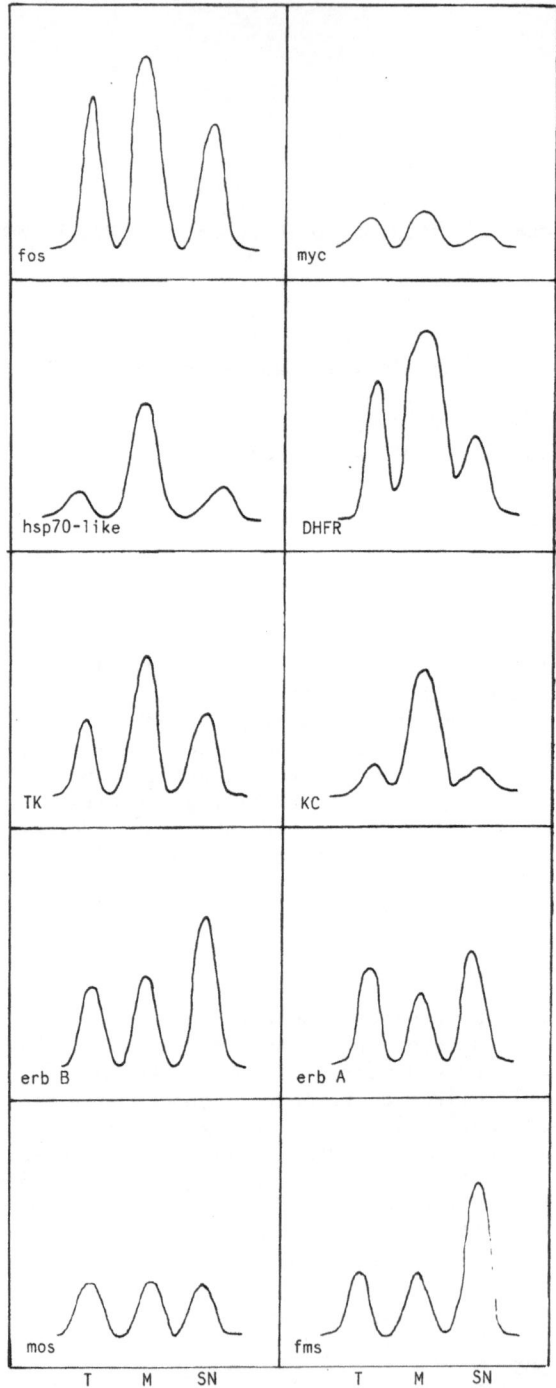

Figure 1.   Densitometry tracings of DNA dot-blot assays of different coding sequences present in total (T), nuclear matrix (M), and bulk chromatin (SN). Each dot contained 1μ of DNA. The sequences used as radioactive probes are marked under each densitogram.

Table I

Gene	Matrix attachment	Transcription in normal rat liver *	Transcription during liver regeneration *
fos	+	-	+
myc	+	-	+
hsp70	+	+	+
DHFR	+	-	+
TK	+	-	+
KC	+	ns	ns
erbA	-	-	-
erbB	-	-	-
mos	-	-	-
fms	-	ns	ns

Relation between matrix attachment and transcriptional activity in normal and regenerating rat liver

*   Biesiada 1989

ns   not studied

suggesting that in rat liver they are present in loops with extended conformation. Other proto-oncogenes such as erbA, erbB, mos, and fms were found in the bulk chromatin fraction, not attached to nuclear matrix. These sequences probably occur in loops of compact configuration.

According to the domain model (Bodnar, 1988), actively transcribed DNA sequences should occur in loops extended into nuclear matrix and attached to it. However, transcripts of c-fos, c-myc, DHFR, TK and KC as well as of erbA, erbB and c-mos were not detected in normal rat liver (Biesiada 1989).

Table I compares the occurance in the 'matrix attached' configuration with transcriptional activity in normal and regenerating rat liver of genes under study. Genes which have been found in matrix attached loops are either transcribed in normal rat liver (as hsp70-like) or induced for transcription during liver regeneration.

Rat hepatocytes start to proliferate in a cynchronized way in response to partial hepatectomy. Regenerating liver exhibits a specific pattern and timing of gene induction. Proto-oncogenes c-fos and c-myc are activated during first hour after hepatectomy (Biesiada 1989). It was postulated that proteins encoded by these genes take part in activation of other genes necessary for proliferation (Kaczmarek 1986). Among genes attached to nuclear matrix, there are the examples of both, early and late activated in liver regeneration. This result could suggest that different 'cascade' events controlling transcription in cell cycle act only on those loops which previously passed the uncoiling and extension step.

REFERENCES

Berezney R., and Coffey D. S. 1974, Identification of a nuclear protein matrix, Biochem. Biophys. Res. Commun., 60:1410.

Biesiada E. 1989, Gene expression in regenerating rat liver, PhD Thesis, Gliwice.

Bodnar J. W. 1988, A domain model for Eukaryotic DNA organization: a molecular basis for cell differentiation and chromosome evolution, J. Theor. Biol., 132:479.

Kaczmarek L. 1986, Protooncogene expression during the cell cycle, Lab. Invest., 54:365.

DNA-FOLDING BY A STABLY DNA-LINKED PROTEIN IN EUKARYOTIC CHROMATIN

Zoya Avramova, Peter Petrov and Roumen Tsanev

Institute of Molecular Biology
Bulgarian Academy of Sciences
1113 Sofia
Bulgaria

Higher levels of structural organization of DNA is achieved through its interaction with different proteins. These interactions are abolished by reagents destroying noncovalent associations, hydrogen bonding and S-S bridges. However, in chromatin of different origin a fraction of nonhistone proteins was found whose association with DNA could not be disrupted by such reagents (Lesko and Emery, 1966; Krauth and Werner, 1979; Neuer et al., 1983; Avramova and Tsanev, 1987). The presence of such a protein fraction was proved by *in vitro* iodination and by *in vivo* incorporation of labelled aminoacids. This firmly bound protein fraction showed several unusual properties: 1) After iodination it could not enter the polyacrylamide gels upon electrophoresis. This made impossible the estimation of its molecular mass; 2) The two-dimensional tryptic peptide map of the iodinated protein isolated from eleven different chromatins - Drosophila, fish, frog and rat liver, chicken erythrocytes, rat and ram sperm, Guerin tumor cells, mouse Friend cells, maize leaves and roots - showed a practically identical pattern (Avramova et al., 1989a, b), revealing its high evolutionary conservation; 3) By *in vivo* labelling of DNA and of the protein it was found that this protein was metabolically stable, transmitted to the progeny like DNA (Avramova et al., 1988a); 4) Chemical and enzymic analysis of the DNA-protein linkage of the stable complex have suggested a bond of a phosphodiester type.

The identity of the peptide maps of the protein obtained from different species, its presence in sperm chromatin devoid of RNP network and most of the nuclear proteins, argue against the possibility that this protein fraction may arise from a randon artifactual trapping of nuclear proteins. It would be of great interest to see how this protein is distributed along the DNA. To this end we have used a somewhat unusual approach. It is known that the presence of proteins associated with DNA would fold the latter into flower-like figures (rosettes) when spread in the presence of cytochrome C (Sonnenbichler, 1969; Leon and Macaya, 1983; Zatsepina et al. 1983). It is still highly disputable whether these formations reflect some structures *in vivo* or are artifacts. However, if they are artifacts they should regarded as 'informative' artifacts, showing the frequency and the location of the sites where DNA is folded into loops of a definite size. It was of interest to see whether extensively deproteinized DNA, containing the stably DNA-linked protein only, would be able to fold into such rosettes when spread according to the routine technique (Kleinschmidt, 1968). Figure 1a,b

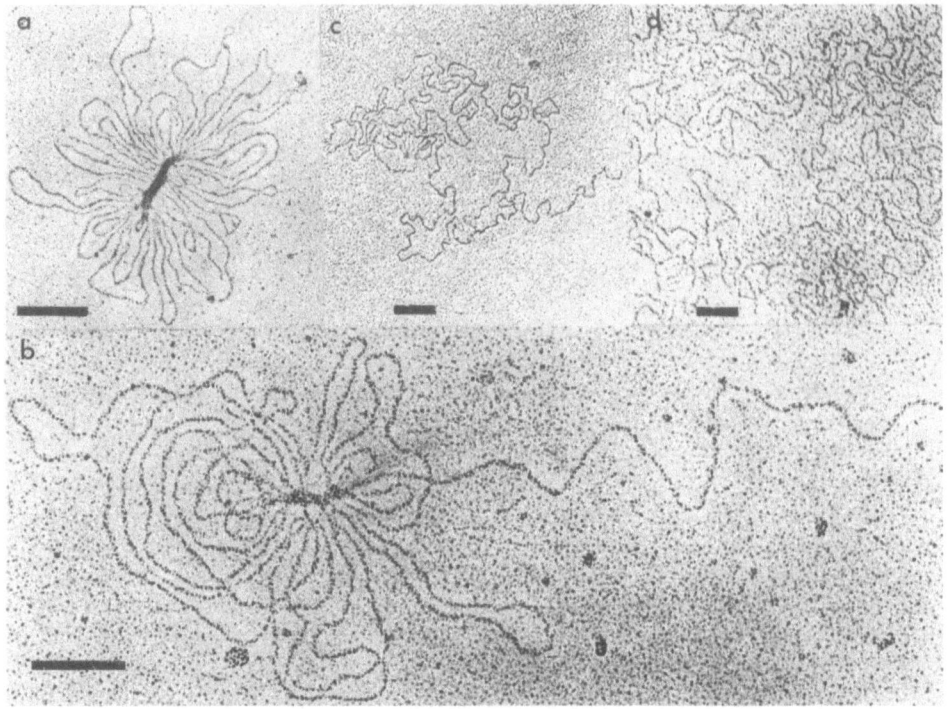

Figure 1.   Electron micrographs of DNA containing the stably sinked
             protein fraction: (a and b) - spread in the presence of
             cytochrome C; (c) - the same DNA as in (a) adsorbed on the
             grid without cytochrome C; (d) - DNA pretreated with pronase
             E and spread on the grid as in (a). Scale bars indicate
             160nm.

show that in the presence of cytochrome C such a DNA folds in the form of
'rosettes' with a well visible central core. Omission of cytochrome C
prevents their formation (Figure 1c). The same was also observed when DNA,
spread in the presence of cytochrome C has been pretreated with pronase
(Figure 1d). The effect of the pronase treatment demonstrates once more
the presence of a residual protein, stably bound to DNA. This confirms the
conclusion that protein-protein interactions are responsible for the folding
of DNA into flower-like structures. Thus, our data show that the presence
of the stably DNA-linked protein is the minimum requirement for folding the
DNA into rosettes. The suggestion (Comings and Okada, 1976) that remnants
of the nuclear matrix are necessary to anchor DNA loops in such structures
is evidently not justified.

    According to the current notion, DNA in interphase nuclei and in
mitotic chromosomes is organized into loops 10-100kbp long, which are not
only separate structural units but represent individually controlled domains
of the genome. However, the size of the individual loops in a rosette is
only 1-10kb. This suggests that the sites of the stably linked protein do
not correspond to the anchorage sites of the chromosomal loops. Long
stretches of linear DNA may be observed between individual rosettes (Figure
1b; Sonnenbichler, 1969), indicating that the sites where the stably linked
protein is located are not evenly distributed along DNA. The average
distance between the stable DNA-protein complexes as indicated by the

flower-like figures is about 3-4kbp. This raises the question whether the protein is really linked to DNA via a phosphodiester bond. Such a bond would lead to nicks in DNA at a frequency which is not revealed experimentally (not shown). Two possibilities should be considered: either the protein is laterally attached through another type of a linkage without breaking the DNA backbone, or whether the protein complexes are much more rarely located but serve as aggregation sites, non-specifically trapping DNA within the large chromosomal loop.

Concerning the old dispute as to whether the rosettes are true structural units corresponding at a cytological level to chromomers (Sonnenbichler, 1969; Comings and Okada, 1978; Okada and Comings, 1980; Zatsepina et al., 1983) our results seem to favour the opposite view (Leon and Makaya, 1983). However, the readiness with which the rosettes are induced by a small fraction of firmly DNA-bound proteins is suggestive of the ease with which such loops may form *in vivo*. This raises the question of a mechanism preventing the strong tendency of DNA for 'collapsing', thereby keeping the chromatin loops in an extended form. The anchorage of a loop to skeletal structures was suggested to keep it in an extended (open) configuration (Gross and Garrard, 1987).

In conclusion, the properties of the stable DNA-protein complexes suggest they have an important role in the functioning of the genome. This may be connected with their ability to form local functional mini-loops within the large chromatin loops.

REFERENCES

Avramova Z., and Tsanev R. 1987, Stable DNA-protein complexes in eukaryotic chromatin, J. Mol. Biol., 196:437.
Avramova Z., Mikhailov I., and Tsanev R. 1988a, Metabolic behaviour of a stable DNA-protein complex, Int. J. Biochem., 20:61.
Avramova Z., Ivanchenko M., and Tsanev R. 1988b, A protein fraction stably linked to DNA in plant chromatin, Plant Mol. Biol., 11:401.
Avramova Z., Mikhailov I., and Tsanev R. 1989, An evolutionarily conserved protein fraction stably linked to DNA in eukaryotic chromatin, Biochim. Biophys. Acta, 1007:109.
Comings D. E., and Okada T. A. 1976, Nuclear Proteins III. The fibrillar nature of the nuclear matrix, Exp. Cell Res., 103:341.
Gross D., and Garrard W. T. 1987, Poising chromatin for transcription, Trends Biochem. Sci., 12:293.
Kleinschmidt A. 1968, Monolayer techniques in electron microscopy of nucleic acid molecules in: 'Methods in Enzymology' 12B, L. Grossman and K. Moldave, eds., Academic Press, New York and London.
Krauth W., and Werner D. 1979, Analysis of the most tightly bound proteins in eukaryotic DNA, Biochim. Biophys. Acta, 564:390.
Lesko S. A., and Emery A. J. 1966, Subunit form of calf thymus DNA, Biochem. Biophys. Res. Commun., 23:707.
Leon P., and Macaya G. 1983, Properties of DNA rosettes and their relevance to chromosome structure, Chromosoma, 88:307.
Neuer B., Plagens U., and Werner D. 1983, Phosphodiester bonds between polypeptides and chromosomal DNA, J. Mol. Biol. 164:213.
Okada T. A., and Comings D. E. 1979, Higher order structure of chromosomes, Chromosoma, 72:1.
Sonnenbichler J. 1969, Nucleoprotein complexes: Possible subunits of chromosomes, Hoppe-Seyler's Z. Physiol. Chem., 350:761.
Zatsepina O. V., Polyakov V. Yu., and Chentsov Yu. S. 1983, Chromonema and chromomere. Structural units of mitotic and interphase chromosomes, Chromosoma, 88:91.

STRUCTURAL CHANGES OF NUCLEOSOMES AT THE SITES OF NON-RIBOSOMAL RNA

TRANSCRIPTION

Roumen Tsanev

Institute of Molecular Biology
Bulgarian Academy of Sciences
1113 Sofia
Bulgaria

The problem of what happens to nucleosomes during transcription is still unsolved. Transcription with RNA polymerase II has been studied mainly by nuclease digestion and electron microscopy, with conflicting results: i.e. absence or presence of nucleosomes (see reviews: Tsanev, 1983; Reeves, 1984; Tsanev and Tsaneva, 1986). These varying results were explained by: the varying organization of different genes; the presence of cells with an active and cells with an inactive gene; the temporary elimination or conformational changes of nucleosomes. Apparent changes in the nucleosomal organization may be induced by the spreading conditions. The observation that under the same conditions neighbouring transcribing and non-transcribing fibers show different nucleosomal organization may be due to a lower stability of transcribed nucleosomes, moreover the effect of stretching forces on the nucleosomes has not been studied.

Thermodynamic considerations (Ausio et al., 1984; Bina et al., 1980) indicate that within the range of ionic conditions used for spreading, the electrostatic interactions between histones and DNA may be easily overcome by stretching forces. We have observed a substantial loss of nucleosomes during centrifugation when heavy particles pull the chromatin fiber (Tsanev and Tsaneva, 1986). Heavy transcripts should exert the same effect, so that fibers bearing transcripts would be stretched more severely and may lose nucleosomes.

The question arises as to exactly what happens when the polymerase interacts with a nucleosome. Experiments *in vitro* have again lead to conflicting results - both preservation (Losa and Brown, 1987) and displacement (Lorch et al., 1987) of nucleosomes has been reported. Nucleosome preservation was suggested to be an exception due to a stronger interaction of histones with some DNA sequences (Lorch et al., 1988). However, our electron microscopic observations have shown that nucleosomes may be present at a normal density even on heavily transcribed fibers (Tsanev and Tsaneva, 1986). Thus, we have to accept that the situation *in vivo* may be quite different from transcription *in vitro*.

The large polymerase molecule prevents us from seeing whether a nucleosomal particle may be present at the base of a transcript. To eliminate the polymerase without losing nucleosomes we have used treatment with urea. Urea is known to induce a reversible unfolding of the

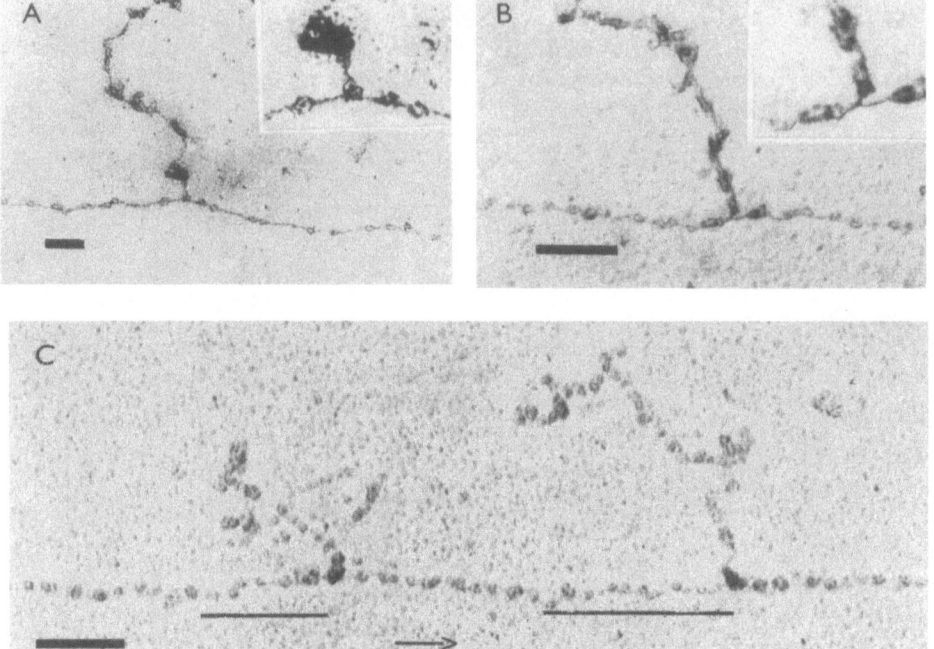

Figure 1.   Electron micrographs of rat liver chromatin (A and B) -
spread through a layer of 6M urea and washed in water. The
insets are enlarged pictures of the anchorage sites of
transcripts - on a nucleosomal particle (A) or on DNA between
particles (B). (C) - chromatin spread in the absence of urea
showing half-nucleosomes behind transcripts (underlined).
Arrow shows the direction of transcription. Scale bars
indicate 74nm.

nucleosomes preserving the contacts of histones with DNA, even in 6M
solutions (Zayetz et al., 1981). On the other hand we have shown that
concentrations of urea higher than 4M lead to dissociation of the enzyme.
We took advantage of this differential effect of urea to eliminate RNA
polymerase II by spreading chromatin through a layer of 4M to 8M urea and
then replacing the urea with water. This procedure revealed at the bases of
some transcripts, small particles identical with the neighbouring nucleosomes. Anchorage of transcripts was observed both on the particles and
between them (Figure 1A, B). Our observation with mammalian chromatin is in
agreement with results obtained with transcribing SV40 minichromosomes
crosslinked with psoralen (De Bernardin et al., 1986). All these data show
that the nucleosomes remain at the sites where RNA polymerase II transcribes
DNA. Thus, we have to consider structural changes which do not eleminate
the nucleosome. Different changes have been suggested: 'unfolding' of the
nucleosome, its opening into two halves or a temporary association of the
histone core with the noncoding strand. The latter possibility was
rejected by experiments with DNase I digestion of the highly transcribed
yeast galactokinase gene, which showed the same 10b repeat in the two
opposite strands (Lohr, 1983). The other possibility, the opening of the
nucleosome into two halves, was shown to occur under some ionic conditions
(Tsanev and Petrov, 1976; Oudet et al., 1977). It was suggested that half-
nucleosomes might be important for transcription (Weintraub et al., 1976).

This hypothesis was rejected by showing that oxidation of the SH groups to form a -S-S- bridge did not prevent *E. Coli* polymerase from transcribing reconstituted H1-free oligonucleosomes (Gould et al., 1980). However, artifacts from the possible generation of free DNA and the use of the much more efficient prokaryotic polymerase make it difficult to accept the validity of this result, especially for the eukaryotic enzyme. Moreover, unpublished experiments by Wu (see also Wu et al., 1979) have shown a low transcription efficiency with dimethylsuberimidate crosslinked nucleosomes.

The mechanism of nucleosome opening is supported by our observation that under normal spreading conditions half-nucleosomes are often present immediately behind a transcript (Figure 1C) or nucleosomes are absent. The latter may be explained by the loss of histone cores when the nucleosome is opened and DNA is uncoiled during centrifugation. These structural changes of the nucleosomes behind transcripts are suggestive for an 'opening-closing' mechanism. Such a mechanism is supported by the finding (Bode et al., 1983) that nucleosomes open as the histones become hyperacetylated (a transcription-related modification) and by the uncovering of the SH groups in active nucleosomes (Sterner et al., 1987). This dynamic model should include four steps: 1. Opening into two halves; 2. Transcription of the first half by a partial local uncoiling-recoiling of DNA; 3. Transcription of the second half; 4. Closing of the two halves.

This model predicts that most of the DNA-histone contacts will be preserved; the DNase I 10b pattern of the two opposite strands would be practically the same; the half-nucleosomes could be attacked internally by micrococcal nuclease; transcribed fibres would more easily lose histone cores during spreading; corsslinking of the two tetramers should prevent transcription at least with RNA polymerase II.

REFERENCES

Ausio J., Seger D., and Eisenberg H. 1984, Nucleosome core particle stability and conformational changes, J. Mol. Biol. 176:77.
Bina M., Sturtevant J. M., and Stein A. 1980, Stability of DNA in nucleosomes, Proc. Natl. Acad Sci. USA, 77:4044.
Bode J., Gomez-Lira M. M., and Schroter H. 1983, Nucleosomal particles open as the histone core becomes hyperacetylated, Eur. J. Biochem., 130:437.
Lohr D. 1983, The chromatin structure of an actively expressed, single copy yeast gene, Nucl. Acids Res., 11:6755.
Lorch Y., LaPointe J. W., and Kornberg R. 1987, Nucleosomes inhibit initiation of transcription but allow chain elongation with displacement of histones, Cell, 49:203.
Lorch Y., LaPointe J. W., and Kornberg R. 1988, On the displacement of histones from DNA by transcription, Cell, 55:743.
Losa R., and Brown D. D., 1987, A bacteriophage RNA polymerase transcribes in vitro through a nucleosome core without displacing it, Cell, 50:801.
Sterner R., Boffa L. C., Chen T. A., and Allfrey V. 1987, Cell-cycle dependent changes in conformation and composition of nucleosomes containing human histone gene sequences, Nucl. Acids Res., 15:4375.
Tsanev R., and Petrov P. 1976, The substructure of chromatin and its variations as revealed by electron microscopy, J. Microsc. Biol. Cell. 27:11.
Tsanev R., and Tsaneva I. 1986, Molecular organization of chromatin as revealed by electron microscopy, in: Meth. Achiev. exp. Pathol., 12: 63, G. Jasmin and R. Simard, eds., S. Karger, Basel.
Weintraub H., Worcel A., and Alberts B. 1976, A model of chromatin based upon two symmetrically paired half-nucleosomes, Cell, 9:409.
Wu H. -M., Dattegupta N., Hogan M., and Crothers D. M. 1979, Structural

changes of nucleosomes in low-salt concentrations, <u>Biochemistry</u>, 18:3960.

Zayetz V. W., Bavykin S. G., Karpov V. L., and Mirzabekov A. D. 1981, Stability of the primary organization of nucleosome core particles upon some conformational transitions, <u>Nucl. Acids Res.</u>, 9:1053.

A FRAGMENT OF CHICKEN NUCLEAR MATRIX–ASSOCIATED DNA CAN MAINTAIN AUTONOMOUS

REPLICATION OF PLASMIDS IN MAMMALIAN CELLS

Y. S. Vassetzky Jr., A. G. Kakandedze, E. G. Kintsurashvili,
K. T. Turpaev, N. F. Grinenko*, A. D. Altstein* and
S. V. Razin

Institute of Molecular Biology,
USSR Academy of Sciences,
32 Vavilov St.,
Moscow 117984, USSR.
* Institute of General Genetics,
USSR Academy of Sciences,
Moscow, USSR.

The data on the existence of different types of association of DNA with the nuclear matrix have been obtained recently (van der Velden et al, 1984, Smith et al, 1984, Jackson, 1986, Razin, 1987). In functionally active nuclei, most specific interactions of DNA with the nuclear matrix elements are connected with transcription (Jackson, 1986, Razin, 1987). However a specific group of DNA associations with nuclear matrix remains in inactive nuclei of mature erythrocytes (Razin et al, 1985). We designated them as permanent attachment sites. Further studies demonstrated that the permanent attachment sites were located near replication origins on the DNA chain (Razin et al, 1986). In the present paper we tried to characterize DNA sequences involved in organization of permanent attachment sites.

CLONING AND SEQUENCING OF SHORT ERYTHROCYTE nmDNA FRAGMENTS

We have previously found that DNA fragments permanently attached to the nuclear matrix (nmDNA) represented only a small fraction of unique sequences of total checken DNA (Razin et al, 1986). One might expect that these fragments have some common features important for their anchorage on the nuclear matrix. We have cloned short erythrocyte nmDNA fragments selected on the basis of prominent hybridization with purified fraction of chicken replication origins (Kalandadze et al, 1988).

Surprisingly, we did not find any pronounced similarity or consensus common to all the sequenced nmDNA fragments (Table I). However, five out of ten sequenced fragments were more or less GC-rich and had multiple imperfect internal repeats (Figure 1) and a number of putative binding sites for different sequence-specific DNA-binding proteins was detected in the sequenced fragments (Table II).

Table I.   Sequences of the Cloned Short Chicken nmDNA
            Fragments.

CO30   160 bp
ATCCGGGGGG GGTCGTGGGG GGGGGGAGTC CCCAATTCCA AGGGGGGGGT CGGAGGGTGG
GGGGGGTCCC CATTCCAAGG GGGGGGCCGG AAGGGGGAGG GTCCCCATTT CTCAGGGGGG
GGTCGGAGGG GGGGGTCCCC ATTCCCAACG GGGGAGGGGG

CO326   489 bp
GGAGGTGCGG GGAAGGGGCC CGAGGAGTTG GGGTGGGGGG GGGGGGAGTT GGGGTTCTTT
GGGGGTGGGT TTGGGGGTCG CGGTCCGCGC TCACCCGTCA GAAGGCCGAG GAGGAGCAGC
GCGGGGCCGC AGAAGGCCAT GGCGGGCCGG GCCCGGTTCG GGCCTCTTCG GGGGGACCCT
TCAGAGCGCT TCGGGGTTCT TCGGGTCCGG TTCGGGTCCG TTCGGCTGAC GTTCGGGTCT
CCGCGACCGC TTCGGGTGTC CTTCGGGTCC GTTCGGCGGC GTTCGGGTCA CTGCAGATCA
CTTCGGGTGT CCTCGGGTCC GGTTCGGGTC CGTTGGCCGG CGTTCGGGTG ACCGCAGAGA
GCTTCGGGTC TCTTCGGGTC CGTTCGGCGG CGTTCGGGTC ACTGCAGACC GCTTCGAGTC
TTTTCGGGTC CGGTTCGGGT CCGTTCGGCG GCGTTCGGGT CACTGCAGAC AACTTCGGAA
CTCTTCGGG

CO33   172 bp
GATCCCTTCT TTTACAAAAT TAAGTCTGCC TAGCTCCATA TCTTTCTTTT GATTAGGTGT
TATTTTAAAG GAGCTTTTCA TGCCACCTAT CCTTCCAATG TTACTCTTTC TGGTTTTTTC
CCCTCTTTGC ACTACTTGCA AAGATGACTT GAAGGAGTTC TATTTTAGAG TG

CO34   173 bp
GAGCTCGGT ACCCGGGGAT CCGGAATCAT AGTTTATCAG CAGCTTCCTC AAAAGCAGCAC
AGGAACTGCT GGGTAATCTA CAGCCTGTTA CTCTAAAGAC ACCCAATGAT TTTTTCTTAC
ACTGTGGACT GCTTACAAAG ACTTTCACAC TACTTCACAT GGATCCTCTA GAG

CO41   140 bp
ACTCACACAT GACAGCTCAG AAATAAAATA AAAACCTGCT TACCCCTGAA AGGCATGTAA
GGACCTTAGA CAGGCAGGGA TCTCTAAACA AGATACTCTT GCTTCCACTG CCAACCCTTT
AATGTTATCT GGGAAGGGGT

CO56   318 bp
GCAGTCCGGT GGCACTGCAG CTGCAGCAGT GGGTGGGAGAG GTAATTATCT CACTGTACTG
GGTGCTCATT GAATCCCGTT TAGAATATTG CATCCGGTTT GGGGCTCTCT GTTGCAACAA
AAATGTCATA AATTGGAGCG AGCTCGGGGG TATGCCCCAA GATGTTTCCC TGAGAGGAAA
GGCTGAGGGA ACTGGGTTTG TTCAGGGTGG GGAGTAGATG GCTTTGGGGA ACCCAAGAGC
AACTTGCCTG TACCTGTGAG GGGGTCAGCA AGGAGTGATG CTCACTGCGC ATCCATTGCC
TGTGGGAGAA GGAGGCCT

CO62   220 bp
GGAACGGAAC GGAACGGAAC GGAACGGAGC GGAGCGGAGC AGAGATGCCA CCGCCGCCGT
CGGCCCGCCG TGACGCCGTG CGCCGGCGGG GGCGGCGGGG AGAGCCGGGG GGGGAACCGC
AGTGCCGCAG CTCCGCCGGT ACCGGAACCC CCGGAGAGGG TGCGTGGCAC GGGGGGGTGC
GGGAGGCGAG GGGGTGGCGG GAGGGCACGG CAGCAGGGGG

CO65   318 bp
CGAGCCGGTA CCCGGGGACC CAAACTGGCA AAGGGTGCAG GTGAGGATTG TCCCAACCAG
CAGTTCTGCG AGCTGTCAGG CGCTGGTAGG CAGGGAGAGT TAAAAAGGAA TTAAGTGATT
AGAAATGCAT AAATATTAAT TGCTTATGGA CTCCCTTTAT GACACTCTCA GGTGGAGGAA
GTTAATAAAT CACTGCTGGC ATTGACCCAC AAAACAATAC AGAGTGTGTC TGAGACACAG
CGAGGGGAGG AGGGGGGAGG CCTTTCTGGA GCACGTGAAA GCGCTCTTTT TAGGTCAGGG
AGTGACACAA CAGAGCTG

CO73   517 bp
CCCCCCAGTG GGGAGGTCCT GTGGGGGGGT TTGGGGGGGT CCTAGGGAGG TTTTGGGGGG
GATAGTGGGG TTTGGGGGGG TCCTATGGGG GATGGGGGGG TTCTGGAGGG GTCCAAGGGG
GGATAGTGGG GTTTGGGGGG GTCCTATGGG GAATGGGGGG GTCCCTGAGG GGTCCAAGGG
GGGATAGTGG GGTTTGGGGG GGCTGGGTCC GTTGGGGATG GGTAGGAATG CGGCCCACTG
TGAGGTCACG GGGTGACTGT GACTCTTCGG TGCCATCACG AAGCACACAG ATAGGTGGAG
GTTTGCTGCC CCGTGGCTAT TGCGTCAGTA TTTTGGGCAT TTGCCTGCCA GAGTGCACCC
GAGGACTTTC AGTGGGCCAT TTTCCCAAGC AGTGACAAAT AAAGGATTCT TGAAGCAAGC
AGTAGAGCAG CTTTGGTGTT GCCGTTGATC AGTGCAGTAG CTGTGACCGT GCCGGTATTT
GATAGTAGTG TCAGCAAAGT AAATGCTGTG CTGGGTA

CO77   297 bp
AAAAAAGAGC CTTCGGTGAT GGCAGGAGGG GAGACGTACG GATGGAGTCT TGCAAGAGGG
ATGCTGCCAT CCCATCCCAT AGAGGCAGAG CAAAGTGGCC ATCTTTCCTC ACCTTGTTAT
TTGAATGTTG ACTAGGTGTT CACTGTTTCA AATTTACTAT CAGGGTAATG ATAATCGTAT
TCATGATTAA GGATCTTTTT TTGGCATTTG AGCTTCTACT CTCTGTGTTA ATGCATCAAA
TCCATAGCAG CAAAGCTTTC AACCTAAGAG TTCTGATGAA TCCACAAGCT GTTACAG

CO 30   CO 73

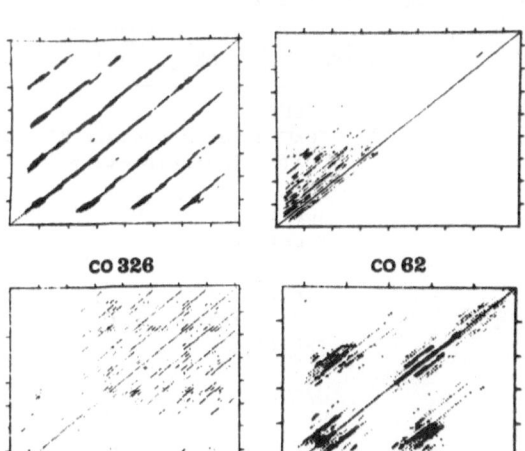

CO 326   CO 62

Figure 1.   Dot matrix map of internal homologies in the cloned nmDNA
             fragments. The frame width is 17/31.

nmDNA CLONES SPECIFICALLY BIND TO PROTEIN EXTRACTS FROM PROLIFERATING CELLS

     The next series of experiments was designed to find whether any
specific DNA-binding protein from actively replicating cells actually could
interact with the nmDNA fragments. Protein extracts from actively growing
(embryonic chicken fibroblasts) and resting (chicken hepatocytes) cells
were incubated with the cloned fragments of nmDNA in the presence of
competitor DNA. Specific binding was monitored using band-shift assay
(Fried and Crothers, 1981). Six of the ten nmDNA fragments, namely CO326,
CO33, CO34, CO56, CO62, and CO77 specifically interacted with protein
extracts from proliferating but not resting cells (data not shown).

     Using the DNase-I footprinting technique we have found a specific
binding site for the CO326 fragment. It turned out to be a novel DNA
recognition pattern:

$$GTC^A_TC^T_CCGCAGA$$

The CO326 insertion contains five similar binding sites (F1-F5) for a
protein factor from chicken embryonic fibroblasts (Figure 2). Computer
analysis revealed recognition sites for this factor in several other cloned
nmDNA fragments, namely CO56 and CO34.

A PLASMID, CONTAINING A FRAGMENT OF NUCLEAR MATRIX-ASSOCIATED DNA CAN
PERSIST AS AN EPISOME IN MAMMALIAN CELLS

     Binding of protein extracts from replicating cells to nmDNA fragments
only indicates possible connection between these fragments and replication.
To obtain more valid data we tested whether the nmDNA fragments were cloned
into a pBR327 plasmid containing truncated HSV thymidine kinase gene
(Holst et al., 1988) and transfected into RAT-1 cells. Clones that could
grow on HAT medium were selected and analyzed. The concept was that only the
presence of ca. 20 plasmids per cell could allow production of a sufficient
amount of thymidine kinase and, hence make a cell viable (Holst et al,
1988). The original report of these authors on the identification of

347

Table II.   Computer-Derived Protein-Binding Sites in the
             Cloned nmDNA Fragments.

CLONE	BINDING SITES FOR:
CO30	C/EBP ~CAC-binding protein (2 sites) ~CATT-binding protein
CO326	ADR-1 T-antigen
CO33	C/EBP TGGCA-protein (CTF) ~NFIII
CO34	C/EBP
CO41	CTF
CO47	CTF (2 sites)
CO56	ADR-1 (2) C/EBP T-antigen binding site CTF
CO62	ADR-1 Sp1 CTF (2)
CO65	ADR-1 T-antigen binding site CTF
CO73	DTF-1 CTF (2)
CO77	ADR-1 ANTP octamer B1a binding site Progesterone response element CTF (3) ~H1 box ~NFIII

The above sites have 100% or at least 85% (~) homology

autonomously replicating sequences turned out to be a misinterpretation of experimental data. However, the approach and the plasmid construct used in their work is above suspicion.

In our experiments, the construct containing a CO77 insertion repeatedly gave positive results. A cell strain that could grow on HAT medium was selected and analyzed. An episome could only be detected in Hirt cell extracts (Figure 3), and no traces of integration could be observed (data not shown). We estimate the plasmid copy number to be *ca.* 10-15 copies per cell.

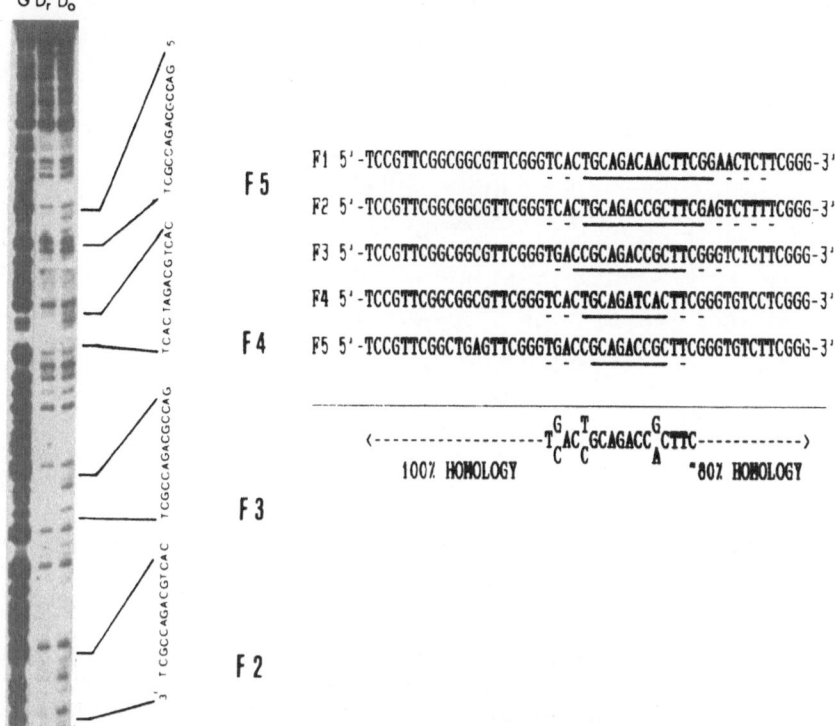

F1 5'-TCCGTTCGGCGGCGTTCGGGTCACTGCAGACAACTTCGGAACTCTTCGGG-3'

F2 5'-TCCGTTCGGCGGCGTTCGGGTCACTGCAGACCGCTTCGAGTCTTTTCGGG-3'

F3 5'-TCCGTTCGGCGGCGTTCGGGTGACCGCAGACCGCTTCGGGTCTCTTCGGG-3'

F4 5'-TCCGTTCGGCGGCGTTCGGGTCACTGCAGATCACTTCGGGTGTCCTCGGG-3'

F5 5'-TCCGTTCGGCTGAGTTCGGGTGACCGCAGACCGCTTCGGGTGTCTTCGGG-3'

$$\langle\text{--------------------}T{}^{G}_{C}AC{}^{T}_{C}GCAGACC{}^{G}_{A}CTTC\text{-----------}\rangle$$

100% HOMOLOGY   ~80% HOMOLOGY

Figure 2. Footprinting of a novel protein binding site on a cloned chicken erythrocyte nmDNA fragment. The end labelled CO326 insertion was incubated with the protein extract from embryonic chicken fibroblasts and then briefly treated with DNase I. Then the protein-associated fragments were separated from free DNA by preparative electrophoresis and applied to sequencing gel. G, standard G cleavage pattern; Dr, DNase I cleavage pattern of protein-bound fragment; Do, DNase I cleavage pattern of the free fragment. Underlined nucleotides in the sequence (right) correspond to protein-protected DNA areas.

Several experiments have been carried out in order to test the nature of this episomal DNA. We tested it for DpnI/MobI sensitivity and found out that it was DpnI-resistant and MboI-sensitive (Figure 3). It means that this episomal DNA replicated in eukaryotic cells. Transformation of *E. coli* with the DpnI-treated Hirt extracts of transfected RAT-1 cells allowed us to obtain bacterial strains carrying the initial plasmid. Hence, the nature of the observed episome leaves no doubts. Besides, the episome could still be observed several months after it was found and it can persist in non-selective conditions for at least several cell generations.

ACKNOWLEDGEMENTS

We would like to thank Drs. V. V. Lobanenkov and V. V. Adler for their valuable help and consultations in gel retardation and footprinting experiments.

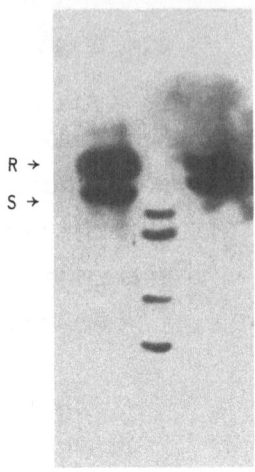

Figure 3. Blot hybridization of Hirt extracts of RAT-1 cells
transfected with the construct, containing truncated HSV
thymidine kinase gene and a CO77 cloned nmDNA fragment, three
months after transfection and growth on HAT medium. 1,
Mixture of supercoiled and relaxed input DNA; 2, Hirt extract
treated with MboI; 3, Hirt extract treated with DpnI. S and
R - positions of supercoiled and nicked circular plasmid DNA,
respectively. The whole (^{32}P)-labelled plasmid was used as
a probe.

REFERENCES

Fried M., and Crothers D. M. 1981, Equilibrium interactions of lac
    repressor-operator interactions by polyacrylamide gel electrophoresis,
    Nucl. Acids. Res., 9:6505.
Grummt F. 1989, Autonomous replication in mouse cells: a correction, Cell,
    56:143.
Jackson D. A. 1986, Organization beyond the gene, Trends in Biol. Sci.,
    11:249.
Holst A., Muller F., Zastrow G., Zentgraf H., Schwender S., Dinkl E., and
    Grummt F. 1988, Murine genomic DNA sequences replicating autonomously
    in mouse L cells, Cell, 52:355.
Razin S. V., Rzheshovska-Volni I., Moreau J., and Scherrer K. 1985,
    Localization of DNA attachment site to the nuclear matrix within the
    domain of chicken α-globin genes in functionally active and inactive
    nuclei, Moleklarnaya Biologia USSR, 19:456 (in Russian).
Razin S. V., Kekelidze M. G., Lukanidin E. M., Scherrer K., and Georgiev G.
    P. 1986, Replication origins are attached to the nuclear skeleton,
    Nucl. Acids Res., 14:8189.
Razin S. V. 1987, DNA interactions with the nuclear matrix and spatial
    organization of replication and transcription, BioEssays, 6:19.
Smith H. C., Puvion E., Buchholtz L., and Berezney R. 1984, Spatial
    distribution of DNA loop attachment and replicational sites in the
    nuclear matrix, J. Cell Biol., 99:1794.
Umek R. M., Linskens M. H. K., Kowalski D., and Huberman J. A. 1988, New
    beginning in studies of eukaryotic replication origins, Biochem.
    Biophys. Acta, 1007:1.
van der Velden H. M., Willigen G., Wetzels R. H. W., and Wanka F. 1984,
    Attachment of origins of replication to the nuclear matrix and the
    chromosomal scaffold, FEBS Lett., 171:13.

HISTONE H1-SPECIFIC AND DNA-ACTIVATED PROTEINASE IS ASSOCIATED WITH THE

NUCLEAR MATRIX FROM RAT LIVER

A. I. Gasiev, M. P. Kutsyi and A. Yu. Martynov

Institute of Biological Physics
Academy of Sciences, USSR
Pushchino
Moscow Region 142292, USSR

INTRODUCTION

It was shown that proteinase capable of degrading histone H1 in the presence of denatured DNA or DNA containing single-strand breaks is associated with the nuclear matrix (NM). The histone H1-specific proteinase is dissociated from the NM by 0.5% Triton X-100 and 0.5M NaCl. This proteinase is inhibited by antipain, leupeptin, phenylmethylsulfonyl floride (PMSF), and dithiothreitol (DTT).

Earlier we have shown that the NM of rat liver possesses a proteinase activity capable of degrading the substrate casein (Kutsyi et al., 1987). In order to understand the functions of this proteinase, we studied the specificity for DNA-binding proteins. Here we examined a histone-specific proteinase, contained in rat liver NM.

MATERIALS AND METHODS

Highly polymeric DNA from calf thymus, circular DNA of the pBR322 plasmid, DNAase I, antipain, leupeptin, PMSF, DTT and Tritron X-100 (Sigma) were used.

Rat liver nuclei were obtained according to Blobel and Potter (1966). NM was isolated from the nuclei by a procedure described by Beresney and Coffey (1977). DNA denatured by heating at 100°C for 10 minutes followed by rapid cooling. To induce single-strand break, DNA was treated with DNAase I (Aposhia and Kornberg, 1962). DNA solutions were γ-irradiated in a ^{60}Co installation at a dose rate of 19.8 Gy/min and UV-irradiated (254nm) at a dose rate of 2.3 j/m^2sec.

The reaction mixture (0.1ml) for analysis of proteinase activity contained 40nM Tris-HCl buffer, pH 8.0, 20μg of NM protein, 100μg of total histones, 50μg of DNA. The mixture was incubated at 37°C for 8h. After incubation, aliquotes of the samples were taken for electrophoresis of histones (Laemmly, 1970).

In order to dissociate the proteins possessing proteinase activity from the NM, NM preparations were treated with a solution containing 0.5% (V/V)

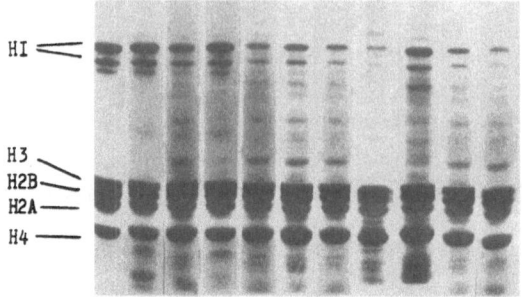

Figure 1.   Electrophoregrams of total histones incubated with NM
preparations in the presence of DNA. 1 - intact histones;
2 - histones incubated without DNA; 3 - in the presence of a
high-polymer thymus DNA; 4 - in the presence of cyclic DNA
pBR322; 5 - in the presence of denatured thymus DNA; 6 - 8 -
in the presence of γ-irradiated DNA (6 - 100Gy; 7 - 250Gy;
8 - 500Gy); 9 - in the presence of UV-irradiated DNA (200j/m);
10 - 11 - in the presence of DNA treated with DNAase I for
5 min and 10 min, respectively.

Triton X-100 and 0.5M NaCl. After treatment, the NM preparations were
centrifuged and the proteinase activity was determined in the supernatant
aliquotes and in the residual NM.

RESULTS AND DISCUSSION

     The NM preparations obtained contained 7.5 to 8.0% of protein and 0.8
to 1.0% of DNA of their initial content in the nuclei, as described by
Kutsyi et al. (1987). We analyzed the histone-specific proteinase using
total histones as a substrate. The analysis results are presented in Figure
1. As seen from these data, incubation of total histones with NM resulted
in partial hydrolysis of core histones H2A, H2B, H4, H3 rather than histone
H1. Addition of native thymus DNA to the incubation mixture somewhat
inhibited the proteolysis of core histones, at the same time, a partial
degradation of histone H1 was observed. DNA plasmid pBR322 did not markedly
affect the hydrolysis of histones by NM-associated proteinase. However,
after adding denatured DNA to the reaction mixture an almost complete
hydrolysis of histone H1 was observed (Figure 1). This points to the fact
that a histone H1-specific and denatured-DNA-dependent proteinase was
associated with the NM. It may be suggested that this proteinase does not
cleave histone H1, either in the free state or in complex with denatured
or structurally-damaged DNA. Therefore, we also used DNA treated with
γ-radiation, UV-radiation, and with DNAase I. The results show that when
γ-irradiated DNA was added to the reaction mixture containing histones and
NM, proteolysis of histone H1 took place. It is seen that here the activity
of the proteinase depended on the dose of γ-irradiation of DNA. However,
the UV-irradiated DNA did not affect the proteolysis of histone H1 by NM-
associated proteinase. At the same time, the DNA containing breaks induced
by DNAase I was able to activate the histone H1-specific proteinase of NM.
In this case the degree of proteolysis of histone H1 in this reaction
increases with increasing time of treatment of DNA with DNAase I (Figure 1).
Analysis showed that the histone H1-specific proteinase of NM has an optimum
pH 8.0 and its activity is not increased by adding Ca^{2+} or Mg^{2+} ions to the
reaction mixture (data not shown).

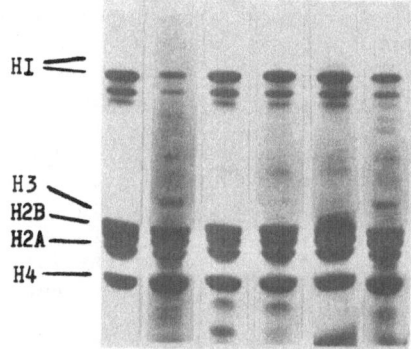

Figure 2.   The effect of inhibitors of proteinases on the proteolysis of
             histone H1 by NM proteinase. 1 - intact histones; 2 - his-
             tones incubated with NM and denatured DNA without proteinase
             inhibitors; 3 - histones incubated in the presence of anti-
             pain; 4 - in the presence of leupeptin; 5 - in the presence
             of PMSF; 6 - in the presence of DTT. All inhibitors were
             used at a concentration of 5mM.

Figure 2 presents the data on the effect of various compounds on the
known to be specific inhibitors of serine proteinases (Powers and Harper,
1986). This proteinase is also inhibited by DTT, a compound promoting
cleavage of protein disulfide bonds.

The histone H1-specific proteinase was not dissociated from NM by 2M
NaCl. However, on combined treatment of NM with 0.5% Triton X-100 and 0.5M
NaCl, part of NM material went to the soluble fraction. This fraction contained some of the DNA and of the NM proteins and possesses histone H1-spe-
cific proteinase activity (data not shown).

Thus, we demonstrated that a histone H1-specific proteinase which is
active only in the presence of denatured DNA or DNA containing single-strand
breaks induced by DNAase I or γ-radiation is tightly associated with the rat
liver NM. It may be suggested that histone H1-specific proteinase is
involved in the regulation of DNA repair and gene expression. It is likely
that the cleavage of histone H1 bound to damaged DNA provides the access on
enzymes to the sites of DNA repair synthesis within chromatin. At present
it is well known that the repair synthesis of DNA in the internucleosomal
linker sites of the chromatin to which histone H1 is bound occurs much
faster compared to that on the DNA within nucleosomal particles (Smerdom and
Lieberman, 1978). It should be also noted that in UV- and γ-irradiated rat
hepatocytes the unscheduled DNA synthesis at the initial stages occurs
essentially at higher rates in NM-bound DNA sites than on the bulk nuclear
DNA (Bezlepkin et al., 1985). It is suggested that the histone H1-specific
proteinase is indirectly involved in the regulation of DNA repair synthesis.

REFERENCES

Aposhia A. V., and Kornberg A., 1962, Enzymatic Synthesis of Deoxyribonucleic
    Acid. IX. The polymerase formed after T2 bacteriophage infection of
    E. coli: A new enzyme, J. Biol. Chem., 237:519.
Bezlepkin V. G., Malinovskij Yu. Yu., Velcovsky V., and Gaziev A. I., 1985,
    The preferential initiation of unscheduled DNA synthesis at the nuclear

matrix of irradiated cells, Dokl. Akad. Nauk SSSR, 283:461.

Berezney R., and Coffey D. S., 1977, Nuclear matrix. Isolation and characterization of a framework structure from rat liver nuclei, J. Cell Biology, 73:616.

Blobel G., and Potter V. R., 1966, Nuclei from rat liver. Isolation method that combines purity with high yield, Science, 154:1662.

Kutsyi M. P., Malakhova L. V., and Gaziev A. I., 1987, The proteinase activity of rat hepatocyte nuclear matrix, Biokhimja, 52:1315.

Laemmly U. K., 1970, Cleavage of structural proteins during the assebly of the head of bacteriophage T4, Nature, 227:680.

Powers J. C., and Harper J. W., 1986, Inhibitors of serine proteinases. In: 'Research Monographs in Cell and Tissue Physiology. Proteinase inhibitors', J. T. Dingle and J. L. Gordon, eds., Elsevier, Amsterdam, New York, Oxford, 12:55.

Smerdon H. J., and Lieberman N. M., 1978, Nucleosome rearrangement in human chromatin during UV-induced DNA repair synthesis, Proc. Natl. Acad. Sci. USA, 75:4238.

HIGH MOLECULAR WEIGHT PROTEINS OF TUMOR NUCLEAR MATRIX

I. B. Zbarsky, S. N. Kuzmina, T. V. Buldyaeva and
T. M. Bazarnova

N. K. Koltzov Institute of Developmental Biology
Academy of Sciences of the USSR
117334 Moscow, 26 Vavilov Street
USSR

In our earlier studies it was shown that nuclei isolated from various tumor cells contained much more alkali-insoluble protein material than those of corresponding normal tissues (Zbarsky and Debov, 1948). Correspondingly, in the nuclear matrices of several hepatomas and other tumors about three-fold more alkali-insoluble polypeptides were revealed than in nuclear matrix of normal liver (Kuzmina et al., 1981).

Electrophoretic study of tumor nuclear matrix showed that it contained more high molecular weight polypeptides than its normal counterpart. This feature was more prominent in solid tumors than in ascites tumor cells (Kuzmina et al., 1984). This pecularity was not revealed in non-tumor proliferating cells. However, in both tumor and non-tumor proliferating cells the nuclear matrix is characterized by predominance of lamin B peak at the expense of decreased or even absent peaks of lamins A and C (Figure 1).

High molecular weight proteins of the nuclear matrix are characterized by high biosynthesis and turnover rate. Thus, after incubation of Zajdela ascites hepatoma cells with labelled *Chlorella* protein hydrolysate the specific activity of the high molecular weight protein group was much higher than that of other nuclear matrix proteins (Figure 2). By comparing incorporation curves after 15, 30 and 60 minutes of incubation the radio-activity of the high molecular weight protein group increased progressively, while that of low molecular weight polypeptides diminished. This result may indicate that high molecular weight proteins could be assembled by some kind of processing, e.g. by polymerisation of low molecular weight polypeptides.

The incorporation of labelled amino acids into the high molecular weight protein group was selectively inhibited by chloramphenicol; the incorporation into most other proteins of the nuclear matrix being practically unaffected with the exception of low molecular weight polypeptides, biosynthesis of which was slightly inhibited (Figure 3).

It is now generally accepted that protein biosynthesis proceeds only in cytoplasm by the ribosomal pathway. However, in earlier studies a considerable incorporation of labelled amino acids into proteins of isolated

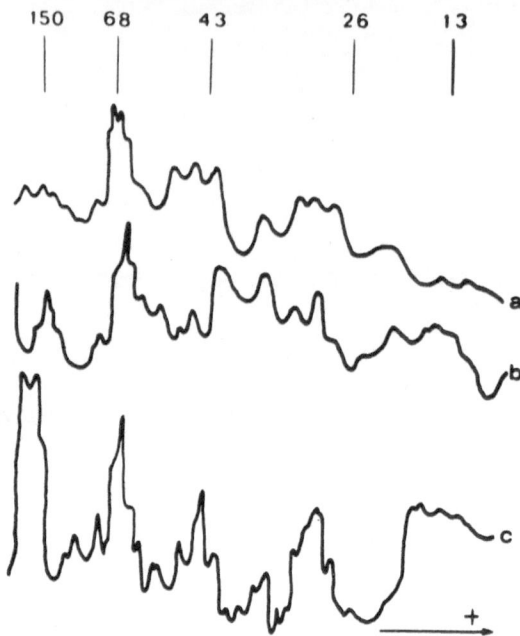

Figure 1.   Densitometric tracing of nuclear matrix proteins: (a) - Rat liver, (b) - Zajdela ascites hepatoma, (c) - Solid hepatoma 27. Above - Molecular weight standards.

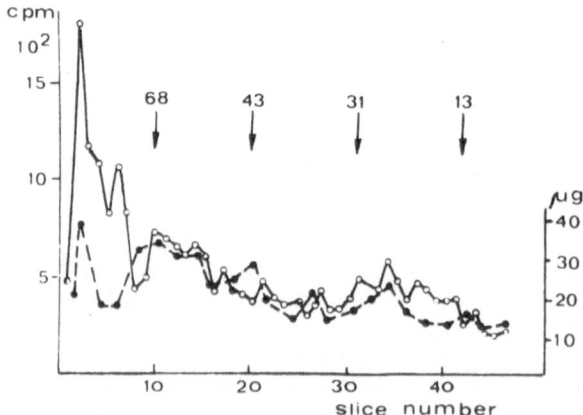

Figure 2.   Radioactivity (o) and protein content (●) in 48 slices of 12% SDS-polyacrylamide gel after electrophoresis of Zajdela ascites matrix proteins. Hepatoma cells were incubated with /^{14}C/ Chlorella protein hydroysate at 30°C for 30 min.

nuclei devoid of cytoplasm was observed repeatedly. Contrary to cytoplasmic eukaryotic protein synthesis, this nuclear protein synthesis proved to be sensitive to chloramphenicol and DNase and relatively unaffected by cycloheximide and RNase (Anderson et al., 1972; Kuehl, 1974; Matinyan and Umansky, 1978). Our experiments differ from these earlier studies by incubation of whole cells and isolation of nuclei and nuclear matrix after

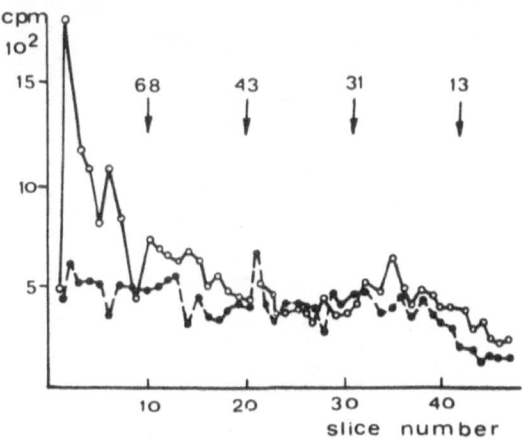

Figure 3.   Zajdela ascites hepatoma cells were incubated in ascitic
fluid with /^{14}C/ Chlorella protein hydrolysate at 30°C for 30
min. o - without chloramphenicol, ● - with chloramphenicol.
12% gel was cut to 48 equal slices and in each slice
radioactivity was measured.

Table I.   Turnover of Solid Hepatoma 27 Nuclear Matrix
Phosphoproteins.

Experiment No.	Initial Radioactivity cpm	Incubation Without ^{32}P min	Postincubation Radioactivity cpm	Residual Radioactiv. %
1	63,440	15	34,680	54.6
2	47,820	30	22,530	47.0
3	44,060	60	9,160	21.0
4	52,720	120	10,180	20.0

Isolated hepatoma 27 nuclear matrix preincubated in a
buffer containing 30 mM Tris-HCl, pH 7.5, 120 mM KCl, 5 mM
MgCl$_2$ and 5 mM dithiothreitol with 2.5 μCi gamma /^{32}P/ ATP
at 30°C for 20 min was postincubated in the same buffer without gamma /^{32}P/ ATP.

the incubation. We suppose that these results may indicate a special type
of biosynthesis or processing of this high molecular weight protein group of
the nuclear matrix.

Hereafter, we studied phosphorylation of non-histone nuclear proteins,
especially those of nuclear matrix, using rat solid hepatoma 27. The
phosphorylation proceeded proportionally to time during the first 20
minutes and was strongly stimulated by sodium vanadate. The phosphorylation
was accompanied by active dephosphorylation of nuclear matrix proteins.
Contrary to cAMP independent proteinkinases associated with the nuclear

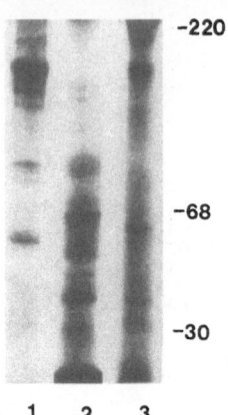

Figure 4.   Autoradiograms of solid hepatoma 27 nuclear matrix labelled
             with ^{32}P at 30°C for 30 min. 7.5% SDS-polyacrylamide gel
             electrophoresis. (1) - Phosphorylation in whole nuclei with
             ^{32}P; the nuclear matrix was isolated in the presence of 1mM
             PMSF and 1mM DTNB. (2) - The nuclear matrix isolated without
             inhibitors and phosphorylated by $\gamma-^{32}P-ATP$. (3) - The nuclear
             matrix isolated in the presence of inhibitors and
             phosphorylated by $\gamma-^{32}P-ATP$.

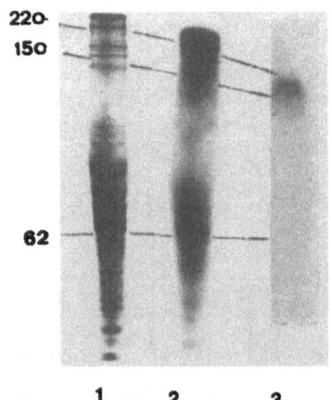

Figure 5.   Visualisation of alkali-resistant (probably phosphorylated on
             tyrosine residues) phosphoproteins in the nuclear matrix of
             rat solid hepatoma 27. (1) - 7.5% SDS-polyacrylamide gel
             electrophoregram, staining with Coomassee blue. (2) - Auto-
             radiogram of a similar gel. (3) - Autoradiogram of a similar
             gel after treatment with NaOH at 40°C for 2 hr.

matrix, protein phosphates are mostly soluble and localized in the nuclear
sap. Nevertheless, even in isolated nuclear matrix the phosphoproteins are
rapidly dephosphorylated; in 15-30 minutes about one half of the phosphoproteins were dephosphorylated and in 1 or 2 hours only about 20% of initial
^{32}P radioactivity remained (Table I).

    When whole nuclei were incubated and nuclear matrix was isolated in the
presence of serine proteinase inhibitors -PMSF and DTNB, phosphoproteins of

molecular mass about 200, 190, 160, 150, 100, 86, 56, and 30kD were revealed. However, without inhibitors of proteolysis, the high molecular weight bands were reduced, the bands of 86, 100 and 67kD became less dense and bands in the region of 45 and 36kD appeared. This shift of radio-activity appears to be due to ATP dependent proteolysis, namely of phospho-proteins. This conclusion is confirmed by similar pattern of phosphoryl-ation in isolated nuclear matrix and in whole nuclei in the presence of the inhibitors (Figure 4). In fact, no protein degradation could be observed both in the presence or absence of PMSF and DTNB when the electrophoregrams were stained with Coomassie blue. In the presence of vanadate, the intensity of phosphorylation increased without alteration of its pattern.

After treatment of electropherograms with 1M NaOH at 40°C for 2 hours only two radioactive bands, at about 190 and 200kD remained (Figure 5). As phosphoserine and phosphothreonine residues are hydrolysed in these conditions it is possible that these phosphoproteins contained considerable amount of phosphotyrosine residues, which may be characteristic of malignant growth.

REFERENCES

Anderson K. M., Slavik M., and Elebute O. P. 1972, Labelling of proteins by isolated rat liver nuclei, Canad. J. Biochem., 50:190.
Kuehl L. 1974, Nuclear Protein Synthesis, in 'The Cell Nucleus', vol. 3 H. Busch ed., Academic Press, New York, p. 345.
Kuzmina S. N., Buldyaeva T. V., Troitskaya L. P., and Zbarsky I. B. 1981, Characterization and fractionation of rat liver nuclear matrix, Eur. J. Cell Biol., 25:225.
Kuzmina S. N., Buldyaeva T. V., Akopov S. B. and Zbarsky I. B. 1984, Protein patterns of the nuclear matrix in differently proliferating cells, Molec. Cell Biochem., 58:183.
Matinyan K. S., and Umansky S. R. 1978, Some properties of nuclear protein synthesis system of eukaryotic cells, Biokhimiya (Russ.) 43:111.
Zbarsky I. B., and Debov S. S. 1948, On proteins of the cell nuclei, Doklady Akad. Nauk SSSR (Russ.) 62:795.

MAP-2-LIKE DNA BINDING PROTEIN: MONOCLONAL ANTIBODY RECOGNIZES UBIQUITOUS

PROTEIN OF INTERNAL FILAMENTS OF NUCLEAR MATRIX

Hiroshi Nakayasu, Kiyoshi Ueda and Ronald Berezney*

Department Med. Biochem.
Shiga University Med. Sci.
Seta,
Otsu 52021
Japan

* Department Biol. Sci.
SUNY at Buffalo
Buffalo, NY 14260
U.S.A.

We have prepared a monoclonal antibody (B3) using isolated rat liver nuclear matrix as immunogen. This antibody recognized a 260kD protein which was one   ubiquitous component (from yeast to man) of nuclear matrices. While this monoclonal antibody showed very high specificity against the nuclear 260kD protein, it did however, also recognize purified MAP-2 (microtubule associated protein-2) on immunoblots. On indirect immuno-fluorescent microscopy, this monoclonal antibody stained the intranuclear meshwork in interphase nuclei, but not nuclear lamina nor nucleolus. We also developed a protein staining procedure using *in situ* biotinylation. This new staining method was very effective in distinguishing each of the intranuclear filaments of nuclear matrix. Double staining (protein staining and B-3 antibody staining) showed that the 260kD protein was located in the overall internal filaments of nuclear matrices (Figure 1). In mitotic cells, this antibody stained mitotic spindle, midbody and chromosomes. Double staining with anti-tubulin antibody revealed that the 260kD protein co-located with tubulin at the spindle and midbody in mitotic cells. These results indicated that the 260kD matrix protein had a MAP-2-like nature. Because purified MAP-2 shows a DNA binding activity, we tried to determine whether this MAP-2 like protein possesses the binding activity or not.

(1) Co-localization of 260kD protein with residual DNA of nuclear matrix and chromosomal scaffold.

In order to determine the localization of residual DNA in nuclear matrix, it was labelled by *in situ* nick translation using biotin-11-dUTP. This method stained the residual DNA in nuclear matrix and in the chromosomal scaffold from mitotic cell ghosts. Double staining showed that the 260kD antigen co-located with the residual DNA in the nuclear matrix and chromosomal scaffold (Figure 2).

Antibody   Biotinylation   Phase

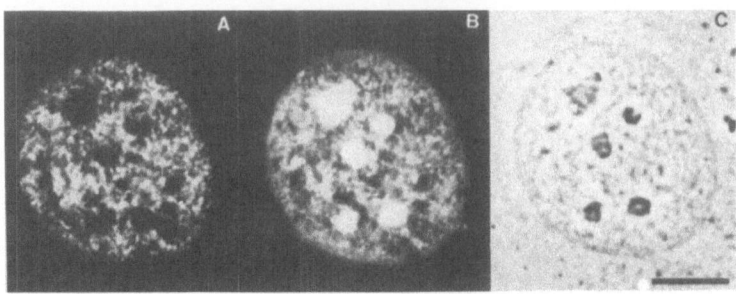

Figure 1.   Distribution of 260kD antigen in nuclear matrix. Mouse 3T3 cells were grown up on cover slips. *In situ* nuclear matrix was prepared by Triton X-100 treatment, DNase I digestion and high salt extraction. The *in situ* nuclear matrix on cover slips was incubated with B3 monoclonal antibody, then with active biotin to stain internal filaments in the *in situ* nuclear matrix (details of this method will be published elsewhere). The bound antibodies and the biotin residues were visualized by TRITC conjugated anti mouse IgM and by FITC conjugated avidine. Biotinylation stained both residual nucleoli and intranuclear filaments. On the other hand, B3 monoclonal antibody stained principally the internal filaments. It must be noted that the antibody stained practically the same filaments which were decorated by biotinylation. The same filaments were also seen in phase microscopy. These results indicated that the 260kD protein existed in the same overall area of the intranuclear filaments. The similar staining pattern by this monoclonal antibody was obtained using paraformaldehyde fixed cells, methanol fixed cells, LIS matrices and reverse matrices.

(2) Co-localization of 260kD protein with DNA binding sites of nuclear matrix and chromosomal scaffold.

The residual DNA was purified from isolated nuclear matrix and labelled by nick translation using biotin-dUTP. Labelled residual DNA was bound to the internal filaments in *in situ* nuclear matrix. Again, double immuno-fluorescent staining (B3 antibody staining and biotinyl DNA binding) showed the co-localization in the nuclear matrix and the chromosomal scaffold (Figure 3).

(3) The 260kD protein showed strong DNA binding activity on Western blots.

Biotinylated residual DNA bound 260kD protein band on western blots after renaturation under suitable conditions. Many nuclear matrix proteins had DNA binding activity, however, this 260kD protein showed the strongest binding among the nuclear matrix proteins under physiological salt concentration and in the presence of competitor RNA. Moreover, the monoclonal antibody inhibited the DNA binding to the 260kD band.

These results suggested that this 260kD protein was a component of the anchoring sites of nuclear DNA loops.

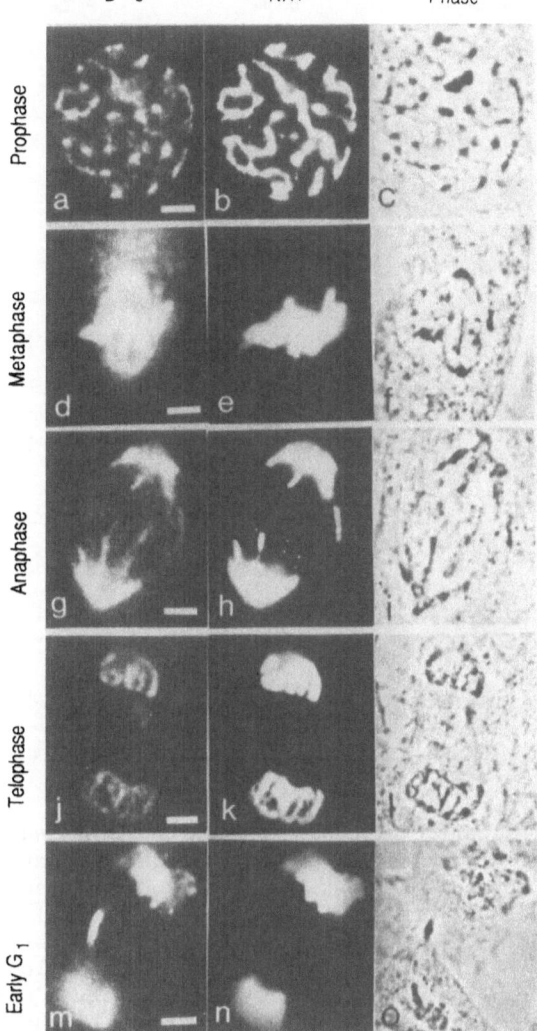

Figure 2.   Co-localization of 260kD antigen with residual DNA. Cells on
            cover slips were treated as described in Figure 1, then the
            residual DNA in nuclear matrix and chromosomal scaffold was
            labelled by *in situ* nick translation using biotin-11-dUTP.
            Without DNA polymerase I, without biotin dUTP, or without
            dNTPs, there was no fluorescence. Moreover, we could see very
            similar staining pattern by Hoechst staining, however, the
            intensity of fluorescence was weaker because of the short
            length of residual DNA. The cover slips were also incubated
            with B3 monoclonal antibody. Biotin residues and bound IgM
            antibodies were visualized as described in Figure 1. The
            staining pattern by the antibody was practically the same as
            that of the residual DNA in interphase nuclear matrix (data not
            shown). In mitotic cell ghosts, the antibody decorated the
            residual chromosomal scaffolds. The staining patterns by
            monoclonal antibody were again very similar to the distribution
            of the residual DNA in the scaffolds. However, the antibody
            also stained mitotic spindle (metaphase) and midbody (early G1)
            besides chromosomes. There was no residual DNA in these
            structures as expected.

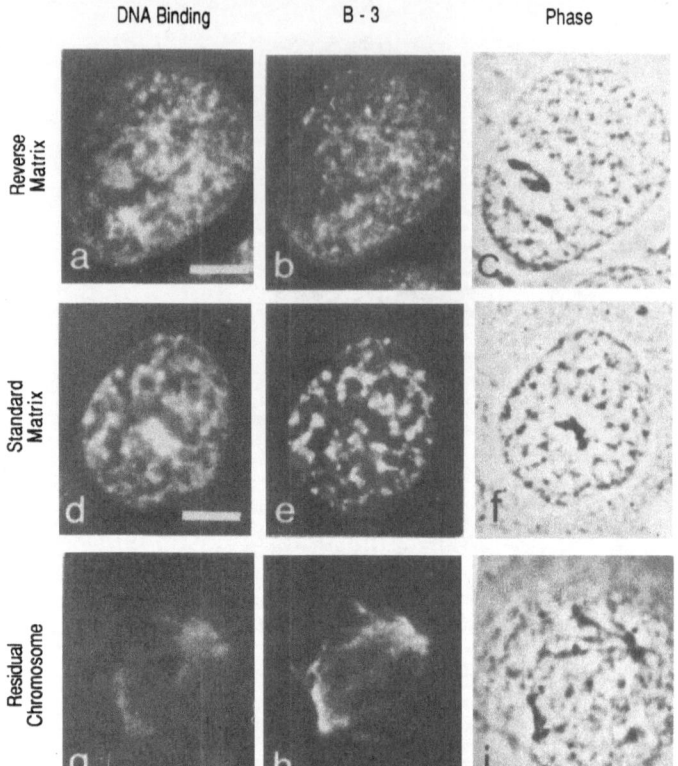

Figure 3.   Co-localization of 260kD protein and DNA binding sites in
            nuclear matrix and chromosomal scaffold. Matrix associated
            DNA (residual DNA) was purified from isolated nuclear matrix.
            The residual DNA was labelled by nick translation using
            biotin-11-dUTP. *In situ* nuclear matrices were incubated
            with biotinylated residual DNA and B3 monoclonal antibody.
            The bound IgM molecules and biotin residues were visualized
            as described in Figure 1. The pattern of the distribution of
            bound DNA was very similar to the distribution of 260kD
            protein in reverse matrix (high salt treatment → DNase I
            digestion), in standard matrix (DNase I digestion → high salt
            treatment) and in chromosomal scaffold.

DISCONNECTION OF DNA DOMAINS IN QUIESCENT AND DIFFERENTIATING CELLS

N. I. Sjakste*, A. V. Budylin* and T. G. Sjakste[+]

* Latvian Research Institute of Experimental and
Clinical Medicine
Ministry of Health of LSSR
Riga, Latvia
USSR

+ Institute of Biology
Academy of Science of LSSR
Salaspils, Latvia
USSR

INTRODUCTION

Nucleoprotein-celite-chromatography reveals three types of DNA-protein
interactions in the cell nucleus: bonds between DNA and the chromatin
proteins and two types of DNA bonds with nuclear matrix proteins. The ratio
of DNA content in chromatographic fractions corresponding to different types
of DNA-protein interactions changes in quiescent and differentiating cells.
The observed changes in chromatographic patterns reflect disconnection of
DNA domains.

MATERIALS AND METHODS

Cultivation of Ehrlich ascites carcinoma cells and transformed
Djungarian hamster fibroblasts (strain 4/21) were described before (Sjakste
and Budylin, 1988; Sjakste et al., 1986). DNA of ethiolated barley shoots
was labelled by incubation of cut coleoptiles in water with 100μCi/ml of ^{3}H-
thymidine. In pulse-chase experiments the labelled coleoptiles were
transferred into medium with 5μg/ml of cold thymidine. Complexes of DNA and
tightly bound proteins were isolated according to Neuer and Werner (1985).
DNA of these complexes was labelled in DNA-polymerase reaction. For NPC-
chromatography nucleoproteins were adsorbed via protein moiety on celite.
DNA was released from the complex with proteins by increasing concentrations
of NaCl (0-3M), LiCl and urea (0-4M, 8M) and heating of the column (0°-
100°C), see Sjakste et al. (1985). All other methods were as described
before (Sjakste et al., 1986; Sjakste and Budylin, 1988; Sjakste and
Blokhin, 1989).

RESULTS AND DISCUSSION

DNA from intact nuclei of proliferating cells is usually presented on
NPC-chromatograms as a sole peak in the end of temperature gradient (DNA II,

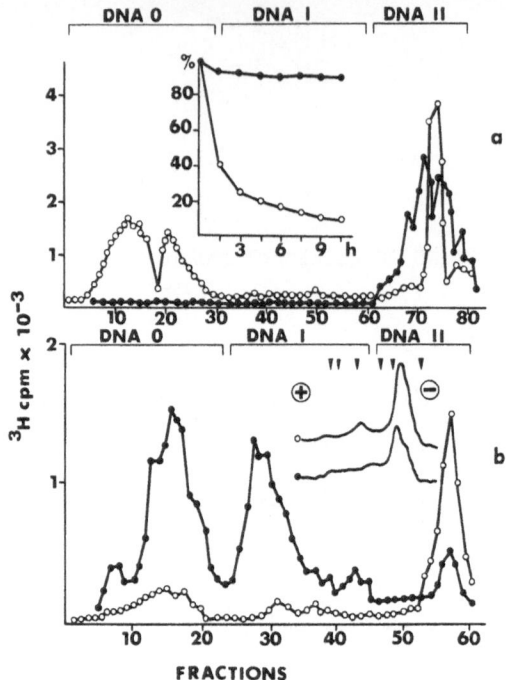

Figure 1.   Changes in NPC-chromatographic patterns of DNA coupled with
cell proliferation and differentiation. a - Djungarian
hamster fibroblasts, proliferating (filled circles) and
quiescent (open circles) cells. Insertion presents profiles
of DNA neutral elution of proliferating and quiescent cells.
b - Barley shoot cells, labelled in S-phase for 30 min (opened
circles) and chased for 10 hours (G_1-phase of the next cycle,
filled circles). Densitometer tracings of the DNA electropherograms of S- and G_1-phase cells are presented in the
insertion. Arrows indicate positions of the molecular mass
standards (23; 9.4; 6.5; 4.3; 2.3; 2.0kbp from right to left).

Figure 1a). Additional fractions appear in quiescent 4/21 cells: a broad
peak in NaCl gradient (DNA 0) and a fraction eluted by LiCl-urea - DNA I
(Figure 1a).

     Shifts in NPC-chromatographic position of DNA, apparently coupled with
differentiation, were also observed in plant cells. NPC-chromatograms of
DNA from synchronously dividing cells from the first leaf of barley shoots
labelled in S-phase are shown in Figure 1b. The DNA II peak prevails on
chromatograms of S-phase cells. In G_1-phase of the next cycle a significant
amount of the label moves to DNA I and DNA 0 (Figure 1b).

     Treatment of isolated nuclei from proliferating cells with DNases
permitted us to model the chromatographic shifts and to characterize each
type of DNA-protein interactions. After treatment with DNase II the nuclear
contents can be divided into fractions of the nuclear skeleton with residual
chromatin and soluble chromatin. As shown in Figure 2a, DNA II prevails on
the NPC-chromatogram of the nuclear skeleton isolated from DNase II-treated
nuclei of Ehrlich ascites cells. The DNA-protein interactions in soluble
chromatin are much weaker and DNA is eluted in NaCl gradient (DNA 0). In
preparation of the nuclear skeleton obtained after treatment of nuclei with
micrococcal nuclease the DNA II fraction is absent, instead of it a large

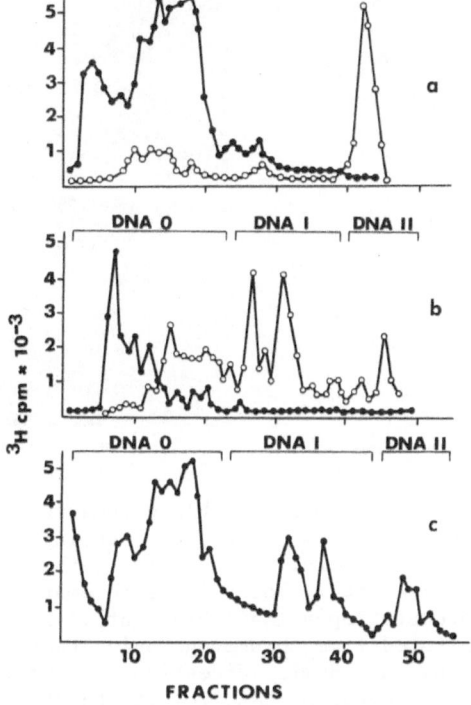

Figure 2.   NPC-chromatograms of subnuclear fractions isolated from
             Ehrlich ascites cells. a - Nuclei were digested with DNase II
             (100 U/ml, 1 hour, 24°C), fractions of soluble chromatin
             (filled circles) and nuclear skeleton with insoluble chromatin
             (open circles) were separated and subjected to NPC-chroma-
             tography. b - Nuclei were digested with micrococcal nuclease
             (100 U/ml, 1 min, 37°C) and treated as in 'a'. c -
             Deproteinized DNA was labelled during the DNA-polymerase
             reaction and subjected to NPC-chromatography.

DNA I peak arises. Soluble chromatin is eluted as DNA 0 (Figure 2b).
Experiments with preparations of isolated nuclear matrices showed that both
DNA I and DNA II bonds are formed by nuclear matrix proteins. Apparently,
DNA II is of topological nature, it is formed by a locally unwound DNA site
and is localized in the replicative complex (Sjakste et al., 1985).

     DNA sites forming the DNA-matrix bonds are not equally sensitive to
treatment with different nucleases. As seen in Figure 2, the bond survives
the DNase II treatment, but it is extra-sensitive to micrococcal nuclease.
The different sensitivity of DNA I and DNA II to nucleases enabled us to
purify the proteins forming these bonds (Sjakste and Blokhin, 1989).
Polypeptide spectra of both preparations are much alike. 180 kD polypeptide, triplet at 65-75 kD , 58 and 50 kD polypeptides are major common
components of the complexes, additional 61-63 kD minor proteins being
present in the DNA II-protein complex. Polypeptide composition of both
preparations resembles the tightly bound proteins that remain attached to
DNA after deproteinization (Neuer and Werner, 1985). NPC-chromatography of
the tightly bound protein preparations revealed heterogenity of these
proteins according to tightness of DNA-protein interactions. The chromatogram reflects all types of DNA-protein bonds characteristic of the nuclear
proteins (Figure 2c).

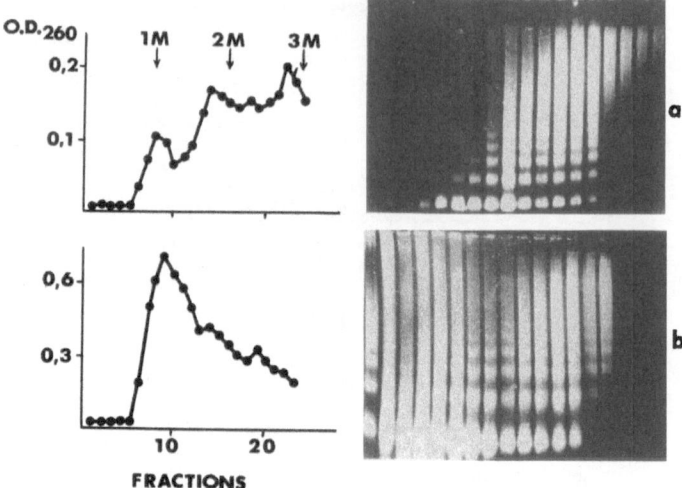

Figure 3.   Size distribution of soluble chromatin particles with
             different tightness of DNA-protein interaction. Ehrlich
             ascites cell nuclei were treated with micrococcal nuclease
             (100 U/ml, 37°C). The soluble chromatin was subjected to
             NPC-chromatography (NaCl gradient). DNA was ethanol-
             precipitated from chromatographic fractions and electro-
             phoresed. a - 3 min. digestion. NPC-chromatogram (left) and
             electropherogram of DNA from 7th to 24th fractions (right).
             b - 10 min. digestion. DNA from 2nd to 20th fractions was
             precipitated and electrophoresed.

     Unlike DNA I and DNA II, the DNA 0 peaks are often heterogenous and
extend throughout the gradient (Figure 2a). Besides, positions of the peaks
are not identical in different subnuclear fractions. DNA 0 of the soluble
chromatin is eluted by a lower salt concentration than DNA of insoluble
chromatin (Figure 2b). The position of DNA 0 on chromatograms is dependent,
to a certain extent, on the size of chromatin particles. Figure 3 shows
the chromatographic distribution of DNA 0 of soluble chromatin obtained
after treating Ehrlich ascites cell nuclei with micrococcal muclease. After
minimal digestion the major part of the DNA is eluted with 3M NaCl, peaks at
1M and 2M being also present (Figure 3a). The peak at 1M prevails in a
chromatin preparation digested to a greater extent, while peaks at 2M and 3M
are still present, but less expressed (Figure 3b). Comparison of the DNA
fragment size from different fractions revealed some greater fragmentation
of DNA eluted at the beginning of the gradient. However, particles of the
same size can be eluted with different salt concentrations (electrophoregrams in Figure 3a, b). Apparently, the particle size alone does not
determine its position on the chromatogram. According to the tightness of
DNA-protein interactions the chromatin particles can be divided into
discrete classes.

     Undoubtedly, DNA strand breaks are responsible for chromatographic
shifts of DNA in intact cells. Insertions in Figure 1 demonstrate that
accumulation of double-strand breaks accompanies chromatographic shifts in
quiescent 4/21 cells (acceleration of DNA neutral elution) and in barley
shoots (increase of DNA electrophoretic mobility).

     A model of a chromatin domain based on the data about three types of
DNA-protein interaction in the nucleus are presented in Figure 4. The loop

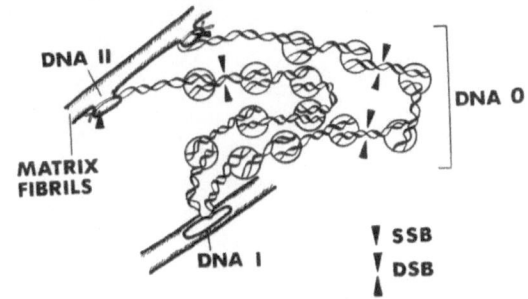

Figure 4.   A model of DNA domain. See the text for detailed
              explanations.

is attached to the nuclear matrix by a topological bond (DNA II) and by
ionic and hydrogen bonds (DNA I). In the proliferating (undifferentiated)
cell the domains are continuous, while transition to quiescence (differentiation) is followed by disconnection of the domains. The topological bond
is destroyed by an incision at the unwound site, thus causing DNA II - DNA
I transition. Several double-strand breaks induced in the loop cause
partial detachment of DNA from nuclear matrix. Apparently, a dynamic
equilibrium exists between induction of DNA breaks and their repair, thus
preserving the integrity of the genome.

REFERENCES

Neuer B., Werner D. 1985. Screening of isolated DNA for sequence released
     from anchorage sites in nuclear matrix, J. Mol. Biol., 181:15.
Sjakste N. I., and Blokhin D. Yu. 1989. Protein composition of DNA-matrix
     complexes with 'tight' and 'loose' type bonds isolated from Ehrlich
     carcinoma cells, Biokhimiya, 54:1217.
Sjakste N. I., and Budylin A. V. 1988. Tightness of DNA-protein inter-
     actions in chromatin fractions differing in sensibility to nucleases,
     Mol. Genetics, Microbiology and Virology, 10:30.
Sjakste N. I., Sjakste T. G., and Zaleskaya N. D. 1986. Reversible
     accumulation of double and one-strand breaks in growth-arrested cells,
     Bull. Exp. Biol. Med., 102:167.
Sjakste N. I., Zaboykin M. M., Erenpreisa Je. A., Lichtenstein A. V., and
     Shapot V. S. 1985. Analysis of cellular nucleoproteins with nucleo-
     protein-celite-chromatography. III. Two types of interactions
     between DNA and nuclear matrix, Molecular Biology (Moscow), 19:1007.

HEAT SHOCK PROTEINS ASSOCIATED WITH NUCLEAR STRUCTURES

S. B. Akopov, T. V. Buldyaeva, L. P. Troitskaya and
I. B. Zbarsky

N. K. Koltzov Institute of Developmental Biology
USSR Academy of Sciences
26 Vavilov Street
117334 Moscow, USSR

INTRODUCTION

Previously, using a cell culture system we have shown that heat shock proteins (HSPs) transported to the nucleus were revealed in association with the nuclear matrix (Akopov et al., 1986; Buldyaeva et al., 1986). We have found as well that whole body heating of rats induced alteration of liver nuclear matrix protein metabolism similar to that in cell culture under heat shock (Akopov et al., 1985). It is known that HSP synthesis leads to thermotolerance i.e. thermoresistancy of organisms or cells to usually lethal heating.

The association of HSPs with nucleoskeleton may be of essential interest for the nuclear matrix, except for its role as a mechanical scaffold, since it may regulate the higher order of chromatin organization, DNA replication and transcription, and RNA transport (Zbarsky, 1981; Berezney, 1984).

We have studied the association of HSPs with liver nuclear matrix, nucleoli, and fibrous lamina following whole body heating of rats.

MATERIALS AND METHODS

Female mongrel rats weighing 100 - 150g were heated in a special positively ventilated automatic incubator at $45^{o}C$ for 30 minutes. Immediately after heating the animals were injected intraperitoneally with $5\mu Ci/g$ ^{35}S-methionine and after 2 hours were sacrificed by decapitation. The nuclei were isolated according Blobel and Potter (1966) with our modification (Buldyaeva et al., 1978). The nuclear matrix was prepared from isolated nuclei by the method of Herlan and Wunderlich (1976) with Triton

Table I.   Incorporation of ^{35}S-Methionine Into Proteins of
Whole Cell Homogenate, Nuclei, Nuclear Matrix and
Nucleoli After Heating[a].

Cellular fractions	Rats	cpm/mg protein	% of incorp.
Whole cells	non heated	261950	100
	heated	209560	80
Nuclei	non heated	39290	100
	heated	30251	77
Nuclear matrix	non heated	117800	100
	heated	135500	115
Nucleoli	non heated	236095	100
	heated	264112	112

[a] Heating at 45°C for 30 min.

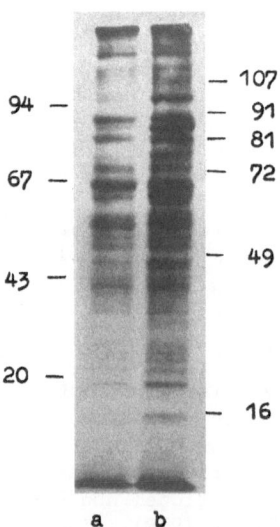

Figure 1.   Fluorogram of ^{35}S-methionine labelled proteins of rat liver
whole cell homogenate; a – without, and b – after whole body
heating at 45°C for 30 minutes. Electrophoresis in 10%
SDS-polyacrylamide gel. Left – molecular mass standards,
right – molecular masses of HSPs.

X-100 concentration raised to 0.5%. Nucleoli were isolated by the method of Chentsov et al. (Stephanova et al., 1988) as follows:

The nuclear pellet was suspended in buffer 1 (30mM sodium phosphate, pH 6.2, 3mM $MgCl_2$, 0.32M sucrose, 0.1% Triton X-100). After homogenization the suspension was overlaid on a cushion of buffer 2 (15mM sodium phosphate, pH 6.2, 1M sucrose, 1.5mM $MgCl_2$) and centrifuged at 3000g for 10 minutes. The pellet was suspended in buffer 3 containing 5mM sodium phosphate, pH 6.2, 0.32M sucrose, and 0.5mM $MgCl_2$, sonicated for 65 seconds overlaid on a cushion of buffer 2, and centrifuged at 1500g for 20 minutes. The pellet was resuspended in buffer 3, sonicated for 20 seconds, overlaid on buffer 2, and centrifuged as above. Preparations of lamina were isolated according to Krachmarov et al. (1986).

The preparations of the nuclear matrix, lamina and nucleoli were studied by SDS-PAAG electrophoresis (Laemmli, 1970) followed by fluorography. The purity of preparations and their morphology were monitored by electron microscopy of ultra-thin sections.

RESULTS

Our results show that heating of rats at $45^{o}C$ essentially changed the rate of ^{35}S-methionine incorporation into proteins of nuclear subfractions (Table I). Thus, in the cell homogenate and in whole nuclei the incorporation was by 20% and 23% lower than in control rats, while in the nuclear matrix and nucleoli it was correspondingly 15% and 12% higher. No essential difference was observed in lamina preparations (not shown).

A fluorogram of liver cell homogenate of rats heated at $45^{o}C$ differed from that of control rats by the presence of HSPs of 107, 91, 81, 70, 49, and 18kD molecular weight bands (Figure 1). In the nuclear matrix an HSP of 70kD was most prominent. In the nucleoli we have found several HSPs of 110, 94, 87, 70, 55, 44, 33, and 20kD. Practically no alteration in polypeptide profile of fibrous lamina was observed (Figure 2).

Electron microscopy of isolated rat liver nuclear matrix (Figure 3) showed that after heating the nucleoli grew very dense, devoid of granular component and contained only the fibrillar dense component. The fibro-granular component of the nuclear matrix grew less dense but exhibited a definite contrast, while no major alterations were observed in the dense lamina (Figure 3).

CONCLUSION

Thus, hyperthermia due to whole body heating of animals induced the synthesis of a specific HSP pattern. These HSPs, transported to the nuclei, were revealed mostly in association with the nuclear matrix. The bulk of the HSPs were observed in association with the nucleoli, while no HSPs were found in isolated fibrous lamina.

It is supposed that the association of HSPs with the nuclear matrix and the nucleoli may be of importance for the survival of nuclear structure during heating and for the restitution of nucleolar morphology, rRNA processing, and rRNA transport to the cytoplasm.

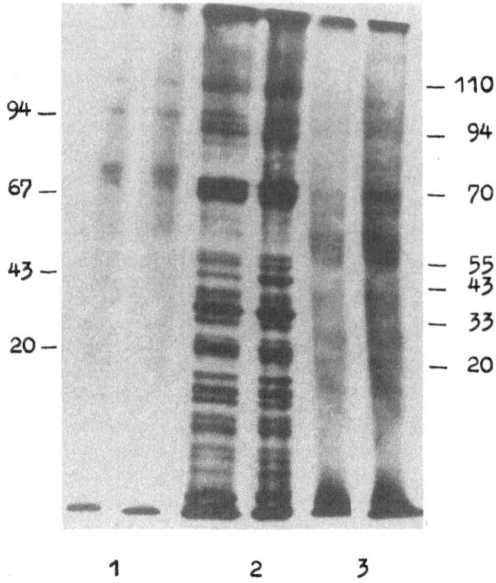

Figure 2.   Fluorogram of ^{35}S-methionine labelled proteins of rat liver
subnuclear fractions; a - without, and b - after whole body
heating at 45°C for 30 minutes.   1 - fibrous lamina,
2 - nucleoli, 3 - nuclear matrix.

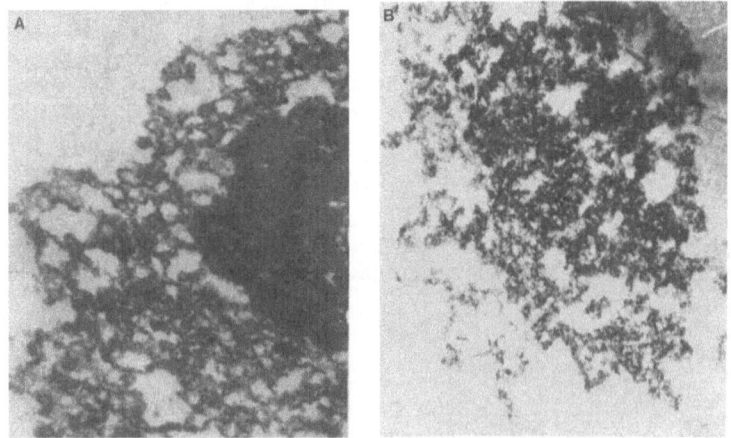

Figure 3.   Electron micrographs of ultrathin sections of rat liver
nuclear matrix.   a - without, b - after whole body heating
at 45°C for 30 minutes.

REFERENCES

Akopov S. B., Buldyaeva T. V., Kuzmina S. N., and Zbarsky I. B. 1985,
    Effect of hyperthermia on polypeptide composition of rat liver nuclear
    matrix (Russ.), Biokhimiya, 50:1127.

Akopov S. B., Kuzmina S. N., Buldyaeva T. V., and Zbarsky I. B. 1986, Heat shock proteins in the nuclear matrix of cultured Zajdela hepatoma cells, (Russ.), Doklady Akad, Nauk SSSR, 287:724.

Berezney R. 1984, Organization and function of the nuclear matrix, In: 'Chromosomal Non-Histone Proteins', vol. 4, L.S. Hnilica ed., CRC Press.

Buldyaeva T. V., Akopov S. B., Kuzmina S. N., and Zbarsky I. B. 1986, Heat shock proteins of Chinese hamster fibroblasts nuclear matrix (Russ.), Biokhimiya, 51:494.

Buldyaeva T. V., Kuzmina S. N., and Zbarsky I. B. 1978, Protein composition of the nuclear matrix and residual protein of rat liver and hepatoma 27 (Russ.), Doklady Akad. Nauk SSSR, 241:1461.

Herlan G., and Wunderlich F. 1976, Isolation of a nuclear protein matrix from Tetrahymena macronuclei, Cytobiologie, 13:291.

Krachmarov Ch., Tasheva B., Markov D., Hancock R., and Dessev G. 1986, Isolation and characterization of nuclear lamina from Ehrlich ascites tumor cells, J. Cell Biol., 30:351.

Laemmli U. K., 1970, Cleavage of structural proteins during the assembly of the head of bacteriophage T4, Nature, 227:680.

Stephanova E., Russanova V., Chentsov Yu., and Pashev I. 1988, Mouse centromeric heterochromatin - isolation and some characteristics, Exp. Cell Res., 179:545.

Zbarsky I. B. 1981, Nuclear skeleton structures in some normal and tumor cells, Molec. Biol. Rep., 7:139.

CHARACTERIZATION OF THE DNA PATTERN IN THE VICINITY OF A REPLICATION ORIGIN

LOCATED UPSTREAM FROM THE DOMAIN OF CHICKEN α-GLOBIN GENES

Sergey V. Razin, Avtandil G. Kalandadze, Eka G. Kintsurashvili
and Yegor S. Vassetzky Jr.

Institute of Molecular Biology
USSR Academy of Sciences
32 Vavilov st.
Moscow 117984 USSR

Substantial progress has been achieved recently in determination of sequences required for initiation of replication in yeast (for review see Umek et al., 1988). This progress became possible with the discovery of so called ARS elements, i.e. short DNA fragments that could ensure autonomous replication of plasmids in yeast cells. Unfortunately, numerous attempts to apply similar techniques to study replication origins in higher eukaryotes yielded highly controversial results (Holst et al., 1988, Grummt, 1989).

In this situation it became important to find other experimental approaches to study DNA sequences involved in initiation of replication in cells of higher eukaryotes. The most direct way is to map positions of replication origins in the cloned areas of genomes and then study these positions in detail. Previously, we have mapped the position of the replication origin to the upstream area of the domain of chicken α-globin genes (Razin et al, 1986).

Our data have recently been confirmed by Leffak who used another experimental approach, run-on replication, to map the position of replication origins (Umek et al., 1988). Both methods suggest that the replication origin is located within a *ca*. 3kb long DNA fragment including both a repetitive element (1kb) and a region containing unique sequence (*ca*. 2kb). In the present paper we report the complete nucleotide sequence and certain functional activities of the second region.

SEQUENCING OF THE FRAGMENT OF THE DOMAIN OF α-GLOBIN GENES CONTAINING A REPLICATION ORIGIN

The studied fragment (we designated it as α5HR) has been subcloned into pUC19 and its sequence has been determined using a modified method of Maxam and Gilbert (Chuvpilo and Kravchenko, 1983). The complete nucleotide sequence of α5HR is shown in Figure 1. This sequence contains multiple stop codons in all reading frames, hence it does not code for any protein. The fragment also contains several short sequences homologous to different regulatory elements (Table I). The region spanning from 500 to 1000bp of the sequenced fragment also contains multiple imperfect repeats.

```
   1 GCGGCACGGG GCGGCCCCGG GCCCGGCGCG CACTTACTGG CCTTGGCGGC GGGGTGCTCG
  61 GCGCCGCGCT GGAAGGGGAA GCGGAAGAGC AGCTTGTTGC CGCGGCTGCC CGAGCTCACA
 121 AGGATAACGC TGATGGGGCT GGTGCTCTCG CCCATGCCGC CGCGCCACAG CGAGCACCGG
 181 GCGGGCAACG ACGGACGCGG CTCCGCGGAA GGCGGCCCGG CCCGCGCGAC TTCCGCTTCC
 241 GCGCCTCCGC CGCCGCCGCC GGTTCCCCCG GGCCGCGGCC GAGCGGCGGG GCGGAGCTGC
 301 GGGCACAGCG CTCCCCGGGC AGGTCGCGCT CAGAGGCCGG GCCGCCGCTT CAGCGCCGTG
 361 CCCTCAGTGC GGCCCAGCGC CGTGCCCGCA GCGCTGCCCA CACGCCCTCG GGGTGCCCCA
 421 CGGCTGCTGC TTGCTCCCGG TGCCCGCCGT TCCTCCCAGC ACCTCGCAGT GCAGCCGTGC
 481 CTGAAGTGCA GCCCAGCACC TCACACCTCA GCCCCGGGCT CCCAGTACGA CCAGCAGGTC
 541 ACGTTGGAGT CTCTTGTCCT CAAGACTGCG CAGTGTCTCA CCTTTGAGCC TTGTGCCCCC
 601 CATTCAGCCC AGCACATCAC ACTGTAGCCC TTACACCCTC ACCACAGCAC AGCACCTCAC
 661 GTTCAGGCCC AGCACGTCA AGATGGAGCC CTGTGCCCCC AGACAGCCAG CATGGAACCA
 721 TCAAATCCTT AGAGTTGGAA GATGTCTGAA TCCTTGTGCC CCCAGTTCAG CCCGGCACCT
 781 CTCACACCCC ACTCAACACT CTTCAGCCAA GAGCCTACAG CTCAACCCAG CACCTCACGC
 841 CACCCAGCAG CACTCCCGCC ATCAGCCCAG TGCCCCCAGT CCGGATCGGT ACCTCTCATG
 901 CCCATGCACA GTGCACCAGA TCAGCCTAGC ACCACTAGTT CATTCCAGCA CCTCACGTGC
 961 CCACAGCCAA CCACTCCAGC ACCCCCGGTG CCCTAGTCAC ACCTCTCCGC TGCCTCAAGG
1021 TTCATTCCCA CCTCTTCCCA CATCCCCTCA CACCCCCTCA TTATTTTCAT GTCTCGCAAT
1081 CTCCTTTGGT CACTTGGAGT CATTCAGTTA TGACAACTCC AGAACTAGAA GCTGCTGGCC
1141 AGCAGCAAGT GCCACAAACT GTGTTCCCCC GGCAGCTCTT CTGGCTCATT TGTCTTATTG
1201 TGTGTCCAGC TGAGATCAGA AAGCTATCGG CAATTATGTC AGAGGATGGC CCAGTTTTTC
1261 ACATAGATTT GTCTGTATTT GATAGCAATA TTTAGTATTT GGTGCTCCGA GTATCCCCAC
1321 TCTGGATTTT TCTCTGCAAG ATTCTTCCCT TGGACTTCAG GCAGAGAAGG GGACTGAAAG
1381 GGAGATGAGC ACCCGCAGTG AGGGCTTAAT CTGCACGGCC ATTCTCTGCA AGGCAGGTGA
1441 TAACAACTGA AGCAAGAGAA GCTGTCATTG AGGGGAGAGA GTTGTTGGTG AGCGATTAAA
1501 GAGCAGTCAC ATTATCACAG CAGAGCATTC ATCGTGGCCC AGTGCTGGGG AGCTACGTTA
1561 GAATTGCCCA GTGTGTCTGC TTCCCAGCAT AACTATGCAT TCTTCAATTA AAAAACTGCA
1621 GGCATGTTTG CCATTTCCAG CTCTCGGAGA TGAGTTAAAG CAAAGCTCTG GAAACCTGCA
1681 AGCTCTCTGA GTGCTAGTAG AATGAAATGA AAGAATAAA
```

Figure 1.   A nucleotide sequence of α5HR fragment of chicken α-globin gene domain.

We have directly assayed the possibility of sequence specific inter-action of different subfragments of α5HR with nuclear protein extracts from nuclei of proliferating and resting chicken cells. All the six subfragments tested in gel retardation experiments specifically interacted with the protein extracts from proliferating (embryonic fibroblasts) but not resting (mature hepatocytes) cells (Figure 2).

The protein binding sites within the α5HR insertion have recently been characterized using methylation interference footprinting experiments (Razin et al., submitted). Ten hitherto unknown regognition sites for sequence-specific nuclear DNA-binding proteins have been identified (Razin et al., submitted).

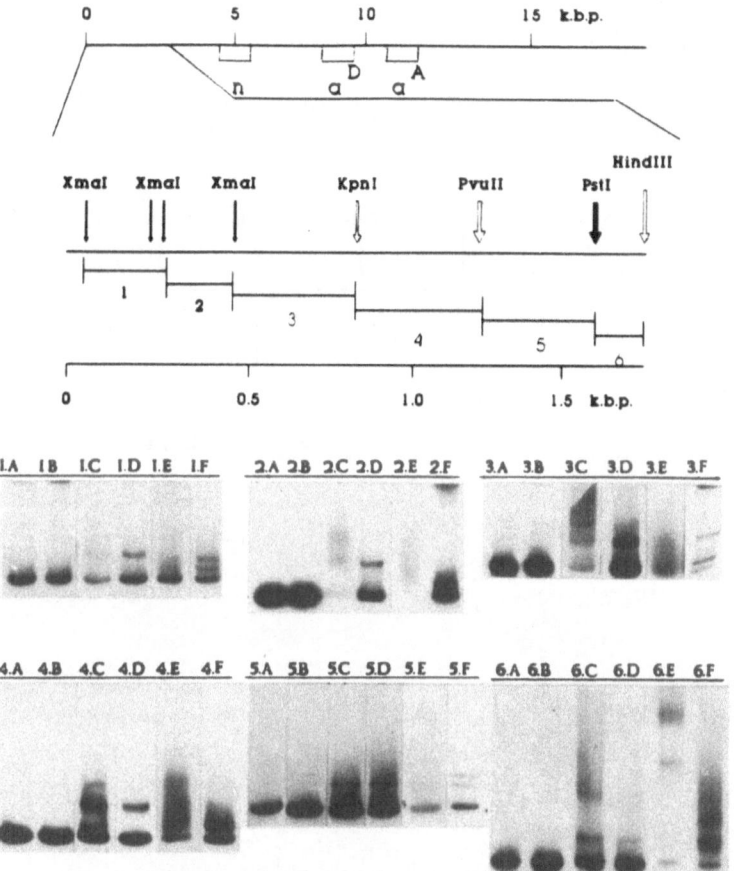

Figure 2. Electrophoretic analysis of complexes of different subfragments of α5HR with nuclear protein extracts from
proliferating and resting cells. Numerals indicate
corresponding subfragments of α5HR (see scheme). A, input DNA
subfragment; B, incubation of a subfragment with purified
nuclear DNA-binding protein extract from hepatocytes (with 50-
fold excess of poly (dAdT); C, incubation of s subfragment
with a crude nuclear protein extract from embryonic fibroblasts (without competitor DNA); D, same as C, but with 50-
fold excess of poly(dAdT); E, incubation of a subfragment with
a purified DNA-binding nuclear protein extract from embryonic
fibroblasts (without competitor DNA); F, same as E, but with
50-fold excess of poly(dAdT).

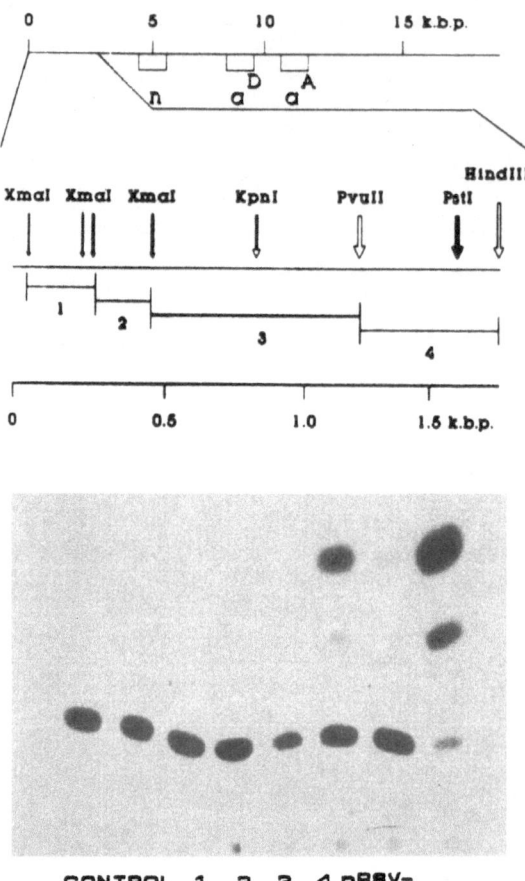

Figure 3.   CAT assay of RAT-1 cells transfected with pUSVL-CAT plasmid
             containing different subfragments of α5HR. Control (mock
             transfection, non-transfected cells, and cells, transfected
             with pUSVL-CAT); cells were transfected with the first XmaI-
             XmaI subfragment (1), the second XmaI-XmaI subfragment (2),
             XmaI-PvuII subfragment (3), and PvuII-HindIII subfragment
             (4); pRSVCAT - transfection with the plasmid containing a
             strong RSV LTR enhancer. The scheme indicates the position
             of α5HR and its subfragments in the domain.

Table I. The DNA Pattern in the
α5HR Fragment of the Domain
of Chicken α-Globin Genes.

PATTERN	α5HR
RETARDATION	+
CTC$_{T}^{A}$GAGA$_{CC}^{GG}$AA Grummt's cons.	3 (87%)
TGGCA (CHICKEN NF I)	1
ATGCAAT (NF III)	2
SV 40 ORIGIN	1
PALINDROMES	1
INTERNAL REDUNDANCY	+

TRANSCRIPTIONAL ENHANCER IS LOCATED IN THE VICINITY OF THE REPLICATION ORIGIN

Several observations made on replication origins of papova and
adenoviruses suggest that transcriptional enhancers can also influence
replication (O'Connor and Subramani, 1988). Keeping this in mind, we have
tested α5HR for possible enhancer function. Constructs containing α5HR and
its subfragments, SV40 late promoter and CAT reporter gene were transfected
into RAT-1 cells. At 48h post infection the cells were lysed and tested
for CAT activity. The vector without insertion and construct carrying RSV
LTR enhancer were chosen as negative and positive controls, respectively.
The whole fragment was found to possess and enhancer activity that was
ca.100 times less than that of the RSV enhancer. Later the subfragments of
α5HR were assayed (Figure 3). The enhancer was mapped to ca.300bp XmaI-
KpnI subfragment (see scheme in Figure 3). However its activity was ca.2
times weaker than that of the whole α5HR. Additional experiments
demonstrated that full activity could be observed in a XmaI-PstI subfragment
containing both minimal enhancer region and a neighboring fragment which was
previously found to include constitutive DNase I hypersensitive site
(Weintraub et al., 1981).

REFERENCES

Chuvpilo S. A., and V. V. Kravchenko. 1983, A solid phase method for
    determination of DNA nucleotide sequence, Bioorg. Khimiya USSR, 9:1634
    in Russian).
Grummt F. 1989, Autonomous replication in mouse cells: a correction, Cell,
    56:143.
Holst A., Muller F., Zastrow G., Zentgraf H., Schwender S., Dinkl E., and
    Grummt F. 1988, Murine genomic DNA sequences replicating autonomously
    in mouse L cells, Cell, 52:355.
O'Connor D. T., and Subramani S. 1988, Do transcriptional enhancers also
    augment DNA replication? Nucl. Acids Res. 16:11207.
Razin S. V., M. G. Kekelidze, E. M. Lukanidin, K. Scherrer, and G. P.
    Georgiev. 1986, Replication origins are attached to the nuclear
    skeleton, Nucl. Acids Res., 14:8189.
Umek R. M., Linskens M. H. K., Kowalski D., and Huberman J. A. 1988, New

beginning in studies of eukaryotic replication origins, <u>Biochem.</u>
<u>Biophys</u>. <u>Acta</u>, 1007:1.

Weintraub H., Larsen A., and Groudine M. 1981, α-Globin gene switching
during the development of chicken embryos: expression and chromosome
structure, <u>Cell</u>, 24:333.

PARAMETERS OF *XENOPUS* rDNA TRANSCRIPTION IN MICROINJECTED OOCYTES

Bärbel Meissner, Michael F. Trendelenburg and
Ansgar Hofmann*

Institute of Experimental Pathology
German Cancer Research Center
Im Neuenheimer Feld 280
D-6900 Heidelberg, Federal Republic of Germany

INTRODUCTION

Microinjection experiments using cloned *Xenopus laevis* rDNA have resulted in a detailed analysis of transcriptional control elements of the rDNA repeat (Busby and Reeder, 1983; Labhart and Reeder, 1984; De Winter and Moss, 1986). However, to facilitate detection of *X. laevis* rDNA transcripts most microinjection experiments were performed in a heterologous system: *X. laevis* rDNA clones were injected into *X. borealis* oocytes. It is clear that this assay has to cope with the phenomenon of nucleolar dominance of *X. laevis* transcripts over *X. borealis* transcripts (Reeder and Roan, 1984). Another line of evidence from oocyte microinjection experiments indicated, that oocyte batches from different females are likely to contain different amounts of rDNA specific transcription factors (Trendelenburg et al., 1978; Sollner-Webb and McKnight, 1982; McStay and Reeder, 1986). From EM-observations of injected rDNA templates it is clear that transcription factors specific for RNA polymerase I compete with those for polII on individual injected templates (Trendelenburg and Gurdon, 1978; Trendelenburg and Puvion-Dutilleul, 1987). In order to allow quantitation of some essential parameters influencing rDNA transcription we used a homologous experimental system: we injected a *X. laevis* rDNA construct of which transcripts could be discerned from endogenous rDNA transcripts into *X. laevis* oocytes and monitored the amount of transcript produced by the S1 nuclease technique. In addition, we co-injected α-amanitin, which leads to a higher yield of RNA polymerase I transcripts by suppressing non-specific RNA polymerase II transcripts (Hadjiolov, 1985; Gurdon and Wickens, 1983).

RESULTS AND DISCUSSION

Detection of rRNA Transcripts Derived from Injected rDNA Templates

In order to be able to discern between endogenous rRNA transcripts and

---

* present address: Institute of Medical Virology, University of Heidelberg, D-6900 Heidelberg, Federal Republic of Germany.

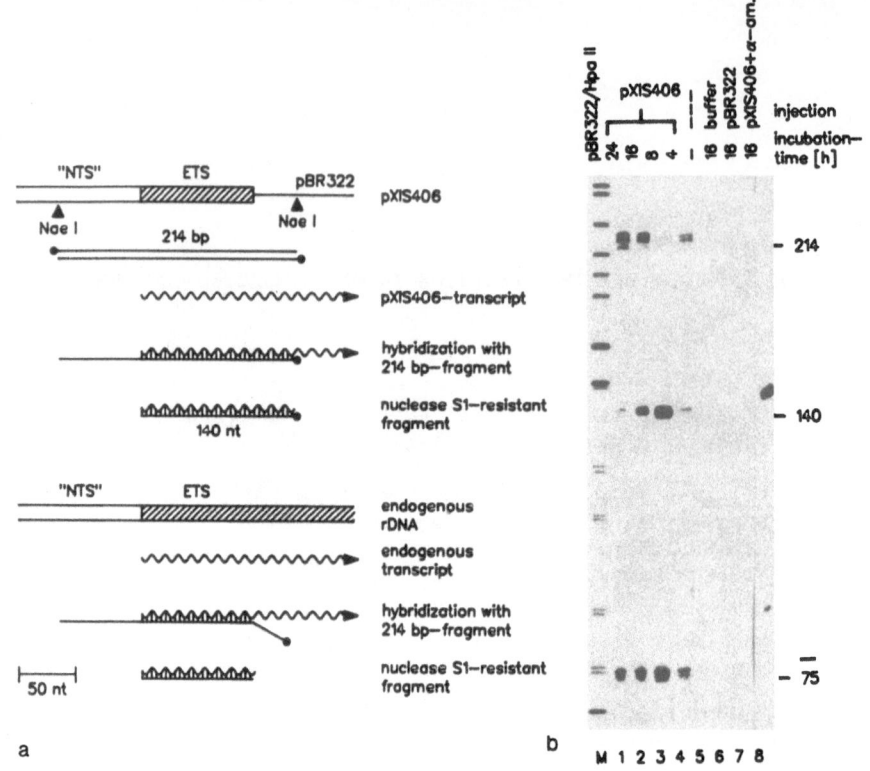

Figure 1.   Transcription of pXIS406 after injection into *Xenopus laevis* oocytes.

Figure 1a.   Only transcripts derived from pXIS406 can protect the labelled 5'-end of the 214bp *Nae* I fragment. Correctly initiated transcripts should protect a 140bp fragment.

Figure 1b.   S1 nuclease analysis of the 214bp probe. Oocytes of adult *X. laevis* females were injected with 6ng of pXIS406 and incubated at 19°C for various times. The RNA was extracted by the proteinase K method (18). The RNA equivalent of 1.25 oocytes was hybridized against 10,000cpm 214bp fragment labe-led at the 5'-end with ^{32}P (0.5-2x10^{6} cpm/pMol) in 20µl 80% formamide, 0.4M NaCl, 40mM PIPES (pH 6.4), 1mM EDTA at 57°C. The samples were digested with 60-80 U S1 nuclease (Sigma, Munich) at 37°C. Nuclease S1 resistant fragments were analysed on 6% sequencing gels. Oocytes were either non-injected (lane 5), buffer-injected (lane 6), injected with 6ng pBR322 (lane 7) or pXIS406 (lanes 1-4, 8). In one case 7.5ng α-amanitin were co-injected (lane 8). After injection the oocytes were incubated for 4 (lane 4), 8 (lane 3), 16 (lanes 2, 5-8) and 24 hours (lane 1).

transcripts of the injected rDNA template we constructed a clone that contains in addition to the rDNA part sequences that are specific for this clone, its transcripts and the S1 probe used. pXIS406 consists of the *Sau*3A I promoter fragment of pXL108 (Boseley et al., 1979) subcloned into pBR322. The clone contains 290bp of the 3'-end of the NTS including the failsafe termination site T3 (Labhart and Reeder, 1986), the complete gene promoter and 116bp of the ETS region (Moss, 1982). After incubation of the oocytes for various time periods the rRNA was extracted and hybridized against a 214bp *Nae* I fragment of pXIS406 (Hofmann et al., 1985) that had been 5'-labelled with ^{32}P (Figure 1a). During the following nuclease S1 digestion correctly initiated pXIS406 transcripts protected a 140bp fragment. Endogenous rDNA transcripts could not protect the labelled 5'-end of this fragment (Figure 1a). In addition to the 140bp signal two signals of 214 and 75bp were reproducibly observed. These bands disappear after co-injection of α-amanitin (Figure 1b, see below).

Modulation of rDNA Transcription in Oocyte Batches from Different *Xenopus* Females

With this experimental set up we first examined in detail wether an individual oocyte batch used would influence transcription of injected rDNA in the homologous system to a major extent. Oocytes of 10 different females were injected with 6ng pXIS406 each, RNA was extracted after 14 hours incubation and RNA equivalent of 3 oocytes hybridized against the 214bp fragment and incubated with 80U nuclease S1 at room temperature. The S1 nuclease resistant fragments were separated on 6% sequencing gels and the density of the signals was measured. The strongest signal was taken as 100% and the other values were calculated accordingly. Average values of two experiments show for each of the oocyte batches tested, great individual variation of the amount of transcripts was found (Table I).

Three of the 10 females exhibited a very reduced transcriptional activity (15-25%), five females had an intermediate activity (31-55%) whereas two females showed a very high activity (95-100%). The above results represent the first detailed analysis in regard to determination of maximal and minimal rDNA transcription rates in oocyte batches of randomly selected *X. laevis* females. They are in line with previous reports which stated differential rates, but did not present detailed quantitation (Trendelenburg et al., 1978; Sollner-Webb and McKnight, 1982; McStay and Reeder, 1986). They also add information on rDNA transcription factors in regard to the so far only well documented case, namely the observed variation in 5S-specific transcription factors (Korn et al., 1982).

Time Course of rRNA Transcription

The next parameter examined was the influence of the length of incubation after injection. The incubation periods were varied as follows: 2, 4, 6, 8, 12, 16 and 24 hours were chosen. Figure 2a shows the relative amount of transcripts calculated from the density of the S1 nuclease resistant 140bp fragment at the different times: during the first hours of incubation, the amount increases until a maximum is reached after 8 hours. Longer incubation leads to a decrease of the amount of transcripts observed. We conclude that about 6 hours after injection the RNA polymerase I reaches its maximal transcriptional activity. During longer incubations the activity decreases again because the injected genes are turned off, probably accompanied by a degradation process involved in the post-transcriptional modifications of the 40S pre-rRNA (Hadjiolov, 1985).

Suppression of Non-Specific Pol II Transcription by Co-Injection of α-Amanitin

In order to identify the RNA polymerases synthesizing the transcript protecting the 214, 140 and the 75bp signal we injected the oocytes with pXIS406 and α-amanitin. Co-injection of α-amanitin has three effects: a) the bands of 214 and 75bp disappear (Figure 1b, lane 1 and 8); b) the amount of transcript increases (Figure 2b); c) the maximal amount of transcript is reached after 12 - 16 hours after injection (Figure 2b).

These results demonstrate that the 214 and 75bp bands are protected by

Table I. Individual Variations of the Amount
of Transcripts Found in 10 Females of
Xenopus laevis.

female no.	exp. 1	exp. 2	average
1	5%	25%	15%
2	10%	35%	22.5%
3	15%	32%	23.5%
4	10%	42%	31%
5	25%	45%	35%
6	20%	55%	37.5%
7	52%	55%	38.5%
8	55%	55%	55%
9	90%	100%	95%
10	100%	100%	100%

transcripts of an α-amanitin sensitive RNA polymerase whereas the 140bp band
is protected by RNA polymerase I transcripts. The 75nt band is probably
derived from a RNA polymerase II transcript initiated within the ETS
sequence of pXIS406 whereas the full length protected 214bp band is caused
by RNA polymerase II read-through transcripts either initiated at the rDNA
gene promoter or at TATA-box-like sequences on the vector DNA (Sassone-
Corsi et al., 1981; Michaeli and Prives, 1987) that do not terminate at the
fail-safe termination site T3 (Labhart and Reeder, 1986). The increased
amount of transcripts and the shift of the maximum are in our opinion due
to two effects: first, the amount of injected DNA templates accessible to
the RNA polymerase I, i.e. the promoters of the templates are not blocked
by RNA polymerase II transcripts, is higher. Secondly, the blocking of RNA
polymerase II leads to a reduction of short-lived RNA polymerase II
transcripts which are thought to be involved in the post-transcriptional
modifications of 40S pre-rRNA transcripts. Therefore the degradation
process is slowed down (Hadjiolov, 1985) resulting in an accumulation of
40S pre-rRNA transcripts.

SUMMARY AND CONCLUSIONS

     Major parameters influencing rDNA transcription were examined by
injection of a *Xenopus laevis* rDNA construct into *X. laevis* oocytes.
Specific transcripts derived from the construct could be detected among the
endogenous rDNA transcripts by nuclease S1 mapping.

I.   First, it could be shown that the amount of transcripts measured
depends on the batch of *X. laevis* oocytes used: of 10 randomly selected
animals 3 showed only little transcriptional activity (15-25%), 5 animals
exhibited an average value of 30-55%, whereas 2 animals showed a very high
activity, probably reflecting excessive amounts of transcription factors
in oocytes of these animals.

II. Secondly, the length of incubation after injection is of major

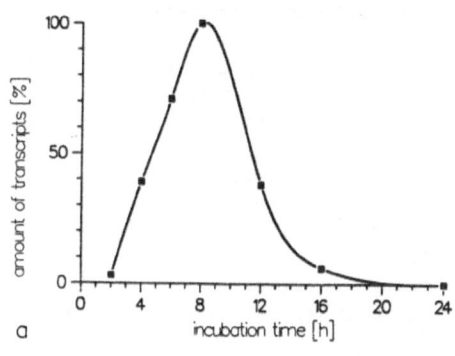 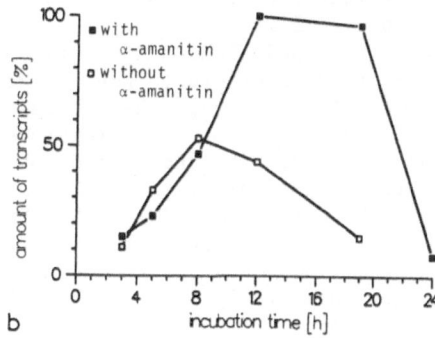

Figure 2.   Dependence of relative amount of transcripts on incubation
               time.
Figure 2a.   Oocytes were injected with 6ng pXIS406 and incubated for 4,
               6, 8, 12, 16 and 24 hours. The RNA equivalent of 3 oocytes
               was hybridized against the 5'-labelled 214bp *Nae* I fragment
               and digested with 80U S1 nuclease at room temperature. The
               nuclease S1 resistant fragments were separated on 6%
               sequencing gels and the intensity of the signals measured
               with a scanner. The highest value was taken as 100% and the
               other values were calculated accordingly.
Figure 2b.   7.5ng of α-amanitin and 6ng of pXIS406 were co-injected
               (intracellular concentration = 15µg/ml accessible cell
               volume = 500nl) leading to an increase of the amount of
               transcripts and a shift of the maximum of the transcriptional
               activity.

relevance: we could show that approximately 8 hours after injection the
amount of transcripts found is highest. Further incubation leads to a
decrease caused by post-transcriptional processing of the 40S pre-rRNA
transcripts.

III. Thirdly, it was shown that co-injection of α-amanitin leads (i) in all
batches examined to a higher yield of RNA polymerase I transcripts by
suppressing specific and unspecific RNA polymerase II transcripts which
play a part in the post-transcriptional modification of the primary
transcript and hinder the RNA polymerase I to initiate new transcripts by
blocking the promoter, respectively. (ii) Interestingly the time course of
maximal expression of rDNA templates was found to be shifted to 12 - 16h
after injection (compare II).

ACKNOWLEDGEMENT

    We thank our colleagues Roger Fischer, Markus Montag, Herbert Spring,
Herbert Steinbeißer and Helmut Tröster for helpful discussions.

REFERENCES

Boseley P., Moss T., Mächler M., Portmann R. and Birnstiel M. L. (1979),
    Sequence organization of the nontranscribed spacer of *Xenopus laevis*
    ribosomal DNA. Cell 17:19.
Busby S. J. and Reeder R. H. (1983), Spacer sequences regulate transcription
    of ribosomal gene plasmids injected into *Xenopus* embryos. Cell 34:989.
DeWinter R. F. and Moss T. (1986), Spacer promoters are essential for

efficient enhancement of *Xenopus laevis* ribosomal transcription. Cell 44:313.

Gurdon J. B. and Wickens M. P. (1983), The use of *Xenopus* oocytes for the expression of cloned genes. Methods in Enzymology 101:370.

Hadjiolov A. A. (1985), The nucleolus and ribosome biogenesis. Cell Biology Monographs, Vol 12, Springer-Verlag, Wien.

Hofmann A., Laier A. and Trendelenburg M. F. (1985), Geninjektion und Transkriptanalyse in der *Xenopus* Oozyte, in: Molekular- und Zellbioogie - Aktuelle Themen (Blin, N., Trendelenburg, M. F. and Schmidt, E. R., eds.) pp. 144-158, Springer-Verlag, Berlin.

Korn L. J., Gurdon J. B. and Price J. (1982), Oocyte extracts reactive developmentally inert *Xenopus* 5S genes in somatic nuclei. Nature 300:354.

Labhart P. and Reeder R. H. (1984), Enhancer-like properties of the 60/81bp elements in the ribosomal gene spacer of *Xenopus laevis*. Cell 37:285.

Labhart P. and Reeder R. H. (1986), Characterization of three sites of RNA 3'-end formation in the *Xenopus laevis* ribosomal gene spacer. Cell 45:431.

McStay B. and Reeder R. H. (1986), A termination site for *Xenopus* RNA polymerase I also acts as an element of an adjacent promoter. Cell 47:913.

Michaeli T. and Prives C. (1987), pBR322 DNA inhibits simian virus 40 gene expression in *Xenopus laevis* oocytes. Nucl. Acids Res. 4:1579.

Moss T. (1982), Transcription of cloned *Xenopus laevis* ribosomal DNA microinjected into *Xenopus* oocytes and the identification of a RNA polymerase I promoter. Cell 30:635.

Reeder R. H. and Roan J. G. (1984), The mechanism of nucleolar dominance in *Xenopus* hybrids. Cell 38:39.

Sassone-Corsi P., Cordon J., Kedinger C. and Chambon P. (1981), Promotion of the specific *in vitro* transcription by excised 'TATA'-box sequences inserted in a foreign nucleotide environment. Nucl. Acids Res. 9:3941.

Sollner-Webb B. and McKnight S. L. (1982), Accurate transcription of cloned *Xenopus* rRNA genes by RNA polymerase I: demonstration by nuclease S1 mapping. Nucl. Acids Res. 10:3391.

Trendelenburg M. F. and Gurdon J. B. (1978), Transcription of cloned *Xenopus* ribosomal genes visualized after injection into oocyte nuclei. Nature 276:292.

Trendelenburg M. F., Zentgraf H., Franke W. W. and Gurdon J. B. (1978), Transcription patterns of amplified *Dytiscus* genes coding for rRNA after injection into *Xenopus* oocyte nuclei. Proc. Natl. Acad. Sci. USA 75:3791.

Trendelenburg M. F. and Puvion-Dutilleul F. (1987), Visualizing active genes, in: Electron Microscopy in Molecular Biology (Sommerville J. and Scheer U., eds.). pp. 101-146. IRL Press, Oxford, U.K.

TRANSCRIPTIONAL ACTIVITY OF THE 5' REGION OF THE *XENOPUS LAEVIS* R-PROTEIN

GENE L1

Elena Beccari and Francesca Carnevali

Centro di Studio per gli Acidi Nucleici
Department of Genetics and Molecular Biology
University of Rome
La Sapienza
00185 P. le A. Moro 5 Roma
Italy

INTRODUCTION

In a previous study we analyzed in detail the 5' flanking region of the gene coding for the ribosomal protein L14 of *Xenopus laevis* (Carnevali et al. 1989) and found that its functionality as a promoter was abolished when shortened beyond a sequence located at -53 from the major capsite of the gene.

Two different proteins were found to bind to the L14 promoter: one, named XrpFI, made specific contacts with the CTTCC motif at -53 and the other, XrpFII, with a motif located at -27 from the capsite.

In this paper we present a preliminary analysis of the promoter region of a second *Xenopus laevis* ribosomal protein gene, also coding for a protein of the large subunit of the ribosome, L1. L1 is the largest of the r-proteins, with a molecular weight of 45kD. A peculiar feature in the biosynthesis of this protein is the regulation at the level of splicing of the precursor RNA (Caffarelli et al., 1987). This does not occur with the L14 gene.

METHODS

Plasmid Constructions

The region from the NdeI site at -307 to the HaeIII site at +16 of the L1 gene was ligated to BamHI linkers after repair of the NdeI site with the Klenow polymerase. The fragment was BamHI digested and cloned into the unique BamHI site of the CAT vector SP65CAT. This vector is a pSP65 plasmid with the repaired TaqI-TaqI region of the bacterial CAT gene cloned into the SmaI site in orientation opposite to the direction of transcription by the SP6 polymerase. The SP66CAT vector alone, injected into *Xenopus* oocytes, is completely negative to the CAT assay after overnight incubation. Sub-fragments of the linked BamHI-BamHI -307 to +16 L1 region were prepared. Plasmid -129SP65CAT contained the L1 fragment, from the Sau3A site at -129 to the downstream BamHI linker, inserted in the BamHI site of SP65CAT.

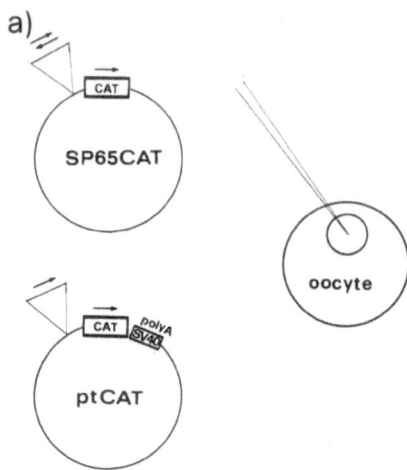

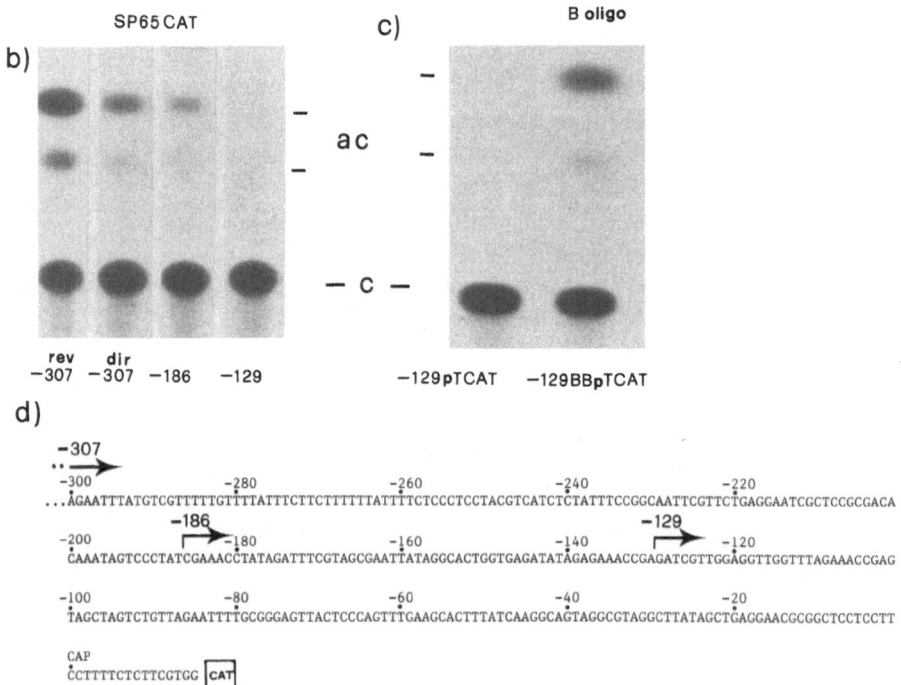

-AGCTTACCACAAACTTCCGGTTATCAGGTGTTCCCA
ATGGTGTTTGAAGGCCAATAGTCCACAAGGGTTCGA-

Figure 1.   Functional assay of promoter regions of the L1 gene. a)
scheme of the CAT constructs. b) CAT assay with the SP65CAT
recombinants: promoters for CAT transcription are the L1 -307
to +16 region in reverse (rev -307) or direct (dir -307)
orientation, and the shortened derivatives of dir -307 maintaining 186 and 129 bases upstream of the capsite. c) CAT
assay after inserting in -129pTCAT (see methods) 2 copies of
the B oligo shown at the top. d) sequence of the L1 region
examined.

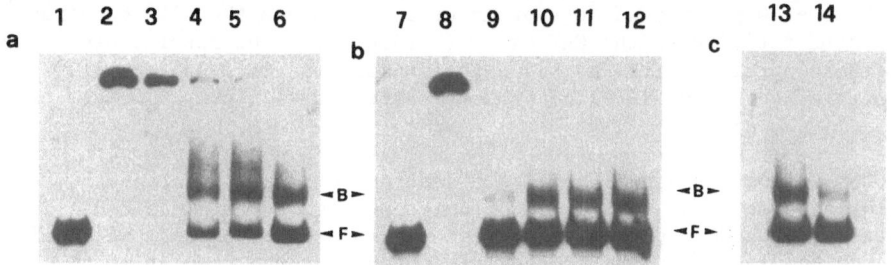

Figure 2.   Gel retardation assay after binding of a crude oocyte nuclear
             extract to the -307 to +16 L1 gene region (a) binding of
             proteins corresponding to 1.5 oocyte nuclei to terminally
             labelled DNA; (1) no proteins added; (2) no competitor; (3-6)
             with the addition of 0.5, 1.0, 2.0, 4.0 micrograms of
             sonicated *E. coli* DNA. (b) time course of complex formation:
             (7-8) as in 1 and 2; (9-12) 5, 10, 20, 30 minutes of
             incubation in the presence of 4 micrograms of competitor DNA.
             (c) specificity of complex B: binding made as in 6, with the
             addition of a 30-fold molar excess over the labelled L1 DNA
             of: (13) a 350bp prokaryotic DNA and (14) the unlabelled
             -307 to +16 L1 region. B=retarded DNA; F=free DNA.

Plasmid -186SP65CAT contained the L1 region from the TaqI site at -186 to
the downstream BamHI site cloned in AccI-BamHI cut SP65CAT.

    In a more recent series of experiments the SP65CAT plasmid was
substituted by another CAT vector, pTCAT (Carnevali et al., 1989), which
contains the SV40 late polyadenylation signal downstream of the CAT gene.
Using this vector, CAT transcripts are stabilized by a poly(A) tail and CAT
assays can be performed shortly after injection. The HindIII-EcoRI fragment
of -129SP65CAT was transferred into HindIII-EcoRI cut pTCAT to form
-129pTCAT. Two copies of a 36bp sequence, called B oligo (Carnevali et al.,
1989), which is a synthetic copy of the L14 -67 to -36 region flanked by
HindIII ends (shown in Figure 1c), were cloned in direct orientation into
the HindIII site of -129pTCAT to give -129BBpTCAT.

CAT Assay, Preparation of the Oocyte Nuclear Extract, Gel Retardation Assay

    The methods employed were described by Carnevali et al. (1989).

RESULTS AND DISCUSSION

    In order to detect sequence motifs important for transcriptional
efficiency, we first constructed recombinant plasmids containing upstream
sequences of the L1 gene cloned in front of a reporter bacterial gene,
coding for the chloramphenicol-acetyltransferase (CAT). The expression of
CAT, directed by L1 promoter sequences, was then assayed after injection
into the nuclei of *Xenopus laevis* oocytes, with a classical CAT assay.
Figure 1 shows a scheme of the CAT vectors used (a), the results of the CAT
assays (b, c) and the sequence of the various fragments of the L1 upstream
region (d).

    The L1 region extending from -307 to +16 relative to the capsite must
contain elements of two divergent promoters, since it promotes CAT
transcription either in the direct or in the reverse orientation (Figure
1b). In the natural orientation of the promoter, the reduction of the

region to -186 greatly affects promoter function and the -129 construct has background activity in the CAT assay (Figure 1b). The results are different from those obtained with the L14 deletion mutants (Carnevali et al., 1989) which retained a high level of promoter strength with only 63 bases of upstream sequences.

Comparison between L1 and L14 5' regions did not reveal obvious homologies except for the region of uninterrupted pyrimidines containing the capsite and the GC rich region between the TATA-like box and the capsite. A sequence homologous to the RPG-box, important for the regulation of transcription of most ribosomal protein genes in yeast, is present at -182 in L1 and at -53 in L14. Deletion of the region from -186 to -129 in L1 severely reduces promoter activity, as in the case of the reduction from -63 to -49 in L14.

The small box CTTCC which binds a specific factor, XrpFI, in the L14 region, is present in L1 only at the level of the capsite; its presence, however, is not sufficient *per se* to support transcription.

We have inserted a small sequence, the B oligo shown in Figure 1c, containing the CTTCC motif, upstream of the -129pTCAT L1 deletion mutant, unable to direct CAT expression. The recombinant -129BBpTCAT, with two copies of the B oligo inserted in the direct orientation, acquired the ability to promote high levels of CAT activity (Figure 1c).

Specific binding of nuclear factors to the upstream region of the L1 gene was detected by gel retardation assay (see Figure 2). A terminally labelled L1 DNA fragment, extending from -307 to +16 relatively to the capsite, was incubated with the crude nuclear extract from *Xenopus laevis* large oocytes. A major shifted band was observed when a specific competitor DNA was included in the binding mixture (lanes 4-6). The formation of the DNA-protein complex (indicated as B in the figure) was not prevented when the competitor DNA was increased up to 200µg/ml (lane 6). The appearance of the shifted species required at least 10 minutes. No further increase was observed beyond 30 minutes. Specificity of the B complex was demonstrated by competition: a 25-fold molar excess of the unlabelled L1 -307 to +16 region (lane 14) decreased the intensity of the B band, but not the same molar excess of a prokaryotic DNA fragment of the same size (lane 13).

REFERENCES

Caffarelli E., Fragapane P., Gehring C., and Bozzoni I. 1987. The accumulation of mature mRNA for the *Xenopus laevis* ribosomal protein L1 is controlled at the level of splicing and turnover of the precursor RNA. EMBO J. 6:3493.
Carnevali F., La Porta C., Ilardi V. and Beccari E. 1989. Nuclear factors specifically bind to upstream sequences of a *Xenopus laevis* ribosomal protein gene promoter. Nucl. Acids Res. in the press.

CORRELATION BETWEEN CHANGES IN SPHINGOMYELINASE ACTIVITY AND LEVEL

OF NUCLEIC ACID SYNTHESIS IN CELL NUCLEI OF REGENERATING RAT LIVER

M. Yu. Pushkareva, O. V. Borovkova and A. V. Alessenko

Institute of Chemical Physics
USSR Academy of Sciences
Moscow
USSR

INTRODUCTION

Sphingomyelin is known to participate in the replication and transcription processes (Alessenko et al., 1984; Alessenko et al., 1983) and the
products of its metabolism control the protein kinase C activity (Hannun and
Bell, 1987). 1,2-Diacylglycerol or phorbol ester, while stimulating cells
to proliferate and activating protein kinase C at the same time, change the
metabolism of sphingomyelin and activates acidic sphingomyelinase (Kiss et
al., 1988; Kolesnick, 1987). The addition of exogenic sphingomyelinase to
cells treated with phorbol ester returns the protein kinase C to an inactive
state (Kolesnick and Clegg, 1988).

In this report we demonstrate the role of sphingomyelinase activity in
DNA and RNA synthesis in the cell nucleus.

RESULTS AND DISCUSSION

There is some data on the presence of sphingomyelinase activity in
practically all subcellular compartments, except the cell nucleus. The
presence of sphingomyelinase both in the intact nucleus and in the nuclear
structures had not been investigated before our experiments.

Our experiments have shown that the nuclei from rat liver cells can
hydrolyse exogenic sphingomyelin. Nuclear sphingomyelinase consists mainly
of the neutral form of the enzyme, with smaller amounts of the acidic
nuclear sphingomyelinase.

Removal of the inner and outer nuclear membranes with 1% Triton X-100
leads to a decrease of nuclear sphingomyelinase activity to 10% of that of
intact nuclei; so it can be concluded that nearly one tenth of total nuclear
sphingomyelinase activity is tightly associated with the chromatin and
nuclear matrix (Table I). Addition of 10mM phosphatidylserine does not
result in an appreciable change in the neutral sphingomyelinase activity
(Table I), from which it appears that sphingomyelinase of the intact nucleus
is localized either mainly in the nuclear membranes and it is the
phosphatidylserine-independent enzyme, unlike the enzyme of the plasma
membrane.

Table I. Effect of Phosphatidylserine on the Sphingomyelinase Activity in the Rat Liver Cell Nuclei With the Nuclear Membrane Removed.

	cpm/mg of protein
Intact nuclei	1762∓96
Nude nuclei	171∓10
Nude nuclei + Phosphatidylserine	183∓12

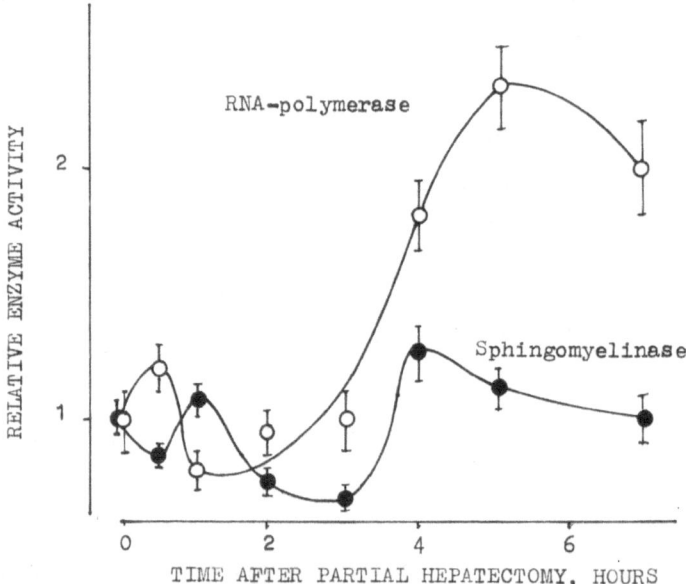

Figure 1.   Changes in the Sphingomyelinase and RNA-synthetase activities in the nuclei of regenerating rat liver cells after partial hepatectomy.

A relationship between the increased sphingomyelin content in nuclei and subnuclear fractions and the synthesis of nucleic acids found earlier (Alessenko et al., 1984; Alessenko and Pantaz, 1983). We suggested that a temporal correlation exists between the changes in sphingomyelinase activity and activation of RNA and DNA biosynthesis in the nucleus. To confirm this assumption a series of experiments has been carried out to determine simultaneously the activities of sphingomyelinase, and of RNA- and DNA-polymerases and the content of sphingomyelin and ceramides in the cell nuclei of regenerating rat liver.

The first 7 hours after operation have been chosen, when the RNA synthesis activation takes place (Figures 1 and 2), and 14-24 hours after operation when the DNA synthesis activation is observed in the cells of the regenerating rat liver (Figures 3 and 4).

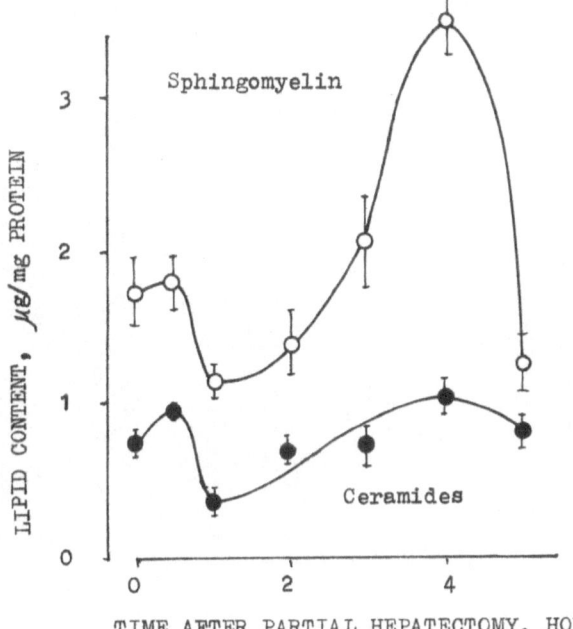

Figure 2.   Changes in the content of sphingomyelin and ceramides in the
              activation of RNA synthesis in nuclei of regenerating rat
              liver cells.

Figure 1 shows changes in the sphingomyelinase activity and transcription in the cell nuclei of the regenerating rat liver in the first 7
hours after partial hepatectomy. Sphingomyelinase activity is markedly
decreased in the nuclei of experimental animals as compared to the nuclei
of control animals, however, by the 4th hour its level increases and
approaches that in the control. The onset of stimulation of RNA-polymerase
activity corresponds to that of sphingomyelinase activity increase.

Oscillation of the sphingomyelinase activity is observed also in the
period 14 to 24 hours after operation, when the DNA synthesis proceeds.
Sphingomyelinase level is considerably decreased at 14-18 hours, as compared
with control and after 18 hours a sharp increase exceeding the control
values is observed. An increase in the sphingomyelinase activity coincides
in time with DNA-synthesis (Figure 3), i.e. a correlation between the
increase of the sphingomyelinase activity and the level of the synthesis of
nucleic acids is observed.

Analysis of the sphingomyelin and ceramide content at the same time
intervals after partial hepatectomy has shown that at 4 and 20-22 hours
after operation the content of sphingomyelin sharply increases from 1.8μg
per mg of protein in the control to more than 3.5μg per mg of protein in the
experiment (Figures 2 and 4). At the same time an increase of ceramide is
observed.

An increase in the sphingomyelin content and accumulation of ceramide
in the nucleus correlates with an increase in the nuclear sphingomyelinase
activity at the time of nucleic acid synthesis activation. That is, the
sphingomyelinase activity in the nucleus increases at a time when its
substrate (sphingomyelin is accumulating. An increase of the ceramide
content is observed in these periods as a result of the enzymatic

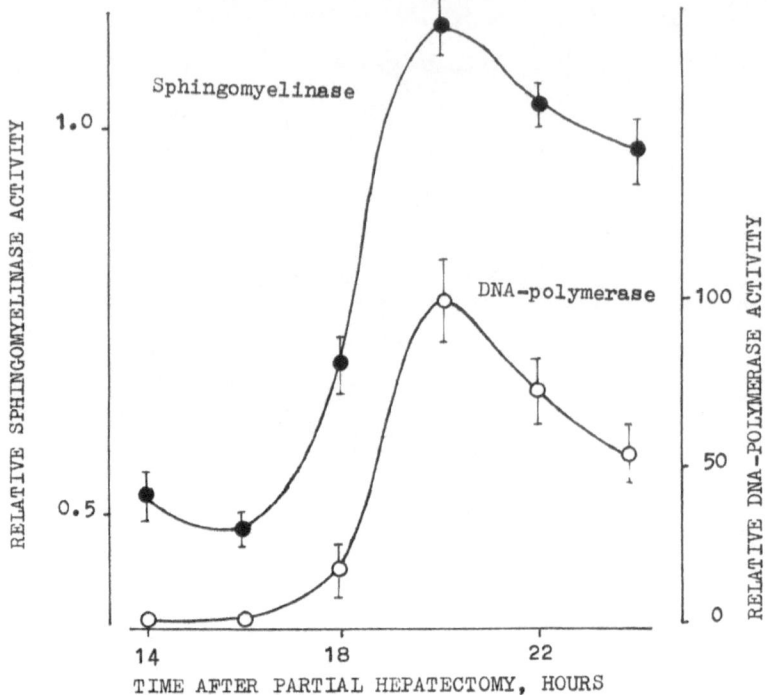

Figure 3.   Changes in the sphingomyelinase and DNA-polymerase activities
              in the nuclei of regenerating rat liver cells after partial
              hepatectomy.

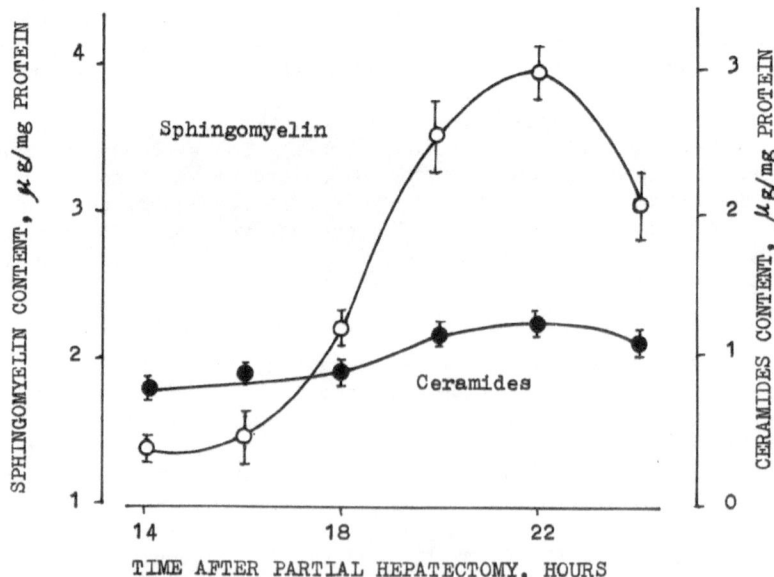

Figure 4.   Changes in the content of sphingomyelin and ceramides in the
              activation of DNA synthesis in nuclei of regenerating rat
              liver cells.

sphingomyelin hydrolysis. We suggest that the amount of ceramide in the nucleus is determined by the sphingomyelin content. It has been shown earlier that sphingomyelin participates in the binding of DNA and RNA to the nuclear matrix (Alessenko et al., 1983, Alessenko et al., 1985), hence an increase in the sphingomyelinase activity during the activation of the nucleic acid synthesis can promote the release of the newly synthesized nucleic acids from the nuclear matrix protein structures due to the enzymatic sphingomyelin hydrolysis.

Other mechanisms of participation of sphingomyelin, the products of its enzymatic hydrolysis and of sphingomyelinase in the regulation of synthetic processes in the nucleus are also possible.

REFERENCES

Alessenko A. V., Boikov P. Ya., Pushkareva M. Yu., and Shevchenko N. I., 1985, RNA polymerase activity distribution in the chromatin and the nuclear matrix and the relation to the sphingomyelin content in the intranuclear structures, in: Abstracts of 9th Nucle(ol)ar Workshop, September 13th-17th, 1985, Crakow-Poland.

Alessenko A. V., Burlakova E. B., and Pantaz E. A., 1984, Effect of sphingomyelin on the RNA-polymerase activity in cell nuclei of normal regenerating rat liver, Biochimia, 49:621.

Alessenko A. V., Krasilnikov V. A., and Boikov P. Ya., 1983, The participation of sphingomyelin in the formation of the bind of DNA with the nuclear matrix during replication, Adv. Acad. Sci. USSR, 273:231.

Alessenko A. V., and Pantaz E. A., 1983, Differences in the composition of nuclear chromatin phospholipids in proliferating rat liver cells after partial hepatectomy. Biochimia, 48:263.

Hannun Y. A., and Bell R. M., 1987, Lysosphingolipids inhibit protein kinase C: Implications for the sphingolipidoses, Science, 235:670.

Kiss Z., Rapp U. R., and Anderson W. B., 1988, Phorbol ester stimulates the synthesis of sphingomyelin in NIH 3T3 cells. A diminished response in cells transformed with human A-raf carrying retrovirus, FEBS Lett., 240:221.

Kolesnick R. N., 1987, 1,2-Diacylglycerol but not phorbol ester stimulate sphingomyelin hydrolysis in GH3 pituitary cells, J. Biol. Chem., 262:16759.

Kolesnick R. N., and Clegg S., 1988, 1,2-Diacylglyserol but not phorbol esters activate a potential inhibitory pathway for protein kinase C in GH3 pituitary cells. Evidence for involvement of a sphingomyelinase, J. Biol. Chem., 263:6534.

ROLE OF LIPIDS IN FUNCTIONAL ACTIVITY OF CELL NUCLEUS

A. V. Alessenko

Institute of Chemical Physics
USSR Academy of Sciences
Moscow
USSR, 117334

INTRODUCTION

Lipids have been found in the interphase chromatin and metaphase chromosomes from a variety of sources. Phospholipids have been also found in nuclear matrix, which is considered to be the sit of synthesis of DNA and RNA. Unfortunately there is little information available on the functional role of lipid and lipid metabolism in cell nuclei.

In our investigation we have charaterized the phospholipid composition of phospholipids of chromatin and nuclear matrix both in the resting and activated nucleus (during RNA and DNA synthesis). An attempt has been made to establish which of phospholipids is functionally the most active in the processes of replication and transcription.

RESULTS AND DISCUSSION

The experiments were carried out according to Alessenko et al. (1983 and 1984). The liver of intact animals, regenerating liver after partial hepatectomy and the liver of the animals injected with CHI at sublethal doses (3mg/kg) was used for studying the composition of phospholipids in nuclei, chromatin and nuclear matrix.

The phospholipid spectra of the nucleus, chromatin and nuclear matrix isolated from the liver cells of intact rats are very different. The chromatin and nuclear matrix phospholipids are enriched in sphingomyelin (Table I, column 1). Sphingomyelin appears to be present in the chromatin and nuclear matrix as lipoprotein. This assumption is supported by the data of Manzolly et al. (1976), who have shown that the acidic proteins from lymphocyte nuclei are enriched in sphingomyelin.

The autonomous behaviour of chromatin and nuclear matrix phospholipids is emphasized by sharp changes in their spectra against minor alterations in the composition of the nucleus during enhanced RNA- and DNA-synthesis processes in liver cells after partial hepatectomy (Table I) and after injection of sublethal doses of CHI (Table II). A sharp increase in the sphingomyelin content during activation of transcription and replication in chromatin and nuclear matrix is characteristic change for the both regenerating liver and inhibition of protein synthesis by CHI.

Nuclear Structure and Function, Edited by J. R. Harris and
I. B. Zbarsky, Plenum Press, New York, 1990

Table I. Content of Phospholipids (in %) in Nucleus, Chromatin and Nuclear Matrix of Normal and Regenerating Liver After Partial Hepatectomy at the Moment of Maximum RNA and DNA Synthesis.

| Phospholipids | control | | | regenerating liver | | | | | |
| | | | | RNA-synthesis | | | DNA-synthesis | | |
	nucleus	chromatin	nuclear matrix	nucleus	chromatin	nuclear matrix	nucleus	chromatin	nuclear matrix
PC	56.3	51.6	30.8	53.0	34.1	10.5	51.8	40.4	14.5
PE	25.6	19.2	12.2	30.3	13.2	7.3	30.5	19.8	3.2
PS	2.3	2.7	0.8	4.4	3.9	--	2.0	4.5	0.3
PI	10.0	7.3	2.9	8.5	10.3	5.3	8.1	6.1	0.9
SPm	3.8	12.7	52.0	3.7	22.5	74.2	1.9	21.0	78.2
CL + PA	2.3	6.7	1.3	3.4	15.6	2.7	2.9	8.2	2.6

Abbreviations : PS - phosphatidylcholine; PA - phosphatidic acid;

CL - cardiolipin; PS - phosphatidylserine;

PE - phosphatidylethanolamine; PI - phosphatidylinositol;

SPm - sphingomyelin

Table II. Content of Phospholipids (in %) in Nucleus, Chromatin and Nuclear Matrix of Rat Liver at the Moment of Stimulated RNA and DNA Synthesis After Injection of a Sublethal CHI Dose.

| PHOSPHOLIPIDS | CONTROL | | | CHI INJECTION | | | | | |
| | | | | RNA SYNTHESIS | | | DNA SYNTHESIS | | |
	NUCLEUS	CHROMATIN	NUCLEAR MATRIX	NUCLEUS	CHROMATIN	NUCLEAR MATRIX	NUCLEUS	CHROMATIN	NUCLEAR MATRIX
PC	53.8	49.2	28.7	56.6	41.1	22.9	57.1	39.8	25.1
PE	28.0	20.8	13.7	25.3	24.3	12.7	25.7	19.5	11.0
PS	2.6	2.9	2.0	2.5	5.2	1.2	2.9	4.5	2.1
PI	9.6	9.3	1.4	9.3	8.3	1.9	8.3	7.8	2.0
SPm	2.5	12.4	48.5	3.5	18.2	59.5	3.2	25.0	56.6
CL + PA	2.6	5.6	5.4	2.5	3.1	1.8	2.7	4.8	2.7

ABBREVIATIONS : CHI - CYCLOHEXIMIDE ;   PC - PHOSPHATIDYLCHOLINE
                 PE - PHOSPHATIDYLETHANOLAMINE; CL - CARDIOLIPIN
                 PS - PHOSPHATIDYLSERINE ;   PI - PHOSPHATIDYLINOSITOL
                 SPm - SPHINGOMYELIN ;   PA - PHOSPHATIDIC ACID

Table III.   The Effect of Phospholipase of the Release of Newly Synthesized RNA (in %) and DNA (in %) From the Nuclear Matrix of Regenerating Rat Liver Cells.

Enzyme	Control		DNA release (in %) inactivated enzyme		RNA release (in %)	
	super-natant	matrix	super-natant	matrix	super-natant	matrix
Sphingomyelinase	3.6	96.4	75	25	88.3	11.7
Phospholipase C	4	96	24	76	10.2	89.8
Sphingomyelinase+ Phospholipase C	3	97	95	5	96	4
Phospholipase A_2	3	97	30	70	--*	--
Phospholipase D	3.5	96.5	6	94	--	--

* -- not detected

An increase in sphingomyelin concentration i correlated with the level of DNA synthesis in liver cells in regenerating liver and after CHI injection. Sphingomyelin is the main component of phospholipids (80%) in the nuclear matrix of regenerating liver, but there is only a small increase in its content upon CHI induction, where DNA synthesis is less marked (Tables I and II

On obtaining the data on a high content of sphingomyelin in the nuclear matrix at the moment of the RNA and DNA synthesis we have considered the possibility that it can be incorporated into the replicative and trans-criptional complexes associated with the nuclear matrix. This has been confirmed by digestion of sphingomyelin in the nuclear matrix with sphingomyelinase, as a result of which 75% of the newly synthesized DNA and 90% of the newly synthesized RNA have been released from the nuclear matrix (Table III). The treatment of nuclear matrix with Phospholipases (phospholipase C without sphingomyelinase activity, phospholipase A_2 and phospholipase D) has not led to the release of such a high yield of newly synthesized nucleic acids into the supernatant. Significantly, in vivo a release of DNA and RNA from the protein structures of the nuclear matrix can be brought about by the endogenous sphingomyelinase, whose activity we have detected in the nucleus.

It was of interest to us to study the participation of phospholipids in the DNA-polymerase activity in the nuclear matrix, where the replicative complexes are formed. We have shown that digestion of phospholipids by phospholipase C + sphingomyelinase in nuclear matrix from regenerating liver cells, at the time of the DNA synthesis, results in a sharp decrease in the activity of the DNA polymerase associated with the nuclear matrix.

DNA polymerases are known to be lipid-depended enzymes (Hachman and Lezuis, 1975). It has been shown that many enzymes participating in replication (topoisomerases, DNA-unwinding proteins, nucleases) have hydrophobic properties, easily form complexes with the lipids and their activity is determined by the lipid surrounding (Yoshida, 1977; Herzberg, 1983). Inhibition of DNA-polymerase in the nuclear matrix and release of

Table IV.   Influence of the Sphingomyelin Injection to
             Animals and its Digestion in the Nuclei by
             Sphingomyelinase on the Activity of Free and
             Template-Engaged RNA Polymerases.

	Injection of sphingomyelin		Digestion of sphingomyelin	
	free *RNA* polymerase	engaged *RNA* polymerase	free *RNA* polymerase	engaged *RNA* polymerase
INTACT RATS	163±45	155±10	65±5	55±5
OPERATED RATS	180±28	160±17	30±8	36±8

newly synthesized DNA from the nuclear matrix when treated with phospholipases indicates a structure-function role of phospholipids in the replic-
active complex. Our results lead us to assume that sphingomyelin is the
main phospholipid involved in the formation of the replicative complex in
the nuclear matrix.

     Sphingomyelin may also be present in the replicative enzymatic complex
formed on the nuclear matrix. As with DNA polymerases, RNA polymerases are
lipid-depended enzymes (Menon, 1972). Our experiments have shown that there
exists a certain coordination between sphingomyelin accumulation in
chromatin, both from the regenerating liver and after the CHI injection, and
an increase in the activity of the template-engaged RNA-polymerase, while
the activity of the free enzyme is conversely related to the sphingomyelin
content in chromatin. These data suggest that sphingomyelin facilitates
the transition of RNA-polymerase from the free to the timplate-engaged
state. Injection of sphingomyelin to the intact and operated animals
resulted in sharp increase in the activity of RNA-polymerases (Table IV).

     Sphingomyelin digestion in the intracellular structures resulted in a
decrease in the activity of the RNA-polymerases of all types (Table IV).
Thus, sphingomyelin is associated with the processes of replication and
transcription in a similar manner. This generally appears to be determined
by the effect of sphingomyelin on the DNA structure, which has been shown
in the work of Manzolli (1972) on the thermal DNA denaturation and in our
studies using spin label and fluorescent probes. The unique nature of
sphingomyelin is determined by the structure of this phospholipid molecule,
which includes three active functional groups (hydroxyl, amide and
trimethylammonia), thereby resembling the periodicity in the arrangement of
the functional groups in polyamines.

REFERENCES

Alessenko A. V., Burlakova E. B., Pantaz E. A., 1984 The influence of
     sphingomyelin on the RNA-polymerases activity in the cell nuclei of
     normal regenerating rat liver, Biochimia, 49:621.
Alessenko A. V., Krasilnikov V. A., Boikov P. Ya., 1983, The participation
     of sphingomyelin in the formation of the bind of DNA with the nuclear
     matrix during replication, Adv. Acad. Sci. USSR, 273:231.
Hachman H. J., Lezuis A. G. 1975, High-molecular-weight DNA polymerases from
     the mouse myeloma. Purification and properties of three enzymes.
     European J. Biochem., 50:357.

Herzberg M., Nathanel T., Bibor-Haroty V., Wreschner E. 1983, Location of RNA-ase activity in nuclear residual structure. Biol. Cell, 49:11.

Manzolli F. A., Coccol L., Facchini A. 1976, Phospholipids bound to acidic nuclear proteins in human B and T lymphocytes. Mol. Cell. Biochem. 12:67.

Manzolli F. A., Muchmore J. H., Bouora B. 1972, Interaction between sphingomyelin and DNA. Biochim. Biophys. Acta. 277:251.

Menon J. A. 1972, A possible role of lipids in RNA polymerase from mammalian cells. Canad. J. Biochem. 50:807.

ROLES OF DNA TOPOISOMERASES I AND II IN DNA REPLICATION, MITOTIC

CHROMOSOME FORMATION, AND RECOMBINATION IN MAMMALIAN CELLS

Martin Charron and Ronald Hancock

Centre de Recherche en Cancérologie de l'Université Laval
Hôtel-Dieu Hospital
Québec
Canada G1R 2J6

INTRODUCTION

DNA topoisomerases I and II effect topological changes in DNA, passing a single or double stranded DNA across a second DNA by reactions in which the enzyme integrates covalently and reversibly into the DNA (Vosberg, 1985; Wang, 1985). Their essential functions in chromosome replication and transcription in mammalian cells are not understood. Identification of these functions is important to understand the cytotoxic action of some important and effective anticancer drugs (Adriamycin; Etoposide; Amsacrine) which trap and arrest the reaction of topoisomerase II molecules integrated into DNA (reviewed in Potmesil and Ross, 1987; Drlica and Franco, 1988; Hancock et al., 1989; Liu, 1989).

We are studying the roles of topoisomerases I and II in chinese hamster ovary (CHO) cells, using such agents to trap and inhibit their target enzyme. Movement along the DNA of other factors such as polymerases may then be blocked, but we reasoned that an inhibitory effect at any period in chromosome replication shows that the topoisomerase is functioning at that time. We used camptothecin to specifically inhibit topoisomerase I (Andoh et al., 1987), and VM-26 (4'-demethylepipodophyllotoxinthenylidene-β-D-glucoside) which, according to current evidence, targets topisomerase II specifically (Yang et al., 1987; Snapka et al., 1988; Sullivan et al., 1989).

TOPOISOMERASE I FUNCTIONS ONLY DURING THE G1 AND S PHASES

DNA replication in synchronously growing cells is slowed by camptothecin, added either at the initiation of growth or during the S phase, and is inhibited by > 80% at 1μg/ml (Figure 1). Inhibition is greater when camptothecin is present in the G1 phase also, and does not recover completely when it is removed at the beginning of the S phase, suggesting that inhibition of processes during the G1 phase, probably transcription, contributes to the inhibition of DNA replication.

The progression of cells to mitosis after DNA replication is completed (2 hr before mitosis) is not affected by camptothecin, even at a

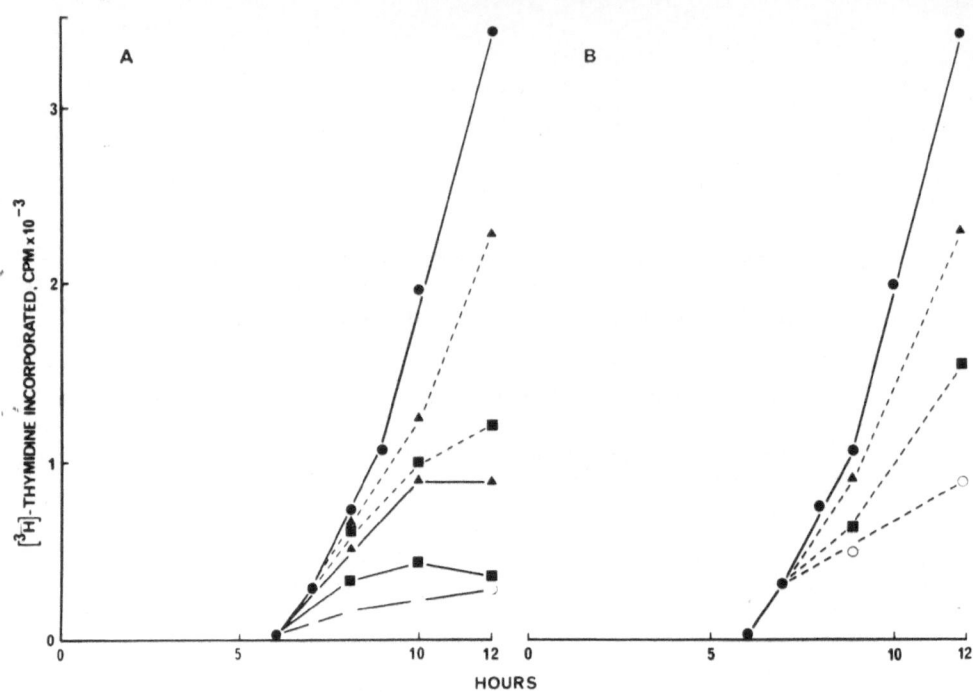

Figure 1.   Inhibition of DNA synthesis when topoisomerase I is inhibited
              by camptothecin in synchronously growing CHO cells.
              Camptothecin was present (A) during the entire growth cycle
              (solid lines) or only until 6hr, the start of the S phase
              (dashed lines), or (B) was added at 7hr, after initiation of
              the S phase at 0 (●), 0.05 (▲), 0.2 (■) or 1 (O) μg/ml.

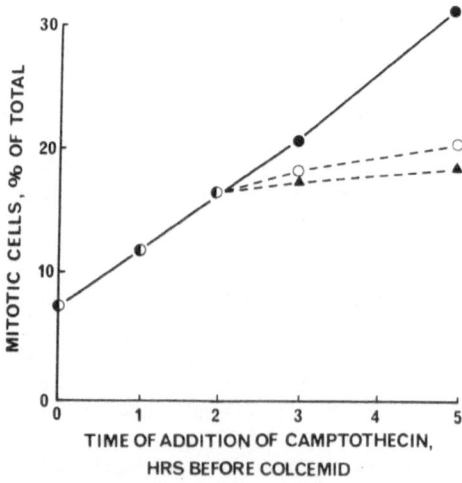

Figure 2.   The progression of cells to mitosis is not affected when
              topoisomerase I is inhibited by camptothecin after the
              completion of DNA replication (< 2hr before mitosis).
              Camptothecin was added at 0 (●), 1 (O) or 5 (▲) μg/ml to
              nonsynchronous cells at different times before adding colcemid
              (0.06μg/ml), and accumulating mitotic cells were counted in
              the phase contrast microscope.

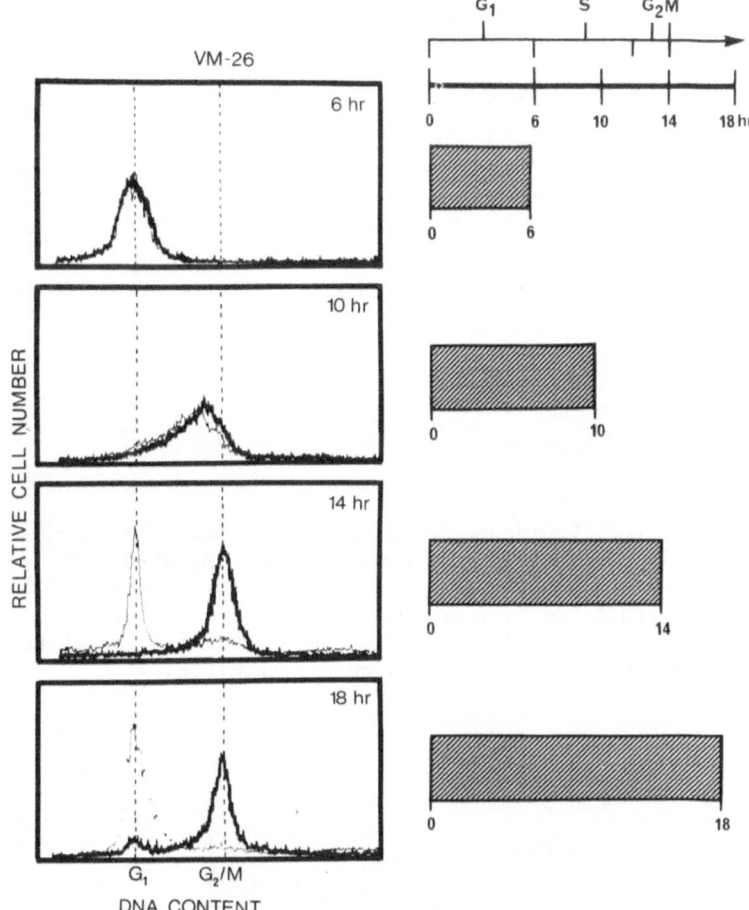

Figure 3.   Inhibition of topisomerase II by VM-26 does not affect DNA
            replication (6 and 10hr) but delays progress through mitosis
            (14 and 18hr). The DNA content of cells growing synchronously
            without (light trace) or with VM-26 (0.05µg/ml; heavy trace)
            was followed by flow cytometry. The schema (right) shows the
            time of sampling relative to the cell cycle periods.

concentration of 5µg/ml (Figure 2), in contrast to its sensitivity to VM-26
(below). Cell division is also not sensitive to camptothecin (unpublished
observations).

TOPOISOMERASE II FUNCTIONS IN MITOTIC CHROMOSOME FORMATION

    DNA replication during the S phase is not affected by VM-26 at a
cytotoxic concentration (0.05µg/ml) (Figure 3). It continues at ≈ 50% of
the normal rate even at a 1000-fold higher concentration (50µg/ml), but
must be measured by flow cytometry because uptake of precursors is inhibited
at > 1µg/ml (Loike and Horwitz, 1976; Richter et al., 1987; unpublished
observations).

    Progress through mitosis is severely delayed by VM-26 at low
concentrations which do not affect DNA synthesis (0.01-0.05µg/ml) (Figure

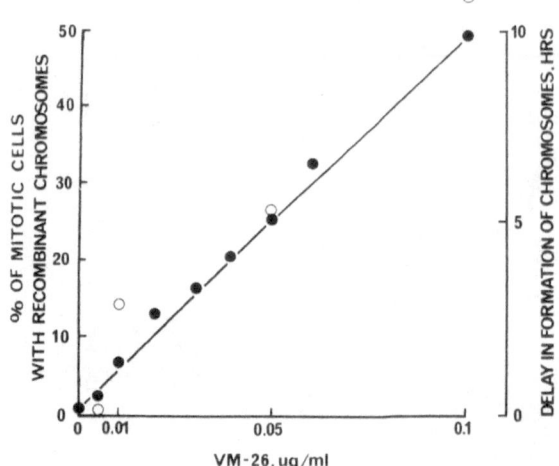

Figure 4.   Delayed chromosome formation and recombinant chromosomes in
             cells grown for 1 cycle with VM-26. Mitotic cells were
             accumulated in synchronous cultures by addition of colcemid
             at 12hr; the time necessary for 10% of the cells to form
             chromosomes (O), and the number containing recombinant
             chromosomes at 16hr (●), were determined.

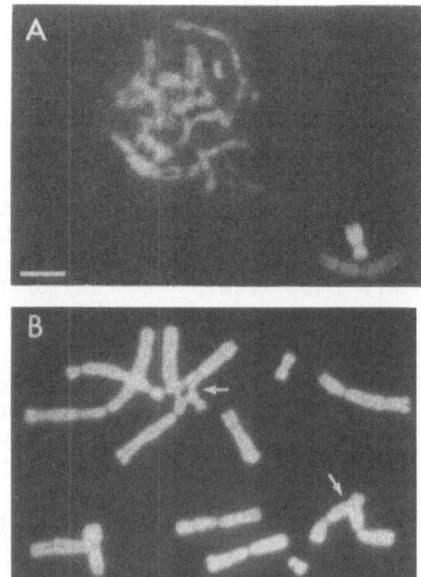

Figure 5.   Chromosomes in mitotic cells after growth with VM-26 (0.05μg/
             ml) in synchronous culture for 1 cycle are (A), incompletely
             formed in ≃ 5% of the cells, or (B), have undergone
             recombination (arrows) in ≃ 25%. Mitotic cells collected
             from 12-16hr were processed for cytogenetic examination and
             stained with the fluorescent DNA stain Hoechst 33258.
             Bar = 2μm.

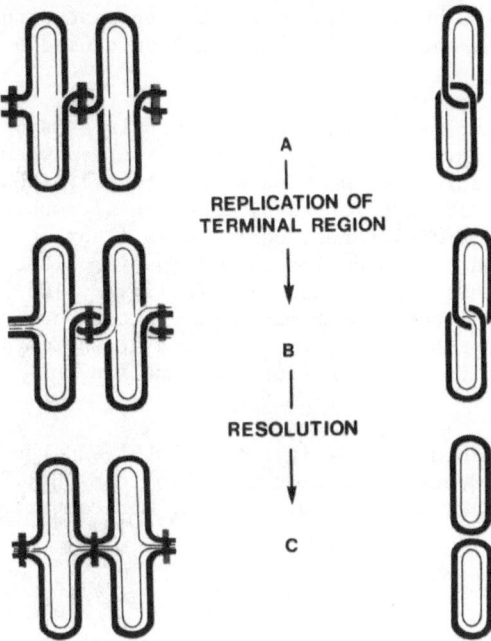

Figure 6.   The termination of replication (A-B) and the resolution into
             long linear molecules (B-C) of adjacent loops of chromosomal
             DNA (left), compared with the topologically analogous steps
             in SV40 DNA (right) in which topoisomerase II functions.
             Parental and new DNA strands are heavy and light; rectangles
             (left) represent sites of attachment and topological
             constraint of DNA in the nucleus.

3). The formation of chromosomes is slowed (Figure 4) and some chromosomes
are incompletely condensed (Figure 5A). More strikingly, many have
undergone recombination (Figure 5B), and at high concentrations of VM-26
some cells contain many chromosomes associated together in complex
configurations.

DISCUSSION

     These experiments show that in mammalian cells, topoisomerase I
functions only during the G1 and S phases, probably to relax DNA super-
helicity to allow DNA polymerase to proceed for replication, and RNA
polymerase for transcription (Giaever et al., 1988). Topoisomerase I does
not function during the G2 phase or mitosis. Topoisomerase II is not
essential during DNA replication, but the slowing by VM-26 at high
concentrations may indicate that it functions facultatively, as in yeast
(Uemura and Yanagida, 1984).

     Topoisomerase II has essential functions in the formation of mitotic
chromosomes. In SV40 virus it decatenates completed daughter DNA molecules
(Richter et al., 1987; Yang et al., 1987; Snapka et al., 1988) and is also
essential for replication of the last $\simeq$ 250 bases (Snapka et al., 1988). In
chromosomal DNA, the topological analogs of these steps are the terminal
replication of DNA loops or domains (Hancock, 1982), and their resolution
into 2 long linear molecules forming the chromatids of a chromosome (Figure
6). Delay in the formation of chromosomes induced by VM-26 suggests that
topoisomerase II functions to form these linear DNA molecules (Figure 6,
B-C) before the chromatin can be correctly folded to form a chromosome.
Topoisomerase II functions in this period in yeast (DiNardo et al., 1984;

Uemura et al., 1987), but in CHO cells the abnormal structure of some chromosomes, and the chromosomal recombination caused by VM-26, can be clearly visualised.

The formation of recombinant chromosomes is a major effect of VM-26 (see also Huang et al., 1973). They are not formed when topoisomerase I is trapped by camptothecin (unpublished observations). Quadriradial chromosomes (Figure 5B) are formed by exchange of double stranded DNA between chromatids of 2 different chromosomes (Therman and Kuhn, 1981). Topoisomerase II can recombine lambda DNA *in vitro* (Bae et al., 1988), possibly by exchange of enzyme subunits (Filipski, 1983), and such exchanges between the DNA of different chromatids may occur when many enzyme molecules are trapped in DNA. This phenomenon focuses attention on the possible role of topoisomerase II in recombination in mammalian cells.

We thank the MRC of Canada (grant MA-9589) and the NCI of Canada for support, the FRSQ (Québec) for a Scholarship to M. C., Bristol-Myers for VM-26, and the Natural Products Branch, NCI for camptothecin.

REFERENCES

Andoh T., Ishii K., Suzuki Y., Ikegami Y., Kusunoki Y., Takemoto H., and Okada K. 1987. Characterisation of a mammalian cell mutant with a camptothecin resistant DNA topoisomerase I. Proc. Natl. Acad. Sci. USA, 84:5565.
Bae Y. S., Kawasaki I., Ikeda H. and Liu L. F. 1988. Illegitimate recombination mediated by calf thymus DNA topisomerase II *in vitro*. Proc. Natl. Acad. Sci. USA, 85:2076.
DiNardo S., Voelkel K. and Sternglanz R. 1984. DNA topoisomerase II mutant of *S. cerevisiae:* topoisomerase II is required for segregation of daughter molecules at the termination of DNA replication. Proc. Natl. Acad. Sci. USA. 81:2616.
Drlica K., and Franco R. J. 1988. Inhibitors of DNA topoisomerases. Biochemistry 27:2253.
Filipski J. 1983. Competitive inhibition of nicking-closing enzymes may explain some biological effects of DNA intercalators. FEBS Lett. 159:6.
Giaever G. N., Snyder L., and Wang J. C. 1988. DNA supercoiling *in vivo*. Biophys. Chem. 29:7
Hancock R. 1982. Topological organisation of interphase DNA. Biol. Cell, 44:201.
Hancock R., Charron M., Lambert H., Lemieux M., Pankov R. and Pépin N. 1989. Topoisomerase II as a target of antitumour agents. *in* 'Molecular aspects of Chemotherapy', E. Borowski and D. Shugar, eds., Pergamon Press, London.
Huang C. C., Hou Y. and Wang J. J. 1972. Effects of a new antitumor agent, epipodophyllotoxin, on growth and chromosomes in human erythropoietic cell lines. Cancer Res. 33:3123.
Loike J., and Horwitz S. B. 1976. Effects of podophyllotoxin and VP-26-213 on microtubule assembly and nucleoside transport in HeLa Cells. Biochemistry 15:5435.
Liu L. F. 1989. DNA topisomerase poisons as antitumor drugs. Ann. Rev. Biochem. 58:351.
Potmesil M., and Ross W. E., eds. 1987. First conference on DNA topoisomerases in cancer chemotherapy. NCI Monograph, 3:1.
Richter A., Strausfeld U. and Knippers R. 1987. Effects of VM-26 (teniposide), a specific inhibitor of type II topoisomerase, on SV40 DNA replication *in vivo*. Nucleic Acids Res. 15:3455.
Snapka R. M., Powelson M. A. and Strayer J. M. 1988. Swivelling and

decatenation of replicating SV40 virus genomes *in vivo*. Mol. Cell. Biol. 8:515.

Sullivan D. M., Latham M. D., Rowe T. C., and Ross W. E. 1989. Purification and characterization of an altered topoisomerase II from a drug-resistant chinese hamster ovary cell line. Biochemistry 28:5680.

Therman E. and Kuhn E. M. 1981. Mitotic crossing-over and segregation in man. Hum. Genet. 59:93.

Uemura T. and Yanagida M. 1984. Isolation of type I and II DNA topoisomerase mutants from fission yeast. EMBO J. 3:1737.

Uemura T., Ohkura H., Adachi Y., Morino K., Shiozaki K. and Yanagida M. 1987. Topoisomerase II is required for condensation and separation of mitotic chromosomes in *S. pombe*. Cell 50:917.

Vosberg H-P. 1985. DNA topoisomerases: enzymes that control DNA conformation. Curr. Top. Microbiol. Immunol. 114:19.

Wang J. C. 1985. DNA topoisomerases. Ann. Rev. Biochem. 54:665.

Yang L., Wold M. S., Li J. J., Kelly T. J. and Liu L. F. 1987. Roles of DNA topoisomerases in SV40 virus DNA replication *in vitro*. Proc. Natl. Acad. Sci. USA 84:950.

EFFECTS OF DRB ON THE ULTRASTRUCTURAL ORGANIZATION AND RNA METABOLISM OF

PLANT NUCLEUS

M. E. Fernandez-Gomez, F. J. Medina and
S. Moreno Diaz de la Espina

Centro de Investigaciones Biologicas
C.S.I.C. Velazquez 144
28006 Madrid
Spain

INTRODUCTION

DRB is an adenosine analogue which preferentially affects the RNA metabolism in animal cells. It acts at the level of ribonucleotide incorporation (Granick, 1975) and interferes with hnRNA synthesis, but not with the labelling of poly A (Shegal et al., 1976). DRB action appears to be mediated by a protein kinase (Zandomeni and Weinmann, 1984). It has been demonstrated that DRB forms covalent complexes with DNA topisomerase II, producing fragmentation of DNA and a reduction of the replicational rate (Löon and Löon, 1987). Its effects appear to be reversible (Tamm and Shegal, 1978; Granick, 1975). In relation to rRNA synthesis, DRB does not impede the attachment of the RNA polymerase I to the template, but it appears to interfere with the processing of the rRNA precursors through its action on the hnRNA synthesis. This gives rise to striking morphological alterations in the nucleolus, known as nucleolar necklaces (Granick, 1975), in which, active rDNA genes become separated from the nucleolar components containing the preribosomal particles (Scheer et al., 1984). In plant cells, the supramolecular organization of the ribosomal genes and their products is in some way different from that in animal cells (Risueño and Medina, 1986). DRB has been reported to reversibly dissociate the nucleolar components in cultured rat cells, establishing a separation between the sites of rRNA synthesis and further processing (Scheer et al., 1984). We decided to test the action of DRB on plant nucleoli as a tool for the morphological dissociation of the nucleolar components and further characterization of their functions.

MATERIAL AND METHODS

Allium cepa meristematic root cells were given a continuous treatment with 200µgr/ml DRB during 24hr. For recovery, after 12hr of treatment, they were transfered to tap water. For RNA synthesis evaluation, ^{3}H Udr at a concentration of 100µCi/ml was given to the roots during the last hour of treatment. Autoradiography was performed on semi-thin sections of roots by the stripping method, and quantified with a semiautomatic image analyzer. Nucleolar segregation in roots was analyzed in the light microscope, using the specific silver staining technique of Fernandez-Gomez et al. (1969).

Nuclear Structure and Function, Edited by J. R. Harris and
I. B. Zbarsky, Plenum Press, New York, 1990

Table I. ^{3}H Udr Incorporation Rates After a Continuous 200µg/ml Treatment, in Relation to Those in Untreated Cells; as Revealed by Light Microscopical Autoradiography.

hrs DRB	0	1	2	3	6	8	12	18	24
nucleoplasm	100	7.6	9	6	15	15	15	13.6	18
nucleoli	100	4.7	17.5	8	19	19	17.5	20	24
cytoplasm	100	3	3	3	3	6.25	6.25	9.37	9.37

RESULTS AND DISCUSSION

RNA Synthesis

DRB has a clear effect on both nuclear and nucleolar RNA synthesis of plant cells, as demonstrated by the autoradiography experiments (Table I). The continuous treatment with 200µg/ml produces a rapid fall of the transcription rate during the first hour of treatment. The nucleolar labelling partially recovers during the treatment, reaching values of about 24% of those in controls at the 24th hr of treatment. Extranucleolar transcription maintain very low levels during the treatment. The effect of the DRB proved to be fully reversible: both nucleolar and nuclear labelling rapidly recover after a 12hr treatment, during the first 6hr of incubation in water (data not shown here). These data agree with previous results in animal cells (Granick, 1975) and also demonstrate that DRB is a clean and useful tool for the study of nuclear RNA metabolism of plant cells as proved to be for animal cells (Puvion-Dutilleul et al., 1986; Scheer et al., 1984).

Nucleolar Morphology and Ultrastructure

The nucleolus of active proliferating meristematic root cells grown in standard conditions shows an ultrastructural organization with three morphological components at the EM: Fibrillar centres; which are known to contain non-transcribing nucleolar DNA; a dense fibrillar component, in which transcription and early maturation of rRNA precursors takes place; and a granular component corresponding to the further steps of maturation of preribosomal particles (Figure 1, Risueño and Medina, 1986: see also Medina et al., this book). In these very active nucleoli the fibrillar centres are numerous and homogeneous in structure, and appear immersed in the patches of the dense fibrillar component which in turn are surrounded by the granular component. The continous treatment with DRB produces not only a sequential rearrangement of the nucleolar components, but also striking ultrastructural variations of them (Figures 2 to 6). In the very early moments of the treatment the nucleolar components segregate in a 60% of the cell population, coinciding with the rapid fall in nucleolar RNA synthesis (Table I; Figures 2 and 8). There is a second peak of segregation later, at about the 9th hour of treatment. At the EM level the segregated nucleolus shows heterogeneous fibrillar centres containing inclusions of condensed chromatin in their interiors, an internal and compact dense fibrillar component, and a peripheral layer of granular component. This ultrastructural organization of the nucleolus has been demonstrated to correspond to a block of nucleolar biosynthetic activity, and has been reported to be induced by other drugs affecting RNA metabolism (Risueño and Medina, 1986; Bernhard, 1971; Fakan and Puvion, 1980). Later the nucleolus loses its organized granular component and shows an ultrastructure typical of those starting activity, like early G1 nucleoli or early reactivating nucleoli from quiescent roots

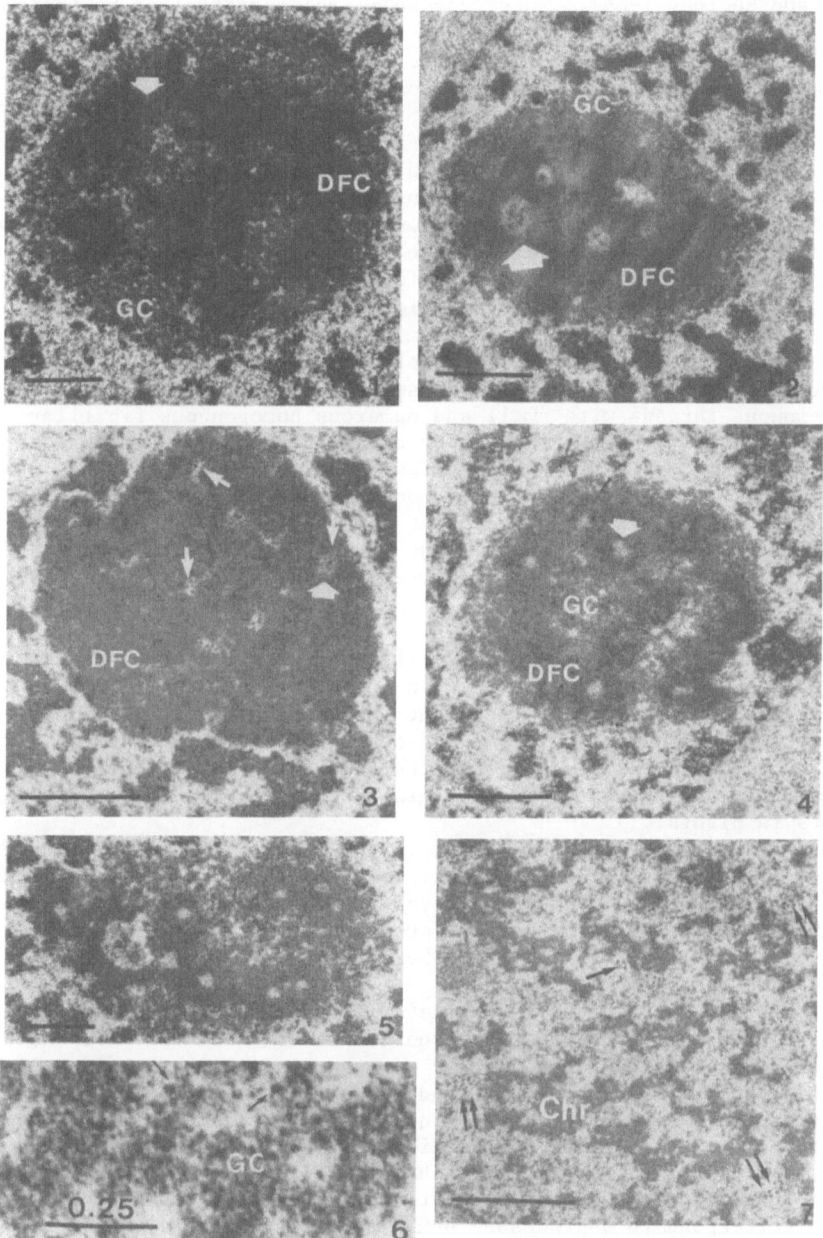

Figure 1.   Untreated, meristematic nucleolus comprised of intermingled
             dense fibrillar components and granular components. Fibrillar
             centres (⬗) are very small and homogeneous.
Figure 2.   1h DRB-treated segregated nucleoli. The DFC has an inner
             localization while the GC is peripherally located. Big
             fibrillar centres (⬗) with inclusions of condensed chromatin.
Figure 3.   2h DRB Fibrillar nucleolus with nucleolar perichromatin
             granules (↑) associated to the fibrillar centres (⬗).
Figure 4.   8h DRB: Nucleolus with aboundant DFC forming a thread with
             numerous homogeneous fibrillar centres (⬗). Scarce GC
             located around the thread of the DFC.

                                          (caption continues ...)

                                                                      415

(Risueño and Medina, 1986). These nucleoli are characterized by their
fibrillar structure and the presence of nucleolar perichromatin granules
(Figure 3), which form at the interphase between the homogeneous fibrillar
centres and the surrounding fibrillar component, where nucleolar transcription occurs (see Medina et al, in this book). Nucleolar perichromatin
granules have also been reported in DRB-treated nucleoli of rat hepatocytes
(Puvion-Dutilleul et al, 1983), but in plant cells they never accumulate in
great numbers within the nucleolus, being either processed to normal
preribosomal particles or degraded. These granules correspond to abnormally
processed preribosomal particles and are considered as a marker of nucleoli
in which processing of rRNA precursors is impaired (Puvion-Dutilleul et al,
1983). After a second peak of segregation when the nucleolar activity
starts, there is a process of rearrangement of the dense fibrillar component,
coinciding with the stabilization of the nucleolar RNA synthesis at the new
lower rates (Figures 4 and 8). The dense fibrillar component appears to
form continuous threads with small and numerous homogeneous fibrillar
centres inside, which are surrounded by a peripheral layer of granular
component (Figure 8). Later these threads disperse forming patches of dense
fibrillar component, observed as interconected dense argyrophylic spots in
the light microscope, while the granular component gradually increases in
size (Figure 5). The latter has a loose structure and shows aboundant
preribosomal particles bound to its fibres, which sometimes appear as
doublets (Figure 6).

We have never observed in plant cells the formation of true nucleolar
necklaces, similar to those described in animal cells (Granick, 1975;
Scheer et al., 1984). The DRB treatment induces a real rearrangement of the
nucleolar components when nucleolar RNA synthesis stabilizes at the new
rates, but the nucleolus never looses its identity. This difference is
probably due to a different supramolecular organization of the nucleolar
components in plants.

Taken as a whole, the nucleolar lesions induced by the drug seem to
indicate that it acts mainly at the level of maturation but it does not
impair the synthesis of the ribosomal precursors, as reported in animal
cells (Puvion-Dutilleul et al., 1983; Scheer et al., 1984).

The main change observed in nuclear ultrastructure, after DRB treatment, apart from that induced within the condensed chromatin mass which is
not included here, is the accumulation of groups of perichromatin granules
which are scarce in untreated cells and in which these structures appear to
be mainly isolated at the borders of the condensed chromatin mass (Figures
1 and 7). Accumulation of perichromatin granules has been correlated with
the impairment of RNA processing (Bernhard, 1987; Fakan and Puvion, 1980);
and according to the actual interpretation of these structures they would
appear to correspond to unprocessed hnRNP particles kept in this state by
the DRB treatment (Fakan and Puvion, 1980).

Figures 5   12h DRB. Nucleolus with a loose structure, typical of high
and 6.   metabolic activity, with homogeneous and numerous fibrillar
              centres and abundant GC.
Figure 6.   High magnification of the GC; its loose structure allows the
              visualization of the preribosomal particles associated to
              fibres (↗).
Figure 7.   4h DRB. The nucleoplasm shows a very high amount of drug-
              induced perichromatin granules mainly forming groups (↗↗)
              but also isolated ones (↗).
Abbreviations: Chr: chromatin; DFC: nucleolar dense fibrillar component;
              GC: nucleolar granular component. When not indicated scale
              bars = 1μm.

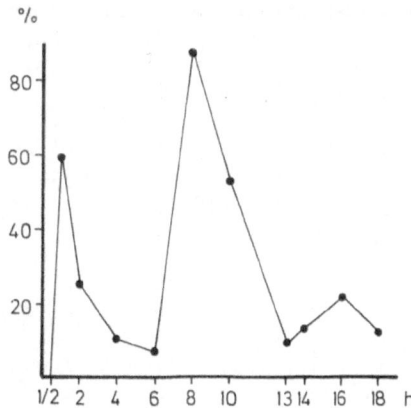

Figure 8.   Nucleolar segregation during a continuous treatment with
             200µg/ml DRB, as estimated in silver impregnated nucleoli at
             the light microscope. Abscissa: hours of treatment.
             Ordinate: Percentage of segregated nucleoli.

CONCLUSIONS

     In general the effects of DRB on RNA synthesis in plant cells are
similar to those described in animal cells. It drastically blocks the
synthesis of nucleoplasmic RNA corresponding mainly to hnRNA. DRB produces
a clear impairment on rRNA synthesis at the beginning of the treatment, but
later on rRNA synthesis recovers and stabilizes at lower rates. The effects
of DRB appear to be rapidly and fully reversible by removal of the drug.

     DRB produces a sequential rearrangement of the nucleolar components
correlating with the variations of rRNA synthesis. The sequence of events
includes: first a transient segregation and degranulation, followed by
formation of nucleolar perichromatin granules. Later, when nucleolar rRNA
synthesis stabilizes at the lower rates, there is a rearrangement and
loosening of the nucleolar components. The nucleolus never loses its
structural identity and true nucleolar necklaces never form. These nucleoli
show an ultrastructural organization similar to that of the very active
proliferating nucleoli (Risueño and Medina, 1986). The ultrastructural
organization of fibrillar centres always correlates with the level of
nucleolar biosynthetic activity, as previously described (Risueño and Medina,
1986).

     The most conspicuous effect on the organization of nuclear RNPs is the
accumulation of groups of perichromatin granules, corresponding to
unprocessed hnRNP particles, a common feature in nuclei treated by drugs
impairing hnRNA metabolism (Fakan and Puvion, 1980).

REFERENCES

Bernhard W., 1971, Drug induced changes in the interphase nucleus. Adv.
     Cytopharmacol., 1:49.
Fakan S. and Puvion E., 1980, The ultrastructural visualization of nucleolar
     and extranucleolar RNA synthesis and distribution. Inter. Rev. Cytol.,
     65:255.
Fernandez-Gomez M. E., Stockert J. C., Lopez-Saez J. and Gimenez-Martin G.,
     1969. Staining plant cell nucleoli with AgNO3 after formalin-hydro-
     quinone fixation, Stain Techn., 44:48.

Granick D., 1975. Nucleolar necklaces in chick embryo fibroblast cells. I. Formation of necklaces by dichlororibobenzimidazole and other adenosine analogues that decrease RNA synthesis and degrade preribosomes. J. Cell Biol., 65:398.

Granick D., 1975. Nucleolar necklaces in chick embryo fibroblast cells. II. Microscope observation of the effect of adenosine analogues on nucleolar necklace formation. J. Cell Biol., 65:418.

Lönn V. and Lönn S., 1987. 5-6-Dichloro-1- -0-ribofuranosylbenzimidazole induces DNA damage by interfering with DNA topoisomerase II. Eur. J. Biochem., 164:545.

Puvion-Dutilleul F., Nicoloso M. and Bachellerie J. P., 1983. Altered structure of ribosomal RNA transcription units in Hamster cells after DRB treatment. Exp. Cell Res., 146:43.

Risueño M. C. and Medina F. J., 1986. 'The nucleolar structure in plant cells'. Barbera-Guillen ed. Serv. Ed. Universidad Pais Vasco. Spain.

Scheer V., Hügle B., Hazan R. and Rose K. M., 1984. Drug-induces dispersal of transcribed rRNA genes and transcriptional products: Immunolocalization and silver staining of different nucleolar components in rat cells treated with 5,6-Dichloro-β-D-ribosuranosylbenzimidazole. J. Cell Biol., 99:672.

Sehgal P. B., Darnell J. E. Jr. and Tamm I., 1976. The inhibition by DRB (5,6-Dichloro-1-β-D-ribofuranosylbenzimidazole) of hnRNA and mRNA production in the Hela cells. Cell, 9:473.

Tamm I. and Sehgal P. B., 1978. Halobenzimidazole ribosides and RNA synthesis of cells and viruses. Adv. Virus Res., 22:187.

Zandomeni R. and Weinmann R., 1984. Inhibitory effect of 5,6-Dichloro-1-β-D-ribofuranosylbenzimidazole on protein kinase. J. Biol. Chem., 259:14804.

REPLICATION TIME OF DNA SEQUENCES RELATED TO G_1 PROGRESSION IN MERISTEMATIC

CELLS

A. González-Fernández, P. Aller, J. Sans* and
C. De la Torre

Centro de Investigaciones Biológicas
C.S.I.C., Madrid
Spain

* Departamento de Biología Celular y Genética
Facultad de Medicina
U. de Chile

ABSTRACT

Two time periods have been detected in G_1 of *Allium cepa* L. meristematic cells where protein synthesis is needed in order to start replication at its due time. The earlier period is located at the telophase/G_1 transition, while the later one is located between one and two hours before the initiation of the S phase.

By bromosubstituting DNA sequences replicated in either early, mid or late S phase, and then irradiating these cells with light of 300-400nm wavelength, it was found that replication of the sequences involved in G_1 progression takes place in both early and late-S phase, but not in mid-S phase.

On the other hand, it was shown in an earlier report that sequences replicated in mid-S phase are involved in a transition point for protein synthesis located in early G_2 phase (De la Torre et al., Eur. J. Cell Biol. 37 (1985) 216).

NUCLEAR LIVER PROTEIN INTERACTION WITH THE RAT TRYPTOPHAN OXYGENASE (TO)
GENE FRAGMENTS INCLUDING THE DNase I HYPERSENSITIVE SITES 'POISED' FOR
TRANSCRIPTIONAL INDUCTION

G. I. Chihirzhina and I. N. Chesnokov

Department of Biochemistry
Leningrad State University
199164 Leningrad
USSR

We tried to detect the trans-acting factors, which bind to fragments
of the 5'-flanking region of the rat tryptophan oxygenase (TO) gene,
containing the tissue-specific DNase I hypersensitive sites 'poised' for
transcriptional induction (Becker et al., 1984). The rat liver nuclear
proteins were found to interact with the fragments from −466 to −292 and
from −292 to −178 of the TO gene. These proteins compete for the NF1 binding
sites. We have also shown that a liver nuclear protein which does not
compete for the NF1 binding sites interacts with the fragment for −292 to
−178 of this gene.

We have used rat liver as a sourse of nuclear extract to identify
DNA-binding proteins which interact with rat TO gene EcoR1 - Taq1 fragment
comprising nucleotides −466 to −178. This fragment contains two DNase I
hypersensitive sites 'poised' for transcriptional induction (Becker et al.,
1984). One site (site 1) is located (Danesch et al., 1987, Shüle et al.,
1988) in the glucocorticoid response element (GRE) and contains the sequences
for binding of the glucocorticoid receptor complex (−472 to −444) and the
CACCC-box binding factor (−455 to −439). A second DNase I hypersensitive
site (site 2) is located in the gene region with an unknown function. To
obtain the EcoR1 - Taq1 fragment containing nucleotides −466 to −178 the
3.7kb EcoR1 fragment to the rat TO gene from λ TO2 (Schmid et al., 1982) was
cloned into pUC 19. Nuclear 0.4M NaCl extract was passed over a DEAE-
cellulose collumn. The DNA-binding proteins only eluted at 300mM NaCl. This
fraction was used in our studies.

In order to study the interaction of nuclear factors with the *EcoR1 -
Taq1* fragment of the TO gene we have utilized an electrophoretic mobility
shift assay which detects the slower migration of DNA-protein complexes with
respect to free DNA. The radiolabelled fragment was incubated with the
nuclear protein fraction in the presence of a 100-1000-fold excess of
unlabelled λ phage DNA as a non-specific competer. The probes yielded two
discrete complexes (data not shown). To ascertain that the bandshifts which
we obtained are due to the specific DNA-protein binding we performed
competition experiments. As shown in Figure 1, the probes which contained a
100-500-fold excess of the unlabelled plasmids (pUC 19, pBR 322), poly(dA)
poly(dT) or TO gene fragment of the coding region, yielded the same two
complexes.

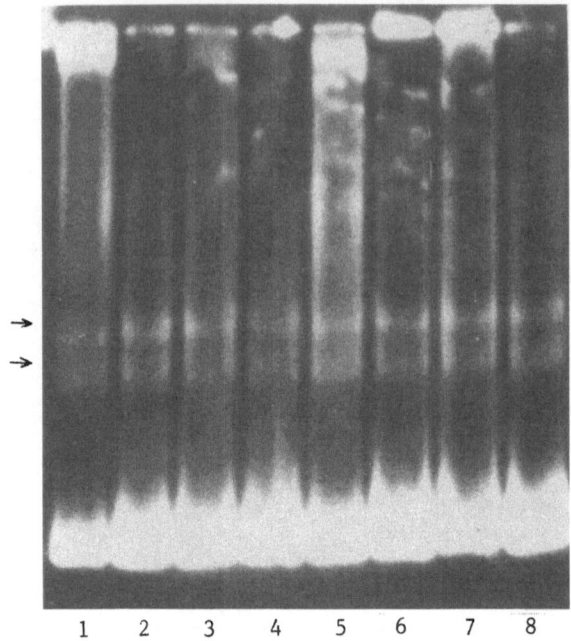

Figure 1.   Binding of the nuclear liver proteins to the -466/-178 EcoR1
             fragment of the rat TO gene in the presence of different
             competitors. The end-labelled fragment was incubated with a
             100-fold and 500-fold excess of various kinds of unlabelled
             DNA as competitors and 4μg of the protein fraction eluted from
             the DNA-cellulose with the 300mM NaCl. Nuclear 0.4M NaCl
             extract was fractionated over the DEAE-cellulose. Lanes 1, 2,
             the TO gene fragment of the coding region; lanes 3, 4, pUC 19;
             lanes 5, 6, poly(dA) poly(dT); lanes 7, 8, pBR 322. Arrows
             denote specific DNA-protein complexes.

        Several observations indicate that the presence of the hormone response
element (HRE) alone is not sufficient for hormone inducibility but that, in
addition, other regulatory elements are required. One of these elements is
a binding sequence for the nuclear factor, NF1 (Cato et al., 1988). We
therefore studied the competition with the plasmid containing the sequences
for binding of the glucocorticoid receptor complex and the nuclear factor,
NF1 (Hynes et al., 1983). These results (Figure 2) demonstrate that the
slower migration complex disappears. Under the present experimental
conditions (Danesch et al., 1987; Schule et al., 1988) the binding of the
glucocorticoid receptor cannot be seen (binding can only be generated by
incubation of purified receptor with DNA in the presence of PEG). We
therefore believe that the liver nuclear protein(s) responsible for the
formation of the slow mobility complex may recognize the sequence for the
NF1 binding. We cannot at present identify the protein(s) responsible for
the formation of the fast mobility complex.

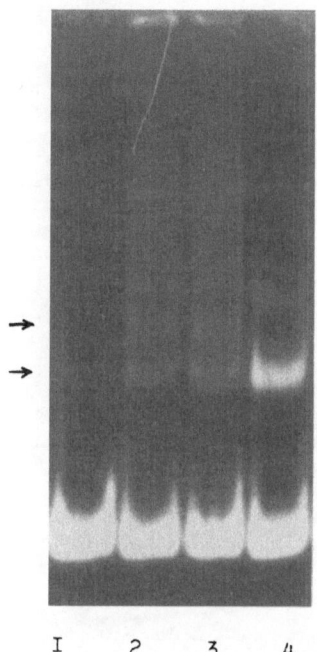

I   2   3   4

Figure 2.　　Binding of the nuclear liver proteins to the −466/−178 EcoRl
fragment of the rat TO gene in the presence of the specific
competitor containing the binding sites for the glucocorticoid
receptor complex and NFl. Lane 1, free DNA; lanes 2, 3, 4
contain a 1−, 2−, and 5−fold molar excess of the competitor,
respectively. Probes were incubated with a 100−fold excess of
the λ DNA. Binding conditions and use of symbols as in
Figure 1.

We studied the specific binding of the nuclear proteins to the −466/
−292 fragment containing hypersensitive site 1 and the −292/−178 fragment
containing hypersensitive site 2 (Figure 3). This and also the results of
the competition experiment (data not shown) indicate that the proteins which
compete for the NFl binding sites interact with the −466/−292 and −292/−178
fragments. Our results also indicate that the proteins which do not compete
for the NFl binding sites interact with the −178/−292 fragment. It is
possible that we have identified the previously unknown trans−acting TO gene
factors.

ACKNOWLEDGEMENTS

We thank Dr G. Schütz for λ TO2 DNA and Dr M. Beato for the plasmid
containing the sequences for binding of the glucocorticoid receptor complex
and the nuclear factor, NFl.

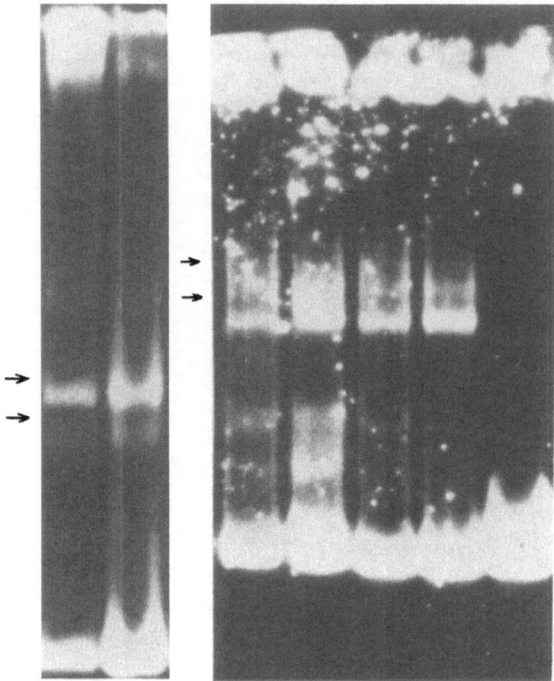

Figure 3.   Binding of the nuclear liver proteins to the -466/-292 EcoR1-
             HindIII fragment (panel A) and the -292/-178 HindIII-Taq1
             fragment (panel B) of the TO gene in the presence of the non-
             specific competitor of the λ DNA. (A) lanes 1, 2 contain a
             100- and 1000-fold excess of the λ DNA, respectively. (B)
             lanes 1 - 4 contain a 20-, 60-, 300-, 600-fold excess of the
             λ DNA, respectively. Lane 5, free DNA. Binding conditions
             and use of symbols as in Figure 1.

REFERENCES

Becker P., Renkawitz R., and Schutz G. 1984, Tissue-specific DNase I
     hypersensitive sites in the 5'-flanking sequences of the tryptophan
     oxygenase and tyrosine aminotransferase genes. EMBO J., 3:2015.
Cato A. C. B., Skroch P., Weinmann J., Butkeraitis P. and Ponta H. 1988,
     DNA sequences outside the receptor-binding sites differentially
     modulate the responsiveness of the MMTV promoter to various steroid
     hormones, EMBO J., 7:1403.
Danesch U., Gloss B., Schmid W., Schutz G., Schule R., and Renkawitz R. 1987,
     Glucocorticoid induction of the rat tryptophan oxygenase gene is
     mediated by two widely separated glucocorticoid-responsive elements,
     EMBO J., 6:625.
Hynes N., van Ooyen A. J. J., Kennedy N., Herrlich P., Ponta H., and Groner
     B. 1983, Subfragments of the LTR cause glucocorticoid-responsive
     expression of MMTV and of adjacent gene, Proc. Natl. Acad. Sci., USA,
     80:3637.
Schmid W., Scherer G., Danesch U., Zentgraf H., Matthias P., Strange C. M.,
     Rowekamp W., Schutz G. 1982, Isolation and characterisation of the rat
     tryptophan oxygenase gene, EMBO J., 1:1287.
Schule R., Muller M., Otsuka-Murakami H., Renkawitz R. 1988, Cooperativity
     of the glucocorticoid receptor and the CACCC-box binding factor,
     Nature, 332:87.

THE snRNA AND REGULATION OF THE EARLY DEVELOPMENT IN TELEOSTS

T. A. Burakova

Koltzov Institute of Developmental Biology
USSR Academy of Sciences
Moscow
USSR

INTRODUCTION

The snRNAs have been described in all eukaryotic cells from mushrooms and protozoa to higher plants and mammals (Busch et al., 1982). The involvement of snRNA in splicing and its role as autocatalyst has been discovered (Maniatis and Reed, 1987). However, at the present time the functional role of these macromolecules has not been investigated in detail.

During early development of amphibia (Forbes et al., 1983), sea urchins (Frederiksen, Hellung-Larsen, 1974), mammals (Lobo et al., 1988) and fishes (unpublished) snRNA synthesis starts concomitantly or prior to initiation of mRNA synthesis. In some animals snRNA synthesis is activated soon after fertilization (at 2 blastomeres in mouse, at 16 blastomeres in sea urchin) and sn RNAs are not abundant in oocytes. In amphibians and fishes snRNA synthesis is activated at mid-blastula transition and snRNA content in oocytes is by several orders of magnitude higher than that in the somatic cells. The presence of such large quantities of snRNAs in cells with low transcriptional level suggests that these molecules are involved in some other cell functions, besides processing.

This paper deals with the study of the DNA content in cells of loach embryos injected with homologous snRNAs, as compared with control embryos during synchronous and asynchronous cleavage.

MATERIALS AND METHODS

Mature loach eggs were obtained and artificially inseminated, as described by Neifakh (1959). The embryos at stages from blastodisc formation to the first cleavage division were microinjected with snRNA preparations from the loach embryos, obtained after phenol deproteinization and subsequent agar electrophoresis of the initial extract (Burkova et al., 1984). 20nl of snRNA preparation (fractions U_1-U_6) in 0.01M Tris-HCl buffer (pH 7.4) were injected into each embryo. The control embryos were injected with the same volume of buffer without snRNA. Injected embryos were grown at 21°C up to the early or mid-blastula stage (5-7h of development). Blastoderms were manually separated from the yolk and dissociated into separate cells by placing into Ca^{2+}-free Holtfreter solution with double concentration. Isolated nuclei were placed onto the glass slides, dried and fixed with 96% ethyl alcohol at 15°C.

DNA content of the nuclei was measured by densitometry at 540nm, of preparations stained according to Feulgen; hydrolysis was performed in 5N NaCl for 15 min at 37°C and staining with Schiff reagent for 1h at room temperature using a scanning densitometer (Vickers M86, England). In each experimental or control series, DNA contents were determined for 100 nuclei of the embryos at the stage of 5h or 7h of development. Loach spermatozoa (1c) and erythrocytes (2c) were used as the ploidy standards. The number of cells in embryos was counted after their dissociation into separate cells in Ca^{2+}-free Holtfreter solution (with doubled concentration) and subsequent staining with a drop of toluidine blue solution. DNA contents was also determined biochemically (Gause, 1984), in isolated nuclei obtained as described earlier (Burkova et al., 1980) from blastoderms of the same stages. Cell divisions at the midblastula stage were synchronized artificially by placing embryos in the cold (3°C) for 2h. Thereafter for 100min embryos were kept at 18°C and at intervals of 10min several of them were prepared and assayed as described above. In order to estimate the degree of synchronization, we counted the number of 4c cells in experimental and control embryos before, placing them in the cold and 100min after their transfer into the normal temperature. The data were treated statistically.

RESULTS AND DISCUSSION

Biochemical determinations have shown that at the early and mid-blastula stage the DNA content of embryos injected with snRNAs increased by 18-20%, compared with the control embryos. According to the results of cytophotometry at the early blastula stage (5h of development) in the control embryos, the proportion of cells in G_1, and G_2 and S phases of the cell cycle was 11%, 25% and 64%, respectively. In embryos injected with snRNAs, this proportion was 9%, 47% and 44%. These data provide evidence that snRNA injection results in considerable changes of the cell cycle at the early blastula stage. It is noteworthy that in experimental embryos as well as in the control ones the total cell number per embryo was equal to 356, i.e. it did not change after snRNA injection. Thus, the overall duration of the mitotic cycle remained unchanged. If we assume that the number of cells which entered a specific phase of the cell cycle is proportional to the duration of this phase, the possible explanation of the results may be as follows. After snRNA injection the duration of G_1 phase did not change while that of the S phase was diminished 1.5-fold. This resulted in an increased synthetic rate and proportional decrease in the duration of G_2 phase. Therefore the increased DNA content in snRNA-injected embryos can be due to an increased number of 4c cells.

The duration of the cell cycle during synchronous cleavage divisions is species-specific. Our results demonstrate that snRNA injection does not affect this parameter but increases the rate of DNA synthesis. As a consequence, the duration of the S phase diminishes and embryonic cells remain at G phase for a longer period of time 'waiting' for the oncoming mitosis which is controlled by cytoplasmic regulatory mechanisms. At present we cannot suggest any other explanations of our experimental data.

At the mid-blastula stage (7h of development) we did not detect any increase in the number of 4c cells. Probably it is due to desynchronization of cleavage divisions and the sharp fall of the mitotic index, character-istic of this stage. In order to further elucidate this question we measured the DNA content and counted 4c cells in embryos with partly synchronized cleavage (see above). In such embryos injected with snRNAs the number of 4c cells was increased by 13-15%, as compared with the control. Therefore, an artificial increase in snRNA content of the embryo results in an increased number of 4c cells at the mid-blastula stage.

Thus, the excessive amount of snRNA in loach embryonic cells results in alterations of the cell cycle not only during synchronous cleavage but also after desynchronization of cell divisions. These observations as well as some disturbances caused by artificially increased snRNA content (multipolar mitoses resulting in cells with decreased DNA content) lead to the conclusion about possible role of snRNAs as endogenous factor controlling cell proliferation during development.

REFERENCES

Burakova T. A., Korzh V. P., Neyfakh A. A. 1984. Injections of low molecular weight RNAs into the loach embryos activate the synthesis of macromolecules. Dokl. Akad. Nauk SSSR (Russian), 274:413.
Burakova T. A., Korzh V. P., Shostak N. A., Neyfakh A. A. 1980. The RNA synthesis activation in loach embryos after injection on 0.35M NaCl extract of the nuclei. Molekul. Biol. (Russian), 14:922.
Busch H., Reddy R., Rothblum L., Choi Y. C. 1982. snRNAs, snRNPs and processing. Ann. Rev. Biochem., 51:617.
Forbes D. J., Kornberg T. B., Kirschner M. W. 1983. Small nuclear RNA transcription and ribonucleoprotein assembly in early Xenopus development.
Frederiksen S., Hellung-Larsen P. 1974. Synthesis of small molecular weight RNA components during the early stages of sea urchin embryo development. Exp. Cell Res., 89:217.
Gause G. G. 1974. Methods of nucleic acid research. In: Research methods in developmental biology (Moscow: Nauka), 365 (in Russian).
Lobo S. M., Marzluff W. F., Seufert A. C., Dean W. L., Schultz G. A., Simerly C., Schatten C. 1988. Localization and expression of u. RNA in early mouse embryo development. Dev. Biol., 127, No 2: 349.
Maniatis T., Reed R. 1987. The role of small nuclear ribonucleoprotein particles in pre-mRNA splicing. Nature, 352:1925.
Neyfakh A. A. 1959. Method of inactivation of nuclei by radiation and its possible applications for the investigation of nuclei functions during early development of fish. Zhurn. Obshch. Biol. (Russian), 20:202.

DIFFERENCES IN TRANSCRIPTION ACTIVITY AND REPLICATION TIME OF rRNA GENES
IN A PAIR OF HETEROMORPHIC NUCLEOLAR ORGANIZING CHROMOSOMES IN TRANSFORMED
KIDNEY CELLS OF THE AFRICAN GREEN MONKEY (RAMT STRAIN)

L. I. Baranovskaya

Institute of Medical Genetics Academy of Medical Science
115478 Moscow
USSR

There is a well known correlation between the transcriptional activity
of chromatin and its DNA replication in the early S-period on the one hand
and genetic inactivity and DNA replication in the late S-period, on the
other. Nevertheless, data obtained by various authors on the replication
time of genes coding for 28S and 18S ribosomal DNA seem to contradict this
regularity. It has been shown that in Chinese hamster cells the bulk of
rDNA is replicated in the first half of the S-period, in mouse and Hela
cells during the whole S-period; in Kangaroo rat cells in late S-period
(for reference see: Balazs and Schildkraut, 1976).

The genes coding for rRNA in mammalian genomes are represented by
clusters which consist of several hundreds of tandemly repeated nucleotide sequences (Long and David, 1980). These are located in chromosomes,
the so called nucleolar organizer regions (NORs). A specific staining of
chromosomes with silver nitrate (Ag-staining) enables us to detect at the
cytological level only transcriptionally active clusters of rRNA genes
(Goodpasture and Bloom, 1975).

In this paper we presented data obtained on the transformed kidney
cells of the African green monkey (RAMT strain), having a pair of
heteromorphic nucleolus organizer (NO) chromosomes (Stobetsky et al.,
1982). We studied: (1) the morphology of nucleolus organizer regions
(NORs) in control cells and those treated with 5-azacytidine (5-AzaC) that
induces reactivation of methylated genes; (2) the relative copy number of
rRNA genes (by *in situ* hybridization); (3) transcriptional activity of
ribosomal genes in the control and the 5-AzaC-treated cells and (4) the
DNA replication time of nucleolus organizer regions.

This particular cell line was chosen as a model because of their two
nucleolus organizer (NO) chromosomes differed from those in parental cells
by the structure of one of the chromosomes and NOR morphology. In parental
kidney cells there were two identical NO chromosomes in which the nucleolar
organizers were manifested by weakly expressed secondary constrictions and
identified by Ag-staining as transcriptionally active. In strain RAMT
cells one NO chromosome turned out to be structurally rearranged (SR),
having in 100% of cells an extended secondary constriction in the middle of
the long arm. In the other NO-chromosome, which is structurally normal
(SN), a less pronounced constriction was observed in less than 40% of cells
(Figure 1a). Following Ag-staining (Howell and Black, 1980), a strong

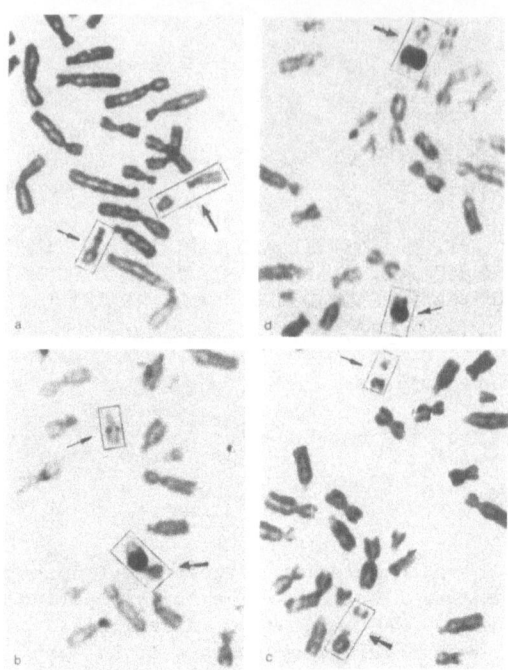

Figure 1.   The metaphase fragments from RAMT-cells. (a,b) - control;
             (c,d) - after 5-AzaC treatment during 34 hours. (a,c) -
             conventional staining; (b,d) - Ag-staining. The NO-
             chromosomes are indicated: thin arrow-structurally normal
             (SN) NO-chromosome; thick arrow-structurally rearranged (SR)
             NO-chromosome.

silver impregnation was found in 100% of cells in the SR chromosome's
second constriction area, while the SN NO-chromosome showed a very weak
impregnation in less than 40% of cells (Figure 1b).

     When hybridized *in situ* with ³H 28S rDNA, both homologues appeared
labelled with the same intensity, which indicates an approximately equal
number of copies of the rRNA gene in their NORs. It means that in SN
NO-chromosome some of the rDNA copies (in 60% of cells, all) are
repressed.

     At present it is accepted that the repression mechanism of ribosomal
genes, as well as other genes, is implemented by DNA methylation (Cooper,
1983). It has also been shown that one can cause gene reactivation by
introducing the analogue of cytozine 5-AzaC into DNA during the replication
(Jones and Taylor, 1981). To reactivate genetically inert rDNA copies in
the SN NO-chromosome we conducted a series of experiments using 5-AzaC
(4-16mkM). Following incorporation of 5-AzaC into DNA during two
replication cycles, a significant increase in the length of the secondary
constriction was detected in the SN NO-chromosome (Figure 1c). The degree
to which constriction lengths were changed varied in different cells. The
secondary constriction in the SN NO-chromosome was observable in 100% of
cells; its extention being in 30% of cells identical to that observed in
the SR NO-chromosome. The NOR morphology in SR NO-chromosome remained
unchanged. In the Ag-stained preparation, a distinct increase in Ag-
staining intensity was noted in SN NO-chromosome, (Figure 1d) and the

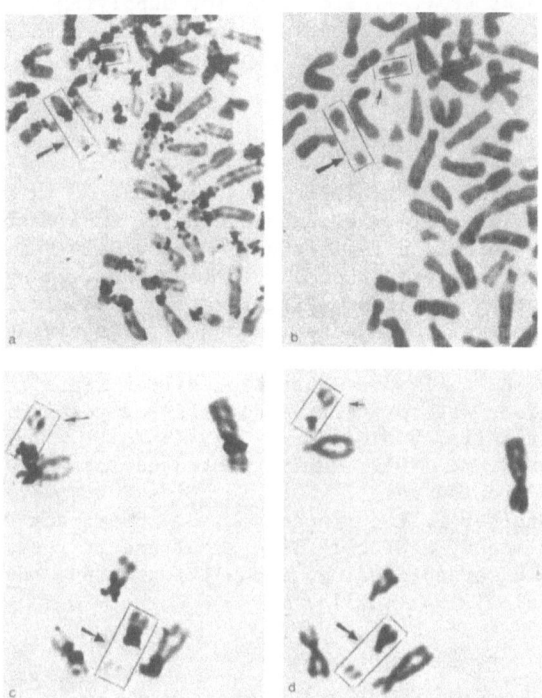

Figure 2.   Fragments of two metaphases. (a,b) - control; (c,d) -
             5-AzaC treated cell. (a,c) - autoradiographs of cells after
             incorporation of ^{3}H-thymidine in late S-phase; (b,d) - the
             same cells after removing label. The arrows indicate the SN
             and SR-NO chromosomes as in Figure 1.

Ag-staining in this particular chromosome was observed in all cells
analyzed. In some cells the Ag-staining intensity in the SN chromosome
corresponded to that detectable in the SR-chromosome.

    The RAMT-strain cells we used proved to be a quite appropriate model
for studying rDNA replication time with regard to its transcriptional
activity. We used the method of autoradiography after incorporation of
^{3}H thymidine to cell cultures at the end of S-period, 6h before cell
harvesting. An autoradiograph analysis revealed the following: (1) In the
SR-chromosome the transcriptionally active NOR (the region of the extensive
secondary constriction) was completely free from the label, while a
transcriptionally inert NOR in the SN NO-chromosome was intensively labelled
(Figure 2a,b); (2) In 5-AzaC-treated cells, however, the SN NO-chromosomal
region carrying the ribosomal genes shaped into an extended secondary
constriction, became free from label in the late S-period, similar to that
observed in the NO-region of the SR-chromosome (Figure 2c,d).

    Thus, in the present investigation the differences in transcription
activity and replication time of rRNA genes in a pair of heteromorphic
nucleolus organizer chromosomes have been revealed. The DNA of transcriptionally active clusters was established to replicate in the first part
of the S-period while that of transcriptionally inert ones, in late S-
period. After 5-AzaC is incorporated into DNA, with the transcription
activity arising in certain clusters, the replication was shifted from late
to the early S-phase.

The author thanks Dr V. I. Stobetsky for supplying the RAMT-strain; Drs I. V. Garkavtsev and T. G. Tsvetkova for helping with hybridization and Dr N. A. Liapunova for discussing the paper.

REFERENCES

Balazs I. and Schildkraut C. L. 1976. DNA replication in synchronized cultured mammalian cells. VI. The temporal replication of ribosomal cistrons in synchronized cell lines. Exper. Cell Res., 101:307.
Cooper D. N. 1983. Eukaryotic DNA methylation. Hum. Genet., 64:315.
Goodpasture C., Bloom S. E. 1975. Visualization of nucleolar organizer regions in mammalian chromosomes using silver staining. Chromosoma, 53:37.
Howell W. M., Black D. A. 1980. Controlled silver-staining of nucleolus organizer regions with protective colloidal developer: a 1-step method. Experientia, 86:1014.
Jones P. A., Taylor S. M. 1981. Hemimetylated duplex DNA prepared from 5-azacytidine treated cells. Nucleic acids Res., 9:2933.
Stobetsky V. I., Grachev V. P., Mironova L. L., Chernicov V. Y. 1982. African green monkey RAMT cell line simultaneous resistant to 8-azaguanine, 6-mercaptopurine, and 6-thioguanine. Bul. Exper. Biol. Med., (in Russian), 94, 10:113.

REACTIVATION OF LATENT NUCLEAR POLYHEDROSIS VIRUS IN SILKWORM

P. Kullyev, N. G. Berdyeva, N. V. Biryukova and N. Agalykov

Institute of Zoology
Turkmen Academy of Sciences
6 Engels St.
Ashkhabad 744000
USSR

Viruses, especially in latent state, are widely held to be beneficial such as in the silkworm, and in insect pests which seriously damage crops and forests. Sometimes a virus can be passed on from one generation to another for a very long period of time, remaining inactive and harmless to an insect. Ecological and other stress factors may activate a latent virus (Aruga, 1963). Sometimes a latent virus is activated spontaneously, which causes a disease in an insect or its death. Thus, latency underlies viral regulation of the insect population.

The present work shows the effect of bruneomycin, an anti-tumour antibiotic, on the activation of a latent virus in a silkworm during its metamorphosis. No spontaneous activity of a latent virus was observed at this developmental stage, as distinct from the larval one. On the third day of their metamorphosis pupae were injected with bruneomycin (0.1, 1.0, 2.5, 5.0 and 10.0µg/g) and the number of mature forms and dead pupae was then registered.

The presence of polyhedral granules caused by the activation of a latent virus under the influence of the antibiotic was detected by microscopy. It turned out that a small dose of bruneomycin (0.1µg/g) does not activate a latent virus and does not influence the morphogenesis. When the dose of the injected antibiotic is increased to 1.0 and 2.5µg/g, the appearance of mature forms is delayed. Polyhedral granules were discovered in dead pupae.

Larger doses (5.0 and 10.0µg/g) completely interrupted the meta-morphosis and activated the latent virus more energetically - the number of polyhedral granules in dead pupae increased manyfold. Similar results were produced by the injection of bleomycin, another anti-tumour antibiotic, in pupae. In this way, dependence was established between the amount of bruneomycin and bleomycin injected and the number of polyhedral granules caused by the activation of the latent virus genome.

The influence of bruneomycin on the DNA molecule was studied by the ultracentrifugation of the nuclear ^{3}H-DNA in alkaline sucrose gradients (0.3M NaOH; 0.7M NaCl; 1mM EDTA; 0.1% sarcosyl). To obtain a labelled ^{3}H-DNA, 20µCi of ^{3}H-thymidine was injected in each of the silkworms two

Nuclear Structure and Function, Edited by J. R. Harris and
I. B. Zbarsky, Plenum Press, New York, 1990

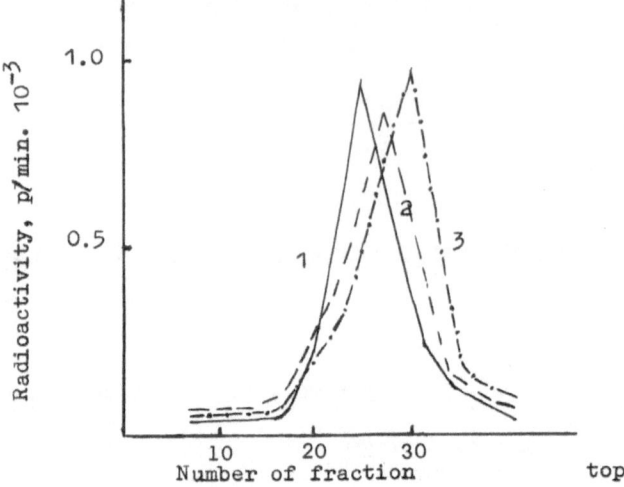

Figure 1.   Analysis of nuclear ^{3}H-DNA preparations from silkworm pupae
(female) by ultracentrifugation in alkaline sucrose gradients
(5-20%w/v). Centrifugation was carried out at +15°C in the
SW - 41 rotor during 4.5 hours at 35,000 RPM. ^{3}H-DNA from
uninfected pupae (1), ^{3}H-DNA from experimental pupae 6h (2)
and 12h (3) after the injection of bruneomycin (10µg/g).

days before they developed into pupae. On the third day of their metamorphosis the pupae were injected with bruneomycin (10.0µg/g). Nuclei were
isolated as described earlier (Mikhailov et al., 1987).

Figure 1 (curve 2) shows that, six hours after the injection of
bruneomycin in pupae, the extracted ^{3}H-DNA sedimented in alkaline sucrose
gradients somewhat slower compared to the ^{3}H-DNA of uninfected pupae
(Figure 1, curve 1). Twelve hours later the ^{3}H-DNA of experimental pupae
sedimented still slower (Figure 1, curve 3). These results indicate that
bruneomycin induces breaks in the DNA chains, and this leads to its
fragmentation.

The infection of cells or the organ of a insect with the nuclear
polyhedrosis virus induces the formation of virus-specific DNA polymerase,
together with other specific proteins and enzymes which bring about the
replication of the viral DNA (Kelly, 1981; Miller et al., 1981; Mikhailov
et al., 1986). The host DNA polymerase and the virally-encoded DNA polymerase are distringuished by their molecular mass, and for this reason they
can be fractioned by ultracentrifugation in glycerol gradients (Mikhailov
et al., 1986).

An analysis of pupal extracts by ultracentrifugation in gradients
showed that only DNA polymerase α was detected in uninfected pupae during
the time of study. Its activity is revealed in gradients by a symmetric
peak with a sedimentation coefficient of 9.6S (Figure 2, curve 1). In the
extracts of experimental pupae which had been injected with bruneomycin
(5.0µg/g) the peak of enzyme activity (Figure 2, curves 2 and 3)
corresponding to the sedimentation coefficient of DNA polymerase α was
detected only during the first three days after the injection. The
difference between uninfected (Figure 2, curve 1) and experimental (Figure
2, curves 2 and 3) pupae lies in the following: the right-hand slope of the
peaks of enzyme activity of experimental pupae is less steep than that of

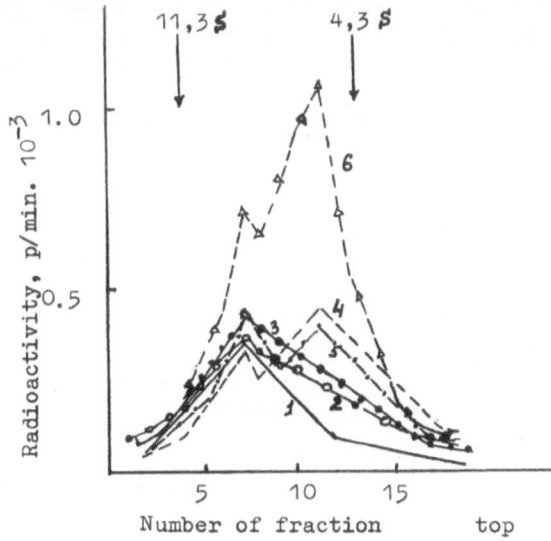

Figure 2. Ultracentrifugation in glycerol gradients (10-30%v/v) of
uninfected pupae extracts (1) and of pupae extracts obtained
a day (2), three days (3), five (4), seven (5) and nine (6)
days after the injection of bruneomycin (5.0µg/g). The DNA
polymerase activity was determined by means of activated DNA.
The arrows indicate the position of sedimentation standards
- BSA (4.3S), catalase (11.3S).

uninfected pupae, which indicates some heterogeneity of the preparation.
This is apparently explained by the appearance of DNA polymerase with a
lower sedimentation coefficient than that of DNA polymerase α. On the
fifth and seventh days after the bruneomycin injection the activity of DNA
polymerase α (Figure 2, curves 4 and 5) is also detected in pupal extracts,
with its level changing slightly compared with uninfected pupae (Figure 2,
curve 1). Meanwhile, one additional enzyme activity appears. It is equal
to the α-polymerase activity and is revealed in gradient in the form of a
peak (Figure 2, curves 4 and 5) corresponding to the sedimentation
coefficient of the virally-encoded DNA polymerase (6.3S). Nine days after
the bruneomycin injection (Figure 2, curve 6) the activity of cellular DNA
polymerase α and of virally-encoded DNA polymerase increased several fold
and the latter polymerase pre-eminated.

The results obtained show that bruneomycin produces breaks in the
chromosome DNA chains of *Bombyx mori* pupae, which leads to the development
of a productive form of the nuclear polyhedrosis virus from its latent
state.

REFERENCES

Aruga H. 1963, Induction of virus infections in :'Insect pathology',
    E. Steinhaus, ed., Acad. Press. N.Y., Vol. 1:499.
Kelly D. C. 1981, Baculovirus replication stimulation of thymidine Kinase
    and DNA polymerase activities in Spodoptera frugiperda cells infected
    with Trichoplusia in nuclear polyhedrosis virus, J. Gen. Virol.,
    52:313.
Mikhailov V. S., Ataeva J. O., Marlyev K. A., Kullyev P. K. 1986, Changes
    in DNA polymerase activities in pupae of silkworm Bombyx mori after

infection with nuclear polyhedrosis virus, J. Gen. Virol., 67:175.

Mikhailov V. S., Marlyev K. A., Kullyev P. K. Krayevsky A. A. 1987, DNA synthesis in isolated nuclei from silkworm pupae infected with nuclear polyhedrosis virus, Biohimiya, 52:24.

Miller L. K., Jewell J. E., Browne D. 1981, Baculovirus induction of DNA polymerase, J. Virol., 40:305.

NEW STRUCTURAL ELEMENTS OF POLY(A)-CONTAINING RNAs IN MOUSE CELLS

D. A. Kramerov, G. P. Shumiatsky and V. V. Svetlov

Institute of Molecular Biology
Academy of Sciences of the USSR
Vavilov Str.
32 Moscow
USSR

Previously, we found (Kramerov et al. 1982, 1985a, 1985b) that RNA polymerase III transcripts of the mouse short B2 retroposon are polyadenylated. This small B2RNA (200-500nt) is the only known example of a polyadenylated RNA polymerase III transcript (poly(A) polIII transcript).

In order to reveal the whole population of poly(A) polIII transcripts in Ehrlich ascite carcinoma cells, we carried out the experiment according to the scheme, shown on Figure 1. The results of the experiment are shown on Figure 2. The data allow us to conclude that:

1.   A great part ($\sim$25%) of newly synthesized poly(A)$^+$RNA of Ehrlich carcinoma cells is poly(A)polIII RNA.

2.   It seems that there are long poly(A)polIII transcripts, but the major part of the poly(A)polIII RNA are small molecules (<500nt).

3.   Apart from pppG and pppA, the unidentified spot X is revealed in the PI-nuclease digests of poly(A)polIII RNA.

4.   Unidentified spots Y and Z are found in PI-nuclease digests of RNA polymerase II transcripts.

Using hybridization and the following fingerprint analysis we showed that 60% to 90% of the small poly(A)polIII RNA is B2 RNA (data are not shown).

The spot X, as well as the spot pppG is revealed in B2 RNA isolated by hybridization (Figure 3). We also detected X in RNA transcribed from the cloned B2 element DNA *in vitro* (Figure 4).

In contrast to pppG, the material of the spot X is resistent to alkaline phosphatase treatment (Figure 3). On the other hand, nucleotide pyrophosphatase hydrolyses X to three products: pG, p (orthophosphate) and an unidentified spot xp (Figure 5). These results suggest that X is a cap-like structure at the 5'-end of B2 RNA: X=xpppG. So, 30-50% of the B2 RNA molecules have a blocked 5'-end. The spot px is located far from the region of nucleotide location on the chromatograms. This fact indicates that the blocking group x has a non-nucleotide nature.

Nuclear Structure and Function, Edited by J. R. Harris and
I. B. Zbarsky, Plenum Press, New York, 1990

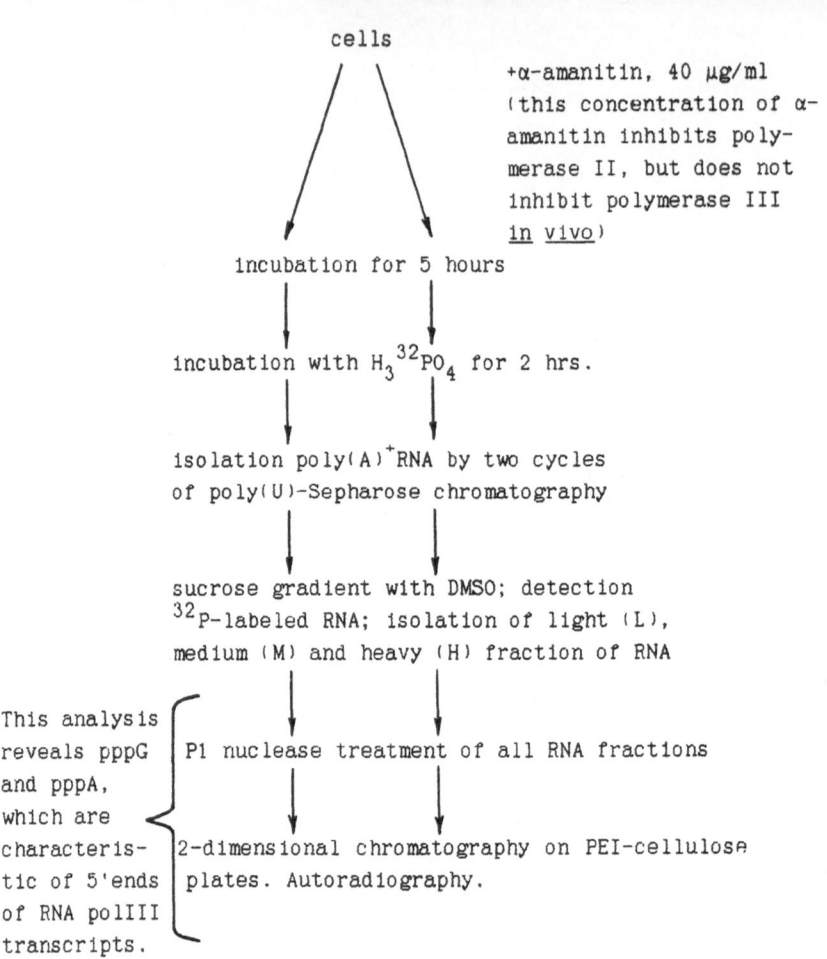

cells

+α-amanitin, 40 μg/ml
(this concentration of α-
amanitin inhibits poly-
merase II, but does not
inhibit polymerase III
in vivo)

incubation for 5 hours

incubation with H$_3$^{32}PO$_4$ for 2 hrs.

isolation poly(A)$^+$RNA by two cycles
of poly(U)-Sepharose chromatography

sucrose gradient with DMSO; detection
^{32}P-labeled RNA; isolation of light (L),
medium (M) and heavy (H) fraction of RNA

This analysis
reveals pppG
and pppA,
which are
characteris-
tic of 5'ends
of RNA polIII
transcripts.

P1 nuclease treatment of all RNA fractions

2-dimensional chromatography on PEI-cellulose
plates. Autoradiography.

Figure 1.   The scheme of isolation and analysis of poly(A) polIII RNA.

A large portion of the molecules of small nuclear U6 RNA also contain
a similar cap-like structure xpppG at their 5'-end, as shown by Reddy et
al. (1987). Thus, B2 RNA is the second example of RNA-polymerase III
transcripts having the cap-like structure at 5'-end. We studied some small
RNAs synthesized by RNA polymerase III and found that only U6 RNA has the
structure X (Figure 6).

The nucleotide sequences of B2 RNA and U6 RNA are not similar. The
only long homologous region in these RNAs is shown on Figure 7. We removed
this region from B2-element DNA by AluI treatment and transcribed the B2-
element *in vitro*. The truncated RNA, as well as the full-length RNA
contained X (Figure 8). This result suggests that the studied box in B2
RNA is not responsible for the synthesis of X. Perhaps, some other
sequence and/or the secondary structure of B2 RNA are responsible for it.

In this work we have also studied the nature of the spots Y and Z. It
has been found that Y and Z were cytosine- and urdine-5'-phosphate-2', 3'-
cyclophosphate, respectively. We believe that they are located at nicks
which are induced in RNA by specific RNAse(s). We have never observed Y
and Z in poly(A)polIII RNA (see, for example, Figure 1). So, it seems
likely that they are characteristic features of RNA polymerase II transcripts. It may be that the formation of pyrimidine-2', 3'-cyclophosphates
is the first step of mRNA degradation *in vivo*.

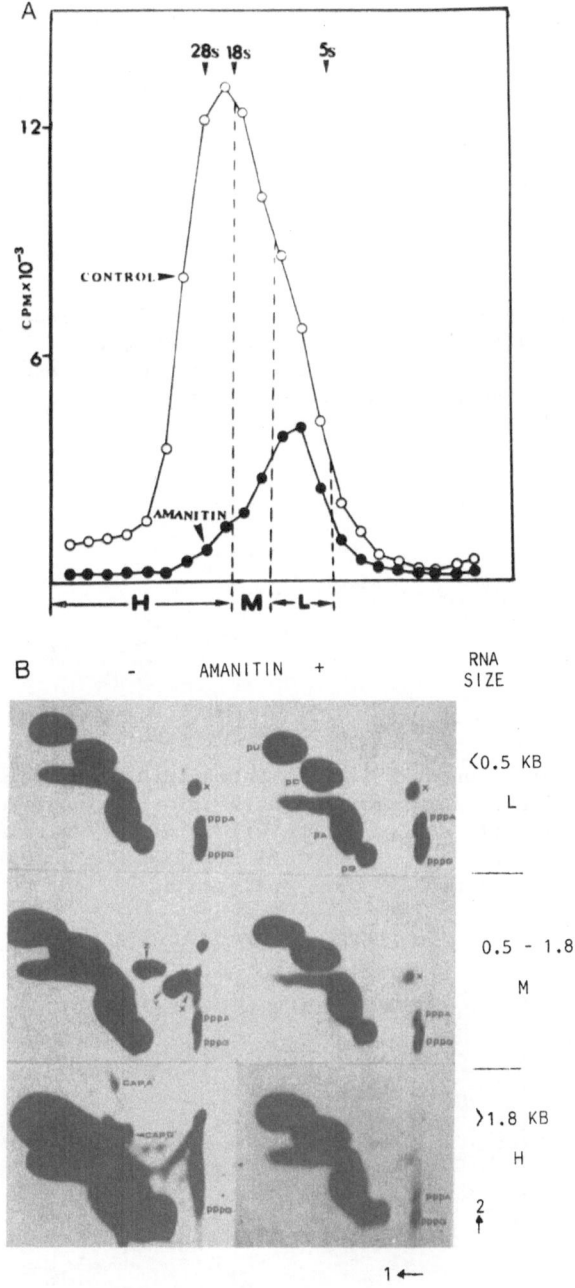

Figure 2.   (A) Sedimentation of (^{32}P)poly(A)$^+$RNA synthesized by cells
incubated with or without α-amanitin.   (B) PEI-cellulose
chromatography of PI-nuclease digests of the light (L),
Medium (M) and heavy (H) poly(A)$^+$RNA after the sedimentation
shown in (A).   TCL chromatography was performed according to
Lycan and Danna (1983).

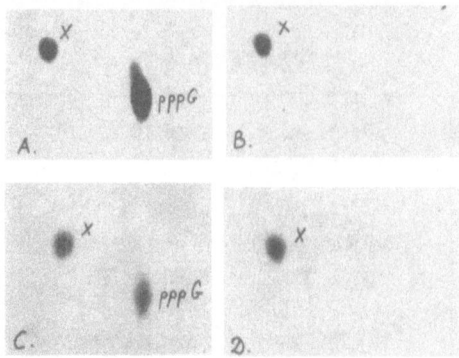

Figure 3.   PEI-cellulose chromatography of PI-nuclease digests of small poly(A)polIII RNA(A,B) and B2 RNA(C,D).   In each case, half of the digest was treated with phosphatase(B,D).   Only part of each chromatogram is shown.

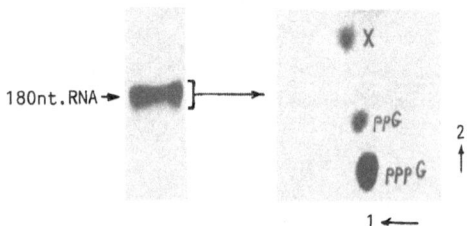

Figure 4.   Detection of X in RNA transcribed from a cloned copy of B2-element *in vitro*.   Transcription was carried out in S100 extracts according to Weil et al. (1979).   After electro-phoresis in polyacrylamide gel B2 RNA (180nt) was digested with P1 nuclease and analyzed by PEI-chromatography.

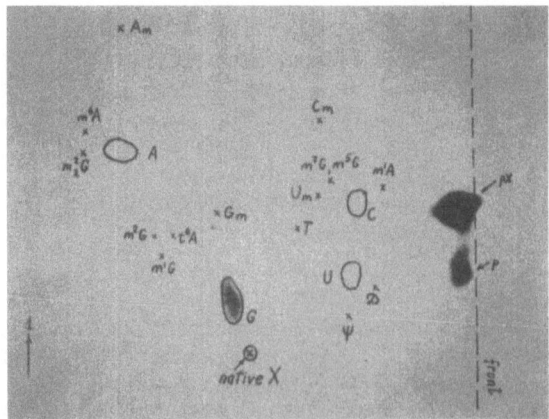

Figure 5.   Cellulose chromatography of X treated with nucleotide pyro-phosphatase.   Chromatography was performed according to Silberklang et al. (1979).   The positions of the modified nucleotides-5'-monophosphates are indicated by crosses.

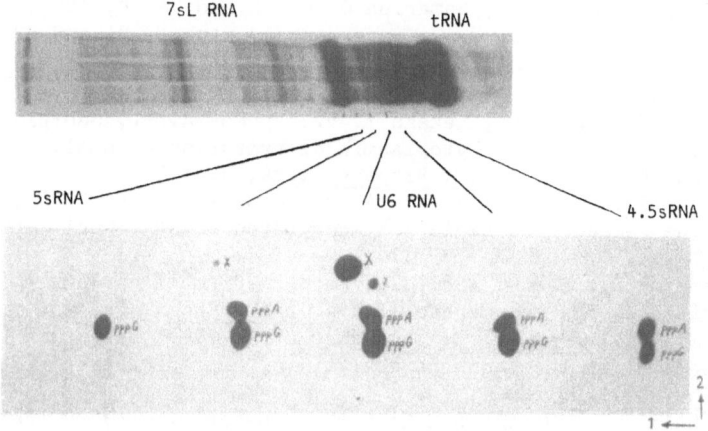

Figure 6.   PEI-cellulose chromatography of PI-nuclease digests of some
             RNAs synthesized by RNA polymerase III *in vivo*.

B2 RNA   5'(x)pppGGGGCUGGUGAGAUGGCUCAGCGGGUAAGAGCACCCGACUGCUCUUCCGA

B2 RNA   AGGUCCUGAGUUCAAAUCCCAGCAACCACAUGGUGGCUCACAACCAUCCGUAACGAGA

B2 RNA   UCUGACGCCCUCUUCUGGAGUGUCUGAGGAC AGCU ACAGUGUACUUACAUAUAAUAAA
                                         AluI   ••••••• ••••

U6 RNA   5'(x)pppGUGCUCGCUUCGGCAGCACAUAUACUAAA

B2 RNA   UAAAUAAAUAAAUAAAUCUUUpoly(A) 3'

U6 RNA   AUUGGAACGAU......

Figure 7.   The region of high homology between B2 RNA and U6 RNA. The
             sequences of the first 40 nucleotides are taken from Epstein
             et al. (1980).

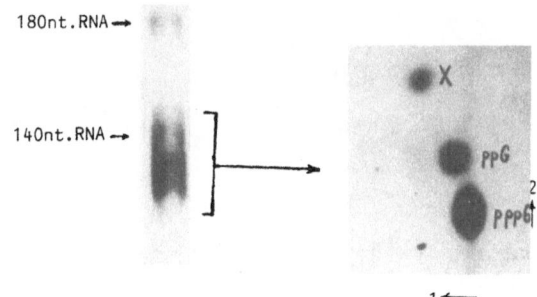

Figure 8.   Detection of X in RNA transcribed *in vitro* from the B2-element
             truncated by AluI treatment (location of the AluI site is
             shown in Figure 7).

REFERENCES

Epstein P., Reddy R., Henning D., and Busch H. 1980, The nucleotide sequence of nuclear U6 (4.7S) RNA, J. Biol. Chem. 255:8901.

Kramerov D. A., Lekakh I. V., Samarina O. P., Ryskov A. P. 1982, The sequence homologous two major interspersed repeats B1 and B2 of the mouse genome are present in mRNA and small poly(A)$^+$RNA, Nucl. Acids Res., 10:7477.

Kramerov D. A., Tillib S. V., Lekakh I. V., Ryskov A. P., Georgiev G. P. 1985a, Biosinthesis and cytoplasmic distribution of small poly(A)-containing B2 RNA, Biochim. Biophys. Acta, 924:85.

Kramerov D. A., Tillib S. V., Ryskov A. P., Georgiev G. P. 1985b, Nucleotide sequence of small polyadenylated B2 RNA, Nucl. Acids Res., 13:6423.

Reddy R., Henning D., Das G. 1987, The capped U6 small nuclear RNA is transcribed by RNA polymerase III, J. Biol. Chem., 262:75.

Silberklang M., Gillum A. M., Raj Bhandary U. L. 1979, Use of *in vitro* ^{32}P labelling in the sequence analysis of nonradioactive tRNAs, Meth. Enzymol., 59:58.

Weil P. A., Segall J., Harris B., Ng S.-Y., Roeder R. G. 1979, Faithful transcription of eukariotic genes by RNA polymerase III in systems reconstracted with purified DNA templates, J. Biol. Chem., 254:6163.

GIBBERELLIN-BINDING PROTEINS OF HIGHER PLANT CELL NUCLEI

D. I. Jokhadze, N. N. Tevzadze and R. I. Goglidze

Institute of Plant Biochemistry
Georgian Academy of Sciences
Tbilisi, 380031
USSR

Investigations carried out in our laboratory on the participation of phytohormones in transcriptional mechanisms of higher plants, in particular pea (Pisum sativum), have shown that gibberellic acid (GA_3) stimulates RNA synthesis in cell nuclei as well as in chloroplasts (Jokhadze and Goglidze, 1977; Tevzadze and Jokhadze, 1986). It was also shown that GA_3 changes the quantity and activity of all three forms A(I), B(II) and C(III) DNA-dependent RNA polymerases. Therefore, it was concluded that the enzyme participates in genome selective functioning (Jokhadze and Goglidze, 1987), since each RNA polymerase form is responsible for the transcription of the corresponding genome and accordingly production of RNA molecular types (Blair, 1988).

Furthermore, specific proteins - mediators participate in the effect of phytohormones on the genetic machinery (Baile et al, 1985; Romanov, 1988). We tried to identify such proteins for GA_3.

While comparing the endogenous ability of cell nuclei and chromatin, isolated from bean plant leaves (Phaseolus vulgaris), to synthesise RNA, it was shown that nuclei have a higher transcriptional activity per DNA quantity than chromatin. The addition of GA_3 to the nuclear or chromatin-containing RNA polymerase systems in both cases stimulated transcription, but to a greater extent with nuclei. Stimulation of RNA synthesis in the system containing chromatin occurs only at high concentrations of GA_3 (Figure 1).

This implies that a mediatory factor, specific for GA_3, exists in cell nuclei and participates in the hormone-genome interaction, but is apparently lost in the process of chromatin isolation from the nuclei.

Experiments, including incubation with ^{14}C-gibberellin, have shown that the higher level of radioactivity was detected in nuclei. It also became clear that the phytohormone is specifically bound with a discrete fraction of the nuclear proteins. This fraction, termed gibberellin-binding nuclear protein (GBP), was identified and partially purified from isolated nuclei by gel-filtration of nucleoplasm on Sephadex G-25 and TSK-HW-55 columns (Figure 2). The molecular mass was approximately 100kD

Experiments on the detection of the GBP effect on RNA synthesis in

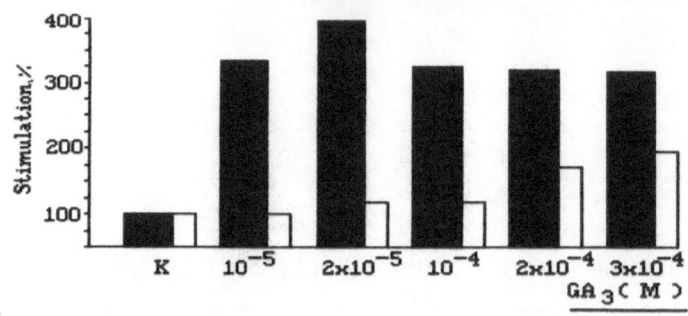

Figure 1.   The effect of GA$_3$ on the endogenic transcriptional activity
of bean leaf cell nuclei and chromatin. Activity was
determined by incorporation of radioactivity into the
insoluble acid material after incubating mixture containing
nuclei or chromatin, ribonucleosidtriphosphates (ATP, GTP,
UTP and ^{14}C-CTP), Tris-HCI, Mg^{+2}, β-mercaptoethanol as it is
described in the article (Jokhadze and Goglidze, 1977). ■ -
nuclei, □ - chromatin, K - control sample activity
(considered as 100%), the other columns represent activity of
samples with GA$_3$, in the final concentrations indicated on
the abscissa. Cell nuclei were obtained according to Jokhadze
and Goglidze (1977) and chromatin according to Haung and
Bonner (1962). Samples with GA$_3$ were incubated with phytohormone for 30 min before the RNA polymerase reaction.

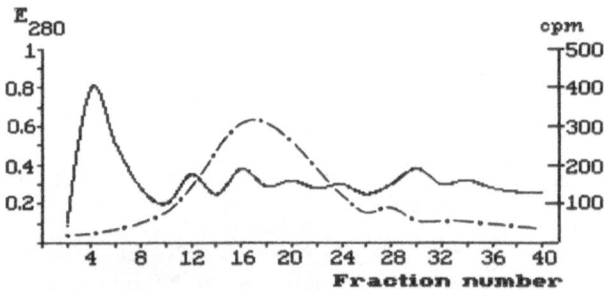

Figure 2.   Fractionation of the bean epicotyl cell carioplasmatic
proteins on the HW-55 TSK - gel tube (− E$_{280}$) and ^{14}C-G
binding to various fractions (−·−).

cell nuclei (Table I) have shown that the addition of GA$_3$ or GBP individually stimulates transcription by 31-44%, and, when added together, the
process is stimulated approximately 2.5 fold. However, GBP added together
with kinetin or IAA stimulates RNA synthesis 25-56%. Similar experiments
performed on choroplasts (Table II) showed that GA$_3$ and GBP separately
stimulate transcription as is the additive effect when they are present
together.

Addition of GA$_3$ without GBP to the RNA polymerase system, containing
either nuclear RNA polymerase (A form) and homologous DNA (Table III) or

Table I. The Influence of GA_3, Kinetin, IAA and GBP on the Endogenic Transcriptional Activity of Bean Epicotyl Cell Nuclei.

Incubation mixture	Incorporation of ^{14}C-CMP, pmol per 30 μg DNA	%
Nuclei (control)	3.2 ± 0.20	100
" ___ "+ GA_3 $(10^{-6}M)$	4.6 ± 0.38	144
" ___ "+ GA_3+GBP (25 μg)	8.0 ± 0.40	250
" ___ "+GBP (25 μg)	4.2 ± 0.21	131
" ___ "+ Kinetin $(10^{-6}M)$	4.6 ± 0.36	144
" ___ "+ Kinetin+GBP (25 μg)	4.4 ± 0.26	138
" ___ "+ IAA $(10^{-6} M)$	4.0 ± 0.28	125
" ___ "+ IAA+GBP (25 μg)	5.0 ± 0.40	156

Table II. The Influence of GA_3 and GBP on the Endogenic Transcriptional Activity of Bean Epicotyl Cell Chloroplasts.

Incubation mixture	Incorporation of ^{14}C-CMP, pmol per 50 μg DNA	%
Total mixture (control)	1.4 ± 0.13	100
" ___ " + GA_3 $(10^{-6}M)$	3.2 ± 0.26	228
" ___ " + GBP (25 μg)	2.2 ± 0.18	157
" ___ " + GA_3 + GBP	5.3 ± 0.48	378

chloroplast RNA polymerase and chloroplast/nuclear DNA (Table IV), does not stimulate RNA synthesis, while the presence of both compounds increases the process 3 to 4 times.

These experimental data enable us to conclude that the fraction-GBP obtained from bean epicotyl cell nuclei interacts selectively with GA_3 and participates in the production of a functional interaction between GA_3 and the genome, not only in nuclei but also in chloroplasts. A detailed study of the nature and functional characteristics of GBP is the subject of our further investigations.

Table III. The Influence of GA_3 and GBP on RNA Synthesis
Into the Nuclear DNA-Dependent RNA Polymerase
System (Form I) *In Vitro*.

Incubation mixture	Incorporation of ^{14}C-CPM pmol per 10 μg DNA	%
DNA + RNA polymerase (control)	2.0 ± 0.10	100
"_____" + GA_3 (10^{-6}M)	1.9 ± 0.11	95
"_____" + GBP (25 μg)	2.7 ± 0.26	135
"_____" + GA_3 + GBP (25 μg)	3.4 ± 0.17	170
"_____" + GA_3 + GBP (50 μg)	3.6 ± 0.17	180
"_____" + GA_3+albium (50 μg)	1.9 ± 0.19	95

Table IV. The Influence of GA_3 and GBP on RNA Synthesis
Into the Chloroplast DNA-Dependent RNA Polymerase
System *In Vitro*.

Incubation mixture	Incorporation of ^{14}C-CMP, pmol per 30 μg DNA			
	nDNA	%	chDNA	%
Total mixture (control)	0.80 ± 0.07	100	0.94 ± 0.07	100
"____" + GA_3 (10^{-6}M)	0.87 ± 0.10	103	0.98 ± 0.08	104
"____" + GBP (25 μg)	1.20 ± 0.10	142	1.30 ± 0.11	138
"____" + GA_3 + GBP	4.00 ± 0.36	476	3.70 ± 0.25	394

REFERENCES

Baily H. M., Barker R. D., Libbenga K. R. 1985, Auxin binding sites in the
    tobacco cells, Biol. Plant., 27:105.
Blair D. G. R., 1988, Eukaryotic RNA polymerases, Comp. Biochem. and
    Physiol., 89:647.
Jokhadze D. I., Goglidze R. I., 1977, Comparative effect of gibberellic acid
    on RNA polymerase activity of cellular nuclei from pea leaves and
    roots, Physiologia Rastenii., (Russian). 24:746.
Jokhadze D. I., Goglidze R. I., 1987, Specificity of transcriptional systems
    of high plant cell organells, N. Vavilov, 5 Meeting, 'VOGIS'.,
    (Russian), 5:77.
Romanov G. A., 1989, Hormone-binding proteins in plants and the problem of
    phytochormone reception, Physiologia Rastenii., (Russian), 36:166.
Tevzadze N. N., Jokhadze D. I., 1986, Stimulation of endogenous RNA
    polymerase activity of chloropalsts with gibberellin acid, Bull. Acad.
    Sci. Georgian SSR,(Rus), 122:141.

DEVELOPMENTAL EXPRESSION OF PROTO-ONCOGENE c-erb-A RELATED GENES ENCODING

NUCLEAR RECEPTORS IN *XENOPUS*

Jamshed R. Tata and Betty S. Baker

Laboratory of Developmental Biochemistry
National Institute for Medical Research
The Ridgeway
Mill Hill
London NW7 1AA
United Kingdom

Recently, genes encoding several receptors for extracellular chemical signals have been cloned and found to fall into a few superfamilies related to cellular oncogenes. Prominent among these are nuclear receptors of steroid and thyroid hormones, vitamin D_3 and retinoic acid which are related to the proto-oncogene c-erb-A (Green and Chambon, 1986; Evans, 1988). Whereas considerable attention is currently being focused on receptor structure, DNA sequences and hormone binding domains, it is not clear how the oncogene-related receptor superfamilies are expressed during ontogenesis and whether a developmental pattern of expression is related to hormonal activity during development. Here, we describe the developmental analysis of expression of c-erb-A related transcripts in *Xenopus*. We show that total c-erb-A related mRNA accumulates in a biphasic fashion before and during metamorphosis. These coincide with two well separated periods when the organism acquires precociously metamorphic and vitellogenic responses to thyroid hormone and estrogen, respectively, well in advance of the physiological developmental processes.

Metamorphosis in insects and amphibia is a dramatic example of a hormone regulated developmental remodelling of the late embryo in which virtually every cell of the organism is a target for the hormone. Thyroid hormones, which control amphibian metamorphosis, can precociously induce the process in frog tadpoles and numerous biochemical studies have been performed on precociously induced metamorphosis (Frieden, 1981; Tata, 1984). Several years ago, we had reported that *Xenopus* larvae exhibited multiple biochemical responses to thyroid hormones (including enhanced overall RNA and protein synthesis, altered permeability to ^{32}P-orthophosphate) at stages well in advance of normal metamorphosis (Tata, 1968). More recently, it was also shown in our laboratory that the response of *Xenopus* larvae to estrogen, as measured by the activation of dormant vitellogenin genes in liver, was first seen in late metamorphic stages (Ng et al., 1984). It is most likely that the *Xenopus* tadpole exhibits response to the two hormones at very different stages of larval development due to acquisition of hormone-binding components or nuclear receptors for thyroid hormone (TR) and estrogen (ER hormones at around developmental stages 42 and 58 respectively.

We first determined the expression of all c-erb-A related genes by measuring the accumulation of total c-erb-A mRNA from the unfertilized

Nuclear Structure and Function, Edited by J. R. Harris and
I. B. Zbarsky, Plenum Press, New York, 1990

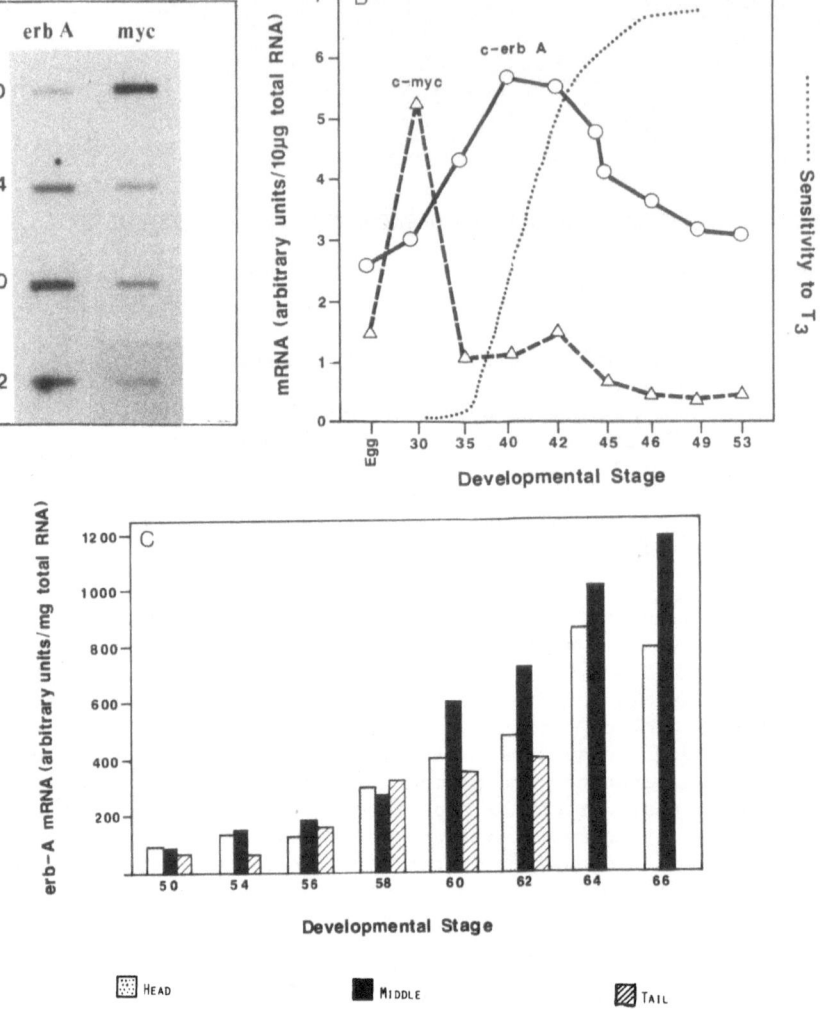

Figure 1.   Biphasic accumulation of c-*erb*-A related mRNAs during
different stages of *Xenopus* development, as compared with
c-*myc* mRNA.
A: Slot-blot hybridization of RNA at early tadpole developmental stages with c-*erb*-A and c-*myc* probes. B: Changes in
concentration of c-*erb*-A related and c-*myc* mRNAs in total
body tissue at different developmental stages of *Xenopus*
tadpoles, in relation to acquisition of sensitivity to T₃.
C: Relative amounts of c-*erb*-A related mRNAs in head (h),
middle (m) and tail (t) regions of tadpoles during and after
different stages of metamorphosis.

eggs, embryos, tadpoles at various developmental stages, including metamorphosis, and in adult tissues, by hybridization under non-stringent
conditions with the full-length c-*erb*-A cDNA. Transcripts of *Xenopus* c-*myc*
oncogene (whose product is also nuclear), albumin and cytoskeletal actin
were monitored as controls. Northern blotting showed a complex pattern of
more that twelve c-*erb*-A related mRNAs, some of which may code for
receptors for thyroid and steroid hormones.

C-*erb*-A mRNA was detected in embryos and larvae at all developmental stages, adult tissues and eggs, albeit in variable amounts. Quantitative slot-blot analysis (Figure 1) showed that the amount of c-*erb*-A related transcripts increased with embryonic development up to stage 42 while *myc* transcripts decreased in the same RNA samples. Thus, the concentration of *myc* RNA was highest in embryos before stage 30 (not shown) and then declined rapidly throughout tadpole development, which is compatible with the association of c-*myc* protein in the cell nucleus with cell proliferation. On the other hand, total c-*erb*-A related mRNA concentrations increased about 4-fold from stages 30 to 42 which is coincidental with the acquisition of T_3 binding and hormone responsiveness (Figure 1A and B). What was surprising was the drop from stages 45 to 54, especially as stage 54 tadpoles are known to retain their sensitivity to T_3 and would undergo accelerated metamorphosis if exposed to the hormone at stages beyond 45. The relatively unchanging concentration of *Xenopus* cytoskeletal mRNA at these developmental stages (data not shown) confirmed that the pattern seen for c-*erb*-A transcripts was unique. A relatively low concentration of *erb*-A transcripts is maintained until stages 55-58 after which it rises again until the froglet stage (stage 66), as shown in Figure 1C. There were no striking variations in *erb*-A transcripts in different regions of the metamorphosing tadpole (head, gut, tail). The onset of this second increase in transcript accumulation coincided with the acquisition by late metamorphic *Xenopus* tadpoles of response to estorgen (May and Knowland, 1981; Ng et al., 1984). Among tissues of adult *Xenopus* the highest levels were found in oviduct, a major target for estrogen with a high ER content in oviparous vertebrates.

We next decided to assay specifically for TR and ER mRNAs in various RNA extracts, particularly around the time of acquisition of hormonal responses. Total RNA samples were probed with cRNA to the hormone-binding domains of TR and ER since ligand-binding sequences are unique and evolutionarily highly conserved (Evans, 1988; Petkovitch et al., 1987; Weiler et al., 1987). Northern blotting detected an α-type TR mRNA of 5kb (there was no detectable signal from the β-type mRNA) as early as stage 45 tadpole which increased substantially at the onset of metamorphosis and until stages 56-62. The two major forms of ER mRNA (1.8 and 3.2kb) could not be clearly detected until stage 57, with only faint signals appearing during mid-metamorphosis. When these determinations were expressed on the basis of cytoskeletal mRNA concentrations, the pattern shown in Figure 2 was obtained. This figure also shows that the differential pattern of accumulation of TR and ER mRNAs was similar to, but not coincidental with, the temporally and ontogenically distinct acquisition of competence to respond to thyroid hormone and estrogen.

To conclude, these studies represent the first quantitative determination of the developmental accumulation of nuclear receptors of thyroid hormone/steroid superfamily related to the proto-oncogene c-*erb*-A. In attempting to correlate receptor gene expression with the developmental acquisition of response to a given hormonal signal, we realize that the accumulation of mRNA for a given receptor may anticipate, if not coincide with, the appearance of functional receptor. In future it will be important to quantify during development of the mRNAs coding for other receptors of this supergene family, such as receptors for progesterone, retinoic acid and vitamin D_3. It would also be of considerable interest to see if one hormone or morphogen activates or regulates the gene encoding another nuclear receptor of this family, as well as to establish by *in situ* hybridization the spatial distribution of individual gene products during development. While awaiting the outcome of such investigations, our present findings should be viewed as one of the many facets of a broad strategy of early establishment during development of intercellular networks of communication via nuclear receptors.

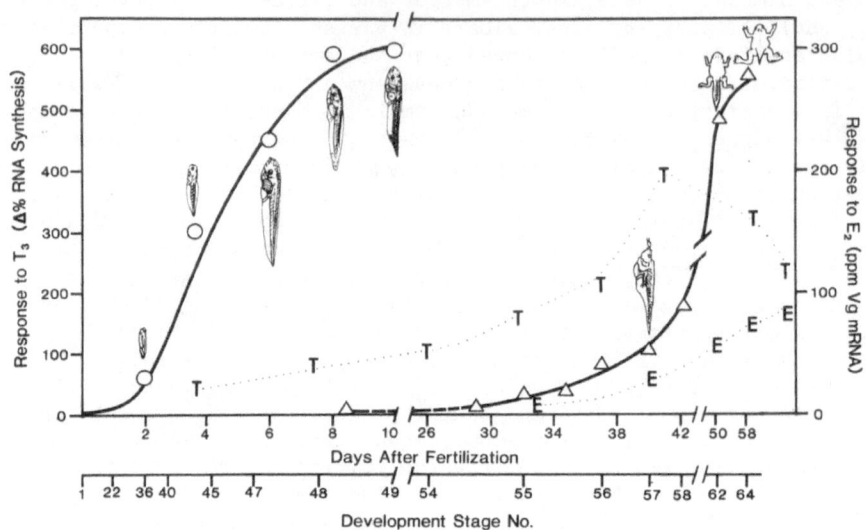

Figure 2.   Accumulation of TR (T) and ER (E) mRNAs at different stages
of *Xenopus* development through metamorphosis. The relative
amounts of TR and ER mRNAs are superimposed on the acquisition
of response to T_3 and E_2 (Tata, 1968; Ng et al., 1984).

We are most grateful to Professor B. Vennström, Karolinska Institute,
Stockholm, for giving us the chicken c-*erb*-A and TR-β cDNAs, Professor P.
Chambon, University of Strasbourg, for the chicken ER cDNA, Dr T. Mohun of
this Institute, for *Xenopus* cytoskeletal actin cDNA, Dr M. Mechali, Institut
Jacques Monod, Paris, for the *Xenopus myc* cDNA, used as hybridization probes
in this study, and Mrs Ena Heather for preparation of the manuscript.

REFERENCES

Evans R. M. 1988, The steroid and thyroid hormone receptor superfamily.
    Science, 240:889.
Frieden E., 1981, The dual role of thyroid hormones in vertebrate develop-
    ment and valorigenesis, in *Metamorphosis,* L. I. Gilbert & E. Frieden
    eds., pp. 545–563, Plenum Press, New York.
Green S. and Chambon P., 1986, A superfamily of potentially oncogenic
    hormone receptors, Nature, 324:615.
May F. E. B. and Knowland J., 1981, Oestrogen receptor levels and
    vitellogenin synthesis during development of *Xenopus laevis,* Nature,
    292:853.
Ng W. C., Wolffe A. P. and Tata J. R., 1984, Unequal activation by estrogen
    of individual *Xenopus* vitellogenin genes during development, Dev. Biol.
    102:238.
Petkovitch M., Brand N. J., Krush A. and Chambon P., 1987, A human retinoic
    acid receptor which belongs to the family of nuclear receptors, Nature,
    330:444.
Tata J. R., 1968, Early metamorphic competence of *Xenopus* larvae. Dev. Biol.
    18:415.
Tata J. R., 1984, The action of growth and developmental hormones.

Evolutionary aspects, in Biological Regulation and Development, Vol. 3B, R. F. Goldberger & K. R. Yamamoto eds., pp. 1-58, Plenum Press, New York.

Weiler I. J., Lew D. and Shapiro D. J., 1987, The *Xenopus laevis* estrogen receptor: sequence homology with human and avian receptors and identification of multiple estrogen receptor messenger ribonucleic acids. <u>Mol. Endocrinol.</u> 1:355.

THE SECONDARY STRUCTURE MODEL OF MOUSE U1 snRNA AS DETERMINED FROM THE

RESULTS OF Pb-INDUCED HYDROLYSIS*

E. Zietkiewicz[1], J. Ciesiolka[2], W. J. Kryzosiak[2],
and R. Slomski[1]

[1] Institute of Human Genetics
Polish Academy of Sciences
Strzeszyńska 32
60-479 Poznań, Poland

[2] Institute of Bioorganic Chemistry
Polish Academy of Sciences
Noskowskiego 12/14
61-704 Poznań, Poland

INTRODUCTION

All eukaryotic cells contain a group of small nuclear RNAs, designated UsnRNAs, that function in mRNA and rRNA processing as ribonucleoprotein snRNPs (Busch et al., 1982). The primary structure of U1 snRNA, the most abundant UsnRNA in animal cells, is known for several closely related as well as distant species, including human, rat, mouse, chicken, frog, fruit fly, peas and beans (Reddy, 1988). The molecules, depending on the organism, are 162-166 nucleotides long and they are thought to form the same secondary structure (Reddy and Busch, 1988) which seems to be also preserved in intact U1 snRNP particles, that are the essential components of spliceosomes (Chabot and Steitz, 1987).

The existing secondary structure model of U1 snRNA (Reddy and Busch, 1988) is based on nucleotide sequence comparisons (Mount and Steitz, 1981) and the results of nuclease digestion studies (Branlant et al., 1981; Epstein et al., 1981). In this paper we verify this model using a novel single strand specific probe - Pb ions, in combination with a computer program designed to generate the lowest energy RNA secondary structure (PC FOLD).

MATERIALS AND METHODS

T4 RNA ligase was obtained from PL-Biochemicals, T1 RNase from Boehringer and other sequencing enzymes from Pharmacia. $5'-(^{32}P)pCp$ (3000 Ci/mmol) was from Amersham. Acrylamide and N, N'-methylene bis-acrylamide were from Serva.

*This work was supported by the Polish Academy of Sciences, grants 03.13.3.1 and 04.12.1.3.

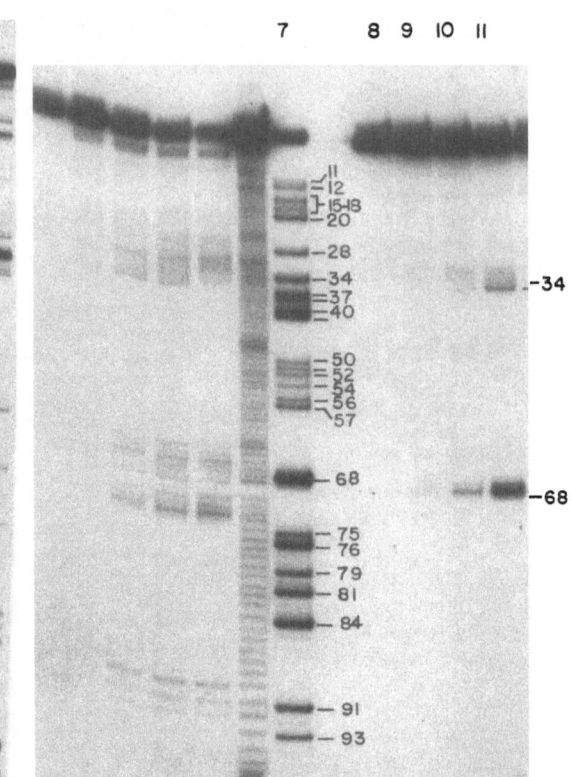

Figure 1.   3'-end labelled U1A1 snRNA analysis. (a) sequencing of U1A1
             snRNA. Lanes: K - control; L - formamide ladder; T1, Ph, U2,
             Bc, CL3 - limited hydrolysis with T1, Ph, U2, Bc and CL3,
             respectively. Side numbers indicate guanosine positions.
             (b) Pb ion-induced hydrolysis of U1A1 snRNA. Lanes: (1,8) -
             controls; (2-5) - reactions with 0.25, 0.5, 1 and 2 mM Pb,
             respectively; (6) - formamide ladder; (9-11) - 0.05, 0.25 and
             1.25 units/ml of S1 nuclease, respectively.

   Isolation of U1 snRNA. Total nuclear RNA was isolated from ASL-1
(strain A spontaneous leukaemia) murine cells according to Scherrer and
Darnell (1962). Briefly, the cells were lysed in 10mM Tris-HCl pH 7.6,
10mM KCl, 1.5mM MgCl$_2$, 5mM 2-mercaptoethanol, 0.35M succrose, 0.12% Triton
X-100, 32µg/ml PVS buffer. The nuclei were centrifuged and homogenized in
50mM sodium acetate pH 5.0, 0.15M NaCl, 10mM EDTA, 0.25% SDS, 32µg/ml PVS
buffer and extracted with: phenol (at 65°C and at room temperature),
phenol:CHCl$_3$:isoamyl alcohol (25:24:1). Ethanol-precipitated RNA was
subjected to polyacrylamide gel electrophoresis - (PAGE), (10% gel in 7M
urea). U1 snRNA was eluted from the gel and precipitated with ethanol.

   3' end labelling of RNA. U1 snRNA was labelled at the 3' end with
(^{32}P)pCp and T4 RNA ligase, a-cording to England and Uhlenbeck (1978).
Labelled RNA was purified by PAGE, eluted from the gel and precipitated
with ethanol.

   Enzymatic sequencing of RNA and digestion with S1 nuclease. 3' end
-labelled U1 snRNA was partially digested with nucleases: T1, PhyM, U2, CL3

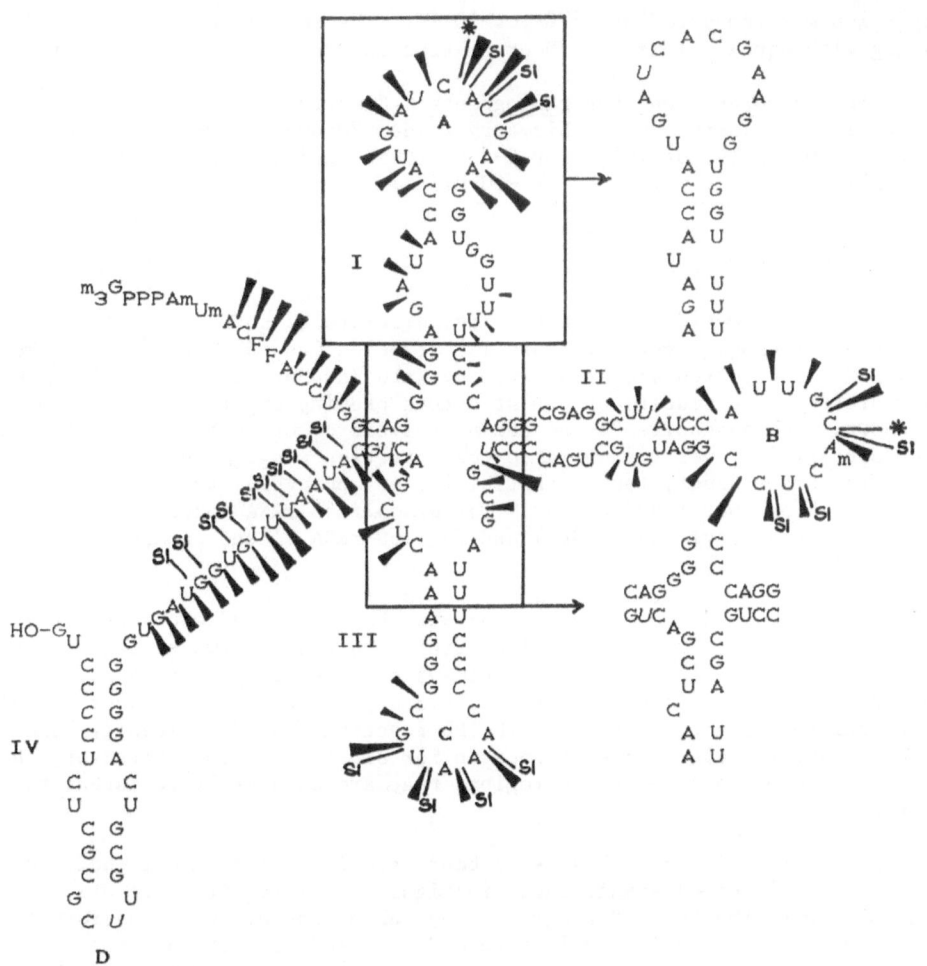

Figure 2. Cleavage sites in U1A1 snRNA from ASL-1 cells presented within the proposed secondary structure. Cuts are indicated by arrows of increasing length corresponding to increasing reactivity of a phosphate bond; bonds reactive towards S1 nuclease are indicated by lines with the letter S and spontaneous hydrolysis sites are indicated by lines with an asterisk. Regions in the proposed structure, that differ from those in consensus U1 secondary structure (shown aside), are indicated by boxes. A, B, C, D - loops; I, II, III, IV - corresponding stems; each tenth nucleotide is printed in italics.

and from *Bacillus cereus,* according to standard procedures. S1 digestion of U1 snRNA was performed as for the Pb-induced hydrolysis, but instead of Pb acetete, S1 nuclease solution containing Zn ions was added (concentrations of the enzyme specified in the Figure legend).

Hydrolysis with Pb ions. Prior to the reaction with Pb ions, the 3' end labelled U1 snRNA was dissolved in 10mM Tris-Cl pH 7.2, 10mM MgCl$_2$, 40mM NaCl buffer containing carrier RNA (total yeast tRNA) at 8μM final concentration of RNA and subjected to renaturation (incubated for 1min. at 55°C and cooled slowly to room temperature). Pb acetate was added and the digestion was conducted at 25°C for 10min. The reactions were quenched by mixing with equal volumes of 7M urea, tracking dyes and 20mM EDTA solution.

Electrophoresis and autoradiography. Electrophoresis was performed in 10% or 12% acrylamide, 0.75% bis-acrylamide, 7M urea, 50mM Tris-borate pH 8.3, 1mM EDTA gels, at 1000V for 3-6 hours in various experiments. Autoradiography was performed at -20°C.

RESULTS AND DISCUSSION

Direct enzymatic sequencing of the discrete U1 snRNA species isolated from mouse leukaemia cells (Figure 1a) revealed its identity with the mouse U1A1 snRNA nucleotide sequence given by Kato and Harada (1985). In Figure 1b representative results of our structure probing experiments with Pb ions is shown. The extensive Pb-induced hydrolysis occurs with practically the same specificity over the whole Pb concentration range used (0.5mM to 2mM). Thus, in this respect, the susceptibility of mouse U1A1 to hydrolysis by Pb ions resembles that found recently for *E. coli* 5S rRNA conformer B (Ciesiolka and Krzyżosiak, 1989) and for 16S rRNA 3'-end domain (Górnicki et al., 1989).

Among the sequences of phosphodiester bonds cleaved are phosphates 25-36 and 65-74. The nuclease S1 cutting within these two regions is shown in Figure 1b, for comparison. The cleavages occur only at phosphates 31-34 and 68-71. It is evident that the S1 hydrolysis, which was conducted under identical solution conditions to the Pb reaction, is significantly more specific and therefore less diagnostic for single strands. Moreover, the Pb cuts take place also in the regions that are completely resistant to S1 hydrolysis.

All Pb-reactive phosphodiester bonds are indicated in our modified structure model of U1 snRNA, shown in Figure 2. This structure was generated with the PC FOLD programme, by taking advantage of the results of Pb-induced hydrolysis. Also in Figure 2, two sections of the molecule are shown with the different base-pairing patterns proposed earlier for U1 snRNA structure by Reddy and Busch (1988). It is clear that our experimental data does not agree with the earlier model and our modified structure can be considered as an attractive alternative.

The modified structure is consistent with results of all earlier U1 snRNA structure probing experiments. The distribution of coordinated base changes in known U1 snRNA sequences from different organisms, that is diagnostic for base pairing, does not rule out the proposed model.

REFERENCES

Branlant C., Krol A., Ebel J-P., Gallinaro H., Lazar E., Jacob M. 1981, The conformation of chicken, rat and human U1A RNAs in solution, *Nucleic Acids Res.*, 9:841.
Busch H., Reddy R., Rothblum L., Choi T. 1982, snRNAs, snRNPs and RNA processing, *Ann. Rev. Biochem.*, 51:617.
Chabot B. and Steitz J. 1987, Multiple interactions between the splicing substrate and small nuclear ribonucleoproteins in spliceosomes, *Mol. Cell Biol.*, 7:281.
Ciesiolka J., Krzyzosiak W. J. 1989, Lead(II) induced hydrolysis of alternative E. coli 5S rRNA A and B forms, *Biochim. Biophys. Res. Commun.*, submitted.
England T. and Uhlenbeck C. 1978, 3'-Terminal labelling of RNA with T4 RNA ligase, *Nature*, 275:560.

Epstein P., Reddy R., Busch H. 1981, Site-specific cleavage by T1 RNase of
U1 RNA in U1 ribonucleoprotein particles, Proc. Natl. Acad. Sci., USA,
78:1562.

Górnicki P., Baudin F., Romby P., Wiewiórowski M., Krzyżosiak W. J. Ebel J.
P., Ehresmann C. Ehresmann B. 1989, Use of lead(II) to probe the
structure of large RNAs. Conformation of the 3' terminal domain of E.
coli 16S rRNA and its involvement in building the tRNA binding sites.
J. Biomol. Struct. Dyn., 6:971.

Kato N. and Harada F., 1985, New U1 RNA species found in Friend spleen
focus forming virus-transformed mouse cells, J. Biol. Chem., 260:7775.

Mount S. and Steitz J. A., 1981, Sequence of U1A RNA from Drosophila
melanogaster: implications for U1 secondary structure and possible
involvement in splicing, Nucleic Acids Res., 9:6351

PC FOLD: Version 3.0, RNA Secondary Structure Prediction, Zuker M.,
National Research Council of Canada, Ottawa.

Reddy R., 1988, Complication of small RNA sequences, Nucleic Acids Res.,
16:r61.

Reddy R. and Busch H., 1988, Small Nuclear RNAs: RNA Sequences, Structure
and Modifications, in Structure and Function of Major and Minor Small
Nuclear Ribonucleoprotein Particles, M. Birnstiel, ed., Springer-
Verlag, Berlin-Heidelberg.

Scherrer K. and Darnell J., 1962, Sedimentation characteristics of rapidly
labelled RNA from HeLa cells, Biochem. Biophys. Res. Commun., 7:486.

AGE-SPECIFIC DIFFERENCES IN THE METABOLISM OF NUCLEAR PHOSPHOLIPID FATTY

ACID COMPONENTS

V. N. Nikitin and N. A. Babenko

Institute of Biology
Kharkov State University
Kharkov, 310077
USSR

Arachidonic acid is the predominant polyunsaturated fatty acid of the nuclear lipids. The level of arachidonic acid sharply changes in the nuclear lipids during the process of postnatal ontogenesis (Nikitin and Babenko, 1987) and following the action of cellular hormonal stimuli (Nikitin and Babenki, 1989). It has been established that cell nuclei are able to acylate endogenous phospholipids in the presence of labelled arachidonic acid (Nikitin and Babenko, 1987). It was reported that phospholipases involved in metabolism of phospholipid acyl components are present in liver cell nuclei of albino rats. Yet the results obtained did not establish a correlation between the change in the level of individual nuclear lipids in ontogenesis and the activity of enzymes involved in their degradation and resynthesis. It was supposed that the endogenous phospholipase A_2 of cell nuclei is involved in the regulation of the level of polyunsaturated fatty acids in nuclear membrane lipids. For this reason the purpose of the present study was undertaken to examine *in vitro* metabolism of the ^3H - arachidonic acid in liver cell nuclei of albino rats of various ages.

It has been established (see Table I) that the addition of 1-acyl-2 ^3H - arachidonyl-sn-glycero-3-phosphorylcholine to isolated nuclei is accompanied by rapid translocation of the residual labelled arachidonic acid to endogenous phospholipids of various types. Similar metabolic conversion of arachidonic acid occurs in the presence of cardimyocyte microsomes (Reddy and Schmid, 1986) and animal brain (Ojima Ayako et al., 1987). Thus, taking into account this evidence as well as our data that about 60% of the phosphatidylcholine and phosphatidylethanolamine of albino rat liver cell nuclei are represented by tetraene molecular species (Nikitin and Babenko, 1987), these are synthesized in liver mainly by a deacylation-reacylation (Misra et al., 1975). Thus, it appears that the cell nucleus has an endogenous transacylation system for phospholipids, providing a unique set of nuclear phospholipid fatty acids. Accumulation of the ^3H - arachidonic acid and ^3H - lysophosphatidylcholine in the incubating medium of cell nuclei of 3 month old rats occurs in parallel with the simultaneous decrease in the level of ^3H - phosphatidylcholine. This confirms the activation of both phospholipase A_1 and phospholipase A_2 under these experimental conditions (see Table I). In the presence of 24 month old rat cell nuclei, in contrast to young rat cell nuclei, a more pronounced degradation of ^3H - phosphatidylcholine and accumulation of its

Table I. [3]H - Arachidonic Acid Distribution in Liver Cell Nuclear Lipids of Albino Rats of Various Ages.

Lipid	Ages of rats, months			
	3		24	
	I	II	I	II
Phosphatidylcholine	41.6±2.6	19.7 ±0.72[b]	29.4±1.4[a]	19.4±2.3
Phosphatidylethanolamine	4.3±0.9	2.72±0.38	4.8±0.51	-
Phosphatidylinositol	6.8±0.2	3.66±0.63[b]	12.0±0.6[a]	-
Lysophosphatidylcholine	6.9±0.4	2.51±0.8[b]	15.2±0.9[a]	-
Fatty acids	22.9±1.9	5.58±0.53[b]	-	-

Incubation conditions: 0.25 M sacchrose, 50 mM tris-HCl, pH 7.6, 2 mM $MgCl_2$, 0.045-0.080 mg nuclear protein, 160 mM 1-
-acyl-2 [³H] -arachidonyl-sn-glycero-3-phosphorylcholine (specific activity - $2.4 \cdot 10^3$ c.p.m/mM). Liposomes were obtained by sounding (3 min, 22 kHz, 20 mA). Control of the [³H] -phosphatidylcholine preparation homogeneity was carried out by a chromatographic method. I - the incubating medium was added with 40 mcm of EDTA, incubating temperature = 4 °C, 20 min. II - 2 mM $CaCl_2$, temperature = 37 °C, 20 min. Extraction of lipids was carried out with chloroform and methanol mixture (2:1, v/v), division of phopholipids into fractions was performed by means of dimetric thin-layer chromatography in the systems: I - chloroform:methanol:water (65:25:4, v/v), 2 - butanol:glacial acetic acid:water (60:20:20, v/v) as well as separation of fatty acids in the system: hexane:diethyl ether:glacial acetic acid (73:25:2, v/v). The results are expressed as a percentage of the [³H] -phosphatidylcholine total radioactivity (mean ± S.D. of 5 separate experiments), a - $P_{3\ months-24\ months} < 0.001$, b - $P_{I - II} < 0.001$.

lysoforms occurs, which is not accompanied by an increase in the fatty acid content. Change of the incubating medium for the cell nuclei (by increasing the calcium ion concentration and temperature) intensifies the [3]H - phosphatidylcholine degradation, especially with the 3 month old rat cell nuclei. This is accompanied by a decrease in the level of endogenous nuclear phospholipids containing the [3]H - arachidonic acid residues. It should be noted that this process is more pronounced in liver cell nuclei of 24 month old rats compared to 3 month old animals. The evidence obtained suggests that in cell nuclei, the arachidonic acid and phospholipids containing it, are subjected to intensive catabolism with increasing age. Presumably in the degradation of exogenous and endogenous phospholipids of 24 month old rat cell nuclei by phospholipase C occurs alongside phospholipase A_1. This presumption is supported by early obtained evidence that a decrease in the level of phospholipids in nuclear membranes in postnatal ontogenesis is accompanied by a considerable increase in the content of the lysophospholipids (Nikitin and Babenko, 1987) and diacylglycerols (Babenko and Nikitin, 1988). It has been also established (Nikitin and Babenko, 1987) that the decrease in the level of phospholipids in albino rats liver cell nuclei with age occurs mainly at the expense of a decrease in the content of molecular species containing arachidonic and other polyene fatty acids. At the same time the level of exogenous arachidonic acid incorporation into nuclear phospholipids actually does not change in postnatal ontogenesis. For this reason one may believe that phospholipases A_1, A_2 and C of nuclei are directly involved in the metabolism of phospholipids containing the arachidonic acid residues. At the same time it can not be excluded that other enzyme systems are involved in the metabolic conversions of nuclear phospholipid acyl components. A decrease in the level of polyunsaturated phospholipids in nuclei in postnatal ontogenesis is apparently the result of activation of phospholipases A_1 and C of nuclei with increasing age. Cell nuclear phospholipases, exhibiting maximum activity under various conditions (pH, ion concentration, etc.), are the targets for various biologically active substances (Nikitin and Babenki, 1989). Thus, it is obvious that the enzyme systems for fatty acids metabolism are an important regulatory mechanism, ensuring alteration of the lipid composition of nuclear membranes in parallel with changes of the functional state of the cell.

REFERENCES

Babenko N. A., and Nikitin V. N., 1988, Arachidonic Acid Metabolism in Albino Rats Liver Cell Nuclei, Reps of the USSR Acad. Sci., 302:460.
Misra R., Misra U. K., and Venkitasubramanian T. A., 1975, Distribution and Synthesis of Molecular Species of Phosphatidylcholine in Weanling Rat Liver, Indian J. Biochem. and Biophys., 12:399.
Nikitin V. N., and Babenko N. A., 1987, Lipids and Lipid Metabolism in Ontogenesis, Advances of Modern Biology, USSR, 104:331.
Nikitin V. N., and Babenko N. A., 1989, Thyroid Hormones and Lipid Metabolism, Phys. J., USSR, 35:91.
Ojima Ayako, Nakagawa Yasuhito, Sagiura Taktynki, Masuzawa Yasuo, and Wakukeizo, 1987, Selective Transacylation of 1-O-alkyglycerophospho-ethanolamine by Docasakhexaenoate and Arachidonate in Rat Brain Microsomes, J. Neurochem., 48:1403.
Reddy P. V., and Schmid H. O., 1986, Coenzyme A-dependent and -independent Acyl Transfer between Dog Heart Microsomal Phospholipids, Biochim. Biophys. Acta: Lipids and Lipid Metab., 879 (L 81) : 369.

CHANGES IN LIPID METABOLISM IN RAT LIVER CELL NUCLEI DURING SUPEREXPRESSION

OF NUCLEAR ONCOGENES

G. N. Filippova, D. D. Spitkovsky* and A. V. Alessenko

Institute of Chemical Physics
Academy of Sciences
Moscow, USSR

* All Union Cancer Research Center AMS USSR
Moscow, USSR

INTRODUCTION

Interest in lipid metabolism relating to oncogene expression has been confined, in the main, to the inositol pathway of regulation. However, there is information that metabolic processes sharply intensify for other phospholipids (PhL), under conditions of increased proliferation (Guy G. R., Murray A. W., 1983).

Lipolytic enzymes, such as, phospholipase C (Besterman J. M. et al., 1986) and sphingomyelinase (Kolesnick R. N., 1987), are the first enzymes to be involved in the cascade of reactions resulting in the transmission of proliferative signals and activation of celluar oncogenes.

In this connection, the present work investigates both changes in the activity of sphingomyelinase and phospholipase C localized in the nucleus, specifically hydrolyzing sphingomyelin (SPhM) and phosphatidylcholine (PhCh), respectively. It also examines changes in the synthesis and contents of PhCh, SPhM and its hydrolysis products, the ceramides, in the nucleus under the conditions of superexpression of nuclear oncogenes induced by inhibition of protein synthesis with sublethal doses of cycloheximide (CHI).

The methods used have been previously described (Alessenko A. V., Filippova G. N., 1989; Filippova G. N. et al., 1989).

RESULTS AND DISCUSSION

The model of oncogene expression, induced by a sublethal dose (3mg/kg) of CHI (Filippova G. N. et al., 1989), used in the investigation, is characterized by reversible inhibition of the protein synthesis (0-6hrs after CHI injection) followed by its activation (12-24hrs) and initiation of the DNA synthesis (48-50hrs after the drug injection). The processes of transcription and translation are clearly separated in time in this model.

A superexpression of c-*fos* and c-*myc* oncogenes has been found in the conditions of strong inhibition of the protein synthesis, against a

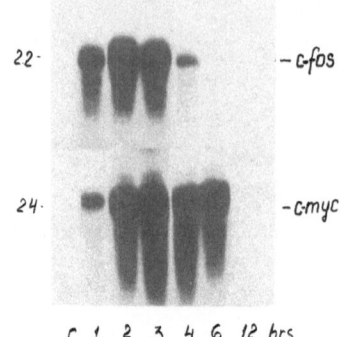

Figure 1.   Blot-hybridization of ^{32}P-v-foc and ^{32}P-c-myc with total RNA samples, isolated from rat liver after CHI injection at a dose of 3mg/kg. Exposure – 12 hours at room temperature. C – control without CHI injection.

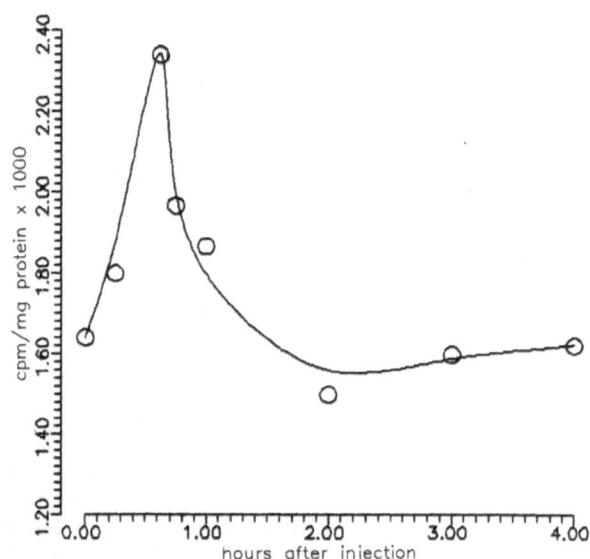

Figure 2.   Changes in sphingomyelinase activity in rat liver cell nuclei after CHI injection.

background of increased RNA-synthesizing activity of nuclei (Figure 1). The figure shows that the most powerful induction of c-*fos* oncogene expression was recorded in 2-3 hours and that the c-*myc* oncogene in 2-4 hours after the drug injection.

The level of the phospholipase C and sphingomyelinase activity was measured in the cell nucleus before and during short-term expression of the c-*myc* and c-*fos* oncogenes. As seen in Figure 2, sphingomyelinase is activated in the nucleus during the first 15-45min after the drug injection, preceding the maximum expression of oncogenes. An increase in the phospholipase C activity was not recorded in the cell nucleus, under conditions of the induction of oncogenes expression (data are not presented).

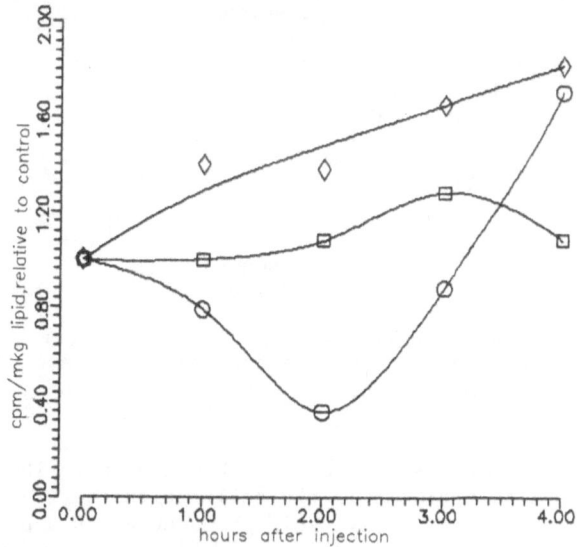

Figure 3.   Changes in the content of PhCh (□), SPhM (O) and ceramides (◇) after CHI injection.

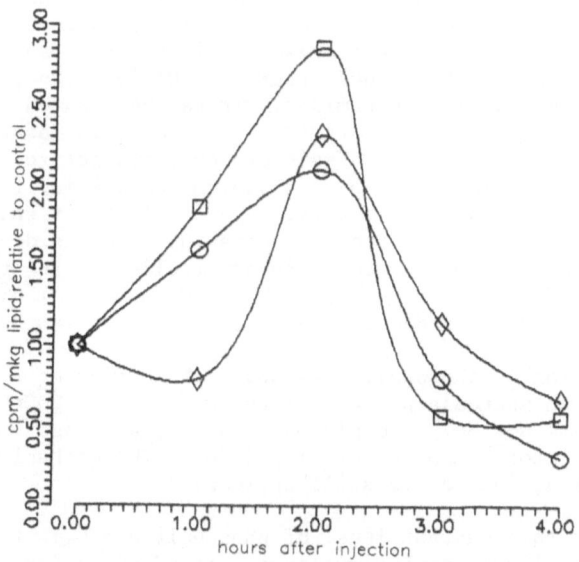

Figure 4.   Changes in the level of the synthesis of PhCh (□), SPhM (O) and ceramides (◇), labelled by sodium acetate-^{14}C, after CHI injection.

Figure 3 gives the data on the contents (relative to the corresponding control) of PhCh, SPhM and ceramides from which it follows that the PhCh level changes very little during activation of oncogenes in the nucleus, while the level of SPhM and ceramides undergo considerable changes. analysis of the data obtained on the change in activity of lipolytic enzymes and level of the corresponding lipids in the nucleus, it is possible to conclude that the stable phospholipase C activity correlates with the almost constant PhCh content, while the transient activation of

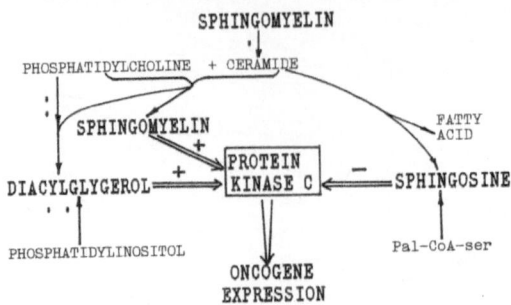

Figure 5.   Scheme of participation of sphingomyelin metabolism products
             in the regulation of nuclear protein kinase C activity (*
             – sphingomyelinase,   ** – phospholipase C).

sphingomyelinase accompanies the fall in SPhM content. The content of
ceramides continues to increase in the cell nucleus during both a decrease
and increase content in the SPhM. This phenomenon has a number of possible
explanations. Firstly, SPhM digestion by sphingomyelinase results in the
accumulation of its hydrolysis product, i.e. ceramides. Secondly, an
increase in the sphingomyelin content, as the reaction substrate, can also
cause accumulation of ceramides in the nucleus.

However, a change in the amount of the lipids can take place not only
because of their enzymatic degradation, but also as a result of de novo
synthesis. The experimental data presented in Figure 4 shows that SPhM
synthesis is activated 1.5-2 hours after the CHI injection, just preceding
the accumulation of labelled ceramides. It can be assumed that an increase
in the quantity of radioactive ceramides (the label is located in the fatty
acid part) is caused by the hydrolysis of newly synthesized SPhM. In
addition, marked PhCh labelling by ^{14}C-sodium acetate was found 1.5-2hrs
after CHI injection. However, the relative stability of PhCh in the
nucleus suggests extensive metabolic processing, such as degradation and
exchange with other lipid fractions, in particular, in the reaction with
ceramides: ceramide + PhCh ======== SPhM + DAG (Lecert J. et al., 1987).
SPhM accumulation was recorded in the nucleus 4hrs after the CHI injection.

The activation of phospholipases under the action of a proliferative
stimulus causes an increase protein kinase C activity, which is transferred
from the cytoplasm not only the plasma membrane, but also into the nucleus
under these conditions (Kiss Z. et al., 1988). The mechanism of the nuclear
protein kinase C activation is still unclear.

Our data on the constant level of phospholipase C in the nucleus and
short-term activation of sphingomyelinase preceding an increase in the
expression of oncogenes, suggests that the appearance of the protein kinase
C activator, i.e. diacylglycerol (DAG), in the nucleus arises not from the
hydrolysis of phosphoglycerides (Besterman J. M. et al., 1986), but as a
result of SPhM digestion by sphigomyelinase. The ceramide, produced in this
reaction can interact with PhCh, bringing about DAG and SPhM formation
(Lecert J. et al., 1987). In its turn SPhM along with DAG can activate
protein kinase C (Maltzeva E. L. et al., 1987). Besides, the product of
intense hydrolysis of SPhM, i.e. sphingosine, is an endogenous inhibitor of
protein kinase C (Hannun Y. et al., 1986).

From our own published data, the mechanism of regulation of nuclear
protein kinase C activity and, as a consequence of this, regulation of the
expression of oncogenes by the products of the sphingomyelin metabolism is

proposed (Figure 5). It should be noted that the proposed mechanism has an advantage, compared with the inositol pathway, of regulation due to the presence of feedback brought about by the products of the metabolic cycle.

REFERENCES

Alessenko A. V., Filippova G. N., 1989, Changes in sphingomyelinase activity and in level of nuclear oncogene expression in reversible inhibition of protein synthesis by cycloheximide, Dokl. Acad. Sci. USSR, 306:486.

Besterman J. M., Duronio V., Cuatrecasas P. 1986, Rapid formation of diacylglycerol from phosphatidylcholine: a pathway for genereation of a second messenger, Proc. Nat. Acad. Sci., 85:6785.

Filippova G. N., Spitkovsky D. D., Bojkov P. Ya., Alessenko A. V. 1989, Oncogene expression in rat liver under conditions of template biosynthesis uncoupling by sublethal doses of cycloheximide, Molekul. Biolog. (in Russian), 23:843.

Guy J. R., Murray A. W. 1983, Effect of a tumor promoter on phospholipid metabolism in Hela cells, Cancer. Res., 43:5564.

Hannun Y. A., Loomis C. R., Merrill Jr., Bell R. M., 1986, Sphingosine inhibition of protein kinase C activity and phorbol dibuturate binding in vitro and in human platelets, J. Biol. Chem., 261:12604.

Kiss Z., Deli E., Kuo J. F., 1988, Temporal changes in intracelluar distribution of protein kinase C during differentiation of human leukaemia HL60 cells induced by phorbol ester, FEBS Lett. 231:41.

Kolesnick R. N. 1987, 1,2-Diacylglycerols but not phorbol esters stimulate sphingomyelin hydrolises in GH$_3$ pituitary cells, J. Biol. Chem., 262:16759.

Lecert J., Fouilland L. Gagniarre J. 1987, Evidence for a high activity of sphingomyelin biosynthesis by phosphocholine transfer from phosphatidylcholine to ceramides in lung lamellar bodies, Biochim. Biophys. Acta. 918:48.

Mal'tseva E. L., Kurnakova N. V., Burlakova E. B., Palmina N. P., 1989, Modification of information signal at the protein kinase C level by lipid peroxidation products in vitro, Studia Biophysica, 132:69.

ROLE OF PHOSPHOLIPIDS ON NUCLEAR STRUCTURE AND FUNCTION

N. M. Maraldi

Ist. Citomorfologia Normale e Patologica
C.N.R.
c/o I.O.R.
Via di Barbiano 1/10 - 40100 Bologna
Italy

Evidence has accumulated in recent years on the presence of phospho-lipids in nuclear substructures, after extensive purification procedures involving detergents (for a review, see Manzoli et al., 1985). Since active chromatin has been reported to contain larger amounts of phospholipids than its repressed form (Rose and Frenster, 1965), these molecules have been thought to play a role in regulating RNA synthesis. Hepatoma nuclei contain increased amounts of phospholipids compared to normal hepatocytes and a considerable increase of sphingomyelin appears to be associated with chromatin fractions (Coetzee et al., 1975). Significant variations of the chromatin-associated phospholipids have been revealed in chronic lymphocytic leukaemia cells, compared to normal B lymphocytes (Manzoli et al., 1977), as well as in rapidly proliferating embryonic tissues compared to resting ones (Manzoli et al., 1982). Chromatin-associated phospholipids found in Ehrlich ascites tumor cells have a fatty acid composition different from that of the corresponding nuclear membrane phospholipids, and are more resistant to diet-induced modifications (Awad and Spector, 1976). In some tumor cells it has been shown that nuclei and chromatin contain lipids with a neutral lipid/phospholipid ratio higher than that in normal cells (Balint and Holczinger, 1978). Among the nuclear matrix components, resistant to both detergent and saline extraction, is a small amount of phospholipids (Berezney and Coffey, 1974). These matrix phospholipids are probably responsible for the hydrophobic interactions between nucleic acids and matrix fibres, as indicated by phospholipase C digestion experiments, which release mostly the newly-synthesized DNA and small RNAs (Cocco et al., 1980).

Phospholipids can interact, in liposome form, with cell-free systems and with isolated nuclei, which constitute a more complete substrate for the expression of their regulatory functions. In this system the following mechanisms have been elucidated: a) phospholipids are transfered from liposomes to the inner nuclear compartments forming lipoprotein complexes with nuclear components different from the nuclear membrane; b) the endogenous RNA synthesis is inhibited by neutral phospholipids and enhanced by negatively charged ones (Maraldi et al., 1984); c) the expression of a single gene can be modified by exogenous phospholipids; d) these effects do not modify the RNA profiles but affect only the efficiency of transcription; e) a possible target for the negatively charged phospholipids is represented by histone H1 which, when removed, causes a solenoid-nucleosome transition in the chromatin arrangement; f) these effects are widely documented by electron microscopy both in thin section and in freeze-fractured samples (Maraldi et al., 1982); g) neutral phospholipids induce an accumulation of the interchromatin granules suggesting that they impair the maturation and transport of the transcripts (Maraldi et al., 1987).

Recently, it has been observed that isolated nuclei can incorporate ATP into inositol phospholipids *in vitro* and that changes occur in this incorporation during cell growth and differentiation (Cocco et al., 1987). For example, in Swiss-mouse 3T3 cells it has been shown that IGF-I pre-treatment of the cell, results in a transient decrease in incorporation from (^{32}P)-ATP into FtdInsP and PtdInsP$_2$ when isolated nuclei are studied *in vitro* (Manzoli et al., 1988), while no significant parallel change is seen in whole cell homogenates. Because of the evidence showing the occurence of PKC at the nuclear level (Capitani et al., 1987), these observations pointed to the likelihood that polyphosphoinositides and PKC could act synergistically in the nucleus, providing an intranuclear signally pathway. Since the occurence of inositol lipids in the nucleus, we have sought to establish the effect of mitogen treatment on these molecules and the relationship between inositol cycle products and hyperphosphorylation of nuclear proteins, via PKC during the lag phase leading to the onset of DNA synthesis.

When Swiss 3T3 cells are labelled for 36 hours with high levels of (^{3}H)-myo-inositol and the radioactivity in nuclear inositol phospholipids is measured, it has been observed that treatment of cells for 2 mins, but not for 4 hours, with a mitogenic concentration of insulin-like growth factor I and bombesin causes a marked decrease in PtdInsP and PrdInsP$_2$. Moreover *in vivo* phosphorylation of some nuclear protein occurs later on. Among these proteins, histone H1 and 0.75M PCA soluble polypeptide with an apparent Mr of a 21,000kD are phosphorylated *in vitro* by protein kinase C in isolate nuclei purified from 3T3 cells treated for 90 mins with IGF-I and bombesin (Manzoli et al., 1989). Since these phosphorylative events follow the earlier changes in nuclear polyphosphoinositide metabolism induced by the same mitogen combination, it seems possible that these two phenoma are related and trigger the synthetic machinary responsible for the replication of DNA.

REFERENCES

Award A. B., and Spector, 1976, Modification of the Ehrlich ascites tumor cell nuclear cell lipids. Biochem. Biophys. Acta, 450:439.
Balint Z., and Holczinger L., 1978, Neutral lipids in nuclei and chromatin fraction of young and old Ehrlich ascites tumor cells. Neoplasma, 25:25.
Berezney R., and Coffey D. S., 1974, Identification of a nuclear protein matrix. Biochem. Biophys. Res. Commun., 60:1410.
Capitani S., Girard P. R., Mazzei G. J., Kuo J. F., Berezney R., and Manzoli F. A., 1987, Immunochemical characterization of protein Kinase C in rat liver nuclei and subnuclear fraction. Biochem. Biophys. Res. Commun., 142:367.
Cocco L., Maraldi N. M., Manzoli F. A., Gilmour R. S., and Lang A., 1980, Phospholipid interaction in rat liver nuclear matrix, Biochem. Biophys. Res. Commun., 96:890.
Cocco L., Gilmour R. S., Ognibene A., Letcher A. J., Manzoli F. A., and Irvine R., 1987, Synthesis of polyphosphoinositides in nuclei of Friend cells: evidence for PIP$_2$ metabolism inside the nucleus which changes with cell differentiation. Biochem. J., 248:765.
Coetzee M. L., Sprangler M., Morris H. P., and Ove P., 1975, DNA synthesis in membrane-denuded nuclei and nuclear fractions from host liver and Morris hepatomas. Cancer Res., 35:2752.

Manzoli F. A., Maraldi N. M., Cocco L., Capitani S., and Facchini A., 1977, Chromatin phospholipids in normal and chronic lymphocytic leukaemia lymphocytes. Cancer Res., 37:843.

Manzoli F. A., Capitani S., Mazzotti G., Barnabei O., and Maraldi N. M., 1982, Role of chromatin phospholipids on template availability and ultrastructure of isolated nuclei. Adv. Enzym. Regul., 20:247.

Manzoli F. A., Maraldi N. M., and Capitani S., 1985, Effect of phospholipids on the control of nuclear DNA restriction in: 'The pharmacological effect of lipids', J. J. Kabara, ed., The American Oil Chemist's Society, Champaign, Illinois, pp. 133-156.

Manzoli F. A., Capitani S., Cocco L., Maraldi N. M., Mazzotti G., and Barnabei O., 1988, Lipid mediated signal transduction in the cell nucleus, Adv. Enzyme Regul., 27:83.

Manzoli F. A., Martelli A. M., Capitani S., Maraldi N. M., Rizzoli R., Barnabei O., and Cocco L., 1989, Nuclear polyphosphinositides during cell growth and differentiation, Adv. Enzyme Regul., 28:25.

Maraldi N. M., Caramelli E., Capitani S., Marinelli F., Antonucci A., Mazzotti G., and Manzoli F. A., 1982, Chromatin structural changes induced by phosphathydilserine liposomes on isolated nuclei, Biol. Cell, 46:325.

Maraldi N. M., Capitani S., Caramelli E., Cocco L., Barnabei O., and Manzoli F. A., 1984, Conformational changes of nuclear chromatin related to phospholipid-induced modification of the template availability, Adv. Enzyme Regul., 22:447.

Maraldi N. M., Galanzi A., Caramelli E., Billi A. M., Ognibene A., Rizzoli R., and Capitani S., 1987, Changes in ribonucleoprotein particle and chromatin organization induced by liposomes in isolated nuclei, Cell Biochem., Funct., 6:165.

Rose H. G., and Frenster J. H., 1965, Composition and metabolism of lipids within repressed and active chromatin of interphase lymphocytes, Biochem. Biophys. Acta, 106:577.

THE CHROMATOID BODY IN SPERMATOGENESIS: NUCLEO-CYTOPLASMIC TRANSPORT OF

HAPLOID GENE PRODUCTS AND ITS CYTOSKELETAL REGULATION

Pekka Mali, Jorma Toppari, Leena-Maija Parvinen and
Martti Parvinen

Institute of Biomedicine
Department of Anatomy
University of Turku
SF-20520 Turku
Finland

INTRODUCTION

All cells of the germ line are characterized by specific cytoplasmic
components, germ cell determinants, that are called polar granules in
insects, germinal plasm in amphibians and nuage in mammals (Beams and
Kessel, 1974). The chromatoid body is a germ cell determinant; it is
characteristic for mammalian spermatocytes and spermatids (Eddy, 1974;
Söderström, 1981, for review, see Sud, 1961). In early spermatids, it is a
lobulated cytoplasmic structure with 1-2μm in diameter (Figure 1). During
the first appearance of the chromatoid material in mid-pachytene spermatocytes (Russell and Frank, 1978; Head and Kresge, 1985), it is often seen
to be intimately associated with mitochondria (Fawcett et al., 1970).
During early spermiogenesis, the chromatoid body is located on the surface
of the nucleus close to the Golgi complex and the developing acrosome (Susi
and Clermont, 1970). During this period, two cytochemically distinct
compartments of the Golgi cortex, characterized by acid phosphatases,
nicotinamide adenine dinucleotide phosphatase (NADPase) and cytidine mono-
phosphatase (CMPase), independently contribute to the formation of the
vesicular component of the chromatoid body (Thorne-Tjomsland et al., 1988).
Towards the end of spermiogenesis, the chromatoid body diminishes in size
and is found in the region of the developing flagellum. It becomes
separated from the spermatozoon at spermiation in the residual body.

Histochemical staining with pyronin suggests the presence of RNA in the
chromatoid body (Daoust and Clermont, 1955), but electron microscope
analysis after RNase digestion failed to confirm this (Eddy, 1970). More
recent analyses have demonstrated basic proteins and polysaccharides in the
surrounding vesicles (Krimer and Esponda, 1980; Paniagua et al., 1985). The
Golgi vesicles containing acid phosphatase have been shown to be associated
with the chromatoid body (Anton, 1983). DNase-gold complex specifically
labels the chromatoid body, but since this marker also binds to actin, the
presence of DNA in the chromatoid body remains unclear (Dadoune et al.,
1987). The chromatoid body reacts specifically with a monoclonal antibody
OX3 to a rat histocompatibility antigen (Head and Kresge, 1985). This
antibody, therefore, should prove useful in determining the chemical
structure, origin and fate of the chromatoid body.

Nuclear Structure and Function, Edited by J. R. Harris and
I. B. Zbarsky, Plenum Press, New York, 1990

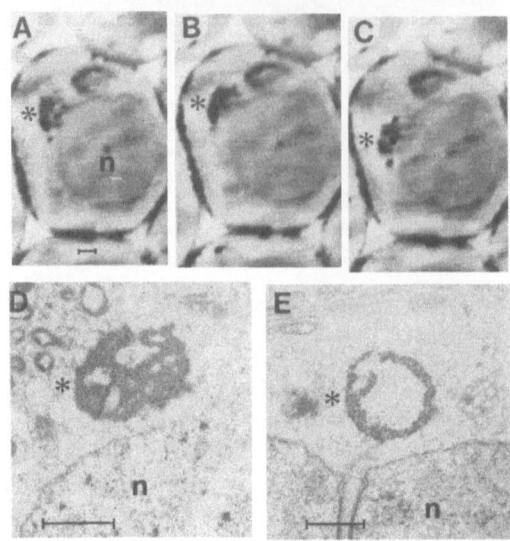

Figure 1.   Phase contrast videomicrographs (A-C) of a living step 4
             spermatid cultured for 24h in Ham's F 12/Dulbecco's MEM (1:1),
             supplemented with Hepes, gentamycin, bovine serum albumin and
             sodium bicarbonate in 5% CO_2 in air at 32°C. A frozen TV-
             screen was photographed at 10s intervals. The movements of
             chromatoid body (asterisk) on the nuclear (n) envelope are
             evident when compared with a stationary Golgi complex and
             acrosomic system (the other dark cytoplasmic organelle). The
             normal lobulated ultrastructure of the chromatoid body
             cultured in above-mentioned conditions (asterisk in D) is
             changed into an abnormal ring-shape when vincristine (20µg/ml)
             was added into the medium (asterisk in E). Scale bars: 1µm.

        The localization of RNase-gold particles suggests the presence of RNA
in the chromatoid body (Walt and Armbruster, 1984). This was also suggested
by incorporation of radioactivity derived from ^{3}H-urdine (Söderström and
Parvinen, 1976). The dependence of the chromatoid body on transcription of
the haploid genome has been suggested, since after specific inhibition of
transcription by actinomycin D, it is changed morphologically (Parvinen et
al., 1978) and remains unlabelled after incubation with ^{3}H-uridine
(Söderström, 1977).

        The chromatoid body moves rapidly on the nuclear surface in a nonrandom
saltatory fashion (Figure 1A-C, Parvinen and Jokelainen, 1974). Two types
of movements have been distinguished that are suggested to play a role in
the incorporation and transport of haploid gene products during early
spermiogenesis (Parvinen and Parvinen, 1979).

        The movements of the chromatoid body are obviously dependent on
cytoskeletal components. Incorporation of gold-labeled anti-actin (Walt and
Armbruster, 1984), and accumulation of microtubules in the vicinity of the
chromatoid body have been demonstrated (Parvinen and Parvinen, 1979). These
microtubules are associated with vesicles suggested to store and release
calcium in association with the chromatoid body (Andonov and Chaldakov,
1989).

        Vincristine, a microtubule inhibitor, influences the movements of the

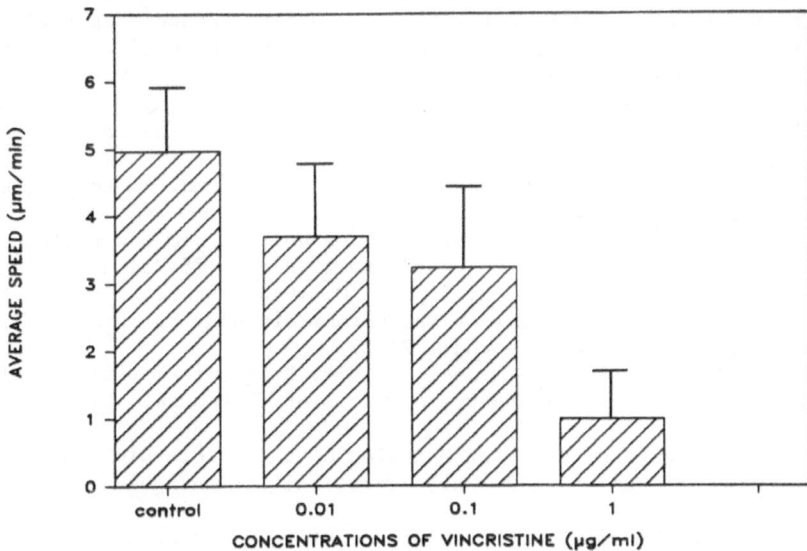

Figure 2.   Quantitative analysis of the average speed of the chromatoid body in control conditions (after 24h culture) and after culture in the presence of varying concentrations of vincristine. The speed was significantly ($p < 0.05$) reduced by 0.01µg/ml of vincristine, and the movements were virtually stopped by 1µg/ml of vincristine.

chromatoid body (Mali et al., 1987). After 24h incubation *in vitro*, 0.01µg/ml of vincristine had a significant inhibitory effect on the movements of the chromatoid body and 1µg/ml slowed the movements to an almost undetectable level (Figure 2). In addition, concentrations of 10 and 20µg/ml vincristine caused marked morphological alterations to the chromatoid body in early spermatids (Mali et al., 1987, Figure 1E).

The chromatoid body obviously has a function in the formation of the acrosomic system. There is evidence that a considerable proportion of the RNA synthesized in pachytene spermatocytes is preserved until late spermiogenesis (Germia et al., 1978). There are no observations about the structural location of this RNA in the spermatids, but the chromatoid body is a clear possibility. Studies on these problems await isolation of the chromatoid body for biochemical analyses.

ACKNOWLEDGEMENT

This project has been supported by the Academy of Finland (Project no. 200 at the Medical Research Council).

REFERENCES

Andonov M. D., and Chaldakov G. N. 1989, Morphological evidence for calcium storage in the chromatoid body of rat spermatids. Experientia, 45:377.
Anton E. 1983, Association of Golgi vesicles containing acid phosphatase with the chromatoid body of rat spermatids. Experientia, 39:393.
Beams H. W., and Kessel R. G. 1974, The problem of germ cell determinants. Int. Rev. Cytol., 39:413.

Dadoune J.-P., Alfonsi M.-F., and Fain-Maurel M.-A. 1987, Marquage ultrastructural du corps chromatide et du corps associe au centriole a l'aide du complexe DNase-or colloidal dans les spermatides de singe. C. R. Acad. Sci., 305:135.

Daoust R., and Clermont Y. 1955, Distribution of nucleic acids in germ cells during the cycle of the seminiferous epithelium of the rat. Am. J. Anat., 96:255.

Eddy E. M. 1970, Cytochemical observations of the chromatoid body of male germ cells. Biol. Reprod., 2:114.

Eddy E. M. 1974, Fine structural observations on the form and distribution of nuage in germ cells of the rat. Anat. Rec., 178:731.

Fawcett D. W., Eddy E. M., and Phillips D. M. 1970, Observations on the fine structure and relationships of the chromatoid body in mammalian spermatogenesis. Biol. Reprod., 2:129.

Geremia R., D'Agostino A., and Monesi V. 1978, Biochemical evidence of haploid gene activity in spermatogenesis of the mouse. Exp. Cell Res., 111:23.

Head J. R., and Kresge C. K. 1985, Reaction of the chromatoid body with a monoclonal antibody to a rat histocompatibility antigen. Biol. Reprod., 33:1001.

Krimer D. B., and Esponda P. 1980, Presence of polysaccharides and proteins in the chromatoid body of mouse spermatids. Cell Biol. Int. Rep., 4:265.

Mali P., Fagerhed R.-M., and Parvinen M. 1987, Cytoskeletal regulation of the chromatoid body. In: 'Development and Function of the Reproductive Organs', Serono Symposia Review no 14, M. Parvinen, I. Huhtaniemi, and L. J. Pelliniemi, eds., Ares Serono Symposia, Rome, Italy, p. 317.

Paniagua R., Nistal M., Amat P., and Rodriguez M. C. 1985, Presence of ribonucleoproteins and basic proteins in the nuage and intermito-chondrial bars of human spermatogonia. J. Anat., 143:201.

Parvinen L.-M., Jokelainen P. T., and Parvinen M. 1978, Chromatoid body and haploid gene activity: Actinomycin D induced morphological alterations. Hereditas, 88:75.

Parvinen M., and Jokelainen P. T. 1974, Rapid movements of the chromatoid body in living early spermatids of the rat. Biol. Reprod., 11:85.

Parvinen M., and Parvinen L.-M. 1979, Active movements of the chromatoid body. A possible transport mechanism for haploid gene products. J. Cell Biol., 80:621.

Russell L., and Frank B. 1978, Ultrastructural characterization of nuage in spermatocytes of the rat testis. Anat. Rec., 190:79.

Söderström K.-O. 1977, Effect of actinomycin D on the structure of the chromatoid body in the rat spermatids. Cell Tiss. Res., 184:411.

Söderström K.-O. 1981, The relationship between the nuage and the chromatoid body during spermatogenesis in the rat. Cell Tiss. Res., 215:425.

Söderström K.-O., and Parvinen M. 1976, Incorporation of ^{3}H-urdine by the chromatoid body during rat spermatogenesis. J. Cell Biol., 70:239.

Sud B. N. 1961, The 'chromatoid body' in spermatogenesis. Quart. J. Micr. Sci., 102:273.

Susi F. R., and Clermont Y. 1970, Fine structural modifications of the rat chromatoid body during spermatogenesis. Am. J. Anat., 129:177.

Thorne-Tjomsland G., Clermont Y., and Hermo L. 1988, Contribution of the Golgi apparatus components to the formation of the acrosomic system and chromatoid body in rat spermatids. Anat. Rec., 221:591.

Walt H., and Armbruster B. L. 1984, Actin and RNA are components of the chromatoid bodies in spermatids of the rat. Cell Tiss. Res., 236:487.

SELECTIVE STAINING OF DNA AT THE ULTRASTRUCTURAL LEVEL AFTER ALKALINE

HYDROLYSIS

P. S. Testillano, M. C. Risueño, M. A. Ollacarizqueta[1] and
C. J. Tandler[2]

Centro de Investigaciones Biológicas
Estructuras Celulares
[1] Servicio de Microscopía Electronica Analítica
CSIC, Velázquez 144
28006 Madrid, Spain

[2]Inst. de Biología Celular
CONICET
Fac. de Medicina
Buenos Aires, Argentina

INTRODUCTION

Cytochemical electron microscope methods have been useful for the
localization of individual molecules or groups *in situ*. Techniques
selective for DNA have been very thoroughly investigated, and have lead to
the different Schiff-like techniques. The Feulgen-like method, with the
osmium-ammine complex according to Cogliati and Gautier (1973) and developed
by Moyone (1980) has been the only method to reveal chromatin fine structure,
because of the specificity and high resolution it offers. Its use has
enabled visualization of nucleosomal and non-nucleosomal (12 and 3nm)
chromatin fibres in both animal and plant cells (Derenzini et al. 1982;
Risueño et al., 1982). The main problem of this technique is the synthesis
of the stain, as it is not commercially available. Recently, the localization of the DNA in the nucleus has been address by immunocytochemical
procedures (Scheer et al., 1987). The use of monoclonal antibodies has made
available a highly sensitive procedure to detect double and single-stranded
DNA. However, this approach requires antibody and the development of tissue
processing to preserve antigenicity.

In this paper we describe a simple method (NaOH-formaldehyde-MA) to
selectively stain DNA-containing structures with uranyl acetate, based on
alkaline hydrolysis followed by methylation and acetylation with methanol-
acetic anhydride (Tandler, 1959; Tandler and Solari, 1982).

MATERIALS AND METHODS

The materials used were root meristematic cells from *Allium cepa* L. and
anthers from *Capsicum annum* L. The onion bulbs were grown under standard
conditions and the *Capsicum* plants were grown in a greenhouse. Root samples
were fixed in 3% glutaraldehyde in 0.025M cacodylate buffer for 3-6 hours;
after washing in the same buffer they were treated with 0.5-1N NaOH in 4%

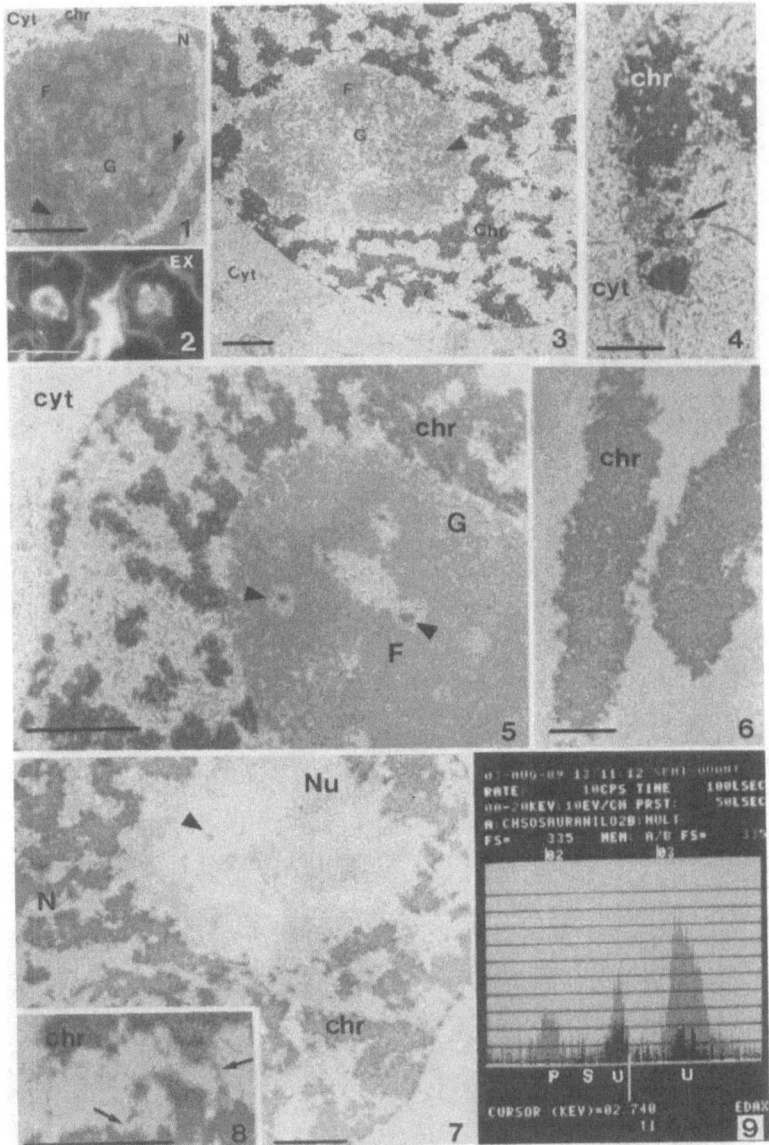

Figure 1.   Glutaraldehyde-fixed, uranyl-lead stained onion root
              meristematic cell. Cyt: cytoplasm, N: nucleus, chr: condensed
              chromatin, F: fibrillar component, G: granular component,
              arrow-heads: het Fcs, arrows: hom Fcs.

Figure 2.   DAPI staining after NaOH treatment of cryosectioned *Capsicum*
              pollen grains. Note the high fluorescence of the DNA inside
              the nucleus. The pollen wall, exine (Ex), shows auto-
              fluorescence. Bar represents 10μm.

Figures 3   Glutaraldehyde-fixed, NaOH-formaldehyde treated, uranyl
and 4.   stained onion root meristematic cells. Figure 3: Interphase
              nucleus. The nucleolus appears less compact and the inter-
              chromatin region presents less fibrilo-granular material than
              in controls; some cytoplasmic structures, such as ribosomes,
              are observed. Figure 4: Mitotic chromosomes. The nucleolar
              material forming the coat in late anaphase (arrows) appears
              less compact than in controls.

formaldehyde for 12-24 hours, washed with distilled water, 0.5 to 1% (v/v) acetic acid, and water. Then, they were divided in two series: one of them was dehydrated in ethanol series and embedded in Epon; the other was dehydrated in methanol and treated with methanol-acetic anhydride (5:1, v/v) freshly prepared (Tandler and Solari, 1982) and finally embedded in Epon. Ultrathin sections were stained only with uranyl acetate and observed in a Philips EM 300. Control samples were gultaraldehyde fixed, Epon embedded, and uranyl-lead stained following the conventional technique; this was done in order to check losses of tissue components and the effectiveness of the blocking reagent MA for suppressing protein stainability. 1μm cryosections from cryoprocessed anthers (for method see Sánchez-Pina et al., in this volume) were mounted on poly-L-lysine coated slides, treated with 0.5N NaOH in 4% formaldehyde for 30 mins. and stained with DAPI (Vergne et al., 1987) to test the DNA presence after alkali treatment. Some ultrathin sections on carbon coated Titanium grids were stained with uranyl and studied by X-ray microanalysis to determine the presence of Phosphorus in chromatin and the nucleolus after the 'NaOH-formaldehyde-MA' technique; this study was carried out in a Philips EM 420 fitted with an energy-dispersive X-ray spectrometer and an EDAX 9100 analyzer computer system.

RESULTS AND DISCUSSION

The weak treatment with alkali of gluteraldehyde fixed material, produce the extraction of some cellular components but the main structural features remain (compare Figures 1 and 3), and DNA preservation is evident from DAPI staining at the light microscope level (Figure 2). After uranyl staining (Figures 3 and 4), the interchromatin region appears less dense in fibrillo-granular material, the nucleolus and mitotic nucleolar material (Figure 4) are less compact than in untreated nuclei (Figure 1). However, condensed chromatin masses (Figure 3), as well as mitotic chromosomes (Figure 4), appear similar to those in control cells. Some cytoplasmic structures are maintained after the treatment (Figure 3).

The application of the methylation and acetylation method (MA) (Tandler and Solari, 1982) in onion root meristematic cells followed by

---

Figure 5.   Glutaraldehyde-fixed, methanol-acetic anhydride (MA) treated, uranyl stained onion root meristematic cell. The electron density of condensed chromatin clumps and chromatin inclusions of het Fcs (arrowheads) is enhanced compared with the nucleolus. F: Dense fibrillar component, G: Granular component.

Figures 6, 7 and 8.   Glutaraldehyde-fixed, NaOH-formaldehyde-MA treated, uranyl stained onion root meristematic cells. Figure 6: Mitotic chromosomes are selectively stained. Figure 7: Interphasic cell. Condensed chromatin masses are selectively stained as well as chromatin inclusions in the het Fcs. The nucleolus and ribosomes in cytoplasm remain unstained. Figure 8: High magnification of a nuclear region. Note the presence of stained fibres (thin arrows) at the periphery of the condensed chromatin and in the interchromatin retion. Bar represents 0.1μm.

Figure 9.   X-ray microanalysis superposed spectra from NaOH-formaldehyde treated cells. Black spectrum corresponds to nucleolus and grey spectrum corresponds to chromatin. Note that there is no Phosphorus peak in the nucleolus, and the Uranium signal is much lower in the nucleolus. Scale bars in electron micrographs represent 1μm.

uranyl provides differential staining of RNP and DNP containing structures, due to the combined blockage of the ammino and carboxyl groups in proteins, avoiding the protein stainability by uranyl. Dense fibrillar and granular components (DFC and GC) of the nucleolus show less electron density than the condensed chromatin masses (Figure 5). The condensed chromatin inclusions of the heterogeneous fibrillar centres (Het FCs) (Risueño et al., 1982) are also densely stained (Figure 5).

When we combine the alkali treatment with the MA method followed by uranyl, the different electron density of RNP and DNP structures is strongly enhanced, giving selective staining of the chromatin and mitotic chromosomes (Figures 6 and 7). Thin fibrils at the periphery of the chromatin clumps and in the interchromatin region were also discerned (Figure 8). The inclusions in the Het FCs appear as stained as the chromatin masses (Figure 7). The DFC and GC of the nucleolus and ribosomes in the cytoplasm remained unstained (Figure 7).

X-ray microanalysis of nuclei after 'NaOH-formaldehyde-MA' technique (Figure 9) indicated that the phosphorus signal appears only over the condensed chromatin masses, while nucleoli and cytoplasm gave no detectable signal, in contrast with untreated cells (Risueño et al., 1987) where in chromatin and nucleoli phosphorus was detected. This data would indicate that RNA and phosphate groups in phosphoproteins were removed or extracted after the treatment. However, the sulphur signal found over chromatin, nucleoli and cytoplasm was as expected, from the protein content in those compartments (Figure 9).

From the results obtained we demonstrate that we have a new selective staining procedure for DNA, the 'NaOH-formaldehyde-MA' technique. This is based on the fact that DNA is not depolymerized by alkali as can be seen by DAPI (Figure 2), while RNA is hydrolyzed and phosphate groups from phosphoproteins are removed (Schmidt and Thannhauser, 1945), together with an effective blocking of stainable carboxyl groups in proteins produced by the MA (Tandler and Solari, 1982). As aldehyde fixatives cross-link tissue proteins, they make them largely resistant to alkali treatment (Tandler, 1959).

In short, the described method is a simple, accessible and reproducible technique, preferential for DNA that may provide a useful tool to study chromatin structural organization *in situ*.

ACKNOWLEDGEMENTS

We wish to thank Dr Sánchez-Pina for her helpful comments and critical reading of the manuscript. Dr C. J. Tandler was supported by a grant from the Ministerio de Educatión y Ciencia of Spain during his stay at the Centro de Investigaciones Biológicas. This work was supported by the project DIGICYT/CSIC 88/91 PB 033201.

REFERENCES

Cogliati R. and Gautier A., 1973. Mise en évidence de l'ADN et des poly-saccharides à l'aide d'un nouveau réactif 'de type Schiff'. C. R. Acad. Sci. (Paris) Sér. D 276:3041.
Derenzini M., Viron A. and Puvion-Dutilleul F., 1982. The Feulgen-like osmiumammine reaction as a tool to investigate chromatin structure in thin sections. J. Ultr. Res. 80:133.

Moyne G., 1980. Methods in ultrastructural cytochemistry of the cell nucleus. Prog. Histochem. Cytochem. 13:1.

Risueño M. C., Medina F. J. and Moreno Díaz de la Espina S., 1982. Nucleolar fibrillar centres in plant meristematic cells: ultrastructure, cytochemistry and autoradiography. J. Cell Sci. 58:313.

Risueño M. C., Olmedilla A., Ollacarizqueta M. A. and Quintana C., 1987. Detection of some elements: Cl, P, S, Ca and Al in the nucleolus of animal and plant cells by X-ray microanalysis. Abs. 10th Eur. Nucle(ol)ar Workshop pp. 127.

Scheer U., Messner K., Hazan R., Raska I., Hansmann P., Falk H., Spiess E. and Franke W. W., 1987. High sensitivity immunolocalization of double and single-stranded DNA by a monoclonal antibody. Eur. J. Cell Biol. 43:358.

Schmidt G. and Thannhauser S. J., 1945. A method for the determination of desoxyribonucleic acid, ribonucleic acid and phosphoproteins in animal tissues. J. Biol. Chem. 161:83.

Tandler C. J., 1959. An alkali-formaldehyde squash technique for plant cytology and cytochemistry. Stain Tech. 34:234.

Tandler C. J. and Solari A. J., 1982. Methanol-acetic anhydride: an efficient blocking agent for electron microscope cytochemistry. Its application to mouse testis and other tissues. Histochem. 76:351.

Vergne P., Delvallee I. and Dumas C., 1987. Rapid assessment of microspore and pollen development stage in wheat and maize using DAPI and membrane permeabilization. Stain Tech. 62:299.

ULTRASTRUCTURAL CHANGES IN NUCLEI WITHIN HIV-1-INDUCED CULTURED CELL

SYNCYTIA

J. R. Harris

North East Thames Regional Transfusion Centre
Crescent Drive
Brentwood
Essex   CM15 8DP
United Kingdom

INTRODUCTION

Cultured lymphoid cells which carry the CD4$^+$ antigen are readily infected by the AIDS virus HIV-1 and in some instances cell fusion occurs. This virally-induced cell fusion results in the formation of multinucleate giant cells or syncytia, generally termed the cytopathic effect of the virus (Montefiori and Mitchell, 1987; Yoffe et al., 1987). The limited ultrastructural data available on HIV-induced syncytiogenesis has placed emphasis upon HIV release from the syncytial surface and into intra-syncytial vacuoles (Dowsett et al., 1988; Harris et al., 1989a). As cell fusion progresses and syncytia of increasing size are formed, it is apparent that marked cytoplasmic and nuclear ultrastructural changes occur (Harris et al., 1989b), which ultimately lead to senescence and syncytial lysis. The data included in the present communication relate primarily to the changes that have been detected within the nuclear envelope and chromatin of nuclei within HIV-1 infected cells and syncytia of increasing size.

MATERIALS AND METHODS

The H9 HIV-1 producer cell line and the permissive C8166 HTLV-1 transformed cell line were cultured in RPM1 1640 medium containing 25mM HEPES buffer supplemented with 15% foetal calf serum. The RF strain of HIV-1 was propagated in the H9 cells and harvested 4 to 5 days postin-fection. The C8166 cells were infected with HIV-1 at a moi of 0.1 to 1.0 and cultured for up to 7 days. Cultures were checked for syncytial formation and aliquots removed daily. Fixation was performed in 3% glutaraldehyde in 0.1N sodium phosphate buffer (pH 7.0). Centrifugally pelleted cells and syncytia were carefully dispersed in low melting temperature agar, which was then allowed to set and was cut into 2mm square pieces for post osmication embedding and sectioning. Thin sections were post stained with potassium permanganate and lead citrate, before being studied at 80kV in the Jeol 100SX and 100CX transmission electron micro-scopes. Electron micrographs were recorded on Ilford technical film, type EM, and printed on Ilfospeed photographic paper, grades 2 and 3.

Nuclear Structure and Function, Edited by J. R. Harris and
I. B. Zbarsky, Plenum Press, New York, 1990

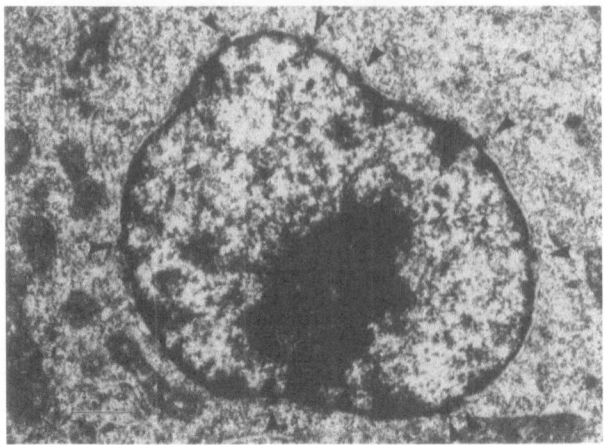

Figure 1.   A nucleus within a HIV-1-infected C8166 cell, showing very pronounced 'plug-like' nuclear pore complexes (arrows). The scale marker indicates 500nm.

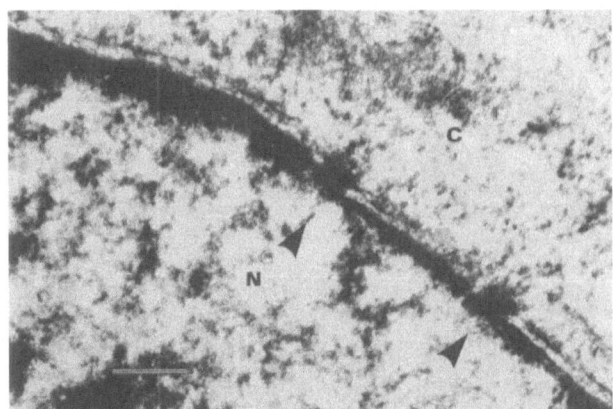

Figure 2.   Part of the nuclear envelope of a HIV-1-infected C8166 cell showing two electron dense nuclear pore complexes (arrows), which appear to have fibrous-like material extending into the cytoplasm. (N, nucleoplasm; C, cytoplasm). The scale marker indicates 200nm.

RESULTS AND DISCUSSION

Following infection of the C8166 CD4$^+$ lymphoid cells with HIV-1, within 2 to 3 days surface budding of developing viral particles can be readily detected and small syncytia are present in the culture. The zones of fusion between adjacent cells are clearly defined, and the two plasma membranes disintegrate as fusion progresses.

The ability of the infected cell to incorporate the HIV-1 genome into its own DNA via reverse transcriptase, with the subsequent production and translocation of viral RNA and mRNP from nucleus to cytoplasm, is known to

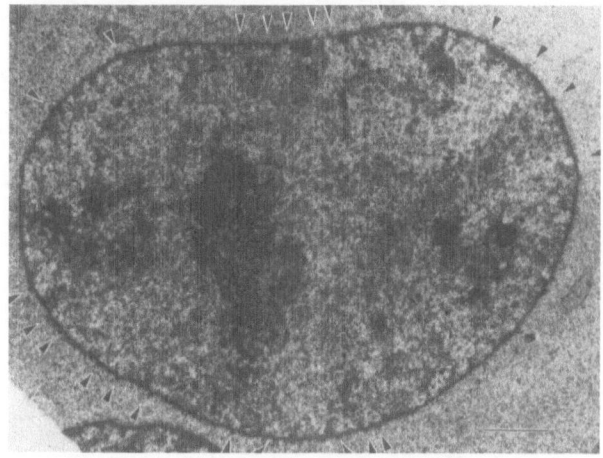

Figure 3.   A nucleus within a small HIV-1-induced syncytium of C8166 cells that is actively releasing virus. Even at this low electron optical magnification the nuclear pore complexes appear as pronounced electron dense structures (arrows). The scale marker indicates 1μm.

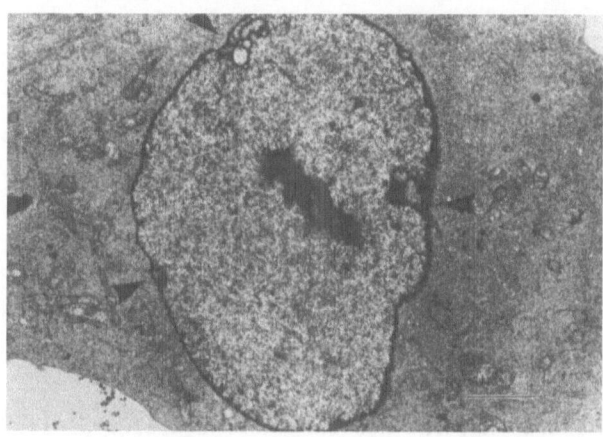

Figure 4.   An intrasyncytial nucleus showing surface undulation and infolding of the nuclear envelope (arrows). The scale marker indicates 2.5μm.

be under the control of the unique retroviral regulatory genes, in particular the tat gene and the rev gene. The latter gene product is believed to activate the sequence-specific nucleocytoplasmic translocation of incompletely spliced HIV-1 RNA species (Malim et al., 1989). It has been found that infected C8166 cells which are actively liberating virus possess very distinct nuclear pore complexes (Figures 1 and 2). The transverse sectioned nuclear pore complexes appear as very pronounced electron dense 'plugs', sometimes with extension of the density into the nucleoplasm and/or cytoplasm. A similar phenomenon has also been detected within the small syncytia, which also liberate large quantities of HIV-1 from their surface (Figure 3). It is proposed that this morphological

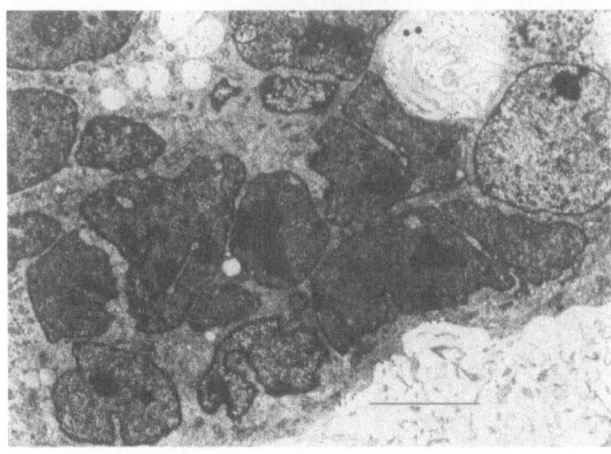

Figure 5.   Intrasyncytial nuclei showing marked distorition of shape.
The scale marker indicates 5µm.

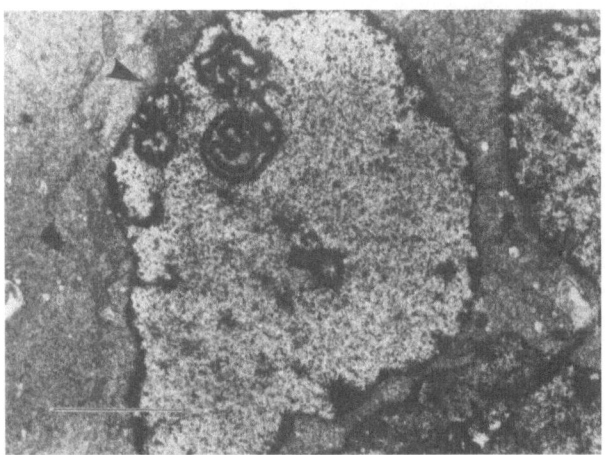

Figure 6.   An intrasyncytial nucleus showing nucleolar reticulation
(arrow). The scale marker indicates 2.5µm.

feature results from the massive activation of intranuclear viral gene
product synthesis with its subsequent translocation across the nuclear
envelope, via the nuclear pore complexes. As the syncytial size increases,
the production of HIV-1 gradually decreases and the nuclei become metabolically less active. Under these conditions the nuclear pore complexes
no longer possess a pronounced electron density. It should be emphasised
that this concept of RNP translocation via the nuclear pore complex has been
present in the literature for many years (Mehlin et al., 1988; Scheer et
al., 1988). Thus, conditions of exceptionally high viral RNA synthesis and
RNP nucleocytoplasmic translocation might reasonable be expected to result
in a more pronounced ultrastructural indication of the functional event.

The nuclei within small syncytia appear to be relatively normal
morphologically, but they have clearly lost the ability to undergo mitosis
and as the syncytial size increases due to the progressive fusion of

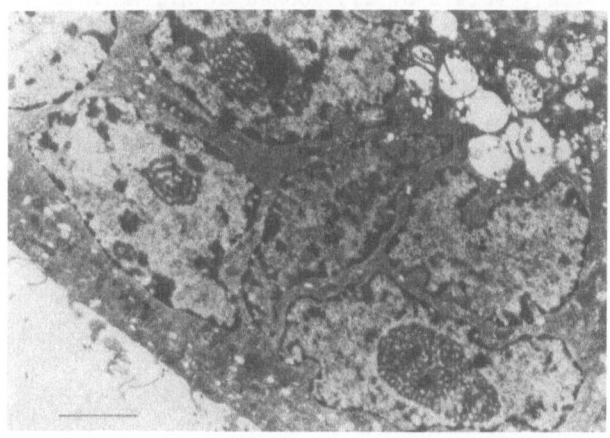

Figure 7.   Shrunken pycnotic nuclei containing reticulated nucleoli,
              within a large vacuolated syncytium. The scale marker
              indicates 2µm.

increasing numbers of cells, degenerative changes become detectable.
Initially, the most pronounced change is the production of an undulatory
nuclear surface, with infolding of the nuclear envelope (Figure 4). This
feature is almost certainly due to nuclear shrinkage as the chromatin
undergoes limited condensation. Within the larger syncytia the initial
approximately spherical shape of the nuclei is lost as the nuclei become
progressively distorted, in a manner similar to that observed for nuclei
within solid tumors (Figure 5). The functionally active 'fibrous' nucleoli
within the nuclei of single infected cells and small syncytia become non-
homogeneous and increasingly reticulated (Figure 6) as the nuclei become
inactive. Ultimately, the syncytia become degenerate or senescent, with
extreme chromatin condensation and shrinkage of the nuclei, which are then
typically described as being pyconotic (Figure 7).

REFERENCES

Dowsett B. A., Roff M. A., Greenaway P. J., Eplphic E. R. and Farrar G. H.
    1987, Syncytia- a major site for the production of human immuno-
    deficiency virus, AIDS, 1:147.
Harris J. R., Kitchen A. D., Harrison J. F. and Tovey G. 1989a, Viral
    release from HIV-1-induced syncytia of CD4+ C8166 cells, J. Med.
    Virol. 28:81.
Harris J. R., Tovey G. and Kitchen A. D. 1989b, Ultrastructural changes in
    HIV-induced cultured cells, in: Proceedings of EMAG-MICRO '89, IOP
    Publishing Ltd, Bristol, p. 273.
Malim M. H., Hauber J., Le S.-Y., Maizel J. V. and Cullen B. R. 1989, The
    HIV-1 rev trans-activator acts through a structured target sequence to
    activate nuclear export of unspliced viral mRNA, Nature 338:254.
Mehlin H., Lönnroth, Skoglund U. and Daneholt B. 1988, Structure and
    transport of a specific premessenger RNP particle, Cell Biol. Int. Rep.
    12:729.
Montefiori D. C. and Mitchell W. M. 1987, Persistent coinfection of T
    lymphocytes with HTLV-II and HIV and the role of syncytium formation in
    HIV-induced cytopathic effect, Virology, 160:372.
Scheer U., Dabauvalle M. C., Merkert H. and Benevente R. 1988, The nuclear
    envelope and the organization of the pore complex, Cell Biol. Int. Rep.
    12:669.

Yoffe B., Lewis D. E., Petrie B. I., Noonan C. A., Melnick J. L. and
    Hollinger F. B. 1987, Fusion as a mediator of cytolysis in mixtures of
    uninfected CD4$^+$ lymphocytes and cells infected by human immuno-
    deficiency virus, Proc. Natl. Acad. Sci. USA, 84:1429.

AN ELECTRON MICROSCOPIC STUDY OF THE STRUCTURAL MECHANISMS OF HYBRID NUCLEUS

FORMATION DURING SPONTANEOUS SPLENOCYTE-FIBROBLAST FUSION

A. Ju. Kerkis

Institut of Cytology and Genetics
Siberian Branch of Scieces of USSR
Novosibirsk
USSR

There are currently many techniques for the production of somatic cell hybrids, by the use of fusion agents and more recently, high voltage electric discharge.

Fusion agents give rise to numerous chromosomal rearrangements in the somatic hybrid cells. This is a disadvantage, because somatic hybrid cells without chromosome rearrangements are more suitable for localization of genes on chromosomes, isolation of individual chromosome, generation of DNA fragment bands and other purposes. Improved techniques for the production of somatic cell hybrids without chromosome rearrangement are clearly needed.

In an attempt to improve the method, an electronmicroscopic (EM) study of splenocyte-fibroblast interaction during their co-cultivation in the absence of fusion agents was carried out.

Cells of the cultured Chinese hamster line deficient in GRFT were used as partners in the fusion experiments. The karyotype (Rubtsov et al., 1981), and the ultrastructure of the nucleus of M15 cells have been previously characterized (Keris et al., 1988); fox splenocytes served as the other cellular partner. The splenocytes were isolated by three washes in sterile Hank's solution containing 100μ/ml penicillin and 100μ/ml streptomycin. The splenocytes were then grown in Eagle's medium containing 1%BSA and 100μ/ml streptomycin. M15 cells were co-cultived in Petri dishes (60mm in dia.) on MEM with 10%FBS in a 5% CO_2 atmosphere (8×10^6 cells/dish). Pretreatment of Petri dishes and the marking of the EM preparations have been described elsewhere (Kerkis and Kristolubova, 1972); 24h later, the medium was discarded, 4×10^6 splenocytes per Petri dish were added to the cell monolayer. The dishes were filled with MEM supplemented with 10%FBS and 240μg/ml FGA.

The material was fixed after 1, 4, 8, 24, 48, and 72h. Ultrathin sections were made from cells that were chosen by light phase contrast microscopy. Thin sections were examined with a JEOL-100 C electron microscope.

The results of this electronmicroscopic study of the spontaneous fusion of splenocytes with fibroblasts during their co-cultivation are as follows. Splenocytes are able to freely enter the cytoplasm of the fibroblast, while

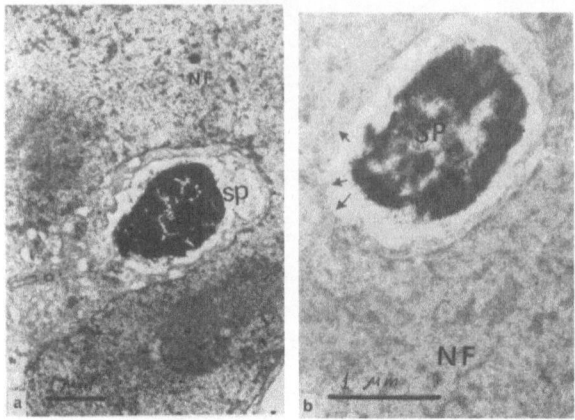

Figure 1.   A fox splenocyte (SP), is seen entering a nucleus (NF) of a
Chinese hamster fibroblast 1-3h after their co-cultivation.
(a) Earliest stage.   (b) Destruction of a nuclear membrane
(arrows) before entrance of SP.

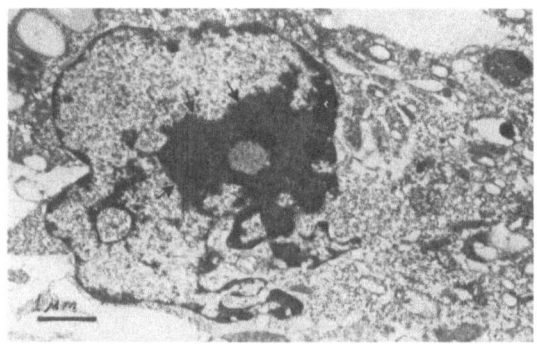

Figure 2.   Chromatin blocks in a hybrid nucleus after 48h of
co-cultivation.

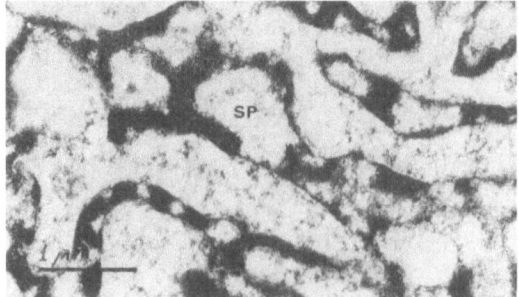

Figure 3.   Fragmentation of the splenocyte nucleus in the fibroblast
cytoplasm. The fragmentate gives rise to distinct chromatin
blocks.

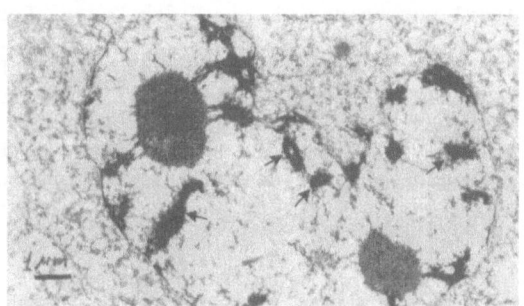

Figure 4.   The structure of the hybrid nucleus 72h after co-cultivation.
             Arrows point to the retained fragments of splenocyte
             chromatin.

the nucleus, being protected from the attack of lysosyme enzymes is spared
and hence shows no morphological changes up to 72h. Some of the splenocyte
nuclei are able to fuse with the fibroblasts at 1-4h. (Figure 1). After
24-48h. blocks of condensed chromatin from the splenocyte nuclei are seen in
the nuclei of the fibroblasts (Figure 2). The amount of condensed chromatin
in the fused nuclei decreases during the process of co-cultivation,
presumably as a result of chromatin activation.

    Based on the data obtained, the sequence of events providing the
formation of somatic cell hybrids may be as follows:

(i) The synkaryon would be formed as a result of the rapid entrance of the
splenocyte into the nucleus of the fibroblast, as a consequence, its plasma
and nuclear membranes would be destroyed.

(ii) The chromatin of the splenocyte nuleus residing in the fibroblast
cytoplasm for 15-20h of co-cultivation, would disperse into chromosome-like
blocks. Each block is seen to be outlined by a double membrane, presumably
formed from the destroyed nuclear membrane of the splenocyte (Figure 3). If
this does occur, the synkaryon would be formed at the time point of mitosis,
when separate membrane-outlined fragments may be integrated into the forming
fibroblast nucleus. As in (i), the condensed chromatin of the splenocyte
are easily identified up to 72h (Figure 4). At this stage the membrane
fragments attached to the remains of the condensed chromatin are still
observed. Our EM study of the interphase cell nuclei of the hybrid clones
of different origin demonstrates that the fragments are invaginations of the
inner nuclear membrane. The investigations provide an additional internal
surface for the nucleus, needed for the attachment of 'alien' chromosomes in
the newly forming hybrid nucleus.

    The data obtained demonstrate the feasility of high frequency pro-
duction of somatic cell hybrids by co-cultivation of fibroblasts with
splenocytes, without the use of fusion agents. This approach enabled us to
produce somatic cell hybrids without chromosome rearrangements, thereby
facilitating investigation of the patterns of hybrid cell formation.

REFERENCES

Kerkis A. Ju., and Khristolubova N. B. 1972, Preliminary choice of labelled
    cell as a method facilitating of EM autoradiography for the study of

the cell functional morphology, Proc. Fifth Europ. Cong. Electron Microscopy, 228.

Kerkis A. Ju., Zhdanova N. S., and Khristolubova N. B. 1988, Nuclear matrix structural elements determining stability of genome of hybrid somatic cell, Inst. Phys. Conf. Ser. 93:3:18:4899.

Rubtsov N. B., Radjabli S. I., Gradov A. A., and Serov O. L. 1981, Chinese hamster x American mink somatic cell hybrids: Characterization of a clone panel and assignment of the mink genes for malatedehydrogenase, NAD-1., TAG., 60:99.

NUCLEAR COMPARTMENTALIZATION IN POLLEN MOTHER CELLS DURING MEIOTIC PROPHASE

M. I. Rodriguez-Garcia[1], J. D. Alché[1], A. Majewska-Sawka[2],
M. C. Fernandez[1] and B. Jassem[2]

[1] Estación Experimental del Zaidin (C.S.I.C.)
Profesor Albareda 1
18008 Granada
Spain

[2] Institute for Plant Breeding and Acclimatization
Weyssenhoffa 11
85-950 Bydgoszcz
Poland

INTRODUCTION

Nuclear vacuoles, which are expansions of the perinuclear space toward the karyoplasm, were first reported in animal cell by Rasmussen (1976) during meiotic prophase in *Bombyx mori*. Some years later, membrane limited spaces in nucleus formed by the invagination of the inner membrane of the nuclear envelope during meiotic prophase in different plant species, were also described by several authors (Sheffield et al., 1979; Karasawa and Ueda, 1983a; 1983b; Sangwan, 1986; Rowley and Walles, 1985; Rodriguez-Garcia et al., 1988; Majewska-Sawka et al., in press). Although these structures have been reported in many different species, the lack of information about the nature and function of the unusual compartments in the nucleus has made many scientists consider them with reservation, or identify them as artifacts, which is a ready explaination for anything new and unknown within a cell.

In this study we will present evidence that supports the existence of nuclear vacuoles as real, dynamic structures which originate in the cell at specific times and under specific circumstances, related to processes of nuclear reorganization during meiotic prophase.

MATERIALS AND METHODS

We based our work on different techniques including standard glutaraldehyde-osmium fixation, cryofixation and cryosubstitution of *Olea europaea* L. and *Beta vulgaris* L.anthers.

Cryofixation. The material was pretreated with the chemical fixation (3% glutaraldehyde at 4°C for 2h), followed by cryoprotection in 2.3M sucrose at 4°C overnight. Specimens were frozen by immersion in liquefied propane at -190°C in the Reichert Jung universal system. Cryosections were

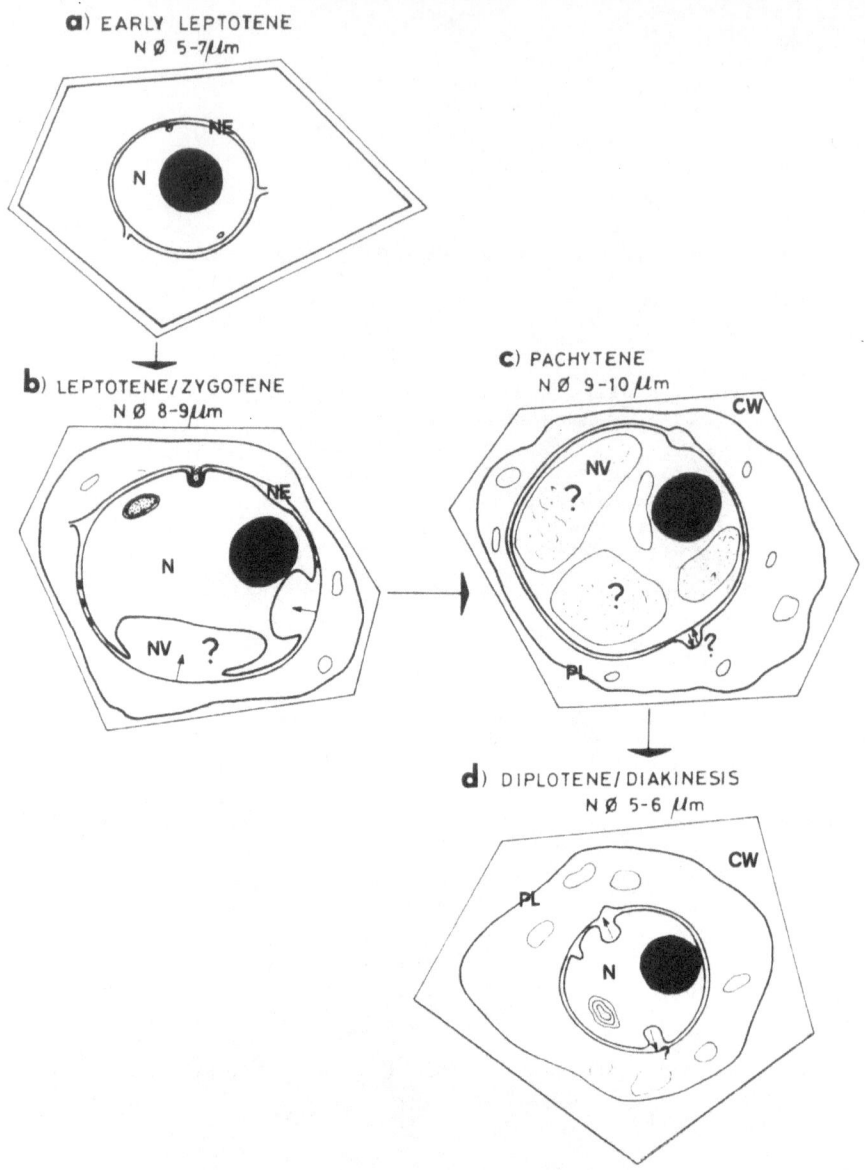

a) EARLY LEPTOTENE
N Ø 5-7μm

b) LEPTOTENE/ZYGOTENE
N Ø 8-9μm

c) PACHYTENE
N Ø 9-10μm

d) DIPLOTENE/DIAKINESIS
N Ø 5-6 μm

Figure 1,   Scheme showing the development of nuclear vacuoles. CW-cell
wall, N-nucleus, NE-nuclear envelope, NV-nuclear vacuoles,
PL-plasmalemma.

obtained using the Reichert Jung FC4, stained by toluidine blue and
observed by light microscopy.

    Cryosubstitution. Following fixation in 3% glutaraldehyde for 3h, the
material was cryosubstituted in 50% at -20°C, and then at -30°C. The
anthers were embedded by low temperature medium Lowicryl K4M.

DEVELOPMENT OF NUCLEAR VACUOLES

    Our present studies in *Olea europaea* show the pattern of formation of

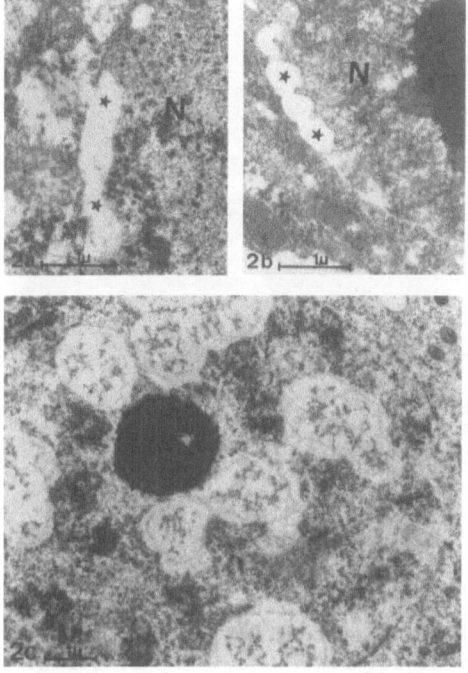

Figure 2,   Leptotene/zygotene. Membrane limited electron clear spaces
(asterisk) at the nuclear periphery. (c): Pachytene.
Nuclear vacuoles (NV) are intercalated with the synaptonemal
complexes (S). (a, c): glutaraldehyde-osmium fixation, (b):
cryosubstitution.

nuclear vacuoles, which agrees with the results obtained earlier, by
electron microscopy (Rodriguez-Garcia et al., 1988; Majewska-Sawka et al.,
in press). We will refer to the scheme in Figure 1a to 1d, which shows the
formation, development, and disappearance of nuclear vacuoles during meiotic
prophase in pollen mother cells (PMCs).

In early meiotic prophase (Figure 1a), the nucleus of the PMC shows a
typical round contour, with no particular features of the nuclear envelope.

In leptotene/zygotene (Figure 1b) local dilations of the perinuclear
space appear to be randomly distributed around the nuclear envelope. As
zygotene proceeds, the perinuclear dilations increase rapidly, growing
toward the karyoplasm, while at the same time the inner nuclear membrane
enlarges. In electron micrographs the nuclear envelope dilations show
membrane limited spaces at the nuclear periphery (Figures 2a, 2b).

In pachytene (Figure 1c), the vacuolar expansions in the perinuclear
space ramify progressively, penetrating deeper into the karyoplasm and
becoming intercalaced with the synaptonemal complexes of the paired
chromosomes (Figure 2c). The increase in size of the nuclear vacuoles is
accompanied by an increase in nuclear volume. Serial sections would be
necessary to determine whether the vacuolar compartments formed in the
nucleus maintain contact with the perinuclear space, or if on the other
hand they are independent units, as the bidimensional images seem to
suggest.

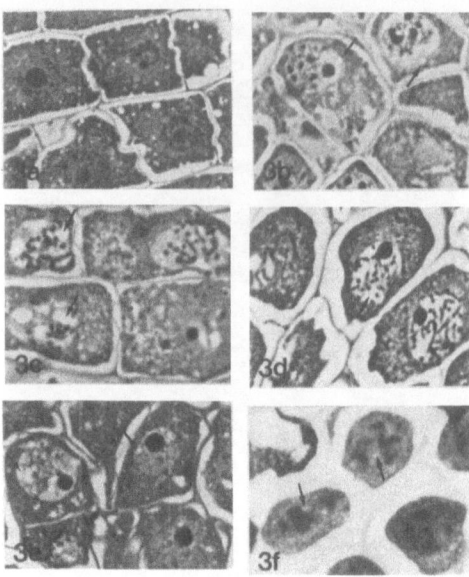

Figure 3.   Pollen mother cells during meiotic prophase. Toluidine blue
staining. (a-d): conventional glutaraldehyde-osmium fixation,
(e): cryosubstitution, (f): cryofixation.

Toward the end of meiotic prophase, diplotene-diakinesis (Figure 1d), nuclear vacuoles are seen less frequently. They become steadily less numerous, and are always seen at the nuclear periphery, while the nucleus becomes noticeably smaller. By contrast, the number of cytoplasmic vacuoles rises. These observations may suggest that nuclear vacuoles are being transferred into the cytoplasm.

CONCLUSIONS

Because of the constant presence of nuclear vacuoles, observed with both conventional electron microscope techniques and with cryotechniques, we can conclude that these structures are real and not artifacts. The nuclear envelope clearly seems to be involved in their formation, together with enlargement of the inner nuclear membrane and the budding out of the perinuclear space into the karyoplasm.

REFERENCES

Karasawa R., and Ueda K., 1983a, Nuclear vacuoles and synizesis during meiotic prophase in *Haplopappus gracilis*. Cytologia, 48:819.
Karasawa R., and Ueda K., 1983b, Occurrence of nuclear vacuoles in meitoic prophase nuclei in *Compositae*. Caryologia, 36:145.
Majewska-Sawka A., Bohdanowicz J., Jassem B., and Rodriguez-Garcia M. I., Development of nuclear vacuoles in microsporocytes of the fertile and sterile sugar beet. Ann. Bot. (in press).
Rasmussen S. W., 1976, The meiotic prophase in *Bombyx mori* females analysed by three-dimensional reconstructions of synaptonemal complexes. Chromosoma, 54:245.
Rodriguez-Garcia M. I., Majewska-Sawka A., and Fernandez M. C., 1988, Why do nuclear vacuoles appear in the prophasic nucleus of pollen mother

cells? Facts and hypotheses, in 'Sexual Reproduction in Higher
Plants. M. Cresti, P. Gori, E. Pacini, eds., Springer-Verlag.

Rowley J. R., and Walles B., 1985, Cell differentiation in microsporangia
of *Pinus sylvestris*. III. Late pachytene. Nord J. Bot. 5:255.

Sangwan R. S., 1986, Formation and cytochemistry of nuclear vacuoles during
meiosis in *Datura*. Eur. J. Cell Biol. 40:210.

Sheffield E., Cawood H. H., Bell P. R., Dickinson H. G., 1979, The
development of nuclear vacules during meiosis in plants. Planta, 146:
597.

INDEX

Nucleic acid
  binding to nucleolar protein B23,
      245,246
Nucleolin, 256
  gene expression
    during embryogenesis, 248, 249
    during oogenesis, 247, 248
  gene promoter, 73-77
  labelling on cryo-sections, 293
  and mRNA accumulation, 248
  and rRNA synthesis, 247-251
Nucleo-cytoplasmic transport, 474,
      475
  via nuclear pore complexes, 486
Nucleolar organiser regions (NORs)
      (*see also* nucleolus and
      silver staining)
  on chromomeres, 220-222
  in fertilization and
      parthenogenesis, 207-209
  silver staining, 227
Nucleolus
  action of DRB, 413-417
  antibody to, 255, 256
  antigens in human tumors, 237-241
  argentaffinity, 259-263
  B23 isoforms
    interaction with nucleic acids,
      243-246
  of bovine embryo, 130,131
  cDNA for protein P120, 239, 240
  chromatin of, 231
  components in onion cells,
      231-234
  dense fibrillar component, 233,
      234
  dominance and hypomethylation
      of NOR-chromosome, 177
  DNA distribution, 149-151
  fibrillarin, 223-225
  functional changes in, 161-164
  genomic clones for protein B23,
      244
  higher order structure of
      chromatin, 203-205
  hypotonic treatment of, 198-201
  immunolocalization of DNA,
      231-234
  inhibition of topoisomerase I,
      153-155
  isolation of fibrillar complexes,
      197-201
  labelling of rRNA and rDNA,
      173-175
  maturation and transcription of
      rRNA genes, 133-136
  monoclonal antibody to protein
      P120, 238, 239
  mRNA level for protein P120, 240
  nucleolin, 243, 247-251
    gene promoter, 73-77

Nucleolus (continued)
  nucleogenesis in cow embryo, 187-190
  of onion cells, 231-234
  oocyte extranucleolar bodies,
      165-168
  of *P. micans,* 293-295
  precursor body, 187-190
  protein B23, 223-225, 243-246, 256
  protein C23, 223-225
  protein localization, 225
  protein P120 gene transcription,
      240, 241
  regulation of activity, 211-214
  and ribosomal gene activation,
      193-196
  ring shaped, 227-229
  satelite (SN), 223-225
  silver staining, 180, 181, 183-185
  structural transitions in, 193-196
  UV-microirradiation of, 211-214
  Xenopus protein NO38, 244
Nucleoprotein mRNA
  expression of in spermiogenesis,
      89-93
Nucleosome
  histone loss, 343
  sedimentation of oligomers, 297-299
  structural changes of, 341-343
  urea treatment of, 341, 342
Nucleus
  antibodies to, 255
  arachidonic acid metabolism,
      459-461
  compartmentalization, 493-496
  /cytoplasmic compartmentalization of
      L2 RNA, 95-97
  development during spermatogenesis,
      99-103
  effects of DRB, 413-417
  extraction procedures and nuclear
      matrix, 317-322
  genes encoding receptors, 447-450
  gibberellin-binding proteins,
      443-446
  glucocorticoid receptor complex,
      421-423
  HIV-1-induced ultrastructural
      changes, 483-487
  hybrid formation, 489-491
  inositol metabolism, 470
  intrasyncytial, 486, 487
  isolation from lymphocytes, 198, 199
  localization of DNA-polypeptide
      complexes, 11-13
  of mouse trophoblast, 273-276
  myosin-like proteins of, 329-331
  nuclearskeleton HSPs, 371-374
  oncogenes and lipid metabolism
      changes, 463-467
  phospholipid fatty acids, 459-461
  reprogramming after transplantation,
      129-131

Nucleus (continued)
  RNA metabolism, 413-417
  70-110S RNPs, 265-271
  role of lipids in, 399-402
  role of phospholipids in, 469, 470
  selective staining of DNA, 477-480
  silver staining of NORs, 179-181
  spatial distribution of DNA and
      histones, 169-170
  sphyngomyelin content, 395, 396
  sphyngomyelinase, 393, 397
  structural proteins of, 237
  ultrastructure, 413-417
  vacuolation of, 493-496

Oogenesis
  of gall midge, 125-128
Oncogene expression
  and lipid metabolism, 463-467
Ouchterlony gels, 237

Phospholipid
  fatty acid metabolism
    age-specific differences in,
        459-461
  role of nucleus, 399-402, 469, 470
Phytohemagglutinin
  stimulation of lymphocytes, 238,
      239
Phytohormone
  gibberellin, 443-446
Plant
  nuclear gibberellin-binding
      protein, 443-446
Plasmid
  autonomous replication, 345-349
Pollen mother cell
  nuclear compartmentalization,
      493-496
Poly(A)-containind RNAs
  structural elements of, 437-441
Polymerase chain reaction (PCR)
  amplification of human insulin
      gene, 119-122
Protamin, 89-93
Protinase
  DNA-activated, 351-353
  histone H1-specific, 351-353
  inhibitors, 353, 359
  treatment of Lowicryl sections,
      294, 295
Protein-A gold, 293, 305
Pulsed field gel electrophoresis
  (PFG), 47, 48

Rat liver
  cytoplasmic low Mr RNA, 115-118
  nuclear lipid metabolism, 463-467
  nuclear matrix, 351-353
  nucleic acid synthesis, 393-397

Rat liver (continued)
  70-110S nuclear RNP complex
    antibody to, 268-271
    characterization of, 265-271
Rattus norvegicus
  DNA-fingerprinting, 31-35
Replication origin
  DNA pattern, 377-381
Retrotransposon, 41-45, 61-64
Ribosome
  gene expression of proteins, 79-81
  5S rRNA genes, 143-146
  R-protein gene L1, 389-392
  translational control, 83-86
Ribosomal RNA (rRNA) gene, 220-222
  transcription and replication,
      429-431
Ribosomal gene expression
  and NOR silver staining, 219-222
RNA
  associated with Poly(A)$^+$, 115-118
  compartmentalization of L1 RNA,
      95-97
  splicing control, 95-97
  DNA-dependent RNA polymerase,
      443-446
  polymerase I, 233, 234, 383-387
  polymerase II, 341-343, 383-387
  polymerase III transcripts, 437-441
  transcription
    non-ribosomal, 341-343

Satelite nucleolus
  and NORs, 223-225
Scanning tunneling microscopy (STC)
  of native DNA, 285-289
Scilla peruviana L.
  pollen grains, 253-256
Silver staining, 125, 126, 128, 198,
      200, 413, 417
  after HCl extraction, 229
  during nucleologenesis, 189, 190
  of extranucleolar bodies, 167
  of NO chromosomes, 429-431
  of NOR proteins, 157-159, 161-164,
      179-181, 219-222
    in leukemic cells, 215-218
  of nucleoli, 157-159, 183-185,
      207-209, 227-229
  of P. micans nucleoli, 291-295
  after ribonuclease digestion, 228,
      229
Small nuclear ribonuclear proteins
  (snRNPs), 269-271
  antibody to, 255, 256
  hydrolysis with lead ions, 453-456
  and regulation of development,
      425-427
  secondary structure model of,
      453-456